NICOTINE, CAFFEINE
AND SOCIAL DRINKING

NICOTINE, CAFFEINE AND SOCIAL DRINKING

Behaviour and Brain Function

Edited by

Jan SNEL
University of Amsterdam, The Netherlands

and

Monicque M. LORIST
University of Groningen, The Netherlands

LONDON AND NEW YORK

First published 1998
by Routledge
2 Park Square, Milton Park, Abingdon, Oxon OX14 4RN
711 Third Avenue, New York, NY 10017

Routledge is an imprint of the Taylor & Francis Group, an informa business

First issued in paperback 2016

Amsteldijk 166
1st Floor
1079 LH Amsterdam
The Netherlands

British Library Cataloguing in Publication Data

Nicotine, caffeine and social drinking: behaviour and
  brain function
  1. Brain — Effect of drugs on   2. Nicotine — Physiological
  effect   3. Caffeine — Physiological effect   4. Alcohol —
  Physiological effect
  I. Snel, Jan   II. Lorist, Monicque M.
  615.7'8

ISBN13: 978-90-5702-218-0 (hbk)
ISBN13: 978-1-138-97719-8 (pbk)

CONTENTS

PREFACE

Regular, moderate use of caffeine, alcohol and nicotine alone or in combination is a well known, widely-established and loved practice of men and women, both young and old. In spite of the attribution of negative health consequences to these three pharmacological agents, people continue, quite persistently, to consume these substances and afford much value to the pleasure of a regular intake.

The assumption is that, in general, nicotine and caffeine have stimulating effects, while alcohol has sedating effects. Surprisingly, in situations of both positive and negative excitement, such as a wedding, the birth of a child, or in a challenging or stressful situation or in boredom or monotony, people may choose the pleasure of either a cup of strong coffee, the smell of a cigarette or the refreshment of a beer or a glass of good wine.

For us, as moderate and regular non-smoking users of caffeine and alcohol, there were several motives to edit the present book.

- People keep asking many questions and have their worries on the effects on behaviour and health. Obviously, there is a shortage of the scientifically based knowledge on the efficacy of these psychoactive substances in the general public.
- There is relatively little information on the effects of the combined use of these substances, which is astonishing in view of the regular use and enjoyment of these substances in varying combinations.
- Relatively little is known about the effects on brain function associated with sensation, perception, attention and memory, although there is a great effort going on in different research centres to get a more precise picture of these effects on information processing as assessed from brain activity.

The emphasis of the book lies on the effects of the normal use of these three substances on behaviour and brain function in humans.

The book consists of three parts dealing with each substance separately. Each part has a similar structure; first, a chapter on the pharmacokinetics and the metabolism of the substance, followed by more theoretical contributions on the psychophysiological mechanisms that underlie the effects of moderate intake of caffeine, alcohol and nicotine. Subsequently, these introductory chapters are followed by chapters dealing with experimental work done on these psychoactive agents.

We wish to thank all authors for their contribution and the patience they must have had in realizing this book. We hope they are satisfied with the final product, such that it was worth the effort they put into it.

A debt of gratitude is also owed to the editorial staff of Harwood Academic Publishers. They did a great job and supported us tremendously in the editorial process.

Finally, we would like to thank Bas, Bert, Hans and Wilfred at the 'helpdesk' of the Faculty of Psychology of the University of Amsterdam for their technical and moral support in the process of producing this book.

NOTES ON CONTRIBUTORS

Dharam P. Agarwal
Institute of Human Genetics, University of Hamburg, Butenfeld 32, 22529 Hamburg, Germany

Maurice J. Arnaud
Institut de l'Eau, Perrier Vittel, BP 101, F-88804 Vittel Cedex, France

Jean-Marc Assaad
Department of Psychology, McGill University, 1205 Docteur Penfield Avenue, Montreal, Canada H3A 1B1

†Karl Bättig
Behavioral Biology Laboratory, Swiss Federal Institute of Technology, Institute of Toxicology, Schorenstrasse 16, CH-8603 Schwerzenbach, Switzerland

Kenneth R. Bruce
Department of Psychology, McGill University, 1205 Docteur Penfield Avenue, Montreal, Canada H3A 1B1

Brenda O. Gilbert
Department of Psychology, Southern Illinois University at Carbondale, Illinois 62901-6502, USA

David G. Gilbert
Department of Psychology, Southern Illinois University at Carbondale, Illinois 62901-6502, USA

Jacques Le Houezec
CNRS URA 1957 – Hôpital de la Salpêtrière 47, Blvd de l'Hôpital – F-75651 Paris, Cedex 13, France

Ypie T. De Haan
Vakgroep Medische Sociologie, University of Limburg, Postbus 616, 6200 MD Maastricht, The Netherlands

Anne Harr
Royal Ottawa Hospital, 1145 Carling Avenue, Ottawa, Ontario, Canada K1Z 7K4

Ronald A. Knibbe
Vakgroep Medische Sociologie, University of Limburg, Postbus 616, 6200 MD Maastricht, The Netherlands

Verner J. Knott
Department of Psychiatry, University of Ottawa/Royal Ottawa Hospital and
Institute of Mental Health Research, 1145 Carling Avenue, Ottawa, Ontario,
Canada K1Z 7K4

Harry S. Koelega
Vakgroep Psychonomie, Heidelberglaan 2, 3584 CS Utrecht, The Netherlands

Adriaan Kole
TNO-Voeding, Postbus 360, 3700 AJ Zeist, The Netherlands

Hartmut Leuthold
Fachgruppe Psychologie, Universität Konstanz, Postfach 5560 [D47], D-77434
Konstanz, Germany

Monicque M. Lorist
Experimental and Work Psychology, University of Groningen, Grote Kruisstraat
2/1, 9712 TS Groningen, The Netherlands

Stacey Lusk-Mikkelsen
Royal Ottawa Hospital, 1145 Carling Avenue, Ottawa, Ontario, Canada K1Z 7K4

Bertie M. Maritz
Volkerakstraat 26III, 1078 XS Amsterdam, The Netherlands

Frances Martin
Department of Psychology, University of Tasmania, GPO Box 252C Hobart,
Tasmania, 7001, Australia

Jennifer M. Nichols
Department of Psychology, University of Tasmania, GPO Box 252C Hobart,
Tasmania, 7001, Australia

Nancy E. Noldy
Playfair Neuroscience Unit, The Toronto Hospital — Western Division,
399 Bathurst Street, MP10-324, Toronto, Ontario, Canada M5T 2S8

Robert O. Pihl
Department of Psychology, McGill University, 1205 Docteur Penfield Avenue,
Montreal, Canada H3A 1B1

Rosanne Pogash
Pennsylvania State University, College of Medicine, 500 University Drive,
Hershey, Pennsylvania 17033, USA

Walter S. Pritchard
Psychophysiology Laboratory, Bowman Gray Technical Center, R&D,
R.J. Reynolds Tobacco Company, Winston-Salem, NC 27102, USA

John H. Robinson
Psychophysiology Laboratory, Bowman Gray Technical Center, R&D, R.J. Reynolds Tobacco Company, Winston-Salem, NC 27102, USA

Andrew P. Smith
Department of Psychology, University of Bristol, 8 Woodland Road, Bristol BS8 1TN, UK

Jan Snel
Department of Psychonomics, Roetersstraat 15, 1018 WB Amsterdam, The Netherlands

Werner Sommer
Fachgruppe Psychologie, Universität Konstanz, Postfach 5560 [D47], D-77434 Konstanz, Germany

Odin van der Stelt
Academic Medical Center University of Amsterdam, Department of Child and Adolescent Psychiatry, PO Box 22700, 1100 DE Amsterdam, The Netherlands

Siegfried Streufert
Pennsylvania State University College of Medicine, 500 University Drive, Hershey, Pennsylvania 17033, USA

David M. Warburton
Human Psychopharmacology Group, University of Reading, Department of Psychology Building 3, Earley Gate, Whiteknights Road, Reading RG6 6AL, UK

Elly Zeef
Foundation Mind at Work, Beeckzanglaan 78, 1942 LT Beverwijk, The Netherlands

Part I — Nicotine

CHAPTER ONE

Pharmacokinetics and Pharmacodynamics of Nicotine

Jacques Le HOUEZEC

INTRODUCTION

Nicotine acts on nearly every physiological system of the human body. The effects of nicotine on the peripheral nervous system have been extensively studied and are now quite well understood. The effects of nicotine on the central nervous system are more complex and our understanding of their effects is limited. This chapter reviews the pharmacokinetics and pharmacodynamics of nicotine with an emphasis on the psychopharmacological basis of nicotine dependence.

In South America, seeking for psychoactive effects of nicotine might be as old as the origin of horticulture, beginning some eight thousand years before present. Ritual tobacco use in Shamanism aimed to achieve acute nicotine intoxication, which induced in the Shamans catatonic states representing symbolic death. The effects of large doses of nicotine on the autonomic and central nervous systems gave the impression of a gradual death of the Shaman, who then returned miraculously to life (Wilbert, 1987).

Nicotine, the major alkaloid of tobacco, was first isolated in a pure form by Posselt and Reimann in 1828. Nicotine and other alkaloids (atropine, muscarine, curare alkaloids) have played a key role in the development of knowledge and understanding of the functional organization of the autonomic nervous system. At the turn of the 19th century, Langley and his colleagues used nicotine to determine the nature of the autonomic innervation and the location of ganglionic synapses for many organs. The concept of receptor arose from Langley's experiments. In, 1914, Dale developed the concept of two different sites of action of acetylcholine, termed muscarinic and nicotinic, based on the relative selectivity of the response to muscarine and nicotine (see Le Houezec and Benowitz, 1991 for references).

Tobacco smoking is a unique and highly addictive form of systemic drug administration in that entry into the circulation is through the pulmonary rather than the portal or systemic venous circulations. Nicotine reaches the brain in less than 10 seconds, faster than after intravenous administration. Nicotine is distributed throughout the brain, with highest concentrations in the hypothalamus, hippocampus, thalamus, midbrain, brain stem and in areas of the cerebral cortex. Nicotine also binds to nigrostriatal and mesolimbic dopaminergic neurons. Brain nicotine concentrations increase sharply after completion of smoking, then decline over 20 to 30 minutes as nicotine redistributes to other tissues (Benowitz *et al.*, 1990). This results in transient high brain concentrations of nicotine which act on many neurotransmitters and produce reinforcing psychological effects (Le Houezec and Benowitz, 1991). Subsequently, venous blood concentrations decline more slowly, reflecting redistribution from body tissues and the rate of elimination (half-life averaging 2 hours). In contrast to inhalation, other routes of absorption result in gradual increase in nicotine concentration in the brain, and a lower brain-to-blood ratio. These routes are considered as less addictive forms of administration.

BASIC AND CLINICAL PHARMACOLOGY OF NICOTINE

Because after tobacco use nicotine is present in the body in very low concentrations, analytic methodology sensitive enough to routinely assay nicotine in biological fluids has been available for only the past 15 to 20 years (Feyerabend *et al.*, 1975; Jacob *et al.*, 1981). The possibility of measuring levels of nicotine in the body gave rise to research on the human pharmacology of tobacco dependence. Many studies have shown that smokers tend to maintain constant their blood nicotine levels from day to day. However, blood nicotine levels are not highly correlated with the nicotine or tar 'yield' of cigarettes, determined with a smoking machine under standard smoking procedures. This body of evidence suggests a 'finger-tip control' of the smoker on his, or her, own nicotine intake.

Chemical properties of nicotine

Research on tobacco dependence has mainly focused on nicotine effects because nicotine is the major alkaloid of tobacco, representing 90 to 95% of the total alkaloid content. In addition to nicotine, there are several related alkaloids in tobacco that may be of pharmacological importance. There is, however, no information on their pharmacological effects in humans as yet. Nornicotine and anabasine have pharmacological activity qualitatively similar to that of nicotine (Clark *et al.*, 1965). In addition, some of the minor alkaloids may also influence the effects of nicotine (Stålhandske and Slanina, 1982). The minor alkaloids represent 8 to 12% of the total alkaloid content of commercial tobacco products, which are mainly derived from *Nicotiana tabacum*. In some other varieties of tobacco, nornicotine (*N. tomentosa*) or anabasine`(*N. glauca*) are present in concentrations exceeding those of nicotine (Wilbert, 1987). Alkaloids are synthesized in the roots and transported to the leaves with a concentration gradient. Leaves that are higher on the stalk have higher concentrations of nicotine than those lower on the stalk. Combining different varieties of tobacco and different parts of the plant is a way to control the nicotine concentration of tobacco products.

The alkaloid content of tobacco products also depends on the way tobacco is processed after being harvested. Blond tobaccos are dried in ovens under specific hygrometric conditions (flue curing). Such treatment makes the tobacco smoke acidic (pH 5–6). Dark tobaccos, such as those used for pipe or cigar tobacco in the United States, are smoked in cigarettes as well in other countries. They are sun or air dried (air curing) after a fermentation period aimed to reduce the alkaloid content, naturally higher in dark than in blond tobaccos. This process makes the smoke more alkaline (pH 6–7 for cigarettes, pH 8 for pipe or cigars).

Nicotine is a tertiary amine composed of a pyridine and a pyrrolidine ring. The natural stereo-isomer of tobacco is the l-nicotine, which is from 5 to 100 times (depending on which specific activity) more potent pharmacologically than the d-isomer (Jacob *et al.*, 1988). The latter is present in tobacco smoke only (up to 10% of the nicotine smoke content), indicating that some racemization occurs during the combustion process. Nicotine is a volatile and colourless weak base ($pK_a = 7.9$), which acquires a brown colour and the characteristic odour of tobacco when in contact with the air. Under atmospheric pressure, nicotine boils at 246°C, and is consequently volatilized in the cone of burning tobacco (800°C). Nicotine in freshly inhaled smoke is suspended in tar droplets (0.3–0.5 μm). Nicotine free base is readily absorbed across membranes because of its lipophilicity.

The nicotinic receptors

The pharmacologically active form of nicotine is a positively charged ion protonated on the pyrrolidine nitrogen. This active form resembles acetylcholine in the spacing of positive and negative charges (Cynoweth *et al.*, 1973). Because of its flexibility, acetylcholine can bind to both nicotinic and muscarinic receptors. Nicotine and muscarine molecules are less flexible and are, therefore, specific agonists of only one type of cholinergic receptors, hence the names.

The nicotinic receptors are member of the super-family of ligand-gated ion channels, and are made up of five subunits. Although the muscle (peripheral) nicotinic receptors present four types of subunits (α, β, γ, and δ in 2:1:1:1 ratio), the neuronal (central) receptors are made of only two types; α and β. The binding site of the receptor is located on the two alpha subunits. When nicotine binds to the receptor, it changes its conformation, opening the ion channel and allowing sodium to enter the cell; depolarization.

Two kinds of nicotinic receptors appear to co-exist in the brain according to their affinity to different ligands. One population is labelled with ^{3}H-nicotine or ^{3}H-acetylcholine and corresponds to a high affinity site, while another population, labelled with ^{125}I-α-BTX (bungarotoxin), corresponds to a low affinity site. The two receptor populations may mediate different effects. Until recently, the mechanism of the nicotinic receptor was only partially known. Many research groups are now studying the receptor extensively (Picciotto *et al.*, 1995). Their results will help to better understand some of its properties which might be directly related to the psychoactive effects and to nicotine dependence.

PHARMACOKINETICS OF NICOTINE

Absorption

The pH is important in determining the absorption of nicotine across cell membranes (pKa = 7.9). In acidic environments, nicotine is in its ionized state and does not readily cross membranes. However, at physiological pH (7.4), about 31% of nicotine is non-ionised such that it rapidly crosses membranes. The methods by which tobacco is processed and used result in considerable differences in the extent and rate of absorption of nicotine. Nicotine is readily absorbed through oral mucous membranes because of the thin epithelium and rich blood supply of these membranes. Nicotine that is swallowed (smokeless tobacco or nicotine polacrilex gum) is absorbed by the small bowel. After absorption into the portal venous circulation, nicotine undergoes presystemic metabolism by the liver so that its bioavailability is relatively low (30–40%). Consequently, buccal (mucous) absorption results in higher blood nicotine levels because nicotine bypasses the liver first pass effect.

The pH of smoke from flue-cured tobaccos found in most cigarettes is acidic. In contrast with some other tobacco products such as chewing tobacco, oral snuff tobacco, pipe or cigar tobacco smoke, there is little buccal absorption from cigarette smoke, even when it is held in the mouth. Inhalation is required to allow nicotine to be absorbed by the huge surface of the alveolar epithelium. Absorption into the systemic circulation is facilitated because pulmonary capillary blood flow is high, representing passage of the entire blood volume through the lung every minute. As we will see in the pharmacodynamics section, the kinetics of absorption of nicotine are important when considering the psychological or subjective effects which may play a role in nicotine dependence. Blood nicotine concentration rises quickly during cigarette smoking and peaks at the completion of smoking. Thus, nicotine absorbed from tobacco smoke can quickly reach various parts of the body, including the brain (Benowitz *et al.*, 1988). In contrast, input from smokeless tobacco has a small lag time, peaks and declines during a 30 minute period of administration, then continues to be absorbed for more than 30 minutes after tobacco is removed from the mouth. The later phase probably reflects delayed absorption of swallowed nicotine. Individual data of this study show that absorption of nicotine varies widely, both in extent and rate, among people.

Smoking behaviour is complex and smokers can manipulate the dose of nicotine delivered to the circulation on a puff-by-puff basis. The intake of nicotine varies considerably with the intensity, duration and number of puffs, depth of inhalation, and the degree of mixing of smoke with air. Because of the complexity of this process the dose of nicotine cannot be predicted from the nicotine content of the tobacco. In one study, the range of intake of nicotine among research subjects was 0.4–1.6 mg per cigarette, and was unrelated to the nominal nicotine yield of the cigarettes (Benowitz and Jacob, 1984).

The same complexity is observed with chewing tobacco or polacrilex gum. The rate of chewing, amount swallowed and local (buccal) factors can influence the absorption of nicotine. A threefold variation was found in a study of gum chewers asked to chew 10 pieces of gum, each for 30 minutes, daily (Benowitz *et al.*, 1987). However, since absorption from chewing gum is slow and persists even after the chewing stops, adjustments of the dose cannot be as precise as when smoking cigarettes (Benowitz *et al.*, 1988).

Because nicotine is readily absorbed through the skin, transdermal delivery systems (nicotine patches) have been developed for use in smoking cessation therapy. Absorption of nicotine from transdermal systems is slow, reaching maximum blood levels in 6 to 8 hours, but allow sustained concentrations of nicotine to be delivered over 24 h. The new generation of nicotine delivery systems is represented by nasal nicotine spray and nasal nicotine aerosol. Absorption of nicotine through the nasal route results in kinetic profiles similar to absorption from tobacco smoke (Sutherland *et al.*, 1992a). Besides their potential value in smoking cessation therapy, these systems present a potential use to deliver safely quantified doses of nicotine to smokers and non-smokers in experimental studies.

Distribution

Smoking is a unique form of systemic drug administration in that it delivers the drug into the pulmonary rather than the portal or systemic venous circulations. Based on physiological considerations nicotine is expected to reach the brain in about 10 seconds (Benowitz, 1990), faster than when nicotine is given intravenously. The drug is then extensively and quickly distributed to other body tissues with a steady state volume of distribution averaging 180 litres (2.6 litre/kg).

Simulation of nicotine concentrations in various organs after smoking a cigarette has been performed, using tissue distribution data derived from experiments in rabbits (Benowitz *et al.*, 1990). Arterial blood and brain concentrations increase sharply following exposure then decline over 20 to 30 minutes as nicotine redistributes to other body tissues, particularly skeletal muscle. In the minutes during and immediately following nicotine absorption, levels of nicotine are much higher in arterial than in venous blood. The discrepancy between arterial and venous blood concentrations has been observed in rabbits after rapid intravenous injection of nicotine (Porchet *et al.*, 1987) and in people after cigarette smoking (Henningfield *et al.*, 1990). Subsequently venous blood concentrations decline more slowly, reflecting redistribution from body tissues and the rate of elimination. The ratio of the concentration of nicotine in the brain to that in venous blood is highest during and at the end of the exposure period, and gradually decreases as the elimination phase is entered.

In contrast to inhalation, the oral and transdermal routes of absorption result in a gradual increase in nicotine concentrations in the brain, with relatively lower brain-to-blood ratio and little arterial-venous disequilibrium. Nicotine nasal spray and aerosol systems are probably in an intermediate situation between inhalation and these slow delivery systems.

Elimination

Nicotine is extensively metabolized in the liver, but also to a small extent in the lungs and kidneys (for a review see Benowitz *et al.*, 1990). Renal excretion of unchanged nicotine depends on urinary pH and urine flow, typically accounting for 5–10% of total elimination. The elimination half-life of nicotine averages 2 hours, although there is considerable interindividual variability. The primary metabolites of nicotine are cotinine (70%) and nicotine-N'-oxide (4%). Seventeen per cent of nicotine is not accounted for by metabolism to cotinine or nicotine-N'-oxide and may be metabolized in nicotine isomethonium ion and nornicotine.

Cotinine is formed in the liver in a two-step process involving cytochrome P-450 and aldehyde oxidase enzymes. Cotinine is further metabolized, with only about 17% of cotinine excreted unchanged in the urine. Trans-3′-hydroxycotinine is the major metabolite of cotinine. Cotinine-N-oxide and 5′-hydroxycotinine have been identified in human urine but little is known about their quantitative importance. Because of its longer half-life (averaging 16 hours), cotinine is commonly used as a biochemical marker of nicotine intake, such as in assessing whether a person has stopped smoking. Trans-3′-hydroxycotinine, the concentration of which in the urine exceed those of cotinine concentrations by 2–3 fold may also prove to be a useful indicator of nicotine exposure when a routine assay method becomes available. However, as nicotine replacement therapy becomes frequently used as an adjunct to smoking cessation, markers of nicotine intake are not useful in determining smoking status. Assays of minor tobacco alkaloids, such as anabasine or anatabine, which are present in tobacco but not pharmaceutical preparations of nicotine, may be of use in this situation.

Chronic administration

The half-life of a drug is useful in predicting the rate of accumulation of that drug in the body with repetitive dosing and the time course of decline after cessation of dosing. Consistent with a half-life of 2 hours, nicotine accumulates over 6–8 h (3–4 half-lives) of regular smoking. The usual blood nicotine levels sampled in the afternoon (plateau) range from 10 to 50 ng/ml. Each cigarette produces an increment in blood nicotine concentration of 5 to 30 ng/ml, depending on how the cigarette is smoked (Benowitz, 1988). During the night, blood levels decline and little nicotine is present in the blood of smokers when they wake up in the morning. Thus, cigarette smoking represents a situation where smokers are exposed to significant concentrations and pharmacological effects of nicotine for 24 hours a day. However, as it will be discussed in the pharmacodynamics section, tolerance to many effects of nicotine develops during the daily exposure. Overnight abstinence allows considerable elimination of accumulated nicotine and resensitization to actions of nicotine.

Individual differences

The level of a drug in the body during chronic administration depends on the rate of intake and the rate of elimination of this drug. There is considerable individual differences of these two parameters in smokers. Since there may be a four-fold difference in clearance of nicotine between individuals, a person with a very high clearance will have blood and body nicotine levels four times lower than a person consuming the same amount of nicotine but having a very low clearance (Benowitz et al., 1990). Smoking is known to accelerate the rate of metabolism of many drugs (Benowitz, 1988). However, brief abstinence (1–2 weeks) in smokers has been associated with increased clearance rather than decreased clearance. A possible explanation for this latter observation is that inhibition of metabolism may be due to a metabolite of nicotine or to any of the more than 3000 other tobacco smoke components (Benowitz et al., 1990). Gender differences in nicotine metabolism have been reported by Benowitz and Jacob (1984). Clearance of nicotine seems to be greater in men than in women. However, Russell and colleagues (1980) have shown in 330 smokers that men and women have similar blood nicotine levels on average, but also that men smoked more than women, which is consistent with a compensation by men for their faster elimination. Finally, since hepatic extraction of nicotine is high (60–70%), clearance of nicotine should be dependent of liver blood flow. Thus, physiological events that modify hepatic blood flow such as exercise, posture, meals or presence of other drugs, may have significant influence on nicotine metabolism. Consumption of a meal has effectively been shown to influence nicotine blood levels 30 minutes after the end of the meal (Lee *et al.*, 1989), but the biological importance of this finding remains to be determined (Benowitz *et al.*, 1990).

The use of subcutaneous administration in clinical studies

Studying the pharmacology of nicotine delivered via the smoking route is difficult because of marked intra- as well as interindividual variability in the absorption of the nicotine, related to differences in the smoking process. Moreover, it eliminates the possibility to compare the effects in non-smokers. Intravenous administration of nicotine allows for more precise control of the dose and time-course of dosing, but intravenous infusions are technically complicated, requiring intensive medical monitoring and severely limiting the mobility of research subjects. The oral route (i.e., polacrilex gum or nicotine tablets) is safe but because the rate of absorption of nicotine is slow, the rate of rise of levels of nicotine in the brain and resultant psychological, subjective, effects are dissimilar to those of smoking or intravenous nicotine (Benowitz *et al.*, 1988). Jones and colleagues (1992) have studied the effects of subcutaneous nicotine (0.4 and 0.8 mg) in Alzheimer patients and control subjects. However, blood nicotine levels were not measured in this study. Russell and colleagues (1990) studied acute tolerance to subcutaneous nicotine (0.75 or 1.0 mg, depending on subject's weight) in non-smokers. However, the brief period (60 minutes) of measurement of plasma nicotine levels was inadequate to characterize the pharmacokinetics of subcutaneous nicotine, which requires measurements for at least 8 to 9 hours (Benowitz *et al.*, 1991). We then decided to evaluate the pharmacokinetics of subcutaneous administration of nicotine as a potential route of administration for research studies (Le Houezec *et al.*, 1993).

The study was realized in 6 highly dependent cigarette smokers (Fagerström dependence score: m = 8.7; s.d. = 1.2) (Fagerström, 1978). The subjects were tested in the morning, after overnight abstinence from cigarette smoking, with 4 different doses (0.4, 0.8, 1.2 or 2.4 mg) on 4 different days separated by at least 2 days. Abstinence was assessed by measuring nicotine levels in baseline blood samples. Because possible side effects of doses over 0.8 mg were unknown, we gave the 4 nicotine doses in increasing order of dose rather than in a randomized sequence. Eighteen blood samples (5 ml) were taken at frequent intervals during the 8 h of each session. Cardiovascular responses and subjective, visual-analog scales, effects were also measured as preliminary data to design future studies on the psychophysiological effects of nicotine.

Subjects tolerated subcutaneous nicotine well, except for one subject who reported mild nausea at the highest dose. Subcutaneous nicotine produced a steady rise in plasma nicotine concentrations which peaked on average between, 19 and 25 minutes after injection, depending on the dose. Despite some interindividual variability, the peak level showed the expected dose effect [F(3/15) = 54.26, p = 0.0001] with a good correlation between peak concentration and dose (r^2 = 0.86). The peak concentration (Cmax) after the highest nicotine dose (2.4 mg) was similar to that reported by Benowitz *et al.* (1988) for smokers smoking one and one third cigarettes and higher than that measured after chewing two pieces of 2 mg nicotine gum. The peak concentration time (Tmax) occurred later compared to smoking (19–25 minutes vs. 9 min.), reflecting a slower absorption after subcutaneous administration, but earlier than that reported after nicotine gum (30–40 .min.). The pharmacokinetic parameters were in general similar to those reported in previous studies (Benowitz *et al.*, 1991, Feyerabend *et al.*, 1985, Lee *et al.*, 1987, Rosenberg *et al.*, 1980). The bioavailability (F) appears to be nearly 100% (equivalent to intravenous infusion). The half-life was similar to what have been reported after intravenous nicotine (Benowitz *et al.*, 1991). Our data suggest that subcutaneous nicotine is an alternative to intravenous dosing to estimate pharmacokinetic parameters in people.

Cardiovascular responses were somewhat variable, but in general were of a lesser magnitude than those seen after cigarettes smoking (Benowitz *et al.*, 1988). However, considering the low dose (0.4 mg) as a placebo, which is supported by the decrease in heart rate observed for 120 minutes after this dose, the nicotine-placebo differences in heart rate after the two highest doses showed an increase of about 7–8 bpm at 15 minutes, similar to that previously reported after subcutaneous nicotine by Russell *et al.* (1990). The effects on heart rate were more transient than the effects on systolic and diastolic blood pressure, which stayed above baseline longer, principally after the 2.4 mg dose.

The subjective effects of subcutaneous nicotine were modest. This might be explained by the slower absorption and distribution to the brain compared to cigarette smoking. However, we must emphasize that the use of a visual-analog scale may have produced more variability in the responses, compared to a rating scale which offers less choices, for example rating on a 5 point scale. Taking into account the small number of subjects (n = 6) this may have prevented us obtaining reliable results. However, our analysis of the subjective data could be of value to other investigators in planning future studies of the effects of subcutaneous nicotine in human subjects. A significant *time* effect was observed on responses to 'I would like a cigarette now' (urge to smoke) and 'I feel high', suggesting that subjects were responding to the effect of nicotine. A significant *time* × *dose* interaction observed for 'I feel high' and 'How strong was the dose of the injection' supports the idea that the effect of the highest dose lasted longer than the effect of lower doses. Despite the lack of interaction between *time* and *dose* for 'I would like a cigarette now', it is interesting to note that urge to smoke remained suppressed for the entire 90 min. measurement after the highest dose while scores returned to baseline for this question by the end of the session after the three other doses. Subjects presented a substantial reduction in anxiety after the first (0.4 mg) nicotine dose. This is because the baseline, pre-treatment-level of anxiety was highest at this time, and quite variable between subjects. Consequently, a biased effect was observed for the response concerning anxiety, which would have probably be attenuated if a counterbalanced order of treatment had been used between subjects. However, we have also observed a great interindividual variability in responses in a study where treatment order was counterbalanced (Le Houezec *et al.*, 1994). It is likely that variability was due to differences in anxiety and expectation from the injection and the experimental procedures. Although subjects may be less apprehensive about subcutaneous compared to intravenous nicotine, some anxiety prior to the first treatment must be anticipated. In planning testing of psychopharmacological effects of nicotine, we suggest including at least one control session to accustom subjects to the experimental procedure and to the administration of nicotine.

In conclusion, considering pharmacokinetic parameters, cardiovascular responses and subjective effects, subcutaneous nicotine seems suitable for studying effects of nicotine in people. Doses of 26–31 μg/kg will produce blood levels similar to those observed in cigarette smokers. Based on these results and those obtained from a study in non-smokers (Le Houezec *et al.*, 1994), doses of 15–20 μg/kg should be suitable for studies in non-smokers.

PHARMACODYNAMICS OF NICOTINE

The term pharmacodynamics refers to the relationship between levels of a drug in the body and its effects. This section will focus on how observations on the pharmacodynamics of nicotine may help us understand some of the consequences of cigarette smoking. Two issues are particularly relevant in understanding the pharmacodynamics of nicotine: its complex dose-response relationship and the development of tolerance with repeated administration.

Dose-response effects

Nicotine is known to have a complex dose-response relationship. Langley's early works on nicotine, at the turn of the century, have shown that low doses of nicotine stimulate ganglion cells while high doses produce a short stimulation followed by a blocking effect. This biphasic response is observed *in vivo* as well, although the mechanisms are far more complex. Smoking has been reported to have both stimulating or calming effects. However, if it is clear that nicotine effects may be interpreted in terms of increased activation of the autonomic and central nervous systems, the evidence that high doses of nicotine obtained from smoking have opposite effects is far less convincing. The relaxing effect of nicotine could be a conditioned response secondary to the peripheral muscle-relaxing effect of nicotine induced by inhibition of the Renshaw cells of the anterior horns of the spinal cord.

Moreover, comparisons of the effects of acute nicotine with effects of chronic tobacco use should be made with caution. And before extrapolating pharmacological observations from animals to humans, blood concentrations should be measured to ensure that the effects are being studied in a portion of the dose-response curve relevant to smokers.

Acute and chronic tolerance

Tolerance is usually defined as when, after repeated doses, a given dose of a drug produces less effect or increasing doses are required to achieve the response observed with the first dose. Pharmacodynamic tolerance can be further defined as when a particular concentration of a drug at receptor site (often approximated by the blood concentration) produces a lesser effect than it did previously. Pharmacokinetic tolerance refers to induction of drug metabolism resulting in a reduced effect after repeated doses of a drug. The latter appears not to be of significance as a mechanism for development of tolerance to nicotine. Individual differences in sensitivity to a drug, often referred to as drug tolerance, should preferably be called acute drug sensitivity. Two kinds of tolerance to nicotine effects are usually observed; acute and chronic. Acute tolerance develops in minutes and is observed in smokers as well as in non-smokers. In smokers, tolerance to nicotine effects builds up during the day as the plasma nicotine levels increase. During the night, due to the relatively short half-life, nicotine levels decrease and very little nicotine is still present when smokers wake up in the morning. Some nicotine effects, like heart rate response, are very sensitive to acute tolerance. This was shown in smokers by Porchet and colleagues (1988) and in non-smokers by Russell (Russell *et al.*, 1990).

Although studies of acute tolerance to nicotine began with the work of Langley and his colleagues about a century ago, little has been known until recently about nicotine effects in relation to its blood concentration. The effects of intravenous nicotine given as a continuous infusion (Benowitz *et al.*, 1982), or in a manner to simulate cigarette smoking, as a series of injections given every 30 minutes for 3 hours (Rosenberg *et al.*, 1980), showed the same pattern of response as after smoking a cigarette. Heart rate and blood pressure increased sharply after the beginning of nicotine administration, but showed little increment with further increase in blood nicotine levels. These data indicated rapid development of acute tolerance, but not complete tolerance in that these effects remained above baseline measurements. In contrast, skin temperature followed inversely the rise and decline of blood nicotine levels, showing no evidence of tolerance.

Some properties of the nicotinic receptor can be related to tolerance, which is believed to play a role in nicotine dependence. The common rule in receptor pharmacology is that chronic exposure to an agonist results in a down-regulation of the receptor, that is a reduction in number of receptors, whereas chronic exposure to an antagonist produces an up-regulation. In contrast, the nicotinic receptor is up-regulated in the presence of nicotine, its agonist. The reason for this seems to be that the configuration of the receptor changes when nicotine binds to it, resulting in a different form unable to respond to nicotine for a period of time. This phenomenon is known as desensitization. However, desensitization *per se* is not sufficient to induce measurable up-regulation. A *single* administration which presumably desensitize nicotinic receptors is without effect on receptor numbers; *repeated* administration is required for up-regulation. So, other mechanisms must be involved which need to be better characterized (Wonnacott, 1990).

Chronic tolerance has been much less studied but certainly plays a critical role in nicotine dependence. Although, it is tempting to parallel acute tolerance to desensitization and chronic tolerance to up-regulation, this is only speculation. There is a need for more studies on both acute and chronic tolerance for a better understanding of tobacco dependence mechanisms.

PHARMACOKINETICS AND PHARMACODYNAMICS OF NICOTINE DEPENDENCE

The reinforcing properties of a drug are expected to be stronger when positive, subjective effects quickly follow the self-administration. If the effects of the drug is delayed due to slower absorption or if tolerance to the drug develops quickly, the abuse liability of this drug is low. Slow delivery systems present consequently a low abuse potential (Henningfield and Keenan, 1993), whereas the abuse potential of nicotine nasal spray and aerosol is probably higher because their kinetic profiles are close to that of smoking.

However, as mentioned previously in discussing the distribution kinetics of nicotine, venous blood levels may not reflect levels at the site of action. To examine the importance of distribution kinetics in the apparent development of acute tolerance to nicotine, Porchet *et al.* (1987) compared effects of rapid- and slow-loading infusions of nicotine, followed by a maintenance infusion. Venous blood nicotine levels, similar in both conditions, did not explain the much greater cardiovascular and subjective effects of nicotine during the rapid- compared to the slow-infusion. Concentrations of nicotine in the brain were simulated using a model incorporating distribution kinetics in arterial blood and the brain derived from studies in rabbits (Benowitz *et al.*, 1990). While pharmacodynamic analysis of nicotine effects based on venous levels, which underestimate brain levels, incorrectly indicated acute tolerance, simulated brain concentrations showed no tolerance. In other words, nicotine brain levels more closely resembled the heart rate acceleration curve than did venous blood levels.

It was observed, however, that a late peak (30 minutes) in venous nicotine concentration was not accompanied by a peak in heart rate. This cannot be explained by distributional tolerance, but rather indicates true tolerance. To characterize the time course of tolerance development and regression, a study in which subjects received paired intravenous infusions of nicotine, separated by different time intervals was conducted (Porchet *et al.*, 1988). Despite higher blood concentrations, heart rate and subjective effects were of much less magnitude after the second compared to the first infusion when separated by 60 or 120 minutes. When 210 minutes separated the two infusions, the response was fully restored. The pharmacokinetic-pharmacodynamic model developed from this study indicated a half-life of development and regression of tolerance of 35 minutes. Thus, after 1.5 to 2 hours (3–4 half-lives) of steady-state exposure to nicotine, tolerance is almost fully developed. This model also suggests that tolerance is not total and that 20 to 25% of the expected pharmacological effect at a similar level of nicotine in the non-tolerant state persists.

These pharmacodynamic studies suggest that nicotine effects in a smoker will persist during the daily smoking cycle despite the development of some acute tolerance. The interval between 2 cigarettes may be influenced in part by the fluctuations of tolerance during this cycle. The first cigarette of the day produces substantial pharmacological effects, but at the same time tolerance begins to build. The duration of time until the next cigarette is smoked may be determined as the time at which there is some regression of tolerance but before severe withdrawal symptoms appear. With succession of cigarettes, there is accumulation of nicotine in the body, resulting in a greater level of tolerance. Transiently, high brain levels of nicotine after each individual cigarette may partially overcome tolerance, but the reinforcing effects tend to lessen throughout the day. In accordance with the elimination half-life of two hours, nicotine is almost totally eliminated from the body after overnight abstinence, allowing resensitization to actions of nicotine (Benowitz, 1990). Full resensitization, however, may require days (Lee *et al.*, 1987). Chronic tolerance to nicotine in human subjects needs to be studied more systematically, this will require to compare regular smokers to non-dependent smokers ('chippers') as well as to non-smokers.

PSYCHOPHARMACOLOGICAL EFFECTS OF NICOTINE AS REINFORCERS OF TOBACCO DEPENDENCE

Nicotine has numerous effects on physiological systems, but its actions on the central nervous system are of particular interest in that such actions presumably reinforce smoking behaviour. There is evidence that smokers can adjust their nicotine intake to *optimize* mental functioning and to control their mood (Le Houezec *et al.*, 1996). There is no doubt that a drug that facilitates shifts toward optimal cognitive performance and positive affect may stimulate reward systems as does any other seeking behaviour (Ashton and Golding, 1989; Le Houezec and Benowitz, 1991). People who are likely to 'use' these reinforcing properties of nicotine are probably at greater risk for dependence and for severe withdrawal symptoms during smoking cessation. The research from our laboratory is based on the hypothesis that the improvement of mood and cognitive function by nicotine is a powerful reinforcer of tobacco dependence.

Cognitive performances

Many studies have found that cigarette smoking or nicotine administration improves mental functioning in abstinent smokers (U.S. Department of Health and Human Services, 1988). It is likely that improvements in cognitive performance could be a major factor in why people smoke and could explain why they find it difficult to quit smoking (Le Houezec and Benowitz, 1991). Interest in nicotine effects on human performance has been growing as can attest the special issue on nicotine published in Psychopharmacology (Vol. 108, No. 4, September, 1992). However, this body of research has serious methodological limitations (Le Houezec and Benowitz, 1991) and despite a lot of interesting data there are still a few questions that need more precise answers.

1 One is whether the enhancement of performance observed after smoking is due to relief of symptoms of abstinence or to a primary effect of nicotine on the brain, or even a combination of both (Hughes, 1991). A study by Snyder and colleagues (1989), illustrating the effect of relief of the symptoms of abstinence, assessed the decrement of performance of smokers during a 10-day period of abstinence. Using a battery of cognitive tests, impairment of performance was

observed within 4 hours after the start of the deprivation period, peaked between 24 and 48 hours, and then returned towards baseline values. However, some performance deficits remained throughout the 10 days of abstinence. Resumption of smoking at the end of this period resulted in return of performance to baseline values within 24 hours. In contrast, a study by West and Hack (1991a) illustrated the direct effects of nicotine by showing an improvement of short-term memory using Sternberg's memory search task (see for details of task Kole *et al.*, Ch. 9; Zeef *et al.*, Ch. 16). Subjects were presented for 3 seconds with a list of 2 or 5 digits, termed the 'memory set'. A series of 16 probe digits were then presented and subjects had to indicate as fast as possible whether or not the probe was a member of the memory set. The authors tested occasional and regular smokers before and after a nicotine or a nicotine-free cigarette and before and after 24 hours of abstinence. The memory search rate was significantly increased in both groups after the nicotine cigarette compared with the nicotine-free cigarette, but only for the larger (5 items) memory set. These results suggest that nicotine acts specifically on memory scanning, an effect never reported before with any other drug. Interestingly, there was no difference in results between the occasional and the regular smokers and no influence of abstinence.

A second question that has not been fully investigated is the role of nicotine itself in determining the effects of smoking on performance. In most of the studies the role of nicotine is generally inferred but has rarely been directly assessed. Neither actual nicotine intake, nor nicotine blood levels (reflecting nicotine intake) have been measured or controlled. Most studies have experimentally manipulated the type of cigarette, comparing cigarettes with low and high nicotine delivery, or the number of cigarettes smoked. Neither of these measures accounts for interindividual variability in smoking behaviour and nicotine absorption (Benowitz and Jacob, 1984; Feyerabend *et al.*, 1985).

Finally, a third question is whether the improvement observed after nicotine is due to changes in specific mental processes (e.g., attention) or reflects a more general effect on all mental processes (e.g., arousal effect). The use of reaction time (RT), the time between the presentation of a stimulus and the subject's response, is not sufficient to answer this question. Analysis of Event-Related Potential (ERP) components, extracted from the electroencephalogram, gives indexes of information processing stages and allows one to distinguish between the perceptive and the motor segment of information processing, and to approach the chronometry of mental functioning (Donchin, 1979; Renault *et al.*, 1988). For example, the appearance of the P_{300} wave (a positive wave occurring 300 to 600 ms after a stimulus presentation) is considered to reflect the end of the stimulus processing, the remainder of the RT representing the response processing. Thus, increasing stimulus complexity will both increase P_{300} latency and RT, while increasing response complexity will only affect RT, with no effect on P_{300} latency. Another component, the N_{100} wave (a negative wave occurring around 100 ms after a stimulus presentation), reflects the physical characteristics of the stimulus as well as some aspects of attention (Luck *et al.*, 1990).

SUBCUTANEOUS NICOTINE AND INFORMATION PROCESSING IN NON-SMOKERS

Because the possibility of nicotine withdrawal effects cannot be excluded in smokers, it is necessary to study the effects of nicotine in non-smokers. In a few studies nicotine has been administered by oral tablets to smokers and non-smokers, but opposite findings, both the presence and absence of improvement, have been reported (Wesnes *et al.*, 1983; Wesnes and Warburton, 1984c; Wesnes and Revell, 1984).

Studies on the effects of nicotine on performance have paid little attention to the neurocognitive processes that mediate performance (Le Houezec and Benowitz, 1991). Edwards *et al.* (1985) for example, reported that nicotine speeded RT and decreased the latency of the P_{300} wave of the ERP. Changes in P_{300} latency are largely under the control of stimulus variables (Van der Molen *et al.*, 1991), suggesting that nicotine improves performance by acting on stimulus variables. This interpretation is consistent with the findings showing that anticholinergic drugs such as scopolamine increase P_{300} latency and slow RT (Callaway *et al.*, 1985; Brandeis *et al.*, 1992). By contrast, amphetamine speeds RT but has only a small effect on P_{300} latency, suggesting that this drug acts mainly on response processes (Naylor *et al.*, 1985; Halliday *et al.*, 1987). There have been few studies using information processing paradigms that have assessed the effect of nicotine on P_{300} and RT. However, the findings of Edwards *et al.* (1985) needed to be replicated because they did not assess earlier components, such as the N_{100}. The P_{300} latency changes could effectively reflect changes that occur earlier in processing.

To address the above issues we examined the effects of a low dose (0.8 mg) of subcutaneous nicotine on multiple measures of information processing in a group of non-smokers (Le Houezec *et al.*, 1994). We assessed performance (RT, accuracy, and speed-accuracy trade-off measures) and the N_{100} and P_{300} components of the ERP. Speed-accuracy analyses were used because drugs may alter performance by improving speed but at the expense of decreasing accuracy (Woods and Jennings, 1976). We also assessed blood nicotine levels during the intervals when the subject performed the task. Our hypothesis was that nicotine would improve cognitive function by acting on attentional or stimulus processes, and so would affect both RT and P_{300} latency.

The choice reaction time task we used, called the Stimulus Evaluation-Response Selection (SERS) task (Callaway *et al.*, 1985; Naylor *et al.*, 1985) discriminates between two processing stages, stimulus evaluation and response selection, by independent manipulation of stimulus and response complexity. This gives four experimental conditions with different levels of difficulty. We tested the effect of nicotine by measuring performance in the task before and after subcutaneous saline or nicotine or no treatment (control) in the same subjects (n = 11) on 3 different days. The most convincing demonstration that nicotine improved performance was found by examining speed-accuracy trade-off function. This function is based on the fact that when subjects are asked to respond faster, they speed their responses but they make a larger number of errors. However, instructions to respond accurately result in an increase in mean RT. In our task both speed and accuracy were emphasized in order to minimize speed-accuracy trade-off effects. The speeding of RT observed after nicotine compared to saline was not statistically significant (less than 10 ms). However, when RT was normalized, by task condition and subject, we observed a significant increase in the number of fast RTs after nicotine compared to saline. This effect was not accompanied by an increase in errors. Thus, the increase in number of fast RTs did not occur because subjects were trading speed for accuracy. It is of interest to note that Jones and colleagues (1992) reported a small non-significant speeding of RT with the same dose (0.8 mg) of subcutaneous nicotine in a different task. The fact that nicotine at this dose acts selectively on the fast end of the RT distribution may be a function of dose. We tested 2 smokers in the same conditions with a higher nicotine dose (2.4 mg subcutaneously) and obtained a mean RT speeding of 50 ms in both of them.

No significant effects were found on the N100 wave. Nicotine speeded P_{300} latency, but only in the hardest of the 4 task conditions. Herning and Pickworth (1985), examining the effects of nicotine gum (0, 4 and 8 mg dose) on P_{300} latency in an auditory oddball task, found a shorter latency after nicotine only in the hardest of two task conditions. These results, like ours, suggest that nicotine might increase stimulus sensitivity when the task becomes more difficult. However, in our study nicotine substantially slowed P_{300} latency in an easier task condition and had no effect on the two easiest ones. This illustrates the importance of assessing nicotine effects on P_{300} latency for a wide variety of task conditions before implicating nicotine as affecting such broad construct as stimulus evaluation processing.

Another important point is that the blood nicotine levels we obtained in the non-smokers were lower than expected from the pharmacokinetic study we did previously in smokers (Le Houezec *et al.*, 1993). This can temper the fact that the cognitive performance changes we measured were not dramatic, and illustrates the importance of assessing nicotine intake in studies looking at nicotine psychopharmacological effects.

Affective effects of nicotine

There is some evidence that smokers may be able to control their mood as well as their cognitive function (Carton *et al.*, 1994a,b; Le Houezec *et al.*, 1995, 1996). There is also some evidence that a history of major depression, even in remission, is associated with greater difficulty in smoking cessation (Anda *et al.*, 1990; Glassman *et al.*, 1990). Occurrence of mood perturbation as a smoking withdrawal symptom may then represent a negative prognostic for smoking cessation.

People who are likely to 'use' these reinforcing properties of nicotine may present specific personality dimensions, like arousal seeking, which may predispose them to dependence. One of our hypotheses links energetic deficit encountered in some psychiatric diseases (depression, schizophrenia, degenerative pathologies) to frontal hypoactivity implicating dopaminergic pathways. Due to its direct action on dopaminergic systems (Balfour, 1994), nicotine could be considered as a self-medication of such deficits.

Illustrating this hypothesis is the result obtained in one subject recruited as an anhedonic person (loss of pleasure) who pretended to use smoking as a psycho-stimulant drug. This subject was not depressed at the time of the tests but had a history of recurrent major depressive episodes. He came twice in the morning after overnight abstinence from smoking. We recorded his ERP in a classical oddball task before and 2 hours after either smoking of 3 small cigars or taking of a single dose of dopaminergic antidepressant drug. ERP traces in both conditions appeared to be very similar. In comparison with each of the pre-drug (abstinence) conditions, a similar shift of the ERP traces toward the left (speeding of information processing) was observed in both post-drug conditions. This result which need to be replicated suggests that nicotine may produce a cortical activation through the stimulation of dopaminergic pathways. If it is confirmed this could be the evidence of a behavioural trait (stimulating smoking with anti-anhedonic effect) which may expose this type of smokers to a greater risk of development of a depressive episode during smoking cessation.

The next step is our study in smokers trying to quit smoking with nicotine replacement therapy (transdermal systems). Evaluation of cognitive function (ERP) and mood changes (clinical evaluations) are made before and during the treatment. The results of this study should help to better understand the role of affective and cognitive perturbations in failures of smoking cessation attempts.

CONCLUSION

Several issues still need to be addressed in order to better understand both the actions of nicotine on the central nervous system and their implication in tobacco dependence. One of these issues concern the tolerance to nicotine effects on human information processing. Most of the studies that have been conducted, including the one in non-smokers presented above, provide some interesting information about the direct acute effects of nicotine, but give no information concerning chronic nicotine use. There is a need for studies in different populations (non-smokers, ex-smokers, regular and occasional smokers) to better understand the role of tolerance. Moreover, as we observed huge differences in nicotine response, it is of importance to study individual effects because predominant effects of nicotine may greatly differ among individuals. In this view, it might be necessary to better characterize the personality and mood of subjects as well as their pharmacological response to nicotine.

SUMMARY

Actions of nicotine on the central nervous system are of particular interest in understanding why people smoke. The complexity of the relationships between nicotine and the neurotransmitters leaves considerable gaps in the understanding of dependence. The current knowledge on nicotinic receptors is fragmentary. The same is true for mechanisms of tolerance, which probably play an important role in nicotine dependence. Many studies have reported improvement of cognitive function. However, most of the results are inconclusive owing to methodological problems. It is still not known if smoking enhances performance by acting directly on the brain or merely by relieving the symptoms of abstinence. It is also unclear whether nicotine improves task performance by affecting specific mental processes or whether it has a more global effect on all mental processes. The research from our laboratory is based on the hypothesis that the improvement of mood and cognitive function by nicotine is a powerful reinforcer of tobacco dependence.

CHAPTER TWO

Effects of Nicotine on Human Performance*

Walter S. PRITCHARD and John H. ROBINSON

INTRODUCTION

In this review, we will focus the attention on the effects of smoking/nicotine on cognitive performance, although we will also review effects on perceptual, motor, and academic performance. We use the term smoking/nicotine to indicate that the majority of the studies that we review investigated the effects of cigarette smoking, with nicotine being the primary psychopharmacological agent in tobacco (Robinson *et al.*, 1992; for data concerning how different smoking regimens can differentially affect performance, see Morgan and Pickens, 1982). However, studies employing alternate routes of nicotine administration will also be considered.

Previous reviews of this area (Koelega, 1993; Parrott, 1992; Sherwood, 1993; Warburton, 1988a, 1988b, 1992; Warburton *et al.*, 1988; Wesnes and Warburton, 1984a) indicate that the acute effects of smoking or nicotine administration on cognitive performance are generally beneficial compared with either pre-administration baseline or a 'no-administration' control group. In a meta-analytic review of nicotine and a wide variety of other psychopharmacologically-active compounds, Hindmarch *et al.* (1991a) concluded: It is apparent from the data that nicotine compares "favourably" with the other substances in that it has small, positive effects on all the dependent variables ... similar to those of caffeine ... (p. 517). In fact, based on Cohen's *d* scores derived from a priori t-tests, Hindmarch *et al.* (1991a) found that nicotine not only produced significant facilitation of both *'decision reaction time'* (the time it takes the subject to release a central button that he/she had been keeping depressed) and *'movement'* (the time it takes the subject subsequently to move to and then press one of a set of response keys) relative to placebo, but produced the greatest facilitation (+0.561 and +0.853 respectively) of any compound tested. This compares with the significant impairments produced by other compounds, e.g., –0.804 and –0.744 for alcohol, –0.751 and –0.897 for morphine, and –1.821 and –1.468 for the benzodiazepine lorazepam. Caffeine did

---

* A set of all tables with references to the notes of each table is available from the senior author upon request.

not differ from placebo effects for either measure. The major *theoretical* question that we will address in our review is the *fundamental* nature of the effects of smoking/nicotine on human performance.

Three specific hypotheses may be formulated regarding such effects.

- The *absolute facilitation* hypothesis. According to this hypothesis, smoking/nicotine results in 'better than normal' performance, that is, an enhancement of performance that is absolute in nature.
- The *withdrawal deficit* hypothesis. According to this hypothesis, pre- to post smoking 'improvements' in performance actually represent the amelioration of performance deficits brought on by abstinence from smoking for a period of time (cf., Heishman *et al.*, 1990; Herning *et al.*, 1990; West, 1990). Many of the studies in the literature have employed smokers who have been tested in the morning following overnight smoking cessation. The implicit goal of this procedure is to maximize the chances of obtaining smoking effects, which is understandable given the premium placed on obtaining statistically significant results.
- However, such a procedure leaves unclear whether effects on performance are either absolute, due to the relief of withdrawal deficits, or due to some combination of the two (the latter possibility being what we term the *combination* hypothesis).

Ruling out the withdrawal deficit hypothesis

Alternatives for ruling out the withdrawal-deficit hypothesis are testing non-abstaining smokers and testing non-smokers. Both non-abstaining smokers and non-smokers 'by definition' would not be in a withdrawal state, and any performance enhancements seen following smoking/nicotine would have to be absolute in nature. An important sub-issue here is the definition of 'non-abstaining'. A smoker who has just finished his or her usual brand of cigarette would of course be 'non-abstaining', but one would not expect a second cigarette in a very short period of time to have any effects (other than possible toxicity). Thus, within an experimental context, 'non-abstaining' would have to be operationally defined as a smoker who had not had a cigarette within a period of time short enough so that any potential withdrawal effects would not be seen, but long enough so that one could reasonably expect to see any potential performance effects. The literature reviewed below suggests that testing 30 minutes to an hour post-smoking qualifies as 'non-abstaining'. Studies employing non-smokers rely on routes of nicotine administration other than inhalation of cigarette smoke, primarily nicotine absorption from the buccal mucusa of the mouth. The gastrointestinal bioavailability of nicotine is low due to significant first-pass metabolism by the liver, so that the swallowing of 'nicotine pills' is not a viable alternative route (Benowitz, 1987). Alternative routes are necessary when testing non-smokers for two reasons. The first relates to ethical concerns about requiring non-smokers to smoke in an experimental setting. The second reason is more practical: it can be difficult to get non-smokers to inhale cigarette smoke so that nicotine is absorbed (because of the acidic nature of cigarette smoke, the nicotine that it contains is not absorbed buccally; buccal absorption is limited to more alkaline media such as cigar smoke or prescription nicotine gum; Benowitz, 1987). An important issue to keep in mind regarding non-smoker studies is that non-smokers generally have a lower *functional* tolerance to toxic effects of nicotine such as feelings of lightheadedness and nausea (Nyberg *et al.*, 1982). A fine line exists in non-smokers between a level of nicotine sufficient to affect performance without producing malaise.

Absolute facilitation versus the combination hypothesis

The demonstration that smoking/nicotine enhances performance in non-abstaining smokers/non-smokers does not distinguish absolute facilitation from the combination hypothesis. Just because nicotine can produce absolute facilitation does not mean that one-hundred percent of the performance enhancement seen in abstaining smokers is absolute in nature. Distinguishing the absolute facilitation hypothesis from the combination hypothesis is more difficult. There are two ways to approach this problem.

The first approach involves comparing the effects of nicotine in non-smokers, abstaining smokers, and non-abstaining smokers. If nicotine facilitates cognitive performance equally in all three groups, then absolute facilitation is supported. If there is no effect of nicotine in non-smokers and non-abstaining smokers, but there is facilitation in abstaining smokers, then the withdrawal deficit hypothesis is supported. If there is an effect in non-smokers and non-abstaining smokers, but an even greater effect in abstaining smokers, then the combination hypothesis is supported, since the greater effect in the abstaining smokers is inferred to come from a withdrawal-relief component added onto an absolute component. Note that the non-smoker group is necessary to experimentally support the combination hypothesis, since merely demonstrating greater facilitation by smoking/nicotine in abstaining than in non-abstaining smokers might be due to tolerance and/or ceiling effects for the task employed. A result where the effect is greater in both non-smokers and abstaining smokers presents a more complicated situation. This would indicate absolute facilitation combined with tolerance and/or ceiling effects with regard to performance of the task in question. Other combinations of outcomes are possible. Note that it is possible to come up with combinations that seem unlikely, nearly impossible, and even outright ludicrous. Others do not have a ready interpretation (at least to us!) and seemingly do not fit within the simple bipolar theoretical framework of absolute facilitation versus withdrawal deficit. Such outcomes would call for a more sophisticated underlying theory.

The second approach to discriminating absolute facilitation from the combination hypothesis is to track the effects of smoking cessation over a period of days relative to pre- and post-cessation 'baselines' as well as relative to a no-cessation control group. Examining the effects of cessation for several days is necessary because, by definition, potential withdrawal symptoms are transient phenomena that should resolve as body physiology re-equilibrates to the absence of the compound in question. Thus, studies that merely compare performance following overnight cessation with the appropriate baselines cannot distinguish possible withdrawal deficits from the removal of absolute facilitation – it always remains possible that, had the subjects been tracked for several days, no attenuation of the initial performance deficits would be seen. A no-cessation group is necessary to control for the mere passage of time. Without such a control, 'recovery' toward baseline cannot be distinguished from either gradual, long-term learning with regard to the cognitive tasks employed or adaptation to the experimental setting. An example of the latter would be conducting a study in which subjects were confined to a clinical ward. Such a procedure is excellent for ensuring compliance with an extended period of smoking abstinence. However, it may also engender unwanted effects on performance. For example, in subjects whose goal is not to permanently stop smoking, performance may deteriorate early in the study as it begins to 'sink in' to the subjects that they still have several days to go until the study is over. Similarly, performance may later improve as the end of the study is anticipated. The importance of a no-cessation control group was emphasized by Hatsukami *et al.* (1985) in a study of the mood effects of smoking cessation. They found that the addition of the control group eliminated findings that would have been significant if only a baseline/ deprivation comparison had been made. For example, restlessness increased significantly from baseline to the deprivation period for subjects in the experimental [cessation] group. However, there was also an increase in this measure for subjects in the control [no cessation] group over the same period of time. Thus, the inclusion of a control prevented an erroneous conclusion from being drawn, as would have happened if data from only the experimental group had been analyzed [pp. 59–60].

To illustrate the points that we have been making, we present three sets of hypothetical results from a cessation study of the sort outlined above. The upper graph of Figure 2.1 illustrates absolute facilitation. Although there is some improvement in the cessation group across cessation days, this improvement is no more than that seen in the control group and may be attributed to factors (such as those discussed above) related to the passage of time. The middle graph of Figure 2.1 illustrates a withdrawal deficit. Performance deteriorates markedly on the first day of cessation, then rapidly improves back to baseline over the next three days; the amount of improvement is much greater than that seen in the control group. Finally, the lower graph of Figure 2.1 illustrates the combination hypothesis.

The cessation group shows greater improvement than the control group, but performance levels off on days four-five without returning to baseline. The resumption of smoking leads to an (absolute) improvement in performance equal in magnitude to the withdrawal deficit. The time course that we have chosen for illustrative purposes is somewhat arbitrary – it may be that subjects have to be 'tracked' through one or even two weeks of smoking cessation to resolve the issue.

THE ISSUE OF NON-SMOKER 'CONTROLS'

Many studies of the performance effects of smoking/nicotine have employed non-smokers who are not administered nicotine as controls (a quite different procedure from investigating the effects of nicotine in non-smokers relative to smokers as outlined in the previous section). The presumption made is that non-smokers would represent 'normal' performance against which not only the acute effects of smoking/nicotine could be judged, but also longer-term effects of smoking on cognitive performance. This presumption is fraught with pitfalls. There is evidence

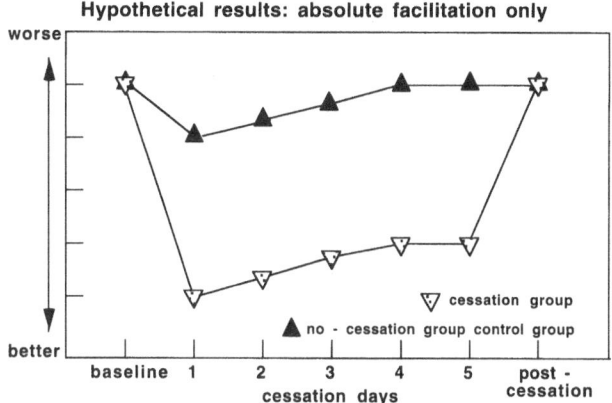

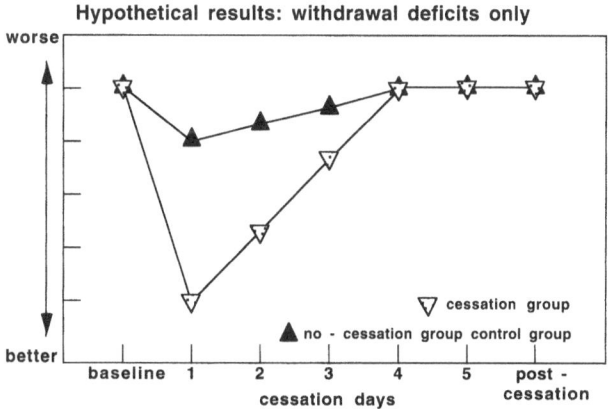

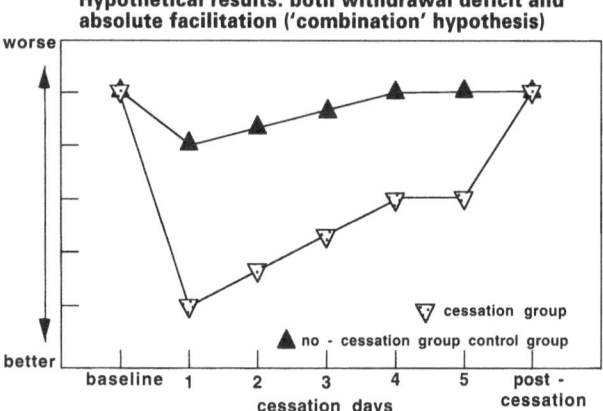

Figures 2.1 – Hypothetical results of absolute facilitation, withdrawal deficit or both combined on performance.

that both the initiation and persistence of smoking are genetically influenced (e.g., Carmelli *et al.*, 1992; Heath and Martin, 1993), and important constitutional differences have been demonstrated between persons who take up smoking and persons who do not. For example, smokers and non-smokers as groups differ along a number of personality dimensions. These differences pre-date the onset of smoking (Cherry and Kiernon, 1976, 1978; Seltzer and Oechsli, 1985; Sieber and Angst, 1990) and thus are largely a function of smoker versus never-smoker differences (as pointed out by Hughes, 1991: "most studies that use a non-smoker control group imply but do not clearly state that the group includes never-smokers but not ex-smokers" [p. 409]). The most reliable smoker/non-smoker personality difference is that smokers as a group score higher than non-smokers on Eysenck's dimension of psychoticism, or P (for a review, see Pritchard, 1991a). Despite its name, P primarily measures a combination of impulsive behaviour along with a cynical, often anti-social, attitude toward the world in general. Both these factors could conceivably impact cognitive performance (greater impulsivity may lead to faster but less accurate responding; a high level of cynicism may lead to less than maximal effort when performing a given cognitive task). Further, smokers on average have lower IQs than non-smokers (Hill and Gray, 1982; Ray, 1985). That is not to say that there are not intellectually brilliant individual smokers (Warburton, 1993), merely that smokers as a demographic group have lower IQs than non-smokers. Regardless of the relative contributions of heredity versus early environment in determining IQ, it is largely fixed at an age that predates the onset of smoking. Thus, when a group of smokers and a group of non-smokers are compared on a more complex cognitive task, and overall the non-smokers perform better, this may reflect the smoking/IQ literature. It certainly says little about the effects of smoking on cognitive performance, either acute or long-term. Matching smokers and non-smokers with regard to personality and IQ would seem to be a minimally necessary procedure for researchers determined to employ non-smoker controls in the investigation of the acute effects of smoking/nicotine. However, even this procedure leaves open questions regarding how representative the subject groups are (e.g., either the non-smokers may have higher levels of P than the population of non-smokers, or the smokers may have lower levels of P than the population of smokers). For making inferences regarding the long-term effects of smoking on cognitive performance, matching prior to the onset of smoking would seem to be necessary. Such a longitudinal study not only would be extremely difficult and expensive to do, but would again raise questions regarding how representative the original samples were. An alternative, which to our knowledge has never been done, would be a cross-sectional study of the cognitive performance of smokers and non-smokers across several years, focusing on the rate of change in cognitive performance. We do note that epidemiologically, smokers have a lower risk of developing Alzheimer's dementia (Smith and Giacobini, 1992), possibly indicative of a long-term protective effect of smoking on neural systems subserving cognitive performance[1]. A final strategy might be to employ ex-smokers as the control group. While using ex-smokers as controls represents an

---

[1] Recently, interest has been focused on the possible use of nicotine as a therapeutic agent in the treatment of Alzheimer's dementia (Jones *et al.*, 1992; see also discussion in Pritchard *et al.*, 1992b; Warburton, 1992), which is associated with a massive loss of brain nicotinic-cholinergic receptors (Giacobini, 1992; Smith and Giacobini, 1992).

improvement over using (presumably largely) never-smokers, this strategy only partially solves the problem of constitutional differences, since ex-smokers also differ with regard to personality from current smokers, falling somewhere in between never smokers and current smokers (see Hughes, 1991).

OUTLINE OF THE REVIEW

We divide our review into five major sections covering the effects of smoking/ nicotine on perception, motor function, cognitive performance, learning and memory, and academic performance. We believe that the separation of 'cognitive performance' from 'learning and memory' is organizationally useful, although we recognize that it is somewhat artificial. All of the studies reviewed in the cognitive performance section involve memory components, and some of the learning and memory studies involve rapid responding, a characteristic of most of the studies in the 'cognitive performance' section. Studies designed to assess the speed of short-term memory scanning by means of the classic Sternberg paradigm (see below) could have been placed in either section; we elected to place them in the 'cognitive performance' section. Unless stated otherwise, – all studies reviewed that had subjects participate in multiple sessions employed counterbalanced designs, and – all studies that administered nicotine by means other than smoking were double-blind, with effects of nicotine being assessed relative to placebo. Our review of the perceptual effects of smoking/nicotine is limited to the visual modality, as we were unable to identify any studies in which other sensory modalities were examined.

EFFECTS OF SMOKING/NICOTINE ON (VISUAL) PERCEPTION

Critical Flicker Fusion Threshold (CFFT)

Critical Flicker Fusion Threshold (CFFT) is the point in Hz ('Hertz', standing for cycles/second) at which a subject is able to perceive that a flickering light is in fact flickering as opposed to perceiving the light to be of constant brightness. CFFT is a measure of the functioning of the visual system; the higher the CFFT in Hz, the 'better' the functioning.

Effects of smoking

Both Larson *et al.* (1950) and Warwick and Eysenck (1963) reported that smoking raised CFFT in overnight-abstaining smokers. Barlow and Baer (1967) reported that smoking elevated CFFT in 1 hour abstaining smokers, while both Philips (1971) and Waller and Levander (1980) reported similar results in 2 hour abstaining smokers. In the Waller and Levander study, raised CFFT was seen as early as the third puff and persisted for 20 minutes following the last puff. Fabricant and Rose (1951) reported that smoking raised CFFT in non-abstaining smokers. In contrast, both Larson *et al.* (1950) and Warwick and Eysenck (1963) reported negative results for non-abstaining smokers. Leigh (1982) conducted four studies of the effects of smoking on critical flicker fusion (CFF) using both the traditional threshold measure (CFFT) as well as two non-threshold measures from signal-detection theory: sensitivity (a 'pure' measure of the subject's ability to discriminate signals [in this case, flicker] from non-signals) and response bias (the criterion or 'bias' adopted by the subject for reporting the presence of a signal).

In the *first* experiment, overnight-abstaining smokers were tested using a signal-detection paradigm that randomly alternated flickering (at the subject's baseline threshold) and non-flickering signals. In three sessions subjects [a] did not smoke, [b] smoked a total of six cigarettes each having a nicotine yield of 0.1 mg, and [c] smoked a total of six cigarettes each having a nicotine yield of 1.2 mg. Perceptual sensitivity was computed based on all responses over the 20 minute test period. Smoking the 1.2 mg cigarettes increased perceptual sensitivity.

The *second* experiment was similar to the first, except that the 0.1 mg cigarette was eliminated, with separate groups of overnight-abstaining subjects either not smoking or smoking four of the 1.2 mg cigarettes. This time, CFF testing was done by making classic CFFT measurements in Hz, which took less time than the signal-detection paradigm of experiment 1. Smoking elevated CFFT (i.e., the visual system was performing 'better').

The *third* experiment was the same as the second except that subjects did not abstain. The results were essentially identical to those of the second study (smoking elevated CFFT).

In the *fourth* experiment, separate groups of non-abstaining and overnight-abstaining smokers were tested. This time, both 'long-block' perceptual sensitivity was measured as in experiment 1 as well as 'short thresholds' as in studies 2 and 3. For perceptual sensitivity, the results replicated experiment 1 for both the abstaining and the non-abstaining group (smoking increased perceptual sensitivity). For the 'short threshold' measurements, the results replicated experiments 2 and 3 for both the abstaining and the non-abstaining groups (smoking elevated CFFT). In sum, all four studies indicated that smoking facilitated visual perception regardless of whether the subject had abstained or not. Dye *et al.* (1991) repeatedly measured CFFTs of female non-smokers and non-abstaining 'light' (1–10 cigarettes/day) and 'heavy' (>15 cigarettes/day) smokers. The smokers did not actually smoke as part of the experimental procedure, but "... it was assumed that subjects were tested at their preferred nicotine level" (p. 528). Across all phases of the subjects' menstrual cycles, the *'heavy'* smokers had lower CFFT thresholds than either the *'light'* smokers or the non-smokers.

Effects of nicotine gum/tablets in overnight-abstaining smokers

Warwick and Eysenck (1963; 1968) reported that buccally absorbed nicotine tablets raised CFFT in overnight-abstaining smokers. Sherwood *et al.* (1992) examined the effects of repeated administrations of 2 mg nicotine gum on CFFT in overnight-abstaining smokers. The first piece of nicotine gum raised CFFT. Elevated CFFT was maintained but not further enhanced by the second and third pieces of gum.

Effects of nicotine gum in non-abstaining smokers

Hindmarch *et al.* (1990a) administered placebo, 2 mg and 4 mg nicotine gum to non-abstaining smokers in three separate sessions. Both the 2 mg and 4 mg gum elevated CFFT; increased visual sensitivity. Sherwood *et al.* (1991) in two sessions studied the effects of 2 mg nicotine gum on CFFT in non-abstaining 'heavy' (>15 cigarettes/day) and 'light' (<5 cigarettes/day) smokers. In contrast to Hindmarch *et al.* (1990a), nicotine did not affect CFFT. Kerr *et al.* (1991) also reported no effect of 2 mg nicotine gum on CFFT in non-abstaining smokers.

Effects of nicotine gum/tablets in non-smokers

Warwick and Eysenck (1968) reported that buccally absorbed 0.1 mg nicotine tablets elevated CFFT in non-smokers. In contrast, Sherwood *et al.* (1991c) reported that 2 mg nicotine gum did not affect CFFT in non-smokers. Similarly, Hindmarch *et al.* (1990a) in separate sessions administered placebo and 2 mg nicotine gum to non-smokers. Again, the 2 mg gum did not affect CFFT. However, the measured blood nicotine concentrations achieved in the 2 mg condition (4.92 ng/ml versus 12.40 ng/ml for the 2 mg, gum-chewing smokers) probably was not sufficient to produce any effects on CFFT. The authors noted that many of the non-smokers reported the gum to be unpleasant to chew and therefore may not have complied with the chewing instructions.

Two-Flash Fusion (TFF)

Tong *et al.* (1974) studied the effects of smoking on two-flash fusion (TFF). TFF involves the measurement of the inter-stimulus interval (ISI) at which a subject can perceive two successive flashes as being separate events. In separate sessions, overnight-abstaining smokers did not smoke and smoked two cigarettes each having a nicotine yield of 1.3 mg. Smoking increased TFF perceptual sensitivity without affecting response criterion.

Visual Perception: Summary

Smoking/nicotine facilitates visual perception in abstaining smokers. In such subjects, smoking consistently raises CFFT, increases CFF perceptual sensitivity, and increases TFF perceptual sensitivity. Similarly, nicotine gum elevates CFFT in such subjects. However, we believe that the overall pattern of results for visual perception must at this time be considered inconclusive regarding absolute facilitation versus withdrawal deficit. In abstaining smokers, both positive and negative results have been reported for both smoking and nicotine gum. Similarly, mixed results have been reported regarding nicotine gum/tablets administered to non-smokers (note that negative results could have resulted from poor nicotine absorption on the part of gum-chewing non-smokers). Based on evidence regarding the time course of effects in smokers, Sherwood (1993) concluded that smoking could produce absolute facilitation. In particular, Sherwood noted that Fabricant and Rose's (1951) positive results in non-abstaining smokers were only significant for subjects who had reported not having a cigarette for at least 30 minutes prior to testing. Similarly, Barlow and Baer (1967) reported positive smoking results in 1 hour abstaining smokers, and Sherwood *et al.* (1992) reported that in overnight-abstaining smokers, CFFT was elevated by an initial piece of nicotine gum but not subsequent pieces of gum administered 1 hour later. This 30 minute to 1 hour time course was "... far shorter than that believed to induce abstinence effects ...

but in accordance with the elimination pharmacokinetics of nicotine in the brain ... suggesting ... an absolute effect rather than a recovery of normal function, albeit rapidly reaching a ceiling for improvement after a single dose" [p. 162].

While we do not dispute Sherwood's analysis, we believe that additional studies demonstrating effects in non-smokers are necessary before concluding that nicotine can produce absolute facilitation (analysis of blood nicotine concentrations would also be necessary in order to demonstrate sufficient nicotine intake). Again, as outlined in the Introduction, the demonstration (from non-abstaining smoker and non-smoker studies) that smoking/nicotine can produce absolute facilitation does not necessarily rule out a withdrawal-deficit component to effects in regular smokers (the combination hypothesis). The type of 'definitive' cessation study outlined in the *Introduction* section has not been done assessing the effects of smoking/nicotine on visual perception.

EFFECTS OF SMOKING/NICOTINE ON MOTOR FUNCTION

In this section we review the effects of smoking/nicotine on a widely used motor task, finger tapping. The rate of finger tapping gives an index of the speed of motor processing in general. Studies of 'athletic' types of performance are not reviewed, although we note that nicotine does not appear to affect brief, high-intensity (anaerobic) performance (Baldini *et al.*, 1992). Studies measuring choice reaction time data, where total reaction time was divided into decision RT (also referred to as 'decision time') and movement reaction time (also referred to as 'movement time') are reviewed in subsequent sections. Reeves and Morehouse (1950) reported no effect of smoking a cigarette (yield unspecified) relative to two-hour abstention on finger-tapping rate. Frith (1967a,b,c) investigated the effect of chewing buffered nicotine tablets (0.1 mg) versus chewing placebo tablets on finger tapping. Subjects were either non-smokers or "occasional smokers who had not had a cigarette the day of testing" (further details not given). Across the three studies, nicotine attenuated a drop-off in tapping performance across the 60-second length of the task. Valeriote *et al.* (1979) studied the effects of smoking on finger tapping in three different sessions: four-hour abstention with no smoking during the experiment, four-hour abstention with two 1.3 mg cigarettes smoked during the experiment, and no abstention with no smoking during the experiment (by the time of testing, these subjects had been without a cigarette for at least 25 minutes). Finger-tapping rate did not differ among the sessions. Neither did foot tapping, which was also measured. West and Jarvis (1986) in a series of five studies examined the effects on finger-tapping rate of nasally-administered nicotine in non-smokers. This method of administering nicotine had been shown to yield plasma nicotine concentrations comparable to those "between a third and a whole cigarette" [p. 727]. Four of the five studies were performed double blind. Nicotine was found to increase finger tapping rate in all cases. Further, this effect was partially blocked by pretreatment with the central nicotinic receptor-antagonist mecamylamine. The duration of the increase was around 30 minutes, and showed no evidence of tolerance with seven administrations across a six-hour period.

Perkins *et al.* (1990a) in two sessions examined the effect of nicotine administered via nasal spray (15 µg/kg body weight) versus placebo on finger tapping rate in overnight-abstaining smokers. Baseline data were collected prior to administration of both types of spray. Nicotine increased finger-tapping rate. Perkins and associates also tested a group of non-smokers. Unlike the smokers, nicotine spray did not increase finger-tapping rate in the non-smokers. This did not appear to be a withdrawal effect, however, as baseline finger-tapping rates did not differ between the abstaining smokers and the non-smokers. The authors speculated that because of their 'smoking history', smokers are able to "... differentially adapt to the various behavioural effects of nicotine in order to take advantage of its 'positive' effects on performance..." [p. 14], in this case, an increase in finger-tapping rate.

EFFECTS OF SMOKING/NICOTINE ON COGNITIVE PERFORMANCE

This section focuses primarily on studies involving reaction time (RT) tasks. In some studies, these tasks have been performed for prolonged periods of time and thus fall under the classification of *vigilance* tasks (also known as *sustained attention* tasks). Studies focusing on tasks specifically designed to assess effects on either short-term memory capacity or learning/long-term memory are reviewed in the next major section.

Before beginning, we will briefly review three tasks that have been used frequently in the study of the effects of smoking/nicotine on cognitive performance.

- The *Rapid Visual Information Processing* task (RVIP) is a sustained-attention task that requires subjects to monitor a video screen where single digits appear, and to respond (usually by pressing a button) to 3 consecutive odd or even digits (variously, the digits either appear at a constant rate of 100/minute or at a rate that increases slightly following correct responses and decreases slightly following incorrect responses).
- The *Stroop task* is a colour-word interference task in which the subject is presented with words printed in colour. The subject is instructed to rapidly name the colour of the letters forming the word. However, the words themselves spell-out a conflicting colour. For example, the word *blue* written in red letters requires the response *'red'*. Since, for most literate adults, reading has become such an overlearned, automatic skill, the colour, spelled-out by the letters, interferes with naming the colour of the letters themselves. Control conditions can include naming the colour of coloured rectangles or the letter colour of words that do not spell-out a conflicting colour.
- The so-called *Sternberg task* (Sternberg, 1966) measures the speed of scanning through short-term memory. The subject is presented with a memory set consisting of a varying number of single digits. A *probe* digit is then presented, and the subject must respond quickly whether the probe was or was not a member of the previously presented memory set. In addition to the 'usual' measures of RT and RT errors, the RT slope across memory set size gives the speed of scanning in digits/unit time.

In evaluating the results of experiments measuring RT, it is important also to consider the accuracy of responding (e.g., number of errors). This is necessary so that the role of possible speed/accuracy-tradeoffs can be evaluated. For example, if smoking in a given study was associated with not only faster RTs but also more errors, then the correct conclusion is that smoking did not produce 'true' performance enhancement, but merely resulted in subjects trading speed for accuracy. Alternatively, if smoking was not associated with faster RTs, but did result in less errors, then the correct conclusion would be that smoking in fact enhanced performance. Many studies, particularly early ones, either do not report error data or do not report RT data.

Overnight-abstaining Smokers

A summary of studies employing overnight-abstaining smokers may be found in Table 2.1. The table also provides additional technical information regarding the studies. In general, smoking/nicotine facilitates performance of a variety of cognitive tasks in abstaining smokers, although results are not uniformly positive. Facilitation by smoking has been reported for choice RT tasks (including the RVIP task) in terms of faster RTs, fewer errors, or both (Edwards *et al.*, 1985; Hasenfratz and Bättig, 1993b; Hasenfratz *et al.*, 1989a; Lyon *et al.*, 1975; Petrie and Deary, 1989; Parrott and Craig, 1992; Parrott and Winder, 1989; Revell, 1988; Sherwood *et al.*, 1992; Smith *et al.*, 1977; Wesnes and Warburton, 1983a, 1984b). In particular, Revell (1988), who had subjects smoke while performing the task rather than prior to it (8 cued puffs spaced 1 minute apart, beginning 5 minutes into the 20 minute task), reported that for the RVIP task, smoking increased correct detections beginning with the second puff on a cigarette; similarly, beginning with puffs three-to-four, RTs were faster. Facilitation of RVIP task performance has also been reported following nicotine gum (Parrott and Craig, 1992; Parrott and Winder, 1989).

Even when overall RTs or error rates are not affected, both smoking and chewing buffered nicotine tablets have been reported to attenuate vigilance decrements (drop offs in performance as a function of time) in sustained cognitive tasks (Frankenhaeuser *et al.*, 1971 [no error data]; Tong *et al.*, 1980 [no RT data]; Wesnes and Warburton, 1978 [no RT data]). Smoking has also been reported to facilitate visual and auditory signal detection (Myrsten *et al.*, 1975 [no error data; low arousal smoking motivation subjects only – see below]; Leigh *et al.*, 1977 [Experiment 1; no RT data]), letter cancellation (Williams, 1980; Williams *et al.*, 1984 [no error data]; Parrott and Craig, 1992), digit-symbol substitution (trend; Petrie and Deary, 1989), and mental math (Pritchard, 1991b). Nicotine gum has been reported to facilitate two-letter search, logical reasoning, and addition/subtraction of two-digit numbers (Snyder and Henningfield, 1989), as well as short-term memory scanning (Sherwood *et al.*, 1992).

Three additional studies that included a group or groups of smokers abstaining overnight are relevant here. These studies are reviewed below in the section on administering nicotine to both smokers and non-smokers; see Table 2.5. Wesnes *et al.* (1983 [no RT data]) reported that chewing buffered nicotine tablets (1 or 2 mg) attenuated the drop off in sensitivity (vigilance decrement) associated with prolonged performance of the Mackworth Clock vigilance task. Grobe *et al.* (1993 [no error data]) reported that nicotine nasal spray lowered RT in a numerical interference task. Finally, West and Hack (1991 [no error data])

Table 2.1 – Studies employing overnight-abstaining smokers and smokeless-tobacco users (the latter marked ST in the first column)

Study	Task(s)	Duration	Design; sex	Sessions/conditions	Effect of smoking/nicotine on speed	Effect on accuracy
Frankenhaeuser et al. (1970)	visual 4-choice RT	four 7 minutes blocks (1 pre, 3 post)	within ss; male	pre/post no/varied-yield smoking	no effect	not reported
Frankenhaeuser et al. (1971)	sustained visual simple RT	80 minutes	within ss; male	no/varied-yield smoking	attenuated vigilance decrement	not reported
Hartley (1973)	complex visual vigilance	36 minutes	within ss; female	smoking/no smoking	not applicable	no stimulus probability matching
Myrsten et al. (1975) [Task 1]	visual signal detection	90 minutes	within ss; male	smoking/no smoking	faster RT in low arousal smokers only (see text)	not reported
Myrsten et al. (1975) [Task 2]	visual signal detection/ choice RT	60 minutes	within ss; male	smoking/no smoking	faster RT in high arousal smokers only (see text)	not reported
Lyon et al. (1975)	visual 8-choice RT	25 minutes	within ss; male	smoking/no smoking	faster decision RT (see text)	too few errors to analyze
Leigh et al. (1977) [Exp. 1]	auditory signal detection	30 and 15 minutes	within ss; male	smoking/no smoking	not measured	fewer errors
Leigh et al. (1977) [Exp. 2]	auditory signal detection	blocks of 10 minutes	within ss; male	smoking/no smoking	not measured	no effect
Smith et al. (1977)	visual 4-choice RT	3 blocks of 50 trials	within ss; male	no/varied-yield smoking	faster decision RT	too few errors to analyze
Tong et al. (1978)	velocity and time estimation	3 blocks of 12 trials	within ss; mixed	smoking/no smoking	underestimation velocity and of both time	underestimation velocity and of both time

Table 2.1 – *Continued*

Study	Task(s)	Duration	Design; sex	Sessions/ conditions	Effect of smoking/ nicotine on speed	Effect on accuracy
Wesnes and Warburton (1978)	Mackworth clock vigilance	80 minutes	within ss; mixed	0-, 1- & 2- mg buffered nic. tablets	not reported	attenuated vigilance decrement
Williams (1980)	letter cancellation	about 3 minutes	within ss; male	pre/post sham /real smoking	faster cancellation	not measured
Tong et al. (1980)	auditory vigilance	72 minutes	between ss; male	smoking/no smoking	not reported	attenuated vigilance decrement
Suter et al. (1983)	Stroop task	15 trials	within ss; male	pre/post varied-yield smoking	no effect	no effect
Wesnes and Warburton (1983a, Study 1)	RVIP task (100 digits/ minute)	30 minutes (10 baseline, 20 post)	within ss; male	pre/post varied-yield smoking	faster RT	more correct detections
Wesnes and Warburton (1983a, Study 2)	RVIP task (100 digits/ minute)	30 minutes (10 baseline, 20 post)	within ss; mixed	pre/post varied-yield smoking	no effect	more correct detections
Wesnes and Warburton (1984b)	RVIP task (100 digits/ minute)	30 minutes (10 baseline, 20 post)	within ss; male	pre/post varied-yield smoking	faster RT	more correct detections
Williams et al. (1984)	letter cancellation	about 3 minutes	within ss; female	pre/post smoking	faster cancellation	not reported
Edwards et al. (1985)	RVIP task (100 digits/ minute)	30 minutes (10 baseline, 20 post)	within ss; male	pre/post no/varied-yield smoking	faster RT	more correct detections

Table 2.1 – *Continued*

Study	Task(s)	Duration	Design; sex	Sessions/ conditions	Effect of smoking/ nicotine on speed	Effect on accuracy
Knott (1985) [Task 1]	visual simple RT (sRT)	2 minutes	within ss; female	sham/real smoking	no effect	not applicable
Knott (1985) [Task 2]	visual sRT with secondary task	2 minutes	within ss; female	sham/real smoking	no effect on either task	no effect on secondary task
Revell (1988)	RVIP task (100 digits/minute)	20 minutes	within ss; mixed	pre/post no/ sham/ real smoking	faster RT	more correct detections
Hasenfratz *et al.* (1989a)	RVIP task (variable rate)	60 minutes (30 baseline, 30 post)	between ss; female	smoking/no smoking	no effect	increased processing rate
Parrott and Winder (1989)	RVIP task (100 digits/ minute)	30 minutes (10 baseline, 20 post)	within ss; male	pre/post 0, 2 and 4 mg gum/smk.	faster RT only following 4 mg gum	more correct detections
Petrie and Deary (1989) [Task 1]	RVIP task (100 digits/ minute)	10 minutes	within ss; mixed	smoking/no smoking	faster RT	no effect
Petrie and Deary (1989) [Task 2]	digit-symbol substitution test	90 seconds	within ss; mixed	smoking/no smoking	trend (p<0.10) toward more correct substitutions	not reported
Petrie and Deary (1989) [Task 3]	inspection time	15–20 minutes	within ss; mixed	smoking/no smoking	no effect on inspection time	not applicable
Snyder and Henningfield (1989) [Task 1]	two-letter search	about 2 minutes	within ss; male	0, 2, and 4 mg gum	faster search only following 4 mg gum	no effect

Table 2.1 – Continued

Study	Task(s)	Duration	Design; sex	Sessions/ conditions	Effect of smoking/ nicotine on speed	Effect on accuracy
Snyder and Henningfield (1989) [Task 2]	six-letter search	about 2 minutes	within ss; male	0, 2 and 4 mg gum	no effect	no effect
Snyder and Henningfield (1989) [Task 3]	logical reasoning	about 2.5 minutes	within ss; male	0, 2 and 4 mg gum	faster RT	no effect
Snyder and Henningfield (1989) [Task 4]	math task	about 3 minutes	within ss; male	0, 2 and 4 mg gum	faster RT	no effect
Pritchard (1991b)	mental math (two tasks)	60 30 seconds	within ss; mixed	pre/post smoking	faster counting for both tasks	no effect
Landers *et al.* (1992) ST [Exp. 1, Task 1]	visual simple RT	42 trials pre and post	mixed design; male	2 g ST/ no ST	no effect	not applicable
Landers *et al.* (1992) ST [Exp. 1, Task 2]	visual 'anticipation'	42 trials pre and post	mixed design; male	2 g ST/ no ST	no effect	not applicable
Landers *et al.* (1992) ST [Exp. 1, Task 3]	visual 2-choice RT	42 trials pre and post	mixed design; male	2 g ST/ no ST	no effect on either decision RT or movement RT	not reported
Parrott and Craig (1992) [Task 1]	RVIP task (100 digits/minute)	30 minutes (10 baseline, 20 post)	within ss; mixed pre/post	0, 2 and 4 mg gum/smk.	faster RT only following gum	more correct detections
Parrott and Craig (1992) [Task 2]	letter cancellation	1350 total letters	within ss; mixed pre/post	0, 2 and 4 mg gum/smk.	faster cancellation following smk. but not after gum	no effect

Table 2.1 – Continued

Study	Task(s)	Duration	Design; sex	Sessions/ conditions	Effect of smoking/ nicotine on speed	Effect on accuracy
Parrott and Craig (1992) [Task 3]	Stroop task	50 trials	within ss; mixed pre/post	0, 2 and 4 mg gum/smoke	no effect	no effect
Parrott and Craig (1992) [Task 4]	'width-of-attention' task	360 trials	within ss; mixed pre/post	0, 2 and 4 mg gum/smoke	no effect	no effect
Hasenfratz and Bättig (1992)	num. interference task	440 trials	within ss; female	smoking/no smoking	faster RT; less interference	not reported
Sherwood et al. (1992) [Task 1]	visual 6-choice RT	20 trials	within ss; mixed	0 and 2 mg gum (3 pieces)	faster movement RT following each piece of gum	too few errors to analyze
Sherwood et al. (1992) [Task 2]	short-term memory scanning	24 trials	within ss; mixed	0 and 2 mg gum (3 pieces)	faster RT following first piece of gum	too few errors to analyze
Sherwood et al. (1992) [Task 3]	analogous to driving (see text)	5 minutes	within ss; mixed	0 and 2 mg gum (3 pieces)	better tracking following each piece of gum	too few errors to analyze
Hasenfratz and Bättig (1993b)	RVIP task (variable rate)	40 minutes (20 baseline, 20 post)	mixed; design; female	pre/post smoke/ no smoking	faster/slower RT in active/passive-coping group	faster processing rate
Pritchard et al. (1995)	RVIP task (paper and pencil)	30 seconds (1 pre, 6 during, 4 post)	within ss; male	pre/during/ post smoking	increased speed (esp. for puffs 1 and 2 of 8)	no effect

The term 'x-mg cigarette' designates a cigarette having a nicotine yield of 'x' mg.

reported that smoking resulted in faster scanning of short-term memory in the Sternberg task.

In contrast to the above, negative choice-RT results have been reported both for smoking (Frankenhaeuser *et al.*, 1970 [no error data]) and smokeless tobacco (Landers *et al.*, 1992). Negative letter-cancellation results have been reported for nicotine gum (Parrott and Craig, 1992) and negative auditory signal-detection results for smoking (Leigh *et al.*, 1977 [Experiment 2; no RT data]). Other tasks, reportedly not affected, include: for smoking, inspection time (Petrie and Deary, 1989), Stroop colour-word interference (Parrott and Craig, 1992; Suter *et al.*, 1983) and width of attention (Parrott and Craig, 1992); for nicotine gum, six-letter search (Snyder and Henningfield, 1989), Stroop colour-word interference (Parrott and Craig, 1992), and width of attention (Parrott and Craig, 1992); and for smokeless tobacco, visual anticipation (Landers *et al.*, 1992). Finally, simple-RT studies have produced negative results both for smoking (Knott, 1985) as well as smokeless tobacco (Landers *et al.*, 1992). We note that Suter *et al.*'s (1983) and Parrott and Craig's (1992) negative results for the Stroop task may have been a function of an insufficient number of trials. As discussed below, effects of nicotine on Stroop performance may require on the order of several hundred trials to emerge. Results employing simultaneous tasks have been more complex. Hartley (1973) employed a sustained visual attention task in which subjects monitored three sources having different signal probabilities. Smoking did not affect task performance other than to block a matching of response frequency to stimulus probability that was seen in a no-smoking control condition. Myrsten *et al.* (1975 [no error data]) reported an interaction between the simultaneous performance of three tasks and arousal smoking motivation (see below). For a combined velocity and time-estimation task, Tong *et al.* (1978) reported that smoking resulted in an underestimation of both velocity and time. Sherwood *et al.* (1992) employed a central visual-tracking task combined with a peripheral RT task (this combination was designed to simulate the major cognitive tasks of automobile driving). They found that nicotine gum resulted in better tracking ('steering') without affecting RT errors.

Interaction with arousal smoking motivation

The study by Myrsten *et al.* (1975 [no error data]) indicated an interesting interaction between the effect of smoking on performance and one aspect of subject smoking motivation. Myrsten *et al.* divided subjects into low- and high-arousal smoking-motivation groups by means of a questionnaire that they developed based in-part on Frith's Situational Smoking Questionnaire (SSQ; Frith, 1971). The SSQ is a psychometric instrument designed to classify subjects as being either 'high-arousal' smokers, who report that their desire to smoke is highest when they feel stressed, anxious, etc., or 'low-arousal' smokers, who report that their desire to smoke is highest when they feel relaxed, bored, tired, etc. They reported that smoking produced faster RTs in low- but not high-arousal subjects engaged in an "easy" sustained visual signal-detection task. In contrast, smoking produced faster RTs in high, but not low-arousal smokers engaged in a "difficult" sustained task that consisted of the simultaneous combination of the visual signal-detection task with visual and auditory choice-RT tasks.

Williams (1980 [no error data]) employed a letter-cancellation task in which subjects crossed out every appearance of the letter E in a list of random letters. Subjects again were divided into high- and low-arousal groups according to the SSQ. Smoking speeded letter cancellation more in the low- than in high-arousal smokers. In a subsequent study, however (Williams *et al.*, 1984 [no error data]), smoking speeded cancellation equally in high- and low-arousal smokers.

Interaction with daily cigarette consumption

In the Williams (1980 [no error data]) letter-cancellation study, subjects were also divided according to daily cigarette consumption, specifically, separate groups of 'light' (<15 cigarettes/day), 'medium' (>15 but <26), and 'heavy' (>26) smokers. Smoking speeded letter cancellation equally in all three groups.

Smokers Abstaining Less than Overnight

A summary of studies employing smokers who abstained for a period less than overnight may be found in Table 2.2. The table also provides additional technical information regarding the studies. As with smokers abstaining overnight, smoking/nicotine generally facilitates performance of a variety of cognitive tasks in smokers abstaining less than overnight, although, again, results are not uniformly positive. Facilitation by smoking has been reported for choice RT tasks (including the RVIP task; Bates *et al.*, 1994 [no error data]; Michel *et al.*, 1987 [no RT data]; Michel and Bättig, 1988;) and auditory signal detection (Mangan, 1982 [no RT data]; Mangan and Golding, 1978 [no RT data]). For a pursuit-tracking task, Heimstra *et al.* (1980) in two studies reported that smoking blocked the vigilance decrement seen in a no-smoking control condition. In a combined choice RT/spatial attention task, Bates *et al.* (1995) reported that smoking resulted in faster decision RT, especially for conditions requiring greater spatial attention (a wider display). Negative smoking results have been reported for the 'classic' pursuit rotor task (Valeriote *et al.*, 1979) and for both warned and unwarned simple visual RT as well as mental math (Heimstra *et al.*, 1980, two studies). Negative nicotine gum results have been reported for the RVIP task (Michel *et al.*, 1988).

A Note on Snyder and Henningfield (1989). We present some brief comments on their study, since a reading of their *Discussion* section provides an illustrative example of the importance of considering the theoretical issues raised in our *Introduction* section. In addition to measuring the effects of nicotine gum following overnight smoking abstention, Snyder and Henningfield also presented pre-abstention baseline data. They reported that, relative to pre-abstinence baseline, performance of all four of their tasks was worse following overnight abstention (similar results for 'reaction time' were reported in a brief convention abstract by Knott and Griffiths, 1992). Snyder and Henningfield interpreted their findings as supporting the withdrawal-deficit hypothesis (although we note that subsequently administered nicotine gum did not affect performance of the six-letter search task, while for the math task it resulted in better-than-baseline performance): "These findings are consistent with the conclusion that performance decrements provide a measure of the nicotine withdrawal syndrome… Consequently, it seems

Table 2.2 – Studies employing smokers abstaining less than overnight

Study	Task(s)	Duration: abstention: task	Design	Sessions/ conditions	Effect of smoking/ nicotine on speed	Effect on accuracy
Mangan and Golding (1978)	auditory signal detection	2 hours; 30 minutes	within ss; mixed	smoking/no smoking	RT not reported	decreased errors of commission
Valeriote et al. (1979)	pursuit rotor	4 hours; 4 minutes	within ss; male	smoking/no smoking	not applicable	basically none
Heimstra et al. (1980) [Study 1, Sub-task 1]	pursuit tracking	no abstention; 3 hours	between ss; female	smoking/no smoking	not applicable	no vigilance decrement
Heimstra et al. (1980) [Study 1, Sub-task 2]	warned simple visual RT	no abstention; 3 hours	between ss; female	smoking/ no smoking	no effect	not applicable
Heimstra et al. (1980) [Study 1, Sub-task 3]	unwarned simple visual RT	no abstention; 3 hours	between ss; female	smoking/no smoking	no effect	no effect
Heimstra et al. (1980) [Study 1, sub-task 4]	mental math	no abstention; 3 hours	between ss; female	smoking/ no smoking	no effect	no effect
Heimstra et al. (1980) [Study 2, Sub-task 1]	pursuit tracking	no abstention; 3 hours	between ss; male	smoking/ no smoking	not applicable	no vigilance decrement
Heimstra et al. (1980) [Study 2, Sub-task 2]	warned simple visual RT	no abstention; 3 hours	between ss; male	smoking/ no smoking	no effect	not applicable

Table 2.2 – Continued

Study	Task(s)	Duration: abstention: task	Design	Sessions/ conditions	Effect of smoking/ nicotine on speed	Effect on accuracy
Heimstra et al. (1980) [Study 2, Sub-task 3]	unwarned simple visual RT	no abstention; 3 hours	between ss; male	smoking/ no smoking	no effect	no effect
Heimstra et al. (1980) [Study 2, Sub-task 4]	mental math	no abstention; 3 hours	between ss; male	smoking/ no smoking	no effect	no effect
Mangan (1982)	auditory signal detection	2 hours; 30 minutes	mixed design; male	no/varied-yield smoking	not reported	greater stimulus sensitivity
Michel et al. (1987)	RVIP task (variable rate)	5 hours; 60 minutes (30 pre, 30 post)	within ss; mixed	pre/post smoking	RT not reported	increased processing rate
Michel et al. (1988)	RVIP task (variable rate)	2 hours; 35 minutes (20 pre, 15 post)	between ss; female	pre/post 0- and 4-mg nicotine gum	no effect on RT	no effect on processing rate
Michel and Bätig (1988, 1989)	RVIP task (variable rate)	2 hours; 40 minutes (20 pre, 20 post)	within ss; female	pre/post sham smk./ smk.	faster RT	increased processing rate
Bates et al. (1995)	visual simple and choice RT	2 hours; 32/64 trials for 1, 2 / 4, 8 choices	within ss; mixed	no/varied-yield smoking	faster decision RT	not reported
Bates et al. (1995) [2 Hours]	choice RT/ spatial attention	2 hours; 32 trials for each task	within ss; female	smk./no-nicotine smk.	faster decision RT, especially for the wider display	not reported

The term 'x-mg cigarette' designates a cigarette having a nicotine yield of 'x' mg.

likely that performance decrements following tobacco abstinence contribute to the high relapse rate observed in stop-smoking programs [p. 21]. However, it is equally plausible that overnight abstention, rather than producing a state of withdrawal, actually resulted in the removal of absolute facilitation. Thus, we again point out that what is needed to unambiguously demonstrate a withdrawal effect is repeated testing over the course of several days, in order to demonstrate that a deterioration in performance associated with longer-term smoking 'cessation' is transient (we note that the importance of such data was acknowledged by Snyder and Henningfield). In fact, a subsequent study by this group (Snyder *et al.*, 1989) did examine cognitive performance during longer-term smoking cessation; we review that study below.

Non-abstaining Smokers

A summary of studies employing non-abstaining smokers appears in Table 2.3. The Table 2.3 also provides additional technical information regarding the studies, which all employed cigarette smoking. The results are largely positive, although the number of studies is much less than the number of abstaining-smoker studies. Facilitation of performance has been reported for simple visual RT in terms of both faster RT 40–55 minutes post-smoking (but slower at five minutes; Cotten *et al.*, 1971) as well as an attenuated vigilance decrement (Myrsten *et al.*, 1972). Positive results have also been reported for choice RT (Myrsten *et al.*, 1972 [trend; no error data]; Frearson *et al.*, 1988 [two types of task; no error data]; Pritchard *et al.*, 1992a). Hasenfratz *et al.* (1989b) reported negative RVIP results for smoking that took place immediately following a standardized lunch. Three additional nicotine-gum studies that included a group or groups of non-abstaining smokers are relevant here (these studies are reviewed below in the section on administering nicotine to both smokers and non-smokers; see Table 2.5). Hindmarch *et al.* (1990a), Sherwood *et al.* (1991c), and Kerr *et al.* (1991) all reported that nicotine gum speeded visual choice RT and improved the tracking portion of a simulated driving task. Although Hindmarch *et al.* (1990a) reported no effect of nicotine gum on RT in the Sternberg task, both Kerr *et al.* (1991) and Sherwood *et al.* (1991a) reported that nicotine gum speeded RT in the Sternberg task, and West and Hack (1991 [no error data]) reported faster short-term memory scanning in the Sternberg task following smoking.

Interaction with IQ

In addition to RT data, Frearson *et al.* (1988 [no error data]) collected IQ data from their subjects (Wechsler Adult Intelligence Scale Verbal and Full Scale, and Raven's Progressive Matrices). Interestingly, the faster decision RTs that they found following smoking for one of their choice-RT tasks (the odd-man-out task) were negatively correlated with IQ scores, that is, lower-IQ subjects 'benefited' more from smoking than higher-IQ subjects.

Repeated administration

Sherwood *et al.* (1992) examined the effects of repeated administrations of 2 mg nicotine gum on a series of three tasks. Since their subjects abstained overnight,

Table 2.3 – Studies employing non-abstaining smokers

Study	Task(s)	Duration	Design; sex	Sessions/conditions	Effect of smoking/ nicotine on speed	Effect on accuracy
Cotten *et al.* (1971)	visual simple RT	6 blocks of 20 stimuli	within ss; male	smoking/no smoking	slower RT at 0 and 5 min; faster at 40 and 55 min	not applicable
Myrsten *et al.* (1972) [Task 1]	visual simple RT	two 25 minutes blocks	within ss; male	smoking/no smoking	prevention of an RT vigilance decrement	not applicable
Myrsten *et al.* (1972) [Task 2]	visual choice RT	two 25 minutes blocks	within ss; male	smoking/no smoking	trend for RT to be faster	not reported
Frearson *et al.* (1988)5 [Task 1]	visual choice RT	blocks of 10 trials	special design; mixed	smoking/no smoking	faster decision but not movement RT	not reported
Frearson *et al.* (1988)5 [Task 2]	visual 'odd- man-out' RT task	60 trials	special design; mixed	smoking/ no smoking	faster decision but not movement RT	not reported
Hasenfratz *et al.* (1989b)	RVIP task (variable rate)	20 minutes	within ss; male	smoking/ no smoking	no effect on RT	no effect on processing rate
Pritchard *et al.* (1992a)	contin. performance task	two 20 minutes blocks	within ss; mixed	smoking/ no smoking	faster RT	no effect

The term 'x-mg cigarette' designates a cigarette having a nicotine yield of 'x' mg.

we reviewed the results of administration of the first piece of gum above in the 'overnight abstaining' section. However, the results for subsequent administrations more properly belong in the present section. They reported that the second and third pieces of nicotine gum (administered one and two hours after the first piece) produced a further speeding of movement RT in their choice-RT task and produced further improvements in tracking performance in their simulated driving task (along with no increase in RT to the peripheral stimuli as was seen in the placebo session). RT in the Sternberg task was not affected by the second and third pieces of gum, although the initial improvement was maintained.

Discussion

Across studies and multiple tasks within studies (including the 'subsequent administration' data of Sherwood *et al.*, 1992, as well as Hindmarch *et al.*, 1990a, Sherwood *et al.*, 1990, and Kerr *et al.*, 1991) 17 of 19 reported beneficial effects of smoking/nicotine on the performance of non-abstaining smokers (omitted from this count is the trend of Myrsten *et al.*'s [1972] second task (no error data) as well as the results of Cotten *et al.* [1971], whose RT effects went 'both ways' depending on when they were measured). This provided an initial indication that smoking can produce absolute facilitation, since, again, non-abstaining smokers 'by definition' would not be in any sort of withdrawal state. However, it will be necessary to review the results of studies in which nicotine was administered to non-smokers before reaching a more firm conclusion regarding nicotine's ability to produce absolute facilitation.

Non-smokers and Occasional Smokers I: Studies not Employing Regular Smokers

A summary of studies in which nicotine was administered only to non-smokers may be found in Table 2.4. The table also summarizes one smoking study that employed 'occasional smokers', and also provides additional technical information regarding all the studies. Several non-smoker results reviewed in Table 2.4 are negative, but in all cases, factors can be identified that offer alternative explanations to the conclusion that nicotine does not affect cognitive performance in non-smokers. For example, Wesnes and Revell (1984) in two studies reported no effect of chewing 1.5 mg nicotine tablets on either RVIP or Stroop-task performance; however, the 60 minute interval between baseline and post-administration testing necessary to allow for full absorption of the other compound investigated in this study (scopolamine) may have been responsible for the lack of an effect of nicotine.

Dunne *et al.* (1986; [no RT data]) reported no effect of 4 mg nicotine gum on either a verbal or a numerical problem-solving task. However, as discussed above, 4 mg nicotine gum usually produces toxic effects in non-smokers (Nyberg *et al.*, 1982) that almost certainly would interfere with cognitive performance. An example of such toxicity may be found in the results of Heishman *et al.* (1990), who reported no effect of nicotine gum on performance of either a letter-search task, a logical reasoning task, or a visual math task. However, Heishman *et al.* also reported that nicotine produced dose-related increases in subjective

Table 2.4 – Experiments in which nicotine was administered to non-smokers/occasional smokers

Study	Task(s)	Duration	Design: sex	Sessions/conditions	Effect of smoking/ nicotine on speed	Effect on accuracy
Wesnes and Warburton (1978)	Mackworth clock vigilance	80 minutes	within ss; mixed	0, 1 and 2 mg nic. tablets	not reported	fewer errors of commission
Wesnes and Warburton (1984c)	RVIP task (100 digits/ minute)	30 minutes (10 baseline, 20 post)	within ss; male nic. tablets	0, 0.5, 1 and 1.5 mg	trend toward an attenuated increase in RT	attenuated decrement
Wesnes and Revell (1984) [Experiment 1]	RVIP task (100 digits/ minute)	30 minutes (10 baseline, 20 post)	within ss; mixed	0 and 1.5 mg nic. tablets	no effect	no effect
Wesnes and Revell (1984) [Exp. 2, Task 1]	RVIP task (100 digits/ minute)	40 minutes (10 baseline, 15 post 1 and 2)	within ss; mixed	0 and 1.5 mg tablets (2 each)	no effect	no effect
Wesnes and Revell (1984) [Exp. 2, Task 2]	Stroop task	400 stimuli	within ss; mixed	0 and 1.5 mg tablets	no effect	no effect
Dunne et al. (1986) [Task 1]	verbal problem solving	6 blocks of 18 items	within ss; female	0 and 4 mg gum	not reported	no effect
Dunne et al. (1986) [Task 2]	numerical problem solving	6 blocks of 20 items	within ss; female	0 and 4 mg gum	not reported	no effect
Heishman et al. ('90)[Task 1]	letter search	about 2 minutes	within ss; male	0, 2, 4 and 8 mg gum	no effect	no effect

Table 2.4 – *Continued*

Study	Task(s)	Duration	Design: sex	Sessions/conditions	Effect of smoking/ nicotine on speed	Effect on accuracy
Heishman *et al.* (1990) [Task 2]	logical reasoning (visual)	about 2.5 minutes	within ss; male	0, 2, 4 and 8 mg gum	no effect	'decreased accuracy'
Heishman *et al.* ('90) [Task 3]	math task (visual)	about 3 minutes	within ss; male	0, 2, 4 and 8 mg gum	no effect	no effect
Provost and Woodward (1991)	Stroop task	3 blocks of 100 stimuli	mixed design; mixed	0 and 2 mg gum	decreasing interference across blocks	not reported
Halliday *et al.* (1992)	visual 2- and 8-choice RT	240 stimuli	within ss; ???	0 and 0.8 mg (subcutaneous)	faster RT; increased number of fast responses	no effect
West and Hack (1991a)	short-term memory scanning	5 minutes	mixed design; mixed	real/ no-nicotine smoking	faster scanning	not reported

The term 'x-mg cigarette' designates a cigarette having a nicotine yield of 'x' mg.

measures of malaise (Total Mood Disturbance and Tension/Anxiety scores from the Profile of Mood States inventory). In contrast to these negative results, Wesnes and Warburton (1978 [no RT data]) reported that chewing 1 mg and 2 mg nicotine tablets resulted in fewer errors of commission in the Mackworth Clock vigilance task (subjects monitor the minute hand of a clock and responded to brief pauses in its motion). Similarly, Wesnes and Warburton (1984c) reported that chewing nicotine tablets produced dose-related attenuation of vigilance decrements in RVIP performance. Provost and Woodward (1991 [no error data]) reported that 2 mg nicotine gum produced increasingly less interference in a sustained Stroop task (3 blocks of 100 stimuli – hence the possibility that negative Stroop results reviewed above may have been a function of an insufficient number of trials). Halliday *et al.* (1992) reported that 0.8 mg of subcutaneously injected nicotine tended to produce faster RTs and produced an increased number of fast responses in a visual choice-RT task. West and Hack (1991, Study 1 [no error data]) examined the effects of smoking on short-term memory in occasional smokers. Occasional smokers were defined as smoking less than 5 cigarettes/day and not smoking every day. West and Hack hypothesized that any smoking effects obtained in such subjects probably would not be a function of withdrawal. The occasional smokers were tested after having abstained from cigarettes overnight. Subjects were randomly assigned to two groups. The testing schedule of the first group was as follows: scanning task; smoke nicotine-yielding cigarette (1.5 mg); scanning task; smoke an herbal, nicotine-free cigarette; scanning task. The schedule of the second group was the same except that the order of the cigarettes was reversed. Smoking the nicotine-yielding but not the nicotine-free cigarette speeded the rate of short-term memory scanning.

Discussion

Overall, the studies reviewed in this section indicate that nicotine can facilitate cognitive processing in non-smokers. Combined with the evidence from the previous section that smoking/nicotine can facilitate cognitive processing in non-abstaining smokers, one is lead to the conclusion that smoking/nicotine can produce absolute facilitation. Again, however, we remind the reader that even if nicotine can produce absolute facilitation, this does not necessarily mean that all of the effects of smoking/nicotine in regular smokers are absolute in nature. Careful examination of the studies in the next section (where the results of studies in which nicotine was administered to both smokers and non/occasional smokers are reviewed) is crucial, as they should provide evidence regarding the first approach outlined in the *Introduction* to distinguishing the *withdrawal deficit* hypothesis from the *combination* hypothesis.

Non-smokers and occasional smokers II: Studies employing regular smokers

A summary of studies in which nicotine was administered to regular smokers as well as non-smokers may be found in Table 2.5. The table also summarizes one smoking study that employed 'occasional smokers', and also provides additional technical information regarding all the studies.

Table 2.5 – Studies in which nicotine was administered to both regular smokers and non-smokers/occasional smokers (Non/OS)

Study (Smoker Abstention)	Task(s)	Duration	Sex; Non or OS	Sessions/conditions	Effect of smoking/ nicotine on speed	Effect on accuracy
Wesnes and Warburton (1978) (???)	Stroop task	2 blocks of 400 stimuli	mixed; Non	0, 1 and 2 mg tablets	less Stroop interference during second block	not reported
Wesnes et al. (1983) (A)	Mackworth clock vigilance	80 min.	mixed; Non	0, 1 and 2 mg buffered tablets	not reported	attenuated drop-off in sensitivity
Hindmarch et al. (1990a) [Task 1] (NA)	visual 6 choice RT	20 trials	female; Non	0, 2 and 4 mg gum	faster movement RT but not decision RT only in smokers	too few errors to analyze
Hindmarch et al. (1990a) [Task 2] (NA)	short-term memory scanning	24 trials	female; Non	0, 2 and 4 mg gum	no effect	too few errors to analyze
Hindmarch et al. (1990a) [Task 3] (NA)	analogous to driving (see text)	not reported	female; Non	0, 2 and 4 mg gum	improved tracking only in smokers	too few peripheral errors to analyze
Sherwood et al. (1990a) [Task 1] (NA)	visual 6 choice RT	20 trials	female; Non	0 and 2 mg gum	faster movement RT	too few errors to analyze
Sherwood et al. (1990a) [Task 2] (NA)	analogous to driving (see text)	not reported	female; Non16	0 and 2 mg gum	improved tracking	too few peripheral errors to analyze
Kerr et al. (1991) [Task 1] (NA)	visual 6 choice RT	20 trials	female; Non	0 and 2 mg gum	faster movement but not decision RT	too few errors to analyze

Table 2.5 – Continued

Study (Smoker Abstention)	Task(s)	Duration	Sex; Non or OS	Sessions/conditions	Effect of smoking/nicotine on speed	Effect on accuracy
Kerr et al. (1991) [Task 2] (NA)	STM scanning	24 trials	female; Non	0 and 2 mg gum	faster scanning	too few errors to analyze
Kerr et al. (1991) [Task 3] (NA)	analogous to driving (see text)	4 minutes	female; Non	0 and 2 mg gum	improved tracking	too few peripheral errors to analyze
Sherwood et al. (1991a) (NA)	STM scanning	72 trials	female; Non	0 and 2 mg gum	positive responses faster scanned	too few errors to analyze
West and Hack (1991b) (Both)	STM scanning	5 minutes	mixed; OS	real/ no-nicot.	faster scanning smoking	not reported
Parrott and Haines (1991) (A)	RVIP task (100 digits/min)	not stated	mixed; OS	0 and 4 mg nicotine gum	not reported	better target detection
Grobe et al. (1993) (A)	numer. interference task	see note	mixed; Non	0, 5, 10 mg and 20 μg/kg	faster RT	not reported

Key: A = Abstaining smokers, NA = Non-Abstaining smokers, B = Both, ??? = Unknown; Effects, if any, are equal in smokers and non-smokers unless otherwise noted. STM = short term memory

The term 'x-mg cigarette' designates a cigarette having a nicotine yield of 'x' mg.

Wesnes and Warburton (1978 [no error data]) reported that chewing 1 mg or 2 mg nicotine tablets produced less Stroop interference in both smokers and non-smokers during the second of two blocks of 400 stimuli. Unfortunately, whether or not the smokers had abstained was not stated. Wesnes *et al.* (1983 [no RT data]) reported that chewing 1 mg or 2 mg nicotine tablets attenuated the drop off in sensitivity seen during performance of the Mackworth Clock vigilance task. This effect was equal in both 'light' and 'heavy' overnight-abstaining smokers as well as non-smokers. Grobe *et al.* (1993 [no error data]) employed a numerical interference task and reported that nicotine nasal spray speeded RT equally in overnight-abstaining smokers and non-smokers. Parrott and Haines (unpublished, reported in Parrott *et al.*, 1991 [no RT data]) reported that 4 mg nicotine gum increased target detections equally in overnight-abstaining regular (>15 cigarettes/day) and occasional (<5 cigarettes/day) smokers (there was a 'near-trend' [p = 0.13] for gum to increase target detections more in the regular smokers). Hindmarch *et al.* (1990a) reported that nicotine gum speeded choice RT and improved the tracking portion of simulated driving in non-abstaining smokers but not non-smokers; short-term memory scanning was not affected in either group. However, as discussed above in the section on visual perception, Hindmarch *et al.* also measured blood nicotine concentrations in their subjects and reported that the level of nicotine achieved in the non-smokers (4.92 ng/ml versus 12.40 ng/ml in the smokers) probably was not sufficient to affect cognitive performance. Both Sherwood *et al.* (1991c) and Kerr *et al.* (1991) reported that nicotine gum speeded choice RT and improved the tracking portion of simulated driving equally in non-abstaining smokers and non-smokers. Both Kerr *et al.* (1991) and Sherwood *et al.* (1991a) reported that nicotine gum speeded short-term memory scanning equally in non-abstaining smokers and non-smokers.

The results of West and Hack (1991 [no error data]) are particularly relevant as they tested both regular and 'occasional' smokers under conditions of both no-abstinence and 24 hour abstinence. Occasional smokers this time were defined as smoking fewer than 20 cigarettes/week and not smoking for at least one day every week. West and Hack hypothesized that any smoking effects obtained in such subjects probably would not be a function of withdrawal. In both groups, both before and after the abstinence periods, short-term memory scanning was equally faster following smoking of a nicotine-yielding cigarette but not following a nicotine-free control cigarette.

Discussion

Omitting the results of Hindmarch *et al.* (1990a), whose non-smokers apparently did not take in sufficient nicotine to affect cognitive performance, and those of Wesnes and Warburton (1978), who did not report whether their smokers were abstaining or not, every study reviewed in Table 2.5 reported equal facilitation of cognitive performance in non-/occasional smokers, abstaining smokers, and non-abstaining smokers (although only West and Hack, 1991, tested both abstaining and non-abstaining regular and occasional smokers in the same study). This pattern seems to fit best with an absolute-facilitation account of smoking/nicotine's effects on cognitive performance. However, further studies employing a wider variety of tasks as well as testing both abstaining and non-abstaining smokers as well as non/occasional smokers in the same study are needed to further resolve this issue. Finally, we note that Hasenfratz and Bättig (1993b, reviewed below as a cessation study) reported less facilitation by smoking in non-abstaining smokers than in overnight-abstaining smokers for the RVIP task, a result inconsistent with absolute facilitation as the sole effect of smoking on RVIP performance (c.f. also the 'subsequent administration' data of Sherwood *et al.*, 1992, for short-term memory scanning). Hasenfratz and Bättig's (as well as Sherwood *et al.*'s) results are consistent with either the combination hypothesis or absolute facilitation plus acute tolerance and/or ceiling effects.

Studies Including Non-Smoker 'Controls'

Groups that are not administered nicotine

In this section we review studies utilizing non-smokers who are not administered nicotine as 'controls'. A summary of studies employing non-smoker 'controls' who are not administered nicotine may be found in Table 2.6. The table also provides additional technical information regarding the studies. Summarizing these studies is difficult, as they represent a wide array of tasks and experimental conditions (e.g., whether or not the smokers abstained prior to testing; whether or not the smokers smoked as part of the testing procedure). We will focus on the issue of whether these studies provide any evidence regarding the hypotheses outlined in the *Introduction*.

Findings that abstaining smokers do worse than either non-smokers or smoking smokers would, assuming that non-smokers make proper 'normal' controls for assessing the effects of smoking/nicotine, indicate a 'normalizing' effect of smoking that would be consistent with the withdrawal-deficit hypothesis. However, as outlined in the *Introduction* section, non-smokers who are not administered nicotine probably do *not* make proper controls due to constitutional differences between non-smokers and smokers. Thus, such findings would not provide unambiguous support for 'normalization', since they would also be consistent with [1] non-smokers having constitutionally higher levels of performance than smokers combined with [2] absolute facilitation on the part of smoking/nicotine. In contrast, findings that smoking smokers do better than either non-smokers or abstaining smokers would indicate that smoking results in performance that is 'better than normal', and thus support absolute facilitation in an unambiguous fashion, since the 'constitutional deck' is stacked against smokers. Since the comparison of abstaining smokers versus smoking smokers is necessary for either of these two outcomes, studies that do not include this experimental manipulation will not be particularly helpful. Thus we will focus on studies whose smokers have undergone some sort of smoking manipulation (this focus excludes the results of the following studies: Ashton *et al.*, 1972 [no error data]; Kucek, 1975; Stevens, 1976; Knott, 1980; Bates and Eysenck, 1994). Tarriere and Hartemann (1964 [no RT data]) employed a simulated driving task. They tested smokers who abstained both prior to testing (overnight) and during task performance with smokers who smoked *ad libitum* prior to testing and during task performance. The smoking smokers had better detection of peripheral signals than the other two groups. Although

Table 2.6 – Studies testing both smokers and non-smoker 'controls'

Study	Task(s)	Duration	Design for smokers; Sex	Smokers: Prior (P) and During (D)	Group differences in speed	Differences in accuracy
Tarriere and Hartemann (1964)	analogous to driving	2.5 hours	within ss; ???	ON abstainers P and D/smoke P and D	not reported	smoking smokers best detection
Heimstra *et al.* (1967) [Task Component 1]	driving: steering the vehicle	6 hours	between ss; male	non-abs. P; smoke/ no-smoke D	not relevant	abstaining smokers less accurate
Heimstra *et al.* (1967) [Task Component 2]	driving: braking to traffic light	6 hours	between ss; male	non-abs. P; smoke/ no-smoke D	no slowing of RT in non-abstaining smokers only	not reported
Heimstra *et al.* (1967) [Task Component 3]	driving: dashboard monitoring	6 hours	between ss; male	non-abs. P; smoke/no-smoke D	not reported ???	deterioration in non-smokers only
Heimstra *et al.* (1967) [Task Component 4]	driving: brake light detection	6 hours	between ss; male	non-abs. P; smoke/no-smoke D	not reported ???	abstaining smokers less accurate
Ashton *et al.* (1972) [Exp. 1, Task 1]	visual choice RT	20 minutes	within ss; mixed	??? P; smoke D	mixed	not reported
Ashton *et al.* (1972) [Exp. 1, Task 2]	choice RT/ simulated driving	20 minutes	within ss; mixed	??? P; smoke D	faster 'slight' steers RT in smokers	not reported
Ashton *et al.* (1972) [Exp. 1, Task 3]	choice RT/ simulated driving	20 minutes	within ss; mixed	??? P; smoke D	mixed	not reported

Table 2.6 – Continued

Study	Task(s)	Duration	Design for smokers; Sex	Smokers: Prior (P) and During (D)	Group differences in speed	Differences in accuracy
Ashton *et al.* (1972) [Exp. 2, Task 1]	choice RT/ simulated driving	20 minutes	within ss; mixed	??? P; high/ low yield smoke D	non-smokers faster for some measures	not reported
Ashton *et al.* (1972) [Exp. 2, Task 2]	choice RT/ simulated driving	20 minutes	within ss; mixed	??? P; smoke D	no differences	not reported
Ashton *et al.* (1972) [Exp. 3, Task 1]	pursuit rotor	5 minutes	within ss; mixed	??? P; smoke D	no differences	not reported
Ashton *et al.* (1972) [Exp. 3, Task 2]	visual 2- and 4-choice RT	10 minutes	within ss; mixed	??? P; smoke D	no differences	not reported
Schori and Jones ('74) [Task 1]	meter monitoring	6 blocks of 30 minutes	between ss; mixed	??? P; smoke/ no smoke D	no group differences	no group differences
Schori and Jones ('74) [Task 2]19	white-light monitoring	6 blocks of 30 minutes	between ss; mixed	??? P; smoke/ no smoke D	no group differences	no group differences
Schori and Jones ('74) [Task 3]19	red-light monitoring	6 blocks of 30 minutes	between ss; mixed	??? P; smoke/ no smoke D22	no group differences	no group differences
Schori and Jones ('74) [Task 4]19	auditory monitoring	6 blocks of 30 minutes	between ss; mixed	??? P; smoke/ no smoke D	non-smokers slower for 'low' complexity/faster for 'high' complexity	no group differences
Schori and Jones ('74) [Task 5]19	mental math	6 blocks of 30 minutes	between ss; mixed	??? P; smoke/ no smoke D	no group differences	no group differences

Table 2.6 – *Continued*

Study	Task(s)	Duration	Design for smokers; Sex	Smokers: Prior (P) and During (D)	Group differences in speed	Differences in accuracy
Lyon *et al.* (1975)	visual choice RT	3 blocks of 50 trials	within ss; male	ON abs. P; smoke/no-smoke D	non-smk. and smk. -smokers faster decision RT	too few to analyze
Kucek (1975)	dual tracking/ mental math	15 minutes	within ss; male	??? P; smoking D	no group differences	smokers poorer
Stevens (1976) [Task 1]	Hunter-Pascal concept task		mixed design; mixed	non abs. P and D	not applicable	heavy smokers poorer
Stevens (1976) [Task 2]	Wisconsin card sort		mixed design; mixed	non abs. P and D	not applicable	no difference
Stevens (1976) [Task 3]	anagram task		mixed design; mixed	non abs. P and D	not applicable	heavy smokers poorer
Stevens (1976) [Task 4]	word-in context test		mixed design; mixed	non abs. P and D	not applicable	heavy smokers poorer
Tong *et al.* (1977)	auditory vigilance	60 minutes	between ss; mixed	3 hours abs. P; smoke/no-smoke D	not reported	improvement in smoking smokers
Knott and Venables (1980)	auditory simple RT	20 trials	between ss; male	ON abs./non abs. P; smoke/no smoke D	no group differences	not applicable
Knott (1980)	visual choice RT	4 blocks of 100 stimuli	between ss; mixed	abs. P (ON) and D	female smokers slower decision time	female smokers more errors
Nil *et al.* (1988)	RVIP task (variable rate)	65 minutes: 35 baseline/ 30 post	between ss; mixed	5 hrs. abs. P; smoke/no-smoke D	RT data not reported	smoking smokers more accurate

Table 2.6 – Continued

Study	Task(s)	Duration	Design for smokers; Sex	Smokers: Prior (P) and During (D)	Group differences in speed	Differences in accuracy
Beh (1989)	visual simple RT	5 minutes	within ss; male	??? P; smoke/ no smoke D	smoking smokers faster decision and movement RT	not applicable
Keenan *et al.* (1989) ST	sustained visual choice RT	23 minutes	between ss; male	24 hours abs./ no abs. P; no ST D	slower RT in abstaining group	no group differences
Hughes *et al.* (1989b)	sustained visual 2-choice RT	45 minutes	between ss; male	24 hours abs./no abs. P; no smoke D	no group differences	no group differences
Spilich *et al.* (1992) [Task 1]	difficult/easy target detection	48 trials	between ss; mixed	abs. P (ON) and D/ non-abs. P and smoke D	no group differences in RT	not reported
Spilich *et al.* (1992) [Task 2]	letter transformation	24 trials	between ss; mixed	abs. P (ON) and D/ non-abs. P and smoke D	non-smoker RT < smoke -smoker RT < abs.-smoker RT	not reported
Spilich *et al.* (1992) [Task 3]	short-term memory scanning	84 trials	between ss; mixed	abs. P (ON) and D/ non-abs. P and smoke D	non-smoker RT faster than other 2 groups	non-smokers fewest errors
Spilich *et al.* (1992) [Task 4]	'computer game' drivin	variable [?]	between ss; mixed	abs. P (ON) and D/ non-abs. P and smoke D	non-smokers games longer (faster driving ?)	non-smokers had more crashes [?]
Landers *et al.* (1992) ST [Exp. 2, Task 1]	visual simple RT	20 trials	within ss; male	ON abs. P; ST/ no ST D	no effect of ST; no ST versus non-ST user differences	not applicable
Landers *et al.* (1992) ST [Exp. 2, Task 2]	visual 'anticipation'	20 trials	within ss; male	ON abs. P; ST/ no ST D	no effect of ST; no ST versus non-ST user differences	not applicable

Table 2.6 – Continued

Study	Task(s)	Duration	Design for smokers; Sex	Smokers: Prior (P) and During (D)	Group differences in speed	Differences in accuracy
Landers *et al.* (1992) ST [Exp. 2, Task 3]	visual 2-choice RT	20 trials	within ss; male	ON abs. P; ST/ no ST D	no effect (no ST-user non-ST user differences)	not reported
Landers *et al.* (1992) ST [Exp. 2, Task 4]	timed mental math	60 trials	within ss; male	ON abs. P; ST/ no ST D	not reported	no effect
Landers *et al.* (1992) ST [Exp. 2, Task 5]	Stroop task	150 trials	within ss; male	ON abs. P; ST /no ST D	not reported	no effect
Bates and Eysenck (1994) [Task 1]	visual choice RT	at least 20–30 correct trials	between ss; mixed	2 hours abs. P; no-smoke D	no group difference	no group difference
Bates and Eysenck (1994) [Task 2]	visual 'odd-man -out' RT task	at least 20–30 correct trials	between ss; mixed	2 hours abs. P; no-smoke D	faster decis. time in 2 hour abstaining smokers	no group difference
Bates and Eysenck (1994) [Task 3]	inspection time	≥9 consecutive correct trials	between ss; mixed	2 hours abs. P; no-smoke D	no group difference	not applicable

Key: Nicotine not administered to non-smokers; 'ST' indicates smokeless tobacco study, 'ON' indicates 'Overnight'. The term 'x-mg cigarette' designates a cigarette having a nicotine yield of 'x' mg.

consistent with absolute facilitation, the fact that results of the other two measures taken (steering and peripheral false detections) were not reported weakens the strength of this conclusion. Heimstra *et al.* (1967 [variously, no RT/error data]) employed a simulated driving task, with half the smokers abstaining during task performance and half being allowed to smoke *ad libitum*. Although the smokers did not abstain prior to testing, the task duration seemingly was sufficient (six hours) to bring out abstention effects. For two of the components of the driving task (steering and brake-light detection), the abstaining smokers did worse than either the non-smokers or the smoking smokers (ambiguous 'normalization'). In contrast, for traffic-light braking, a slowing in RT across the six hours was seen in both non-smokers and abstaining smokers, but not in the smoking smokers (support for absolute facilitation). However, an alternative possible account of these results is that smoking caused subjects to allocate more attention to the traffic light task at the expense of steering and brake-light detection. Schori and Jones (1974) tested smokers who either smoked *ad libitum* (minimum of three cigarettes) or did not smoke during testing. The task was a complex one consisting of five sub-tasks. Although degree of pre-testing abstention was not stated, the duration of the task (three hours) seemingly was sufficient to bring out abstention effects in the no-smoking group. No results supporting either (ambiguous) 'normalization' or absolute facilitation were obtained. Lyon *et al.* (1975) employed a visual choice-RT task. Separate groups of overnight-abstaining smokers either smoked or did not smoke prior to the task. (Ambiguously) supporting 'normalization', non-smokers and smoking smokers had faster decision RTs than the smokers who did not smoke. Tong *et al.* (1977 [no RT data]) employed auditory vigilance task. Separate groups of three-hour abstaining smokers either smoked or did not smoke prior to the task. Smoking smokers showed improvement across the sixty minutes of the task that was not seen in non-smokers or smokers who did not smoke. Knott and Venables (1980) employed an auditory simple RT task. Their design for the smokers crossed overnight abstention/no abstention prior to testing with smoking/ no smoking prior to the task. No group differences of any sort were obtained.

Whether this was a function of the low complexity of the task or the rather intense smoking schedule employed resulting in possible toxic effects is unclear. Nil *et al.* (1988 [no RT data]) employed the RVIP task. Separate groups of five-hour abstaining smokers either smoked or did not smoke during testing. Across the sixty minutes of the task, smoking smokers showed an increase in the number of digits processed per minute that was not seen in non-smokers or smokers who did not smoke. Although these data support absolute facilitation, the strength of these results is weakened by the fact that RT data were not reported. Beh (1989) employed a visual simple-RT task. In two sessions, smokers either smoked or did not smoke during testing. Degree of pre-testing abstention was not specified. In the smoking session, smokers had faster RTs than they did in the no smoking session and faster RTs than the non-smokers, a result supporting absolute facilitation. Keenan *et al.* (1989) conducted a smokeless tobacco study. They tested non-users of smokeless tobacco (who were also non-smokers), 24 hour abstaining smokeless-tobacco users, and non-abstaining smokeless-tobacco users using a sustained visual choice RT task. The abstaining group had slower RTs than the other two groups, a result (ambiguously) supporting 'normalization'. Landers *et al.* (1992) conducted a similar study with the exception that all smokeless tobacco users abstained overnight prior to testing and participated in two sessions: one in which smokeless tobacco was chewed and another in which it was not. A battery of five cognitive tasks was employed. In contrast to Keenan *et al.*, no group differences were obtained for any of the tasks, a result supporting neither (ambiguous) 'normalization' nor absolute facilitation. Hughes *et al.* (1989b) tested non-smokers, 24 hour abstaining smokers, and non-abstaining smokers using a sustained visual choice-RT task. There were no group differences in either RT or RT errors, supporting neither (ambiguous) 'normalization' nor absolute facilitation.

Discussion

In sum, across these studies we find two that (ambiguously) support 'normalization', three that support absolute facilitation, four that support neither, and one (Heimstra *et al.*, 1967) that, depending on the sub-task, supports both. Thus, even assuming for the moment that non-smokers do make proper 'normal' controls against which to assess the effects of smoking nicotine, we find only two of ten studies supporting a 'normalizing' effect of smoking.

Finally, we will focus in some depth on the results of Spilich *et al.* (1992), since they have received widespread publicity (Adler, 1993; Bower, 1993; West, 1993). This series of studies generated such interest because

1 Spilich *et al.* claim their data demonstrate that smoking has a long-term, deleterious effect on cognitive performance,
1 for two tasks, apparent withdrawal *facilitation* was obtained, unique results in the smoking/cognitive performance literature (one of the studies employed a memory task and thus 'properly' belongs below in that section; however, since it produced evidence of withdrawal facilitation, we review it here), and
1 Spilich *et al.* claim their results indicate that smoking impairs automobile driving, a very important task of everyday living in industrialized countries.

Spilich *et al.* (1992) conducted a series of experiments comparing non-smokers (NON), three-hour abstaining smokers who did not smoke during the study (ABS), and three-hour abstaining smokers who did smoke during the study (SMO; different subjects were tested in each experiment). The SMO group smoked a cigarette having a nicotine yield of 1.2 mg just prior to task performance according to an experimenter-driven schedule.

In the *first* experiment, subjects scanned 8 × 12 arrays of letters for target letters that either were similar to the non-target letters (target *X* embedded in other angular letters such as *K*; target *O* embedded in other rounded letters such as *Q*) or were dissimilar to the non-target letters (target *X* embedded in rounded letters; target *O* embedded in angular letters). A total of 48 trials equally balanced between target *X/O* and similar/dissimilar non-targets were run. RT did not differ significantly among the three groups (error data were not reported).

In the *second* experiment, subjects monitored arrays of 20 identical letters, one of which, after a variable period, transformed into a different letter (e.g., one of 20 *B*s transforming into an *E*). Subjects made RT responses to the transformations. A total of 24 trials were run. Results indicated that the NON group was faster than the SMO group, which was in turn faster than the ABS group (error data were not reported).

In the *third* experiment, subjects performed a version of the Sternberg (1966) short-term memory-scanning task, with memory set size varying between one and six (84 trials, 14 for each set size). Results indicated that the NON group has faster RTs than the two smoker groups, which did not differ. The NON group also made fewer errors than the ABS group, which in turn made fewer errors than the SMO group. RT slope across set size (an index of the speed of STM scanning) showed the following significant pattern: NON > ABS > SMO.

In the *fourth* experiment, subjects read a text passage and were tested for comprehension of the passage. Results indicated that the NON group had better overall recall of the content of the passage than the ABS group, which in turn had better recall than the SMO group. For recall of content especially relevant to the central 'point' of the passage, the NON group did better than the two smoker groups, which did not differ from each other.

In the *final* experiment, subjects performed a 'computer-game' driving simulation. The length of time that subjects 'stayed in' the game (one of the measures assessed) was directly related to driving speed and inversely related to the number of rear-end collisions (the other measure assessed). The primary task of the subjects was to stay in the game as long as possible. Specifically, Spilich *et al.* instructed their subjects to "… stay in the game as long as possible by maintaining a quick pace and avoiding accidents" [p. 1321]. Results for the primary task given to the subjects indicated a trend for the SMO group to 'stay in' the game longer than either of the other two groups. Since the SMO group had more accidents than the other two groups, their staying in the game longer must have been a result of a higher overall driving speed.

Withdrawal Facilitation   In both Spilich *et al.*'s short-term memory scanning study (error and RT-slope results) and their text-comprehension study (recall results), abstaining smokers did better than non-abstaining smokers. Such findings are indicative of withdrawal facilitation. Since, other than the results of Roth *et al.* (1992) reviewed below, no group has reported similar results[2], one must look to Spilich *et al.*'s methodology for clues as to why these outcomes may have been obtained; their methodology may also account for the lack of an effect of smoking in the letter-target and 'driving' studies.

It should be noted that different smokers typically exhibit a range of puffing and inhalation behaviours that result in a wide range of plasma nicotine concentrations (Robinson *et al.*, 1992). However, Spilich *et al.* did not allow their SMO groups to smoke *ad libitum*. Instead, smoking by the SMO groups followed an experimenter-driven schedule: subjects took 12 puffs on a cigarette having a nicotine yield of 1.2 mg, with the smoke being held in the lungs for five seconds. Such a regimen could have resulted in many subjects taking in more nicotine than customary. Even in regular smokers, 'over-smoking' can produce symptoms of malaise such as lightheadedness and nausea (Gilbert *et al.*, 1992). Such an effect would hardly be conducive to the performance of complex cognitive tasks

---

[2] Indeed, Spilich *et al.*'s short-term memory performance results are at variance with virtually the entire body of literature, which, as reviewed above, clearly indicates that nicotine facilitates scanning in both abstaining and non-abstaining regular smokers (Kerr *et al.*, 1991; Sherwood *et al.*, 1991; Sherwood *et al.*, 1992; West and Hack, 1991), while occasional smokers (West and Hack, 1991), and non-smokers (Kerr *et al.*, 1991; Sherwood *et al.*, 1991a) have minimal physical tolerance to nicotine.

(indeed, such 'over-smoking' has been utilized as a method of smoking cessation by forming an association between smoking and subsequent feeling of malaise).

Even in the absence of malaise, it is doubtful such artifactual smoking conditions represented the smokers' normal smoking behaviour. Data such as completion of questionnaires assessing feeling states (e.g., dysphoria) and indices of nicotine absorption (tidal-breath carbon monoxide, plasma nicotine concentrations) that would have helped clarify this issue were not collected.

Chronic impairment? A good example of the pitfalls involved in smoker/non-smoker comparisons is Spilich *et al.*'s conclusion, outlined in various portions of their 1992 article, that smoking is associated with chronic decrements in performance. This claim was based on results in which smokers did worse than non-smokers regardless of whether they had smoked recently or not. For example, in discussing their short-term memory scanning study, Spilich *et al.* stated that their results were due to "residual effects of smoking" [p. 1318].

To reiterate, such a conclusion is weak in that it ignores well-documented constitutional differences between smokers and non-smokers with regard to both personality and IQ (*Introduction*). Thus, when large groups of smokers and non-smokers are compared as was done by Spilich *et al.* in their short-term memory scanning study (20 non-smokers versus 40 smokers), and overall the non-smokers perform better, this merely replicates the smoking/IQ literature. Again, equating subjects with regard to IQ prior to the onset of smoking would be necessary in order for the Spilich *et al.* 'residual effect' hypothesis to be tenable. Indeed, Spilich *et al.* themselves addressed this point: "... pre-experimental differences in SES [socio-economic status], IQ, personality, and other variables might differentiate our smoking and nonsmoking subjects" (p. 1315). No such differentiation was attempted, however[3].

Smoking and driving In discussing their driving-game data, Spilich *et al.* focused on their accident findings (again, avoiding accidents was not the primary task given to the subjects: staying in the game as long as possible was). Again, Spilich *et al.* found that non-abstaining smokers had more accidents than either abstaining smokers or non-smokers, and took this as an indication that smoking has a negative effect on driving "performance" (see Spilich *et al.* Figure 6, p. 1322). In actuality, the fact that the non-abstaining group tended to stay in the game longer than the other two groups indicates that smoking may have affected subjects' speed/accuracy trade-off strategy (increasing the willingness to trade increased accidents for increased speed) rather than their overall performance. Since the accidents in the game were not 'real', whether this effect would generalize to real driving is questionable: a stronger experiment would have penalized accidents by subtracting some amount of 'bonus money' available to subjects. At any rate, the Spilich *et al.* findings are at odds with those of Heimstra *et al.* (1967). Again, Heimstra *et al.* reported that, across six hours of a simulated-driving task that was much more sophisticated than the Spilich *et al.* driving game, smokers who

---

[3] In their *Discussion* section, Spilich *et al.* reviewed literature indicating that smokers have impaired regional cerebral blood flow, offering this as evidence supporting their 'residual effect' hypothesis. In our opinion, these findings are irrelevant to the Spilich *et al.* study since they come from elderly smokers, while the average age of Spilich *et al.*'s subjects was 19.2 years.

smoked maintained superior 'steering' performance relative to the smokers who did not smoke. Further, recall that for a comparable measure (tracking in a dual tracking/peripheral-stimulus RT task) both Sherwood *et al.* (1990) and Kerr *et al.* (1991) reported that nicotine produced performance facilitation in both non-abstaining smokers and non-smokers.

Effects of Smoking Cessation

The effects of smoking cessation on cognitive performance is summarized in Table 2.7. The table also provides additional technical information regarding the studies. Since data from cessation studies have the potential to address the hypotheses outlined in the *Introduction*, we review them in detail, with comments presented following the description/results of each study.

Elgerot (1976) tested smokers in two sessions: one following normal smoking and one following 15 hours of smoking abstinence. A battery of tasks was used that included [1] a mental-math task; [2] Raven's Matrices; [3] the 'Bourdon Test'; [4] a letter-series task, and [5] a proofreading task (see Table 2.7 for details). In the abstinence condition, smokers did worse on the mental-math task, Raven's Matrices, and the letter-series task. No effect of abstinence was noted for the Bourdon Test or proofreading.

[1] As discussed in the *Introduction*, the absence of time-course data in acute cessation studies such as Elgerot's makes it impossible to differentiate a withdrawal deficit from the mere removal of absolute facilitation.

Myrsten *et al.* (1977) studied two matched groups of smokers in a 15 day experiment. One group smoked *ad libitum* throughout the study. The other smoked for five days, abstained from smoking for five days, then resumed smoking for the final five days. Subjects were given a battery of tests once during each phase of the study. The battery included (see Table 2.7 for details) [1] a mental-math task; [2] a 'correction test'; [3] a letter-cancellation test; and [4] the Stroop task. The only result clearly related to cessation was a trend for performance of the Stroop task to be poorer during abstinence. Error data were not reported.

[1] Although it was not specifically stated when Myrsten *et al.* tested their subjects during abstinence, their Figure 1 would seem to indicate that testing took place on the fourth day. If this is the case, then the lack of 'cessation' effects on performance may have been due to performance returning to 'normal' following a withdrawal deficit. However, in the absence of actual data from the first day of abstinence demonstrating deteriorations in performance, this conclusion is speculative, especially since error data were not reported.

Elgerot (1978) tested smokers in two sessions: one following four days of normal smoking and one following four days of smoking cessation. A battery of tasks was used that included [1] a mental-math task; [2] Raven's Matrices; [3] the 'Bourdon Test'; [4] a 'geometric patterns' task, and [5] a 'higher-number' task (see Table 2.7 for details). No significant effects were obtained for any of the tasks.

Table 2.7 – Studies investigating smoking/nicotine cessation

Study	Task(s)	Duration	Cessation from …	Length of cessation; Sex	Effect of cessation on speed	Effect on accuracy
Myrsten *et al.* (1977) [Task 1]	mental math	5 min.	smoking	15 days; mixed	no effect	not reported
Myrsten *et al.* (1977) [Task 2]	visual correction	2 min.	smoking	15 days; mixed	no consistent effect	not reported
Myrsten *et al.* (1977) [Task 3]	letter cancellation	1.5 min.	smoking	15 days; mixed	no consistent effect	not reported
Myrsten *et al.* (1977) [Task 4]	Stroop task	12 blocks of 21 stimuli	smoking	15 days; mixed	trend toward slower processing during cessation	not reported
Elgerot (1976) [Task 1]	mental math	???	smoking	15 hours; mixed	poorer performance	
Elgerot (1976) [Task 2]	Raven's Matrices	???	smoking	2 days mixed	poorer performance	
Elgerot (1976) [Task 3]	'Bourdon' test	???	smoking	2 days mixed	no effect	
Elgerot (1976) [Task 4]	'letter series'	???	smoking	2 days mixed	poorer performance	
Elgerot (1976) [Task 5]	proof- reading	???	smoking	2 days mixed	no effect	
Elgerot (1978) [Task 1]	mental math	???	smoking	2 days mixed	no effect	

Table 2.7 – *Continued*

Study	Task(s)	Duration	Cessation from …	Length of cessation; Sex	Effect of cessation on speed	Effect on accuracy
Elgerot (1978) [Task 2]	Raven's Matrices	???	smoking	days mixed	no effect	
Elgerot (1978) [Task 3]	'geometric patterns'	???	smoking	2 days mixed	no effect	
Elgerot (1978) [Task 4]	'highest number'	???	smoking	2 days mixed	no effect	
Elgerot (1978) [Task 5]	'Bourdon test'	???	smoking	2 days mixed	no effect	
Snyder *et al.* (1989) [Task 1]	two-letter search	about 2 min.	smoking	17 days male	slower RT at 4, 8 and 24 hours into cessation	no effect
Snyder *et al.* (1989) [Task 2]	six-letter search	about 2 min.	smoking	17 days male	slower RT at 24 and 48 hours into cessation	no effect
Snyder *et al.* (1989) [Task 3]	logical reasoning	about 2.5 min.	smoking	17 days male	slower RT at 24 and 48 hours into cessation	no effect
Snyder *et al.* (1989) [Task 4]	math task	about 3 min.	smoking	17 days male	slower RT be-tween 8 and 144 h. into cessation 21	see note
Hasenfratz and Bättig (1991)	RVIP task (variable rate)	30 min.	smoking	9 days female	no effect	no effect
Hatsukami *et al.* (1991) [Part 1]	visual 2-choice RT	20 min.	smoking	1-3 days mixed	no effect	fewer errors

Table 2.7 – *Continued*

Study	Task(s)	Duration	Cessation from …	Length of cessation; Sex	Effect of cessation on speed	Effect on accuracy
Hatsukami *et al.* (1991) [Part 2]	visual 2-choice RT	20 min.	nicotine gum	1-3 days mixed	no effect	more commis-sion errors [4-mg only]
Parrott and Roberts (1991)	letter cancel-lation	1350 letters (105 targets)	smoking	12 hrs; mixed	slower processing	fewer targets detected
Hatsukami *et al.* (1992) [Exp. 1]	visual 2-choice RT	23 min.	smokeless tobacco	5 days; male	slower RT	no effect
Hatsukami *et al.* (1992) [Experiment 2]	visual 2-choice RT	23 min.	smokeless tob. (gum repl.)	5 days; male	no effect of gum	no effect of gum
Roth *et al.* (1992) [Task 1]	word- recognition task	2 blocks of 30 stimuli	smoking	over-night; female	varied as a function of 'early' vs. 'late' smokers	
Roth *et al.* (1992) [Task 2]	Austin maze learning	5 learning and 3 retention trials	smoking	over-night; female	varied as a function of 'early' vs. 'late' smokers	
Gross *et al.* (1993)	Stroop task	26 stimuli	smoking	12 hrs.; male	increased time to name smoking- related words	not measured
Hasenfratz and Bättig (1993a)	RVIP task (variable rate)	40 min. (20 baseline and 20 post	smoking	over-night; female	no effect	slower processing rate

The term 'x-mg cigarette' designates a cigarette having a nicotine yield of 'x' mg.

- Three of the tasks employed by Elgerot were also employed in her previous 15 hour abstention study (Elgerot, 1976). Keeping in mind that the data come from different groups of subjects (e.g., policemen in the 1976 study versus teachers in the 1978 study), by comparing the two studies, some information regarding the time course of effects can seemingly be obtained. For both the mental-math task and for Raven's Matrices, poorer performance after 15 hours versus no difference after four days indicates that smoking cessation may produce some component of withdrawal deficit for these tasks (the lack of an effect at either time point for the 'Bourdon Test' indicates that smoking simply does not affect performance of that task).

Snyder *et al.* (1989) conducted a two-phase study that took place on a clinical ward. During an initial week-long baseline period subjects smoked *ad libitum* and were trained on the Snyder and Henningfield (1989) cognitive test battery (descriptions of the tasks provided in Table 2.7). At the end of the week, baseline performance on the test battery was measured. The subjects then stopped smoking for ten days, with the test battery being administered at 1, 4, 8, and 24 hours of the first day of cessation, then at 24 hour intervals thereafter. Subjects then resumed smoking, with the battery being administered at 1, 4, 8, and 24 hours following resumption. Relative to pre-cessation baseline, RT slowed in the two-letter search task at 4, 8, and 24 hours post-cessation, then returned to baseline for the remainder of the cessation period. For both the six-letter search and logical-reasoning tasks, RT slowed at 24 and 48 hours post-cessation before returning to baseline for the remainder of the cessation period. Finally, in the serial-addition/subtraction task, RT slowed between 8 and 144 hours post-cessation before returning to baseline. No effects of cessation on RT errors were obtained for any of the tasks.

- Snyder *et al.*'s (1989) report of deteriorations in RT performance during the early post-cessation period followed by returns to baseline would, at first glance, seem to support the withdrawal-deficit hypothesis. Again, however, these returns to baseline are difficult to evaluate in the absence of a group controlling for the mere passage of time – with the single experimental group employed, recovery toward baseline cannot be distinguished from gradual, long-term learning and/or adaptation to confinement in a clinical ward for an extended period (see *Introduction*).

Hasenfratz and Bättig (1991) employed a version of the RVIP task in which stimuli are presented at a variable rate depending on response accuracy. Separate groups either did or did not stop smoking for nine days. Subjects in the cessation group performed the RVIP before cessation (baseline) and at three and nine days following cessation. Subjects in the control group performed the RVIP at comparable intervals. There were no cessation/control differences in RVIP performance at either baseline, three days post-cessation, or nine days post-cessation.

- The results of Hasenfratz and Bättig (1991) do not fit with a withdrawal-deficit interpretation. At first glance, they also do not seem to fit with an absolute-facilitation interpretation either. However, at the time of testing, subjects in non-abstaining sessions would have been without a cigarette for approximately one hour, during which time it is conceivable that any absolute facilitation of RVIP performance may have dissipated.

Hatsukami *et al.* (1991) examined effects on a choice RT task of smoking cessation. Following one week of smoking cessation, subjects were placed on a placebo-controlled trial of nicotine gum (separate groups receiving 0 mg, 2 mg, and 4 mg gum). This was in turn followed by (for the active-gum group) one week of cessation from nicotine gum. For both smoking and gum cessation, the choice-RT task was administered on the two days immediately prior to cessation (baselines), and on days one, two, and three of cessation. Relative to pre-cessation baseline (average of the two baseline measurements), initial smoking cessation (average of the three cessation measurements) was associated with decreased errors of both commission and omission but no RT changes. Relative to cessation from placebo gum, cessation from nicotine gum was not associated with either changes in RT or errors of omission. Cessation from 4 mg but not 2 mg gum was associated with an increase in errors of commission.

- The smoking-cessation portion of the Hatsukami *et al.* study lacked a 'no-cessation' control group, leaving it open to the same criticisms applied above to the Snyder *et al.* (1989) study. The increased errors of commission seen during cessation from 4 mg gum can be accounted for by either a withdrawal deficit or the removal of absolute facilitation, since any time-course data were lost by the procedure of averaging across the three cessation days.

Parrott and Roberts (1991) tested smokers in four sessions. Sessions one and four were no-abstinence 'baselines'. Half the subjects abstained from smoking for 12 hours prior to session two and half prior to session three. Within each session, a letter-cancellation task was administered before and after smoking. Abstinence was associated with poorer performance that was reversed by subsequent smoking.

- Again, the absence of time-course data makes it impossible to differentiate a withdrawal deficit from the mere removal and restoration of absolute facilitation.

Hatsukami *et al.* (1992) conducted two studies. The first involved cessation from smokeless tobacco and employed a choice-RT task. Subjects were then randomly divided into two groups; one group abstained from smokeless tobacco for five days while the other did not. For the cessation group, the RT task was administered on the two days immediately preceding cessation and on days one through five of cessation. Subjects in the control group performed the task at comparable intervals. RT in the cessation group (average of the five cessation days) increased relative to the baseline average while RT in the control group did not (neither errors of commission nor omission were differentially affected).

- The slower RT seen during cessation can be accounted for by either a withdrawal deficit or the removal of absolute facilitation, since any time-course data was lost by the procedure of averaging across the five cessation days.

Hatsukami *et al.*'s second study was similar to the first, with the exception that all subjects underwent smokeless-tobacco cessation and were divided into three groups in a placebo-controlled nicotine-replacement trial during the five cessation days (0 mg, 2 mg, and 4 mg nicotine gum). Cessation effects were similar to the first for all groups, indicating no effect of nicotine gum on performance during cessation from smokeless tobacco.

- Hatsukami *et al.* did not actually present performance data – they merely stated that there were no differences among the groups during cessation. Assuming that cessation did produce a performance decrement as in the first study, ascribing this to either a withdrawal deficit or loss of absolute facilitation is problematic, since it was not affected by nicotine replacement.

Roth *et al.* (1992) employed two tasks in a study of overnight-abstaining versus non-abstaining smokers. In one session, subjects abstained from smoking overnight and throughout the session. In the other, they were free to smoke *ad libitum* prior to the session, and smoked a total of 4 cigarettes during the session. The two tasks employed by Roth *et al.* were a word-recognition task (WRT) and maze learning (Austin maze; see Table 2.7 for a description of these tasks)[4]. Sessions proceeded as follows: smoking/resting, maze task (5 learning trials), smoking/resting, WRT (30 words), smoking/resting, WRT (30 words), smoking/resting, and maze task (3 retention-testing trials). Smoking abstinence did not affect performance of the maze task. For the WRT task, words not recognized after 20 presentations were greater in the abstinence session (other WRT measures were not affected). However, when Roth *et al.* divided subjects into 'early' and 'late' smokers (early smokers having a cigarette within 1 hour of rising in the morning), they found that the 'early smokers' had better maze performance (less errors and greater speed) in the no-abstinence/smoking session while 'late smokers' had better maze performance in the abstinence session. Similarly, for the WRT, cumulative exposure time was less for 'early smokers' in the no-abstinence/smoking session and for 'late smokers' in the abstinence session.

- Again, the absence of time-course data makes it impossible to differentiate a withdrawal deficit from the mere removal and restoration of absolute facilitation with regard to the effect of overnight abstinence on WRT performance. The better performance on both tasks of Roth *et al.*'s 'late smokers' in the abstinence session is the only report other than Spilich *et al.* (1992) that provides evidence of withdrawal facilitation. One may speculate that Roth *et al.* obtained this effect for a similar reason that Spilich *et al.* may have: nicotine toxicity. One may presume that 'late smokers' are more sensitive to smoking's effects than 'early smokers', and thus may have had a toxic reaction to smoking four cigarettes in such a short time frame.

Gross *et al.* (1993) employed the Stroop task (26 stimuli) and tested smokers in four sessions representing a crossing of 12 hours/no smoking abstinence with two types of word lists: smoking-related words and words unrelated to smoking. The overall time needed to name the colour of words was not affected by

---

[4] The Austin maze involves perceptual-motor learning that is thought to involve different neural systems from those employed in so-called explicit or declarative memory (e.g., memory for verbal material; Baddeley, 1992). Hence, the Roth *et al.* study is reviewed here rather than in the learning/long-term memory section.

abstinence. However, in the abstinence sessions, the time needed to name the colour of the words was greater when they were smoking-related than when the words were unrelated to smoking. In the no-abstinence sessions, the opposite pattern was obtained. Error data were not presented.

- The lack of an effect of abstinence on overall Stroop performance is consistent with studies reviewed above indicating that large numbers of trials are needed for Stroop facilitation to be obtained. The interaction of abstinence with the smoking-relatedness of the words is very interesting and indicates that "... nicotine abstinence decreases the ability to ignore the meaning of smoking-related information" (p. 333; again, in the Stroop task, word meanings are task irrelevant to naming the colour of the ink in which the words are printed). The authors concluded that "The obtained pattern of results cannot be explained by a general decrement in cognitive functioning associated with nicotine abstinence" (p. 335).

Hasenfratz and Bättig (1993b) employed the RVIP task and tested smokers in four sessions representing a crossing of overnight/no smoking abstinence with smoking/no smoking during a session. In each session, subjects performed the RVIP task for baseline 20 minutes, then either smoked or rested, and then performed the RVIP for an additional 20 minutes. Smoking increased processing rate (without affecting RT) in both the overnight-abstaining and non-abstaining sessions, but did so to a greater degree in the abstaining session.

- Although this study did not yield time-course data, the greater effect of smoking following overnight abstinence than following no abstinence is inconsistent with absolute facilitation as the sole effect of smoking on cognitive performance, but consistent with either the combination hypothesis or absolute facilitation plus acute tolerance and/or ceiling effects.

Discussion

The study coming closest to the 'ideal' cessation study presented in the introduction was that of Hasenfratz and Bättig (1991), who reported no effect of abstinence on RVIP performance! If one assumes (from the numerous studies that did report positive effects of smoking/nicotine on RVIP performance) that the lack of effect of cessation in this study was due to *de facto* abstinence of at least one hour in the 'no cessation' condition, then one can conclude that the results are at best 'not incompatible with' absolute facilitation (and do not support the withdrawal-deficit hypothesis). Most other studies fail to present time-course data due to either

[1] taking only one measurement after overnight abstinence (Elgerot, 1976; Hasenfratz and Bättig, 1993a; Gross *et al.*, 1993; Parrott and Roberts, 1991; Roth *et al.*, 1992),
[2] taking only one measurement 'well into' abstinence (Elgerot, 1978; Myrsten *et al.*, 1977),
[3] collecting time-course data but then 'destroying' it by averaging it together (Hatsukami *et al.*, 1991, 1992).

Snyder *et al.* (1989) did collect time course data, but the lack of a no-cessation control group greatly weakens the support of their data for a withdrawal deficit. By combining the two Elgerot studies, there is some indication of a withdrawal deficit for mental math and Raven's Matrices.

Effects of Smoking/Nicotine on Cognitive Performance: Summary

Data from non-abstaining smokers, abstaining occasional smokers who subsequently smoke, and non-smokers administered nicotine, clearly indicate that nicotine is capable of producing absolute facilitation, ruling out the withdrawal deficit hypothesis in its 'strong form'. The question then becomes whether absolute facilitation is the sole basis of smoking's effect, or whether in regular smokers a withdrawal component co-exists with absolute facilitation (the combination hypothesis). Data comparing occasional/non-smokers, abstaining smokers, and non-abstaining smokers, generally indicate equal facilitation in all three groups for numerical interference, choice RT, the tracking portion of simulated driving, and short-term memory scanning. Although this pattern is consistent with absolute facilitation, only one of the studies reviewed (West and Hack, 1991) tested all three types of subjects in the same experiment, a procedure that clearly needs wider application in future studies. Further, West and Hack's short-term memory scanning results, obtained with smoking, seem at variance with the RT results of Sherwood *et al.* (1992). Again, Sherwood *et al.* reported that in overnight-abstaining smokers, a first but not second or third administration of nicotine gum speeded RT in the Sternberg task. Also, Hasenfratz and Bättig (1993b) reported less RVIP facilitation in non-abstaining smokers than in abstaining smokers, a result (like Sherwood *et al.*'s short-term memory scanning data) compatible with either the combination hypothesis or absolute facilitation plus acute tolerance and/or ceiling effects.

Data from cessation studies are basically inconclusive regarding absolute facilitation versus the combination hypothesis due to methodological problems (see that section's *Discussion*). Studies employing non-smokers who are not administered nicotine, even assuming that such subjects make proper controls, give little evidence of a 'normalizing' effect of smoking.

In (very tentative) sum, the evidence as it presently stands indicates that, while there may by a withdrawal-deficit component to some of the effects of smoking/ nicotine, it would seem to be small relative to absolute facilitation (for some tasks, ceiling and/or tolerance effects may exist in conjunction with absolute facilitation). Clearly, however, the old saw 'more work needs to be done' applies to the area of delineating the nature of the effect of smoking/ nicotine on cognitive performance.

EFFECTS OF SMOKING/NICOTINE ON LEARNING AND MEMORY

Most studies of the effect of smoking/nicotine on learning and memory have been interpreted within the context of the Atkinson and Shiffrin (1968, 1971) model of memory (see Hindmarch *et al.*, 1990b; Baddeley, 1992). This model posits two separate memory stores: short-term memory (STM) and long-term memory (LTM). STM (also referred to as 'working memory') is viewed as a rapidly-decaying buffer into which new information is placed. The information is lost unless transferred, or 'consolidated' into LTM. LTM is viewed as being of unlimited capacity and (barring neurological damage) relatively permanent, although the ability to access or retrieve information may decay.

Short-Term Memory (STM) Capacity

The standard neuropsychological way of assessing STM capacity is so-called 'digit span', in which the subject is serially presented with lists of digits of varying length, and is required to repeat the list in its proper order. The list length at which an error is made is the digit span, taken to reflect the capacity of the STM 'buffer'. We were not able to identify any studies that used this procedure. We were able to identify studies that assessed STM capacity in a slightly different manner, which is to present a set of items that clearly exceeds STM capacity (generally held to be 'Miller's Magic Number' of 7 ± 2 items) and assessing speed and accuracy of STM access. This procedure differs from that of the Sternberg STM-scanning paradigm, where an upper limit of six items that is within the STM capacity of most individuals is employed (Klatzky, 1980).

Williams (1980) reported that in overnight-abstaining smokers, real but not sham smoking produced an *increase* in errors for lists of nine digits presented auditorily. This increase tracked the nicotine yields of the three cigarettes employed (0.6 mg, 1.3 mg, and 1.8 mg). The increase in errors was similar in high and low-arousal smokers as classified by the SSQ as well as similar in light, medium, and heavy smokers. Williams did not measure response speed. As reviewed above, Snyder and Henningfield (1989) conducted a study in which smokers were tested three times: baseline, following overnight abstention, and then following a placebo-controlled administration of nicotine gum. In addition to the tasks reviewed above, they also employed a task in which a memory set of nine digits was presented for one second, followed three seconds later by eight of the original nine in a random order. Subjects were instructed to name the missing digit. Relative to pre-abstention baseline, response speed slowed in the placebo session. In seeming contrast to the results of Williams (1980), both the 2 mg and 4 mg gum (but not placebo gum) restored response speed to baseline levels without affecting response accuracy[5]. In their ten-day cessation study, Snyder *et al.* (1989) reported that for their nine-item STM task, response speed slowed at four, eight, and 24 hours, then swung back to baseline. However, response speed again slowed relative to baseline at 96 and 192 hours post-cessation. Similarly, response accuracy was lower at four hours, then swung back toward baseline.

---

[5] In fact, the only response-accuracy trend (p < .10) that Snyder and Henningfield reported in their *Results* section was for correct responding to increase, relative to pre-deprivation baseline, following 4-mg nicotine gum (with no trends relative to baseline following placebo or 2-mg gum). Curiously, however, in the *Discussion* section of their paper Snyder and Henningfield summarize this measure as indicating that "... there were trends in the accuracy data which suggested that tobacco deprivation was associated with decrements which were reversed by nicotine replacement ..." (p. 21).

Again, however, it was lower at 96 and 120 hours post-cessation. In general, performance of the digit-recall task appeared poorer throughout the later portion of the cessation period, indicating that the decrement may not have been a withdrawal effect. Again, the results of this study are difficult to interpret, since it lacked a no-cessation control group.

Heishman *et al.* (1990) administered placebo, 2 mg, 4 mg and 8 mg nicotine gum to non-smokers. The 4 mg and 8 mg gum affected performance of a 'digit-recall task' (presumably this was again the Snyder and Henningfield, 1989, nine-item STM task) in a mixed manner: responding was faster but accuracy decreased (although the latter effect tended to diminish with repeated administration). Again, it should be kept in mind that Heishman *et al.*'s subjects experienced unpleasant mood effects from the gum.

Short-Term Memory capacity: discussion

These few studies provide no consistent evidence regarding the effects of smoking/nicotine on STM capacity. We also note here that the lack of an STM-mediated recency effect in the serial-learning study of Mangan and Golding (1978; reviewed below) indicates no effect of smoking by one-hour abstaining smokers on 'consolidation' from STM to LTM (which presumably would be · more efficient if smoking had increased STM capacity). In contrast, Warburton *et al.* (1992a; also reviewed below) reported that in overnight-abstaining smokers, smoking facilitated the recency effect immediately following free-recall learning; however, it did not do so when testing took place following a 10 minute distractor task. Further studies of the effects of smoking/nicotine on STM capacity as well as 'consolidation' from STM to LTM are needed.

Learning/Long-Term Memory

In laboratory rats, nicotine in low doses has been reported to facilitate acquisition and retention in a variety of learning tasks (for a brief review, see Mangan and Golding, 1978). As early as 1924, Hull (1975 reprint) had reported that actual smoking, as opposed to sham smoking, slowed the rate of paired-associate learning (see below) of both simple geometric shapes and nonsense syllables. Following Hull's report, a vacuum existed in the literature until the mid-1970s. Table 2.8 provides a summary of learning/LTM studies employing overnight-abstaining smokers. Table 2.9 provides a summary of learning/LTM studies employing smokers abstaining less than overnight. The tables also provide additional technical information regarding the studies.

Smoking/Nicotine prior to/during Learning: Overnight-abstaining Smokers

Generally, a mixed pattern emerges regarding the impact of smoking/nicotine prior to or during learning on learning/long-term memory in overnight abstaining smokers. Across studies, experiments within studies, tasks within experiments, and measures within tasks (omitting Andersson and Post's [1974]

Table 2.8 – Learning/long-term memory studies testing overnight-abstaining smokers

Study	Task(s)	Learning-testing interval	Design: Sex	Sessions/conditions	Effect of smoking/nicotine on learning rate	Effect on subsequent memory
Andersson and Post (1974)	serial lrn. of nonsense syllables	testing was during learning	within ss; male	real/nic.-free smoking	slower/faster learning after 1st/2nd cig.	not tested
Andersson (1975)	serial lrn. of nonsense syllables	during learning and 45 minutes	within ss; male	smoking/no smoking	slower learning	trend toward better recall
Andersson and Hockey (1977) [Task 1]	serial recall (location irrelevant)	immediate	within ss; female	smoking/no smoking	not measured	no effect on serial recall; poorer incidental memory
Andersson and Hockey (1977) [Task 2]	serial recall (location relevant)	immediate; overnight	within ss; female	smoking/no smoking	not measured	no effect on recall; trend toward better spatial recall
Gonzales and Harris (1980)	free recall of words	immediate and 45 minutes	between ss; mixed	smoking/no smoking	not measured	poorer recall of words studied prior to smoking
Peters and McGee (1982)	free recall/recognition	immediate and 1 day	mixed design; mixed	high/low-yield smoking	not measured	day 1: no effect; day 2: state effect for recall
Warburton et al. (1986) [Exp. 1]	non-verbal recognition memory	1 hour	between ss; mixed	state-dependent design	not measured	smoking state-dependency
Warburton et al. (1986) [Exp. 2]	free recall of words	1 and 65 minutes	between ss; mixed	state-dependent design	not measured	better recall at 1 minute; state dependency at 65 minute

Table 2.8 – *Continued*

Study	Task(s)	Learning-testing interval	Design: Sex	Sessions/conditions	Effect of smoking/nicotine on learning rate	Effect on subsequent memory
Parrot and Winder (1989)	free recall of words	immediate and 5 minutes	within ss; male	smoking: 0, 2 and 4 mg gum	not measured	no effect on either immediate or 5 minutes recall
Warburton *et al.* (1992a) [Exp. 1]	free recall of words	immediate	within ss; mixed	high/low-yield smoking	not measured	facilitation of recency but not primacy effect
Warburton *et al.* (1992a) [Exp. 2]	free recall with distraction	10 minutes smoking	within ss; mixed	high/low-yield smoking	not measured	facilitation of primacy but not recency effect
Warburton *et al.* (1992b)	free recall of words	0 minute	within ss; male	0 and 1.5 mg nicotine tablets	not measured	better recall

The term 'x-mg cigarette' designates a cigarette having a nicotine yield of 'x' mg.

Table 2.9 – Learning/long-term memory experiments testing smokers abstaining three hours or less

Study	Task(s)	Learning-testing interval Abstention	Design: Sex	Sessions/conditions	Effect on original learning	Effect on subsequent memory
Carter (1974)	serial lrn. of nonsense syllables	7 days; ???	between ss; mixed	smoking/no smoking	not reported	no effect
Mangan and Golding (1978) [Experiment 1]	paired-associate learning	30 minutes; 1 hour	within ss; male	smoking/no smoking	no effect	superior recall
Mangan and Golding (1978) [Experiment 2]	serial word-list learning	30 minutes; 1 hour	within ss; male	smoking/no smoking	not measured	facilitation of primacy effect
Houston et al. (1978)	free recall of words	immediate and 2 days; 3 hours	within ss; mixed	real/nicotine-free smoking	not measured	poorer recall (both immediate and at 2 days
Mangan (1983) [Experiment 1]	paired-associate lrn.	30 minutes; 1 hour	within ss; male	smoking/no smoking	no effect	superior recall
Mangan (1983) [Experiment 2]	serial word-list lrn.	30 minutes; 1 hour	within ss; male	smoking/no smoking	not measured	enhanced primacy effect
Mangan and Golding (1983)	paired-associate learning	30 minutes to 1 month; 1 hour	mixed design; male	see note	not measured	facilitation for the 0.8 and 1.3 mg cigarettes
Peeke and Peeke (1984) [Experiment 1]	free recall of words	0, 10 and 45 minutes; 2 hours	within ss; mixed	smoking/no smoking	not measured	enhanced recall at 10 and 45 minutes ('prior' smoking only)
Peeke and Peeke (1984) [Exp. 2]	free recall of word	24 hours; 2 hours	between ss; mixed	smoking/no smoking	not measured	enhanced recall

Table 2.9 – *Continued*

Study	Task(s)	Learning-testing interval Abstention	Design: Sex	Sessions/ conditions	Effect on original learning	Effect on subsequent memory
Peeke and Peeke (1984) [Exp. 3]	free recall of word	10 and 45 minutes; 2 hours	within ss; mixed	smoking/ no smoking	not measured	enhanced recall
Peeke and Peeke (1984) [Experiment 4]	'levels of processing' paradigm	immediate; 1 hour	within ss; mixed	smoking/ no smoking	not measured	no effect
Jubis (1986) [Condition 1]	free recall/incidental memory	immediate; 1 hour	within ss; male	smoking/ no smoking	not measured	trend toward enhanced free recall
Jubis (1986) [Condition 2]	free recall/word-colour memory	immediate; 1 hour	within ss; male	smoking/ no smoking	not measured	enhanced free recall; no effect on colour memory
Mangan and Colrain (1991)	recall of prose passages	immediate; 2 hours	within ss; ???	no/sham/varied-yield smoking	not measured	enhanced recall for the 0.62-mg cigarette
Colrain *et al.* (1992)	paired-associate learning	10 minutes; 2 hours	within ss; mixed	no/varied yield smoking	not applicable	faster re-learning and fewer first-trial errors
Rusted and Warburton (1992)	free recall of words	1 week; 1 hour	within ss; mixed	real/sham smoking	not measured	enhanced free recall in no-task cond. only

The term 'x-mg cigarette' designates a cigarette having a nicotine yield of 'x' mg.

second-cigarette data, for which the subjects would no longer be overnight abstaining; omitting state-dependent learning results, which will be discussed separately; and counting statistical trends as 'half'), there are 4.5 cases of facilitation, four cases of impairment, and six cases of no effect. Searching for a general pattern in Table 2.8, serial learning of nonsense syllables appears to be impaired (although later recall may be better; Andersson, 1975; Andersson and Post, 1974 [first-cigarette data]), incidental memory appears to be impaired (Andersson and Hockey, 1977), serial recall appears not to be affected (Andersson and Hockey, 1977), free recall has been reported to be either facilitated (Rusted and Eaton-Williams, 1991 [trend for 10-item list]; Warburton *et al.*, 1986, 1992a [Experiment 2], 1992b), not affected (Parrott and Winder, 1989; Peters and McGee, 1982; Rusted and Eaton-Williams, 1991 [30-item list]; Warburton *et al.*, 1992a [Experiment 1]), or impaired (Gonzales and Harris, 1980), and spatial memory shows a trend toward being facilitated (Andersson and Hockey, 1977).

Smoking/Nicotine prior to/during Learning: Smokers Abstaining ≤ Three hours

In contrast to the mixed results in overnight-abstaining smokers, studies employing smokers abstaining three hours or less generally report facilitation of memory by smoking/nicotine prior to learning. Across studies, experiments within studies, tasks within experiments, and measures within tasks (including Andersson and Post's [1974] second-cigarette data, for which the subjects would no longer be overnight abstaining; and counting statistical trends as 'half'), there are 10.5 cases of facilitation, one case of impairment, and four cases of no effect. The one case of impairment (Houston *et al.*, 1978) was for free-recall, and is offset by four cases of facilitation (Peeke and Peeke, 1984 [Experiments 1-3]; Jubis, 1986 [Condition 2]) and one trend toward facilitation (Jubis, 1986 [Condition 1]). Besides free recall, facilitation was reported for paired-associate learning and serial recall (Mangan and Golding, 1978; Mangan, 1983), serial learning of nonsense syllables (Andersson and Post, 1974 [second-cigarette data]), and recall of prose passages (Mangan and Colrain, 1991). Cases of no effect involved two for incidental memory (Jubis, 1986; Peeke and Peeke, 1984), one for the serial learning of nonsense-syllables (Carter, 1974 [presumably non-abstaining smokers]), and one for memory of word colours (Jubis, 1986).

Smoking/Nicotine prior to/during Learning: Comparison of Overnight Abstaining Smokers and Smokers Abstaining ≤ Three hours

Comparison of the results from overnight-abstaining smokers and smokers abstaining three hours or less is summarized in Table 2.10. Although the volume of data summarized in the table is not overwhelming, a clear pattern seems to emerge: the effect of smoking/nicotine on learning/LTM seems, for several types of memory tasks, to move 'toward' facilitation in going from

Table 2.10 – Learning/long-term memory results as a function of abstention

Task	Effect of smoking/nicotine in overnight-abstaining smokers	Effect of smoking/nicotine in smokers abstaining ≤ 3 hours
serial learning of nonsense syllables	impairment	no effect
incidental memory	impairment	no effect
serial recall	no effect	facilitation
free recall	generally facilitation or no effect	generally facilitation
paired-associate learning	???	facilitation
recall of prose passages	???	facilitation

overnight-abstaining smokers to smokers abstaining three hours or less. It is further interesting to note that the one report of impairment of free recall listed in Table 2.9 was for a group that abstained three hours, versus two hours or less for the other free-recall results. This pattern, if anything, could be interpreted as indicating that the facilitation of memory by smoking/nicotine displays *sensitization*, is an outcome that is 'unlikely' when paired with absolute facilitation and 'impossible' when paired with withdrawal deficit! As discussed below, there is evidence that smoking/nicotine can facilitate memory both by directly acting on mnemonic processes as well as indirectly by facilitating the ability to sustain attention. As reviewed above in the section on cognitive performance, smoking/nicotine clearly facilitates attentional processes in overnight-abstaining smokers. This suggests that it is the 'direct mnemonic' component that is lacking in memory tests of overnight-abstaining smokers. Perhaps the 'first cigarette of the day' engages an increase in arousal that is greater than optimal for the facilitation of direct mnemonic processes, a hypothesis that needs further exploration.

Smoking/Nicotine after Learning

Gonzales and Harris (1980) tested overnight-abstaining smokers, and reported that smoking after learning had no effect on free recall. Facilitation by smoking after learning was reported in two paired-associate studies employing smokers abstaining two hours or less (Mangan and Golding, 1983; Colrain *et al.*, 1992). Rusted and Warburton (1992) tested one-hour abstaining smokers and reported that smoking after learning facilitated free recall in a condition where subjects were idle during the 10 minute learning-testing break (during which they smoked), but not when subjects both smoked and engaged in the RVIP task during the same interval. In contrast, Peeke and Peeke (1984), in two experiments employing two-hour abstaining smokers, reported no effect of post-learning smoking on free recall.

Non-smokers

As outlined above in the section on cognitive performance, Dunne *et al.* (1986) examined the effects of 4 mg nicotine gum on problem solving in non-smokers. Two tasks were employed, one verbal and one numerical. Following task performance (results summarized in Table 2.4), subjects were given memory tests for items employed in the tasks in the following order: [1] immediate recall (test right after finishing the task); [2] 60-sec distractor task; [3] 'delayed' recall immediately following the distractor task ('delayed' in quotes because, relatively speaking, this was an extremely short delay); and [4] recognition. Dunne *et al.* reported that nicotine impaired memory performance for both verbal and numerical problem solving tasks. Again, however, Dunne *et al.* did not address possible toxic effects of administering 4 mg gum to nonsmokers.

Memory and Attention

Rusted, Warburton, and colleagues (Rusted and Eaton-Williams, 1991; Rusted and Warburton, 1992; Warburton *et al.*, 1992a) have proposed that attention may mediate at least some of the effects of smoking/nicotine on memory. For example, in discussing why they found that nicotine administered prior to learning facilitated free recall of a 30-word list better than a 10-item word list, Rusted and Eaton-Williams (1991) stated: "As list length increases, so do the attentional requirements of the task, and the observed improvement may therefore be related to the attention-enhancing qualities of nicotine. However, increasing list length increases overall task difficulty as well, and it is difficult to separate these factors" [p. 365]. Rusted and Eaton-Williams proposed that one way to separate attention from difficulty would be to vary exposure time, since "... increases in list length would require attention over a longer presentation period, while increased exposure duration per item would reduce task difficulty by allowing longer for processing of individual items. [Unpublished] data from our laboratory indicate that ... nicotine improves recall of longer lists ... [equally for two- and six-second exposure times]. This is consistent with the notion of nicotine-induced improvements in attention rather than encoding" [p. 365]. Similarly, Rusted and Warburton (1992) accounted for their findings (that post-learning smoking facilitated free recall ten minutes after learning if subjects had no task during the ten minutes other than smoking, but not when they both smoked and performed a 'distractor' RVIP task) as follows: "The result is consistent with the suggestion that post-trial. That is, post-learning, nicotine effects depend on elaborate processing or rehearsal of some sort... This would have been possible only in the no distractor condition ... the absence of nicotine-induced improvements in the distractor condition suggests that overt or conscious attention to the list items was essential for memory enhancement by nicotine" [pp. 454–455].

Finally, recall the results of Warburton *et al.* (1992a): for immediate free recall, smoking facilitated the recency effect; for delayed free recall (where subjects performed a task during the delay), smoking facilitated the primacy effect. Warburton *et al.* interpreted their results as indicating that nicotine can improve memory both by enhancing attention [facilitation of the recency effect] as well as via some effect on mnemonic processing [facilitation of the primacy effect], perhaps by enhancing the formation of associations.

State Dependency Effects: Results

Four studies have examined the issue of nicotine state-dependent learning. For free recall, Peters and McGee (1982) tested overnight-abstaining smokers using high (*H*; 1.4 mg) and low (*L*; 0.2 mg) nicotine-yield cigarettes in a state-dependency design. They reported state-dependent learning for smoking the higher-yield ciga- rette prior to memorization in that *H-L* free recall was poorer than *H-H*. However, no state-dependent learning for the (relative) absence of nicotine prior to memo- rization was obtained in that the *L-H* group actually did better than the *L-L* group. This 'asymmetrical' state dependency is consistent with the administration of nico- tine at the time of memory testing (as opposed to prior to learning) facilitating recall. Kunzendorf and Wigner (1985) employed smoking (*S*; nicotine yield not specified) versus not smoking (*NS*) in conjunction with a task in which subjects memorized the content of an article on education. Degree of pre-experimental abstention was not specified. Results indicated symmetrical state dependency, with the group that smoked twice (*S-S*) and the group that did not smoke at all (*NS-NS*) having better memory performance than either of the other two groups (*S-NS* and *NS-S*). There was also a trend for the *S-S* group to have better performance than the *NS-NS* group. Warburton *et al.* (1986) tested overnight-abstaining smokers in two state-dependency studies.

The first experiment employed smoking (*S*; 1.3 mg) versus not smoking (*NS*) in conjunction with a test of memory for abstract shapes (Chinese ideograms). They reported basically symmetrical state-dependent learning in that *S-NS* free recall was poorer than *S-S* and *NS-S* free recall was poorer than *NS-NS*. Similar to Kunzendorf and Wigner (1985), *S-S* had better recall than *NS-NS*.

The second study employed nicotine (*N*) versus placebo (*P*) tablets and verbal free recall. They reported basically symmetrical state-dependent learning in that *N-P* free recall was poorer than *N-N* and *P-N* free recall was poorer than *P-P*. Similar to Kunzendorf and Wigner (1985) as well as the results of their first study, *N-N* had better recall than *P-P*.

State Dependency Effects: Discussion

The clearest trend to emerge from the four studies seems to be that smoking/ nicotine prior to learning and prior to later recall results in memory performance that is superior to the three other possible permutations.

SMOKING AND ACADEMIC PERFORMANCE

The academic performance of high school students who smoke has been reported to be lower than that of students who do not smoke (Borland and Rudolph, 1975). However, as pointed out by Clarke (1987), this difference is probably a function of lower socioeconomic status and IQ levels of smokers as a group (Hill and Gray, 1982; Ray, 1985). A comparison less biased by these pre-existing demographic differences would be that of smoking and non-smoking groups of successful college students. Warburton *et al.* (1984) examined a sample of 467 students from Reading University (UK). They reported that smokers had better examination and essay marks than did non-smokers, with the percentage of examination failures (unsuccessful students?) being 15 in both groups. In contrast, Clarke (1987), based on a sample of 326 students from Deakin University (Australia), found no difference in introductory-psychology and social-psychology test scores between smokers and non-smokers. A similar negative result was reported by Radovanovic *et al.* (1983) for first-year medical students in what was formerly Yugoslavia. One possible explanation for these different findings is that students in the Warburton *et al.* (1984) study were tested during the traditional British examination week, when the students reported smoking more as well as inhaling deeper compared to non-examination periods (Warburton and Walters, 1989). Although Warburton and Walters interpreted this as "an attempt to maximize nicotine absorption for attentional processing" [p. 229], an alternative hypothesis would be that students may have been smoking more to cope with the stress of examination week and that the better examination and essay marks reflect this affective coping response as much as an improvement in attentional processing. However, West and Lennox (1992) recently compared British college students 1 month before exams with a separate group of students 1 day before exams. They found that, relative to the 1 month group, in the 1 day group, 'stimulant' smoking motivation increased, while 'sedative' motivation did not. This finding would seem to lend support to the 'attentional' hypothesis of Warburton and Walters.

SUMMARY AND CONCLUSION

Smoking/nicotine has been demonstrated to have a variety of beneficial effects on performance in regular smokers tested following overnight smoking abstention. The literature is examined regarding three hypotheses accounting for these results:

- absolute facilitation (smoking/nicotine produces 'better than normal' performance),
- withdrawal deficit (smoking/nicotine abstinence produces performance decrements due to withdrawal that are relieved by subsequent smoking or nicotine administration), and
- the combination hypothesis (effects in regular smokers are a combination of absolute facilitation and the relief of withdrawal deficits).

There is strong evidence from studies in which nicotine is administered to non-abstaining smokers and/or non-smokers that the beneficial effects of smoking/nicotine on performance are not due solely to the relief of withdrawal deficits. Based on the evidence as it currently exists, the answer to the question of whether smoking/nicotine's effects in regular smokers represents absolute facilitation or a combination of absolute facilitation and the relief of withdrawal deficits tentatively appears to be largely absolute facilitation. However, further studies using a wider variety of tasks are needed that either

- administer nicotine to non-/occasional smokers, abstaining smokers, and non-abstaining smokers in the same experiments, or
- track the time course of deficits in performance seen during cessation using a no-cessation control group (for useful methodological recommendations, see Ney *et al.*, 1989).
- The use of non-smokers who do not smoke as 'controls' is not recommended[6].

ACKNOWLEDGEMENTS

The authors thank Drs Michael Houlihan and Carr Smith for comments on an earlier draft.

---

[6] Some other factors that should be attended to in future studies include:

–matching subjects for consumption of caffeine and alcohol, and screening for drug use;

–age matching subjects to better control for age-related declines in attention and memory;

–use of denicotinized cigarettes as a control (e.g., Robinson *et al.*, 1992) rather than sham or no smoking, as the former permits a double-blind design;

–more use of either direct (e.g., blood nicotine levels) or indirect (tidal- breath CO) measure of actual nicotine intake, and

–screening subjects for psychiatric disorders that are more prevalent in smokers and are linked with cognitive performance decrements (e.g., depression and schizophrenia).

CHAPTER THREE

Nicotinic and Non-Nicotinic Aspects of Smoking: Motivation and Behavioural Effects

†Karl BÄTTIG

INTRODUCTION

The 1988 report of the U.S. Surgeon General has determined that nicotine is the substance which is responsible for the continued or even compulsive use of cigarettes. However, the scientific world did not accept this statement as the final word on this issue. On the contrary, scientific research on the putative role of nicotine as the main reinforcer of smoking has greatly expanded since 1988.

Ideally, according to the nicotine concept, non-smoke nicotine should abolish the desire to smoke as is the case when inhaled heroin is replaced by intravenous heroin. Research on the smoking need as well as on smoking cessation in subjects treated with non-smoke nicotine reveals, however, a different picture. West *et al.* (1984a) observed already early that nicotine chewing gums failed to clearly reduce the smoking need of heavy smokers and suggested that craving to smoke may have its roots to a considerable degree in a non-pharmacological dimension. This is consistent with the relatively modest success rates reported from many nicotine replacement trials. Silagy *et al.* (1994) reviewed in a meta-analysis 42 nicotine gum, 9 transdermal nicotine patch, one nasal spray and one inhalation study including a total of about 17,000 persons. The conclusion was that both nicotine gum and transdermal patch treatment are about twice as effective as placebos, whereas nasal spray according to Sutherland *et al.* (1992b) might be somewhat more effective. Quantitatively the success rates after 6 months of treatment have been estimated in another meta-analysis by Fiore *et al.* (1994) for transdermal nicotine. Out of more than 5,000 subjects, 22% treated with nicotine patches as opposed to 9% treated with placebos remained abstinent after 6 months. Therefore, less than about one out of four treated smokers has a chance of quitting. These rather modest results justify a critical assessment of the putative nicotinic and non-nicotinic dimensions of the motivations for and the benefits from smoking.

CIGARETTE PUFFING: NICOTINE OR SMOKE TITRATION?

Modern light cigarettes provide the smoker with air diluted smoke obtained through ventilation holes situated either in the filter itself or in front of it. Filtration is therefore not specific and reductions in smoking tar yields are accompanied by proportional reductions in the nicotine and CO yields. It is commonly accepted today that machine smoking simulates human smoking, as assessed by measuring respiratory CO, plasma nicotine and cotinine, to a limited degree only, because the yield reductions can be compensated in part by intensified puffing. Which of the constituents of smoke are responsible for such compensatory smoking has been the subject of a great number of studies in the past.

Low yield versus high yield smokers

For a cross-sectional study carried out at our laboratory, Höfer *et al.* (1991) recruited 72 male and 72 female smokers equally distributed across consumers of the four nicotine yield classes of 'regular' (1.0–1.2 mg), 'medium' (0.7–0.9 mg), 'light' (0.4–0.6 mg) and 'ultra light' (0.1–0.3 mg) cigarettes. Smoking abstinence was not required and the testing program involved smoking the habituated cigarettes both through a puffing flow meter or by direct lip contact and both when smoking the habituated way or by taking 30 puffs from successive cigarettes. The dependent variables included the different puffing parameters, the number of cigarettes smoked per day, respiratory CO, plasma nicotine and cotinine and in addition a number of additional physiological parameters and subjective assessments. A nearly linear reduction of baseline plasma nicotine before test smoking was seen across the descending yield classes. However, reduction of the machine smoking nicotine yields by 80% resulted in significant reductions of plasma nicotine by only about 50%. Plasma cotinine declined significantly by about 30%, whereas the reduction of respiratory CO by about 25% failed to reach significance. A highly similar picture was obtained for the pre/post test-smoking boosts of these parameters and no differences were seen between lip and flow meter smoking. The partial upregulation for lowered yields appeared rather clearly to be a consequence of increasing the number and volumes of the puffs, whereas the number of the cigarettes smoked daily did not differ across the yield classes in any way.

   In their report, Höfer *et al.* (1991) also included a review of 23 earlier cross-sectional studies. Although these studies differed in manifold methodological aspects, they nevertheless revealed important common findings. No or at best minimal compensation for lowered yields was seen for the number of cigarettes smoked daily across all studies. Plasma nicotine and cotinine indicated partial upregulation, and the few studies which used the puffing flow meter method also observed greater total puff volumes for the lighter cigarettes.

   A subsequent multiple regression analysis of these data by Höfer *et al.* (1992) revealed that nicotine yield predicted plasma nicotine concentration to different degrees. The highest correlations were obtained for measurements obtained immediately after taking 30 puffs, followed by measurements taken immediately after habitual smoking, whereas those taken at baseline correlated more strongly with the number of the cigarettes smoked on the testing day.

In contrast to the upregulation with low yields, downregulation appeared when the smokers required to take 30 puffs had to increase the number of puffs by nearly 150%. The resulting increase in the pre/post-smoking boosts of plasma nicotine then amounted to only 40%. This suggests that above a certain level of machine yields the absorption of nicotine does not further increase, as was shown already in 1980 by Russell *et al.* who failed to detect differences in plasma nicotine with cigarettes of different yields above 1.0 mg nicotine. This may explain why Benowitz *et al.* (1983) failed to detect a reduction of nicotine absorption with light cigarettes. In that study smokers of cigarettes with nicotine yields between 0.8 mg and 2.0 mg represented about half of the 272 subjects and smokers of ultra-lights were underrepresented.

Yield switching

It is of great interest to compare nicotine absorption not only between smokers habituated to different yield classes, but also in the same smokers when they are required to switch to low yield classes. Quantitatively such studies revealed a partial upregulation similar to that seen in cross-sectional comparisons. The corresponding decrease of plasma nicotine and cotinine appears acutely and persists for long periods if not permanently. Guyatt *et al.* (1989) tested their subjects weekly. After switching to moderately lighter cigarettes the initial decline of plasma cotinine amounted to about 40% of the expected decrease and persisted for 36 weeks. Similar reductions with ultra-light cigarettes were obtained by Zacny and Stitzer (1988) 5 days after switching, by West *et al.* (1984b) on the days 1, 3 and 10 after switching and by Jacober *et al.* (1994a) on the days 1 and 2 after switching.

The rather consistent overall picture obtained from such studies has led to the widely popular slogan of 'nicotine titration'. This concept postulates that smokers compensate in part for the dilution of smoke by intensified puffing and that they do this in order to correct for the lowered nicotine yields. However, this concept can be questioned for several reasons. The most pertinent concern can be seen with the fact that the CO, nicotine and tar yields of modern cigarettes are highly correlated across all yield classes. This would call for experimentation with cigarettes by manipulating these different yields independently. Nil and Bättig (1989b) reviewed 12 earlier laboratory studies using this approach. The resulting picture was highly inconsistent, reaching from no compensation to partial nicotine or even 'tar titration'. The same review also considered 10 studies which manipulated nicotine pharmacologically by giving intravenous injections, nicotine chewing gums, transdermal nicotine or the nicotine antagonist mecamylamine.

Generally, no or only modest compensation phenomena were seen and these concerned different parameters (cigarettes/day, puffing, CO-absorption and heart rate (HR)), but plasma nicotine was not measured in any of these earlier studies. However, Conze *et al.* (1994), demonstrated that oral nicotine, although it raised plasma nicotine considerably, failed to affect in continuing smokers any parameter of smoking significantly in comparison with placebo-preloads. The smoking induced increases of plasma nicotine were simply put on top of the increases due to the nicotine preloads.

Nearly nicotine-free cigarettes

A new cigarette made of denicotinized tobacco appeared was presented by Philip Morris for a short period on test markets in the US under the brand name 'Next'. The tar yield amounted to 9.3 mg, a value relatively common for regular cigarettes. The nicotine yield amounted to 0.08 mg, resulting in a nicotine/tar ratio of 1:116 instead of a ratio of between 1:8 and 1:15, characteristic for commercial cigarettes regardless of the actual total smoke yields. Robinson *et al.* (1992) compared this cigarette with a cigarette of identical tar yield but a nicotine yield of 0.6 mg. They found no differences in any puffing parameter between the regular cigarette and 'Next'. In contrast to the control cigarette, 'Next' failed to induce any significant EEG changes and increased heart rate only minimally. The pre/post smoking increases of plasma nicotine amounted to less than 4 ng/ml with 'Next' as opposed to more than 25 ng/ml with the control cigarette. Thus, the ratio of nicotine absorption for the two cigarettes approached the ratio of the machine smoking yields, indicating a total absence of any compensation for the lowered nicotine yield of the 'Next' cigarettes.

Several experiments then were done in our laboratory in order to possibly extend this interesting and remarkable result. Hasenfratz *et al.* (1993a) included in a comparison between these two types of cigarettes also typical ultra light cigarettes with an average tar yield of 1.8 mg and nicotine yield of 0.22 mg in a crossover with 12 female habitual smokers of regular cigarettes. The average pre/post smoking boosts of plasma nicotine reached 13.3 ng/ml with the control cigarette, 7.8 ng/ml with the ultra-lights but only 1.7 ng/ml with 'Next'. Thus, a roughly fourfold decrease of the machine smoking nicotine yield from the control to the ultra light cigarettes lowered the plasma nicotine boosts by only 40%, confirming the compensation phenomenon commonly seen with such cigarettes. On the other hand, the tenfold decrease in machine smoking nicotine yields from the control cigarettes to 'Next' resulted in a nearly tenfold decrease of the nicotine plasma boost. As in the Robinson *et al.* (1992) study, there were no changes in the EEG and only a minimal increase of HR with 'Next'.

A great surprise was obtained, however, with the subjective ratings. Strength, taste and enjoyment were rated lower with the ultra light than control cigarettes and lowest with 'Next'. The post smoking reduction of the desire to smoke, however, was nearly identical for all three types of cigarettes and the same was also the case for the intervals until the subjects wanted to light the second cigarette of the experiment.

In a following experiment, Baldinger *et al.* (1995a) used the same experimental design but included the puffing flow meter method while omitting the plasma nicotine measurements. All results of the previous studies (HR, EEG, cravings) were confirmed with this experiment. In addition, it was shown that in comparison with the control cigarettes only the ultra light cigarettes but not the 'Next' cigarettes were oversmoked with more and greater puffs.

In a further experiment, Baldinger *et al.* (1995b) compared in a field study smoking regular control cigarettes with smoking 'Next' and with complete smoking abstinence. Each condition was required for two complete days separated by a week. The subjects were equipped with HR/activity recorders, with event recorders to mark the time points of lighting the cigarettes and with saliva tubes for collecting saliva every evening. Electronic diaries were used for rating several aspects of the subjective state throughout the day. Saliva cotinine decreased similarly by about 50% in comparison with the habituated cigarettes on the days when smoking 'Next' and on the days when remaining completely abstinent. The daily number of cigarettes and the inter-cigarette intervals remained nearly identical for the control cigarettes and 'Next', although the latter were liked considerably less. Finally, the subjective ratings of the need to smoke increased considerably on the abstinence days but not when smoking 'Next'.

These experiments with the conventional tar but ultra-low nicotine cigarette 'Next' seriously question the nicotine titration hypothesis of smoking behaviour. Compensatory puffing with light and ultra light cigarettes appears therefore rather as a response to the smoke dilution than as a response to the concomitant reduction of the nicotine yields. The possibility that differences in draw resistance might explain the more intense puffing of light cigarettes also appears unlikely in the light of extended experimentation by Woodson and Griffiths (1992). Finally, and perhaps most importantly, compensation with the ultra light cigarettes remains consistently incomplete and this is not corrected by smoking additional cigarettes. Therefore it appears that smoke *per se*, rather than its nicotine content, might govern smoking behaviour to a much greater extent than hitherto believed.

THE TASTE OF SMOKE OR THE TASTE OF NICOTINE?

Tobacco smoke provides the smoker with stinging, burning or scratching sensations in the nose, the oropharynx and the throat. The chemosensory quality of smoke depends not only on the type of tobacco and the process of curing and blending with different ingredients, but most importantly also on the nicotine content of smoke. Not very many studies have tried in the past to disentangle these multiple factors. Some studies attempted to compare different blends of tobacco or to find chemosensory substitutes for reducing the desire to smoke, while other studies attempted to evaluate directly the sensory effects of nicotine.

Sensory substitutes

The importance of the sensory stimulation for the smoking motivation was demonstrated already early by Rose *et al.* (1984). They required the smokers to use lidocaïne solutions for rinsing the mouth or for gargling or to inhale a lidocaïne aerosol, in order to anesthetize the mouth, the oropharynx or the trachea. The subsequently assessed desire to smoke was smaller, the greater the anesthetized area. Later, Rose and Hickman (1988) demonstrated that the inhalation of citric acid aerosols simulates in part the tracheo-bronchial sensations associated with cigarette smoke. Levin *et al.* (1990) conducted a laboratory study with a small handheld citric acid inhaler and observed a short-term reduction of the desire for cigarettes. The same instrument was used by Behm *et al.* (1993) in a clinical smoking cessation trial lasting 3 weeks. The citric acid inhaler was given to the subjects during the first two weeks and the third week served as a follow-up period. The citric acid aerosols reduced smoking in smokers who according to their respiratory CO-levels were deep inhalers of cigarette smoke, but less so in the moderate inhalers. This suggests that deep inhalers are more habituated to tracheal stimulation than light inhalers and may therefore have benefited more from the inhalation of citric acid. In a new study, Rose and Behm (1994) required one group of subjects to puff on a device delivering a vapour from essential oil of black pepper. A second group had to puff on the device with a mint/menthol cartridge and a third group was given the same device with an empty cartridge. After three hours of *ad libitum* puffing on these devices, craving was reduced only with the pepper condition, which quite remarkably also alleviated negative affect and anxiety ratings.

Nicotine versus taste

An experiment by Nil and Bättig (1989a) compared the separate effects of changing the type of tobacco and the machine smoking yields in a crossover design. The experimental design included mentholated, dark and blond cigarettes. Each type of tobacco was presented in two versions, one with a smoke yield corresponding to the habitually smoked cigarettes and the other one with a reduction of the smoke yield and thereby also of the nicotine yield by 50%. In contrast to the habituated cigarettes, satisfaction and taste ratings decreased considerably and similarly for all experimental cigarettes, regardless of the smoke yields. On the other hand, lowering the smoke yields induced compensatory intensification of puffing regardless of the type of tobacco.

With a sophisticated experimental device, Rose *et al.* (1993) compared more directly the impacts of nicotine and other sensory constituents of smoke. Their device allowed them to trap with a Cambridge filter either one eighth of the particulate matter (high nicotine/high sensory condition) or seven eighths (low nicotine/low sensory condition). A third low nicotine/high sensory condition was achieved by trapping all particulate matter and adding with a nebulizer a nearly nicotine free vapour from a regenerated smoke aerosol, which previously by Behm *et al.* (1990) was shown to produce sensations similar to those of smoke from a high nicotine cigarette. Under these conditions the low nicotine/low sensory smoke induced intensified puffing, whereas the subjects puffed and inhaled the low nicotine/high sensory aerosol and the high nicotine/high sensory smoke in a similar manner. Further, the decrease of craving with both of these high sensory conditions was similar and greater than with the low nicotine/low sensory smoke. These results underscore therefore the notion that smokers may regulate smoke intake rather in response to peripheral sensory cues than in response to the bio-availability of nicotine.

Nicotine alone

Using the technique of averaging sensory evoked EEG potentials, Hummel *et al.* (1992a) delivered distinct nicotine boli into the nasal cavity. The sensation of the odour, strongest at medium concentrations, and the sensations of stinging and burning, strongest at the highest concentrations, followed a different time course and produced different evoked potentials in the EEG. This is in accordance with the demonstration by Hummel and Kobal (1992) that from the nose odourous signals are transmitted by olfactory receptors and stinging or burning signals by trigeminal sensory receptors. Comparing smokers and non-smokers with the same method, Hummel *et al.* (1992b) demonstrated further that smokers as well as non-smokers were able to discriminate between the R(+) and the S(−) enantiomer of nicotine. However, whereas both groups disliked the R(+) enantiomer, the S(−) enantiomer, which is the main isomere in cigarette smoke, was liked by the smokers as a sensory reward, but disliked by the non-smokers. In this light, the recent popularity of using nasal nicotine sprays or inhalers as smoking substitutes in laboratory studies, as reported by Perkins *et al.* (1986a, 1992b, 1994a), Sutherland *et al.* (1992a), by Pomerleau *et al.* (1992), and by Hansson *et al.* (1994) is understandable.

These studies show that nasal spray nicotine more closely mimics nicotine pharmacokinetics than nicotine chewing gums or transdermal nicotine patches. The physiological effects are comparable to those of smoking, and finally, the reduction of the need to smoke may also be superior to that achieved with other forms of nicotine substitution.

A placebo cigarette?

The combined evidence from these studies demonstrates that for the smoker the chemosensory stimulation by tobacco smoke might be an important constituent of the smoking reward. However, the complex of substances responsible for the characteristic sensations induced by tobacco smoke includes not only the multitude of non-nicotinic substances but most importantly nicotine as well. In order to disentangle the relative importance of the chemosensory reward obtained with nicotine and the reward value of the systemic effects of nicotine one would need a real placebo cigarette. This cigarette would have to be nicotine-free and indistinguishable for the smoker from a regular cigarette. How difficult this might be follows from neurophysiological studies. Sekizawa and Tsubone (1994) recorded in guinea pigs from the anterior ethmoidal nerve single fiber firing in response to nasal instillations of nicotine, capsaïcine and ammonia. Most fibers fired specifically to one of these substances but some fibers also responded to different combinations of two of the three substances. Furthermore, systematic differences were obtained between these stimuli for the duration and refractory periods of the discharges. Hansson *et al.* (1994) analyzed the cough response of human non-smokers to the inhalation of nicotine and capsaïcine. Although both substances elicited coughing and increases of respiratory resistance, there were differences in the time courses and some subjects coughed in response to capsaïcine only but not in response to nicotine.

Such findings suggest that the chemosensory sensations elicited by tobacco smoke are mediated by a complex network of interacting neuronal systems.

Taken together, all these results underline the importance of smell and taste in the reward value of tobacco smoke. However, it should not be overlooked that none of the different substitutes, neither citric acid nor nicotine-free tobacco smoke regenerates, nor nicotine alone was shown to reduce the smoking need as much as regular cigarettes. Surprisingly, this ability has been seen so far only with the nearly nicotine-free cigarette 'Next', as mentioned above. This cigarette delivers to the smoker the entire tar fraction of smoke but nearly no nicotine and as a consequence its taste was rated poorly by the study subjects. Nevertheless the fact that in comparison with regular cigarettes it was puffed similarly, reduced similarly the smoking need and was smoked at similar intervals across the day, suggests that smoke taste, although important, can hardly be the main decisive factor in the maintenance of the smoking habit.

SEPARATE EFFECTS OF ANTICIPATION AND LIGHTING AND OF NICOTINE ON HEART RATE AND ACTIVITY

Most of the studies on smoking and nicotine carried out so far have been done in the laboratory under highly standardized conditions. Field observations have been restricted mostly to having the subjects use written diaries, to measure cardiovascular state intermittently, etc. In order to extend this approach, we developed a multi-parametric method for the assessing of the behaviour of smokers under field conditions.

The methodological framework

The subjects were equipped with different devices. A combined event recorder and recorder of physical activity and HR with a storage capacity of up to three days was used in the first studies. It was later replaced by separate recorders for activity and HR with storage capacities of up to ten days. Both systems detect heart beats with electrocardiogram electrodes and store the number of beats for every 30 second interval. Activity was measured in the first system by the electric current produced in a coil by a magnetic ball freely moving in a hollow ball and in the second system by using piezo crystals. With both systems, activity counts were also summed up and stored for each 30 second interval. Cross-comparison for reliability did not show any advantage of one system over the other. The event marker was activated in both cases with simple button pressings. The second device the subjects were given, was a preprogrammed pocket computer (Psion type II organizers). These were to be used at fixed occasions such as upon getting up, after breakfast, etc., until before going to bed. At each of these occasions the device asked the subjects to answer particular questions, to perform analog ratings of different aspects of the subjective state and to perform mental tasks, etc. Furthermore, at these occasions the subjects were also reminded to follow particular instructions such as performing self-measurements of blood pressure (BP) with an Omron BP monitor, with which the subjects were also supplied. Finally, the subjects received saliva tubes for collecting saliva with cotton rolls for later cotinine analysis. The main advantages of these systems consisted in the easiness of the data analysis with adequately programmed PCs and in the close control of compliance, as with each entry into the pocket computers running time was also stored.

The averaging procedure

It was expected that the all-day plots of activity and HR might reveal post-smoking changes attributable to the absorption of nicotine and/or changes predicting the lighting of a cigarette. Lowered pre-lighting levels of activity and HR would be expected, if smoking were to serve the smoker to reestablish his arousal level after a smoking free interval, and post-smoking cardio-acceleration was to be expected as a characteristic effect of nicotine. However, the all day plots showed only gradual increases of HR across the first morning cigarettes and thereafter a near stabilization at a level elevated by about 10 beats per minute (bpm). Moreover there was no apparent and consistent relation to the time points of smoking the cigarettes.

As a next approach, the procedure commonly used in the EEG analysis for calculating averaged evoked potentials to a series of identical stimuli was used. The collected data for the 10-minute intervals before and after lighting a ciga- rette were averaged across all cigarettes smoked over the day. The emerging picture was rather surprising. Both activity and HR started to increase gradually and simultaneously from between 3 to 4 minutes before lighting a cigarette. This fluctuation reached its peak at the moment of lighting and then decreased there- after rather abruptly to or even slightly below the pre-lighting level. This shift was therefore termed the 'lighting-anticipatory response' of activity and HR. After lighting, activity remained unchanged for the 10-minute post lighting interval, whereas HR increased gradually and stabilized after about 5 minutes or even tended to decline again toward the end of the 10 minutes. A previous study by Hasenfratz *et al.* (1992) has shown that all-day activity and HR recorded with these instruments failed to be highly correlated in only one subject out of 10 smokers and 10 non-smokers. This allowed the calculation of a pulse/activity index or activity adjusted HR as a third parameter to be averaged across the 10 minutes before and the 10 minutes after lighting. This parameter remained unchanged for the pre-lighting period, it was not affected by lighting but increased in parallel to the post lighting increase of the non-adjusted HR. This indicated an excess of heart beats above the all-day activity/HR ratio and was therefore considered as the 'nicotinic response' of HR. Figure 3.1 shows an example of this picture averaged for 48 subjects and 2 smoking days.

The averaged 'nicotinic responses'

This response averaged over the whole day amounted to about 5 bpm, but its magnitude was clearly affected by several manipulations. Bättig *et al.* (1993) observed that it was identical for workdays and days off. Jacober *et al.* (1993) demonstrated that it was rather pronounced for the first five cigarettes of the day but completely absent for the last five cigarettes of the day, indicating therefore a pronounced intraday development of acute tolerance to the cardio- accelerative action of nicotine. Jacober *et al.* (1994a) further revealed that this nicotinic response was qualitatively and quantitatively almost identical when smoking regular or ultra light cigarettes, although the ultra light cigarettes sig- nificantly decreased all-day HR, respiratory CO and saliva cotinine. Finally, Baldinger *et al.* (1995b) demonstrated that the response was completely absent when smoking the nearly nicotine-free, regular-tar cigarette 'Next', or with imaginary smoking on abstinence days, when the subjects had to press the light- ing event recorder whenever they felt that they would like to light a cigarette right at this moment.

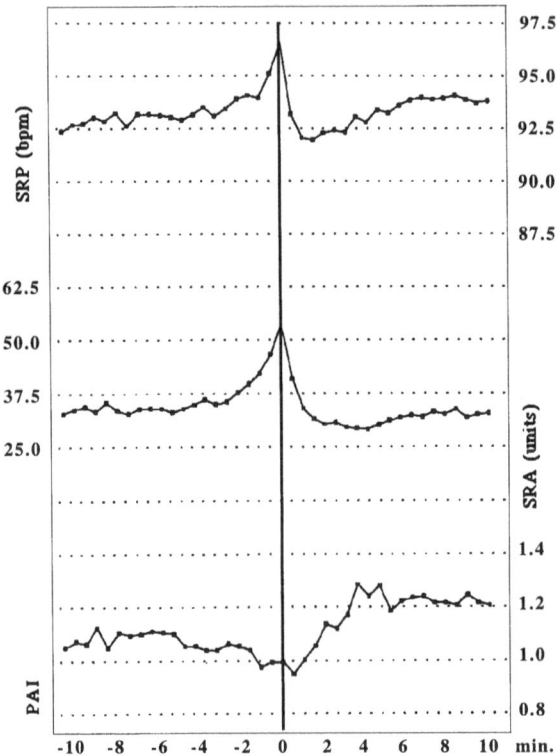

Figure 3.1 – Heart rate and actometer counts in relation to cigarette lighting. Zero denotes the time point of lighting. Top: Heart rate. Middle: Actometer counts. Bottom: Heart rate/ activity index. The index of 1.0 represents the all-day ratio of heart rate: activity. Data averaged across 48 smokers and two smoking days.

The averaged 'lighting-anticipatory response'

When this response was first seen it raised the suspicion that it might represent an artifact due to the motor activity involved in the search for the packet and the lighter, or the search for a place were smoking is allowed. However, in the comparison by Bättig et al. (1993) between workdays and days off which was carried out with hospital employees, it remained nearly identical for the two conditions. More importantly, however, this response also appeared with imaginary smoking on abstinence days in the study by Baldinger et al. (1995b). Furthermore, it remained unchanged under all circumstances mentioned above which showed to affect or suppress the 'nicotinic response', for the first as well as for the last cigarettes of the day, for regular as well as for ultra light cigarettes, and finally also for the nearly nicotine-free cigarette 'Next'.

Rhythmicity and specificity

In the study by Baldinger et al. (1995b), the intraday distribution of imaginary smoking followed a surprisingly similar time course to that of smoking regular and the nearly nicotine-free cigarettes 'Next'. Some individual cumulative day plots of pressing the event recorders for smoking regular cigarettes, 'Next' cigarettes or imaginary smoking are shown in Figure 3.2. The plots were selected so as to represent the smokers from the study with the highest and the lowest frequencies of smoking. The striking similarity for the intraday development of pressing the event recorders across the three conditions was apparent in most subjects. The exceptions concerned further either only the abstinence condition or only the 'Next' condition, but in no case both of these no nicotine conditions simultaneously. The plots seem to indicate therefore that smoking is initiated not only by external challenges or a need for nicotine, but rather by an 'internal clock' representing the individually acquired habit of smoking.

These observations were taken by Jacober *et al.* (1994c) as the rationale for performing a circadian analysis of the activity and HR data of 12 subjects for

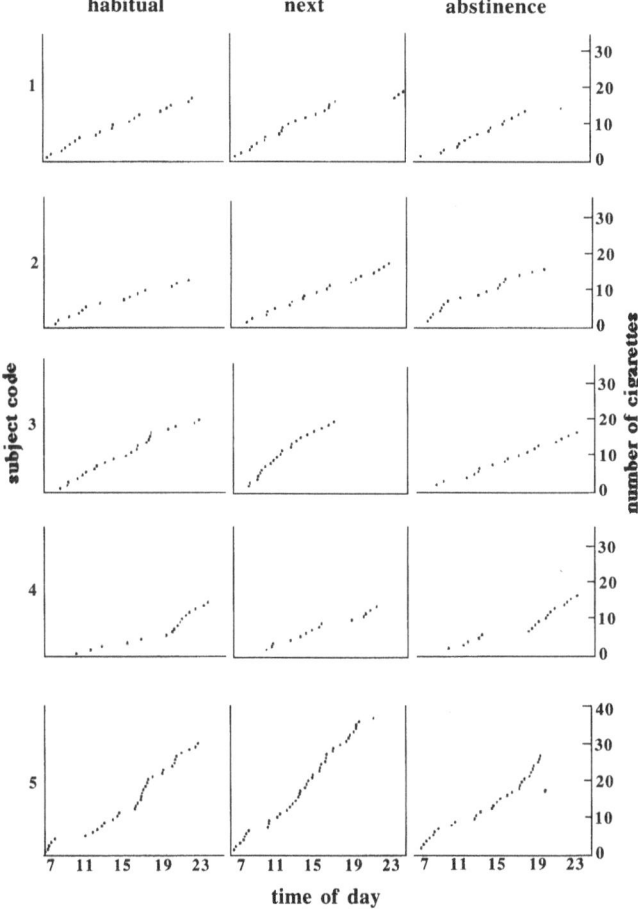

Figure 3.2 – Cumulative records of event marking by 5 subjects for lighting the habituated cigarette (left column), lighting the nearly nicotine free cigarette 'Next' (middle column) and imaginary lighting on an abstinence day (right column).

two consecutive days and nights. In more than half of these individual analyses, using the individual averaged inter-cigarette intervals as the time base revealed a significant periodicity for simultaneous increases of HR and activity, representing thus the lighting-anticipatory activations. On the other hand, for HR alone such periodicities which might then represent 'nicotinic responses' were obtained only exceptionally.

As a next question one might ask as to whether the lighting anticipatory shifts of activity and HR are specific to smoking or appear with other habits as well. The well-known fact that anticipatory arousal shows up shortly before feeding in laboratory animals which are fed at fixed times of day favors the idea that anticipatory arousal may not be specific. In order to test this hypothesis, Jacober *et al.* (1993) applied the same technique in non-smokers who were instructed to use the event recorder of the system whenever they took the first sip of a cup of coffee. A pronounced first sip-anticipatory activation emerged from this analysis which closely resembled the lighting-anticipatory shifts seen with cigarette smoking. However, and quite unexpectedly, after the first sip a gradual increase of the pulse-activity index was also seen, similar to that for smoking. Since caffeine is known not to raise HR, the reason for this shift perhaps the intake of a hot beverage or social communication remains to be explored.

Smoking, an instrumental reward?

Increasing activation towards the moment of lighting up a cigarette or taking a first sip of coffee followed by immediate relief from overactivity could be seen as a gradually increasing need followed by an immediate need reduction upon initiation of a consummatory act. This raises the question as to what extent the consummatory act of lighting up might act as a reinforcement *per se* or represents a phenomenon conditioned to the effects of the psychoactive substance nicotine.

In order to see whether puffing might be rewarding *per se*, Hasenfratz *et al.* (1993b) analyzed the psychophysiological changes related to single puffs for two 5 second intervals before lip contact and for two such intervals after lip contact. Several EEG parameters suggested that lip contact is followed by tranquilization, which, however, already develops in anticipation from the first to the second 5 second block before lip contact. For control, the experiment also included coffee drinking, and the same procedure was applied to the single sips as to the single puffs. The sip related changes were almost identical to the puff related changes, suggesting a non specific attribute of a rewarding act. This is all the more likely, as the physiological changes before and after an entire cigarette and an entire cup of coffee were clearly differentiated in the expected direction.

The theoretically unconventional view of considering an instrumental response as a reinforcement *per se* regardless of the nature of its unconditioned consequences has a parallel in experiments done by Ettenberg *et al.* (1981) in the rat. The animals were allowed to self-stimulate hypothalamic structures for 90 seconds after which they received stimulations passively applied by the experimenter. By activating a second response lever the rats regained the possibility of administering the stimulations by themselves. This was learned very quickly and efficiently by the animals. In a second experiment, the experimenters made use of the well-known fact that tastes associated with rewarding brain stimulation will also become rewarding. Drinking coffee was thus associated with the stimulations. In a subsequent water/coffee free choice trial the rats which associated coffee with active self-stimulation consumed nearly twice as much coffee as the rats which never experienced pairings of coffee and brain stimulation or the rats which associated the taste of coffee with passively administered brain stimulations.

A further support for this view can also be seen in the light of a theoretical model proposed by Tiffany (1990) for drug use behaviour. According to this concept, smoking is initiated by automated action schemata rather than by substance withdrawal. Automatic action schematas organize behaviours which are highly autonomous and may occur without intention, may be difficult to suppress, are carried out without effort and do not need to be controlled by conscious awareness. Several automatic schemata can be executed at the same time, such as coffee drinking and smoking, whereas non-automatic voluntary actions can be carried only one at a time, are slow, depend on intention and attention and require a conscious effort. For the case of smoking it can be assumed that smoking the first cigarette in life represents a non-automatic conscious act. Through repetition not only the act becomes automated but an automated intraday program also develops gradually for smoking at particular occasions or at particular intervals. This would explain why such programs are maintained even when a smoker has to switch to nicotine-free cigarettes or to imaginary smoking.

NICOTINE AS A COGNITIVE ENHANCER

According to extensive research in animals and humans, cognitive enhancement is the most widely documented effect among the central nervous actions of nicotine. In his review, Warburton (1992) concluded that nicotine induced improvements include attention, mental processing, short- and long-term memory across a wide variety of tasks and independent of whether the substance was obtained through smoking or through other routes of administration. In parallel, nicotine also increases electrocortical arousal, as reviewed by Knott (1991), and Roth and Bättig (1991) showed that EEG arousal and tapping speed, a behavioural function consistently improved after smoking, were correlated quantitatively. Furthermore, recent studies have demonstrated that the facilitation of the formation of memories is not merely a consequence of increased attention but also of an improvement of post-learning consolidation. This was seen by Warburton *et al.* (1992a), who observed that nicotine input through smoking after rather than before the input of the task items facilitated memory recall, and the same result was obtained in another study by Warburton *et al.* (1992b) with oral nicotine. Similar evidence for the consolidation hypothesis was further presented by Rusted and Warburton (1992) and by Colrain *et al.* (1992). However, the question has also been raised in a review of the relevant literature by Hughes *et al.* (1990) as to whether such effects in the smoker represent an objective improvement or merely a recovery from withdrawal. A more recent review by Sherwood (1993) remarked that most earlier studies failed to test critically for this possible bias but, based on newer studies, he concluded that nicotine may well have positive effects beyond the mere alleviation of withdrawal. There are different ways to test for this ambiguity.

Nicotine effects in smokers and non-smokers

Non-smokers do not suffer from withdrawal and should therefore not be improved by nicotine according to the withdrawal alleviation concept. Such a negative result was obtained by Heishman *et al.* (1993) in non-smokers given nicotine chewing gum. However, the study did not include a group of smokers as controls and it was also reported that the subjects complained considerably about nausea and dizziness, and this particularly with the higher dose.

Perkins *et al.* (1990a) obtained with nasal nicotine spray in the smokers a greater increase of the speed of finger tapping than in non-smokers, as seen already earlier by West and Jarvis (1986). On the other hand, with subcutaneous application, Foulds *et al.* (1994) observed in smokers and non-smokers very similar activations of the EEG, and Le Houezec *et al.* (1994) confirmed this result and demonstrated in addition improvements of attention and stimulus processing in non-smokers.

Nicotine effects in deprived and non deprived smokers

An extended study by Hasenfratz and Bättig (1993a) compared 20 regular smokers in a crossover 2×2 design of abstaining or smoking before the experimental session and smoking or not smoking between the two trials of the sessions. Pre-session abstinence as opposed to *ad libitum* smoking reduced performance, electrocortical and cardiovascular arousal in the first trial and induced in addition subjective symptoms of withdrawal with increases of anxiety and craving to smoke. Smoking between the two trials, however, increased performance in the second trial both after pre-session smoking and pre-session abstinence. The increases with the pre-session abstinence condition were only slightly and not significantly greater than with the pre-session *ad libitum* smoking condition. Similar interactions were also obtained for the cardiovascular and electrocortical arousal in the second trial. The resulting conclusion that smoking does more than alleviate withdrawal finds further support from additional other studies.

Sherwood *et al.* (1992) demonstrated similar improvements in a series of different performance tasks with one as well as with repeated nicotine doses given in chewing gum to regular smokers who had abstained from smoking overnight. Pritchard *et al.* (1992a) did not require his subjects to abstain and compared their performance between two testing days, one with and the other one without immediate pre-task smoking. Pre-task smoking improved performance compared with no pre-task smoking, although the subjects were not deprived. Warburton and Arnall (1994) conducted two experiments to test the withdrawal relief hypothesis. In the first experiment, deprived smokers and non-smokers failed to differ in their performance. In the second experiment, performance was highly similar both after 1 and 10 hours of deprivation regardless of whether the subjects were allowed to smoke or not in the experimental session.

Intraday acute tolerance?

The cardio-accelerative effects of nicotine seem to vanish across a complete smoking day, as was shown by Jacober *et al.* (1993) with the method of all day recording of HR and activity. Only partial tolerance was seen, however, in laboratory experiments by Hasenfratz *et al.* (1990) across the first few cigarettes of the day, and by Perkins *et al.* (1994b) across a few repeated doses of nasal spray. The results of the studies comparing cardiovascular nicotine effects in deprived and non deprived smokers raise the question whether similar tolerance might also develop for the effects on mental performance. Ideally one would need all-day recordings of performance and perhaps EEG in order to answer this question. It should be kept in mind also that the results reported from comparisons between deprived and non deprived subjects leave open the possibility that withdrawal as a performance deteriorating factor might arise very rapidly, perhaps even within a half hour after the last cigarette.

Several attempts have been made to approach this question. Perkins *et al.* (1994b) compared in their study the responses of smokers and non-smokers to challenge doses of nicotine given via measured nasal sprays. In smokers as opposed to non-smokers the dose effect curves for most of the subjective effects but less for cardiovascular and behavioural effects demonstrated chronic tolerance. A similar pattern was observed for subsequent challenge doses in comparison to the first dose, suggesting thus that acute tolerance resembles chronic tolerance and again that subjective effects are more affected by tolerance than cardiovascular and behavioural effects.

Parrott (1994b) and O'Neill and Parrott (1992) required the subjects to complete brief feeling state questionnaires before and after each cigarette throughout the day. The questionnaires assessed anxiety/stress and arousal. Post-smoking decreases of anxiety/stress and increases of arousal were greatest after the first morning cigarette but remained unchanged thereafter at a reduced level throughout the day. Taken together, these results suggest that tolerance to the central effects of nicotine is at least not complete.

The specificity and reinforcement value of the nicotinic action

Numerous studies with non-smoke nicotine have shown dose dependent positive effects on cognitive functions. There remains, however, the possibility that the sensory/manipulatory reward of smoking might contribute to the facilitating effects of nicotine. This possibility seems unlikely in view of the results of a recent study by Baldinger *et al.* (1995c). In this experiment performance in an information processing task presented in two versions was measured before and after smoking regular or the nearly nicotine-free cigarette 'Next'. For both task versions post-smoking increases of performance, cardiovascular and electrocortical arousal were obtained with regular cigarettes but not with 'Next'. This makes it unlikely that the sensory/manipulatory reward of smoking might contribute to the positive effects of the nicotine content of a cigarette.

Such results raise the question as to what extent nicotinic cognitive enhancement may be one of the primary reinforcers of smoking. On the basis of common sense this would be difficult to imagine. A youngster hardly smokes his first cigarettes in order to perform better in school, and an intellectual smoker also does not smoke preferentially to improve his mental work but quite probably on a great number of other occasions as well. Finally, there are hardly any smokers who smoke in order to lower their risk of Alzheimer's and Parkinson's diseases. These two not infrequent conditions in elderly people have been documented by

the reviews of Shahi and Moochhala (1991) and Lee (1994) to develop less frequently in smokers than in non-smokers, a finding which was confirmed for
Alzheimer's disease in a large prospective study by Grandinetti *et al.* (1994).
Furthermore, Nordberg (1994) has outlined the growing evidence obtained from
molecular biology that nicotine may have a protective effect against these two
degenerative diseases, and Jarvik (1991) included these two diseases into the list
of conditions which may react positively to nicotine treatment.

An indirect answer to the question about the possibility that cognitive
enhancement through smoking might act as a reinforcer for smoking, which dis-
affirms the common-sense skepticism expressed above, comes from the study by
Baldinger *et al.* (1995c). As already mentioned they compared the effects of
'Next' cigarettes and regular habituated cigarettes on mental performance efficiency. The cigarettes were smoked in this experiment both between the two
trials and throughout the second trial of the experiment. The regular cigarettes
but not the 'Next' cigarettes improved performance in information processing.
Surprisingly, 'Next' cigarettes, which in all earlier experiments reduced the need
to smoke as much as regular cigarettes did, failed to do so in this case. The
reduction of the need to smoke was clearly smaller than with the regular cigarettes. The main difference between this and the earlier laboratory studies was
that hard mental work was required in this study but not in the earlier ones. This
raises the intriguing possibility that satiation of the smoking need may be greater
when smoking helps than when smoking is of no help.

PARADOXICAL OR SENSORY-HEDONIC MOTIVATIONS FOR SMOKING?

Paradoxical motivations for smoking can be assumed when smoking does not
provide any objective benefit besides the subjective sensory/manipulatory enjoyment of smoking. Intensified smoking occurs rather individually for many
different occasions ranging from car driving to watching TV and social interactions. However, so far only a few such smoking behaviours have been investigated systematically.

Stress smoking

It is commonly known that a smoker under stress smokes more cigarettes, and
this has also been verified empirically. Perkins and Grobe (1992) reported a
greater desire to smoke after a stressful computer task than after a non stressful
task. West and Lennox (1992) observed increased smoking in students before
examinations, and the increases paralleled the increases in hours spent reviewing
and the anxiety about passing the exams. Several studies focused on questions
concerning the effects of smoking on mood, on the physiological and endocrinological effects of stress and on performance.

The picture resulting from these studies remains in part controversial.
Smoking was seen to dampen slightly or mask the cardiovascular responses to
stress when the stressor had to be sustained passively such as with stressful
video scenes (Fuller and Forest, 1977; Gilbert and Hagen, 1980; Gilbert *et al.*,

1989b), noise bursts (Schachter, 1978; Woodson *et al.*, 1986) and electric shocks (Schachter, 1973). On the other hand, when subjects were required to actively perform tasks, the cardiovascular effects of smoking and the mental stress were disappointingly additive without any interaction. This was seen with video games (Dembroski *et al.*, 1985; Perkins *et al.*, 1986b) and with subject paced information processing (Hasenfratz *et al.*, 1989a; Michel and Bättig, 1989). Perkins *et al.* (1992c) included in their study a high and a low challenge computer task and the smokers smoked regular cigarettes and sham smoked during the task, whereas the non-smokers sham smoked only. Also in this study the cardiovascular effects of smoking and of the tasks were purely additive without any interaction.

Hasenfratz and Bättig (1991) made an attempt to contrast active and passive coping in continuing and abstaining smokers on the days 1, 3 and 9 after quitting by the group of abstainers. This was done by requiring one half of the subjects of both groups to passively sustain electric shocks while performing a computer task. The other half of the subjects had to try, according to the instructions but not in fact, to avoid shocks by optimally performing the same task. The cardio-vascular tonus decreased gradually in the abstainers, but remained elevated in the continuing smokers. The cardiovascular stress responses were highly similar in both groups and generally greater with active than passive coping. Both conditions raised BP similarly but HR increased with active coping only. Craving to smoke remained elevated in the continuing smokers but decreased rapidly and about threefold in the abstainers. Subjective pain increased gradually with passive coping and decreased with active coping. Neither EEG parameters nor task performance differed between the two conditions. A subsequent study by Hasenfratz and Bättig (1993b) made use of the same experimental paradigm, but all subjects were abstinent before the sessions and smoking or sham smoking were required between two presentations of the task. The results were similar with the exception that smoking facilitated performance under both stress coping conditions. The main difference between the two experiments was that in the first study the last cigarette smoked by the continuing smokers preceded the task by about one hour due to the time needed for the preparation of the experiment and the fixing of the multiple electrodes. In the second experiment, in contrast, two performance runs were required and smoking immediately preceded the second run. The two experiments indicate therefore that the beneficial effects on performance last less than an hour, that they are not mediated by any alleviation of vegetative stress and finally that smoking withdrawal from between an hour to ten days does not aggravate any consequences of stress.

On the other hand the two experiments failed to support the notion of a differential effect of smoking in active and passive stress conditions, although these two conditions produced clearly different profiles of the cardiovascular stress responses.

As a second aspect, the effects of smoking on mood and subjective stress perception merit particular attention. Several studies which did not include physiological measurements reported in part conflicting results. Cutler and Barrios (1988) failed to obtain stress related mood changes attributable to smoking or smoking deprivation. Jarvik *et al.* (1989) obtained smoking induced anxiety reductions with an auditory anagram task, but not with the cold pressure test nor with white noise exposure. On the other hand, the studies mentioned above which included physiological measurements reported, as far as they assessed mood, beneficial effects of smoking, with the exception of the Hasenfratz and Bättig (1991) study in which the interval between smoking and stress was an hour or more. Perkins *et al.* (1992c) assessed stress ratings repeatedly during the stress tasks and observed decreases after smoking which appeared almost immediately, but dissipated within less than 10 minutes.

All-day subjective stress ratings before and after the smoking of each cigarette were analyzed by O'Neill and Parrott (1992) and they reported significant immediate post-smoking decreases which remained of similar magnitude throughout the testing days. Parrott (1994a) further observed with the same technique personality differences in the post-smoking changes. Sedative smokers, as classified by a smoking motivation questionnaire, reported more stress modulation, whereas stimulant smokers reported more arousal modulation. Further, the picture he obtained, that the pre-smoking stress ratings were elevated and the post-smoking ratings normalized and the reverse for the arousal ratings, was seen as a suggestion that smoking may merely reduce withdrawal symptoms. An additional argument for this view was seen in the fact that stress ratings across the day were similar for smoking smokers and non-smokers but elevated in abstaining smokers. However, a report by Parrott and Joyce (1993) also showed that the stress ratings of the smoking smokers were higher than those of the non-smokers before the first morning cigarette and lower than those of the non-smokers toward the end of the day, suggesting an anxiolytic effect of the gradual accumulation of nicotine. Evidently such results would gain more weight if comparisons between regular and nicotine-free cigarettes were included in the design.

The study of endocrinological responses has been a third object of the research on smoking and stress. Pomerleau and Pomerleau (1990) found an additive effect of smoking and a mental stressor. Gilbert *et al.* (1992b) presented evidence that post-smoking increases of plasma beta-endorphin and cortisol can hardly be seen as a consequence of normal smoking since they appeared only after smoking cigarettes with an extreme nicotine delivery (2.4 mg) and no correlations between mood ratings and plasma concentrations of these neuromodulators were obtained. Kirschbaum *et al.* (1992) on the other hand claimed that normal cigarettes (1.0 mg delivery) increase plasma cortisol and later Kirschbaum *et al.* (1993) reported that smoking attenuated the cortisol responses to stress, but not those to an injection of corticotropin-releasing hormone.

Given the subtlety, complexity and in part also inconsistency of the results of the research on smoking and stress, it is not surprising that attempts have been made to test the possibility that sensory pleasure rather than nicotine might mediate the smoking induced reduction of anxiety and stress. Levin *et al.* (1991) provided smokers with nearly nicotine-free aerosols from smoke regenerates and found that these reduced stress ratings to a similar extent as the aerosol with real cigarette smoke added.

Certainly, smoking during stress appears as a paradoxical smoking motivation, as it increases rather than decreases cardiovascular stress and perhaps also endocrinological responses to stress. However, this does not exclude the possibility that smokers use smoking as a hedonic sensory/manipulatory self-reward to reduce feelings of distress. This speculation is further supported by the apparently very short duration of the subjective stress reduction and also by the absence of a development of intraday tolerance. If this could be demonstrated by future work, the term 'paradoxical' might become obsolete.

Food intake, body weight and smoking

Among the reasons for the maintenance of the smoking habit, the 1988 US Surgeon General Report attributed an important role to the putative effects of nicotine on body weight mediated by effects on energy metabolism and appetite. Further it was assumed that these effects also contribute to the relapse tendency in quitting smokers.

The evidence, however, has revealed that the smoking-body weight issue is much more complex than initially believed. An excellent review of the relevant literature was presented by Klesges and associates (1989). Twenty-four out of 29 cross-sectional studies cited in the review reported non-smokers to be heavier than non-smokers and 31 out of 41 prospective studies following up weight after quitting reported weight gains in the quitters. Surprisingly, smoking status as assessed by the number of cigarettes per day, by years smoking or by biochemical measurements was considered only in the minority of these numerous studies. Among the studies considering the number of cigarettes per day the great majority obtained a curvilinear rather than a linear relation. Only moderate smokers appeared to weigh less than non-smokers, whereas heavy smokers with more than 20 cigarettes per day failed to differ from the non-smokers. In addition, the five studies which compared food intake in continuing smokers and in non-smokers, did not find a difference.

However, there is abundant evidence that food intake increases when smokers quit smoking, although there exist considerable discrepancies between the different reports with respect to the magnitude and duration of this effect as well as with respect to intervening variables such as physical activity, possible changes in food preferences and metabolic rate.

Perkins (1993) reviewed studies on the weight gain after smoking cessation comes to the conclusion that there is no easy explanation for the effects of smoking on body weight and energy balance. Also his own numerous studies indicate that the thermogenic effects of smoking or nicotine are too short lived and too small to explain the epidemiological data and the same holds for possible anorectic effects. Furthermore, the increase of food intake after smoking cessation is transient and dissipates after a new and higher level of body weight is established. The author proposes therefore that smoking in humans may lower the set point of body weight regulation as nicotine does similarly in laboratory animals.

Effects on appetite could be seen as a first candidate to explain the differences in body weight. As shown earlier shown with smoking, Perkins *et al.* (1992a) demonstrated also with nasal nicotine spray that this treatment failed to affect subsequent food intake. However, the meals increased craving for smoking independently of the nicotine dose, and earlier Hasenfratz *et al.* (1991) observed that such post-meal increases of craving to smoke are also independent of the meal size.

With respect to the cardiovascular effects of nicotine, they were less than additive to those of a meal, as reported by Perkins *et al.* (1990b). Thermogenic effects of nicotine were additive to those of caffeine and more than additive to those of low intensity bicycle riding, as also shown by Perkins *et al.* (1994c). However, the effects of caffeine and nicotine combined were again too modest to explain the weight differences between smokers and non-smokers. In addition, smokers hardly profit from the more than additive effect of nicotine and physical activity, as they are known to engage less in physical activities than non-smokers.

A methodologically new approach to study the differences between smokers and non-smokers and the effects of acute smoking abstinence was chosen by Bättig *et al.* (1994). Instead of manipulations under laboratory conditions, the subjects were equipped with portable devices for continuous recording of HR and activity, with pocket computers used as electronic diaries for entering all foods and beverages, the results of intermittent self-measurements of BP, the answers to self-rating questions and finally the performance on a numerical version of the Stroop task. Saliva tubes were handed out for the later analysis of cotinine content on the experimental days.

The weights of the smokers and non-smokers did not differ significantly, a fact that was not surprising, as the subjects were relatively young (30 years) and as, according to the epidemiological review by Klesges *et al.* (1989), smoker/non-smoker differences appear mostly after the age of 40 years when smokers stabilize their weight, whereas non-smokers continue to gain some weight. Caloric intake was greater in the non-smokers than smokers, but this difference disappeared when controlling for sport and gross physical activity, which as assessed by questionnaire was considerably more frequent in the non-smokers than in the smokers. Casual activity, as measured continuously by the portable actometers, did not differ between the two groups and was also unaffected by smoking abstinence. Coffee and alcohol consumption were the only dietary variables which were (positively) correlated with saliva cotinine. On one of the 6 recording days the smokers had to abstain from smoking and for that day their food intake increased by about 500 Calories without changes of physical activity, subjective state ratings or mental performance but with a decrease of HR of the order to be expected when not smoking. It was suggested that the excess of caloric intake on the abstinence day may reflect more the use of snacks as a hedonic substitute for missing the cigarettes than any specific pharmacological action. This view gained further support in an additional study by Kos *et al.* (1997) which used the same method to compare young subjects and subjects between 55 and 65 years old. In this study it was also considered by post hoc comparison whether staying at home or being away from home during the manipulation might have an effect. It turned out that this was the most important variable for explaining the excess food intake on the abstinence day, and a re-analysis of the Bättig *et al.* (1994) data revealed the same result. If this could be confirmed in further studies it might suggest that the distance to the refrigerator might turn out to be a crucial variable for the magnitude of excess eating on a day of smoking abstinence.

Coffee drinking and cigarette smoking

Commonly it can be observed that smokers drinking coffee also tend to light one or more cigarettes. In their review, Swanson *et al.* (1994) cited 9 large studies (n > 1000), all of which demonstrated the association between coffee and cigarette consumption. Not only do smokers usually report about 50% more cups per day than non-smokers, but the absence of any coffee consumption is also more frequent among non-smokers than among smokers. Surprisingly, however, ex-smokers are nearly as frequent coffee consumers as smokers. It remains open whether this interaction is based on behavioural or pharmacological actions.

Kozlowski (1976) reported no differential effects of adding 0, 75, 150 or 300 mg caffeine to decaffeinated coffee on subsequent cigarette smoking, and adding no caffeine at all even increased smoking. The same negative result was obtained with 200, 400 and 800 mg caffeine by Chait and Griffiths (1983). As these and other studies with similar negative results were done with observation periods of less than 4 hours, Brown and Benowitz (1989) studied the effects of multiple daily doses of caffeine across 4 days with total daily loads of caffeine of 0, 6 or 12 mg/kg per day. However, even with the highest dose, which is quite high, cigarette consumption tended to increase by only about 10%. Marshall *et al.* (1980) observed that not only coffee drinking (regardless of its caffeine content) but also water drinking increased subsequent cigarette smoking. In an interesting study, Bickel *et al.* (1992) investigated what price, in terms of the number of responses required, the subjects would pay for coffee and cigarettes. For both the cups of coffee and the number of cigarettes, the increase of the price from 100 to 1,000 and 2,500 responses decreased consumption price dependently. In a choice situation between one substance at a fixed price (100 responses) and the other substance at increasing prices, the price increase for cigarettes decreased both coffee and cigarette consumption, whereas price increases for coffee reduced coffee consumption but not cigarette consumption.

A pharmacological interaction could be presumed, based on the well-documented fact that smokers metabolize caffeine (like many other substances) nearly twice as rapidly as non-smokers. Benowitz *et al.* (1989) measured plasma caffeine in a smoking cessation study at baseline, 12 and 26 weeks after cessation. Coffee consumption was unchanged after 12 weeks and slightly decreased after 26 weeks, but plasma caffeine was similarly elevated by about 260% after 12 and 26 weeks of smoking abstinence. On the other hand, only a nonsignificant increase was found by Oliveto *et al.* (1991), but coffee consumption as well as cigarette smoking were lower in the subjects of that study.

Several studies examined the interaction of caffeine and nicotine on physiological parameters and performance. Smits *et al.* (1993) showed that the effects of intravenous caffeine and nicotine chewing gum on systolic and diastolic BP were similar in magnitude and additive when both substances were combined. HR increased with nicotine alone and similarly with the combination but not with caffeine alone. Plasma catecholamines also increased in a purely additive fashion with the combination of the two substances. With slight physical work (standing up) the pressor effects of the two substances were less than additive and the same was the case for the forearm vasodilator response when performing mental arithmetic. A similar picture was obtained by Perkins *et al.* (1994c) for nicotine given via nasal spray. The physiological effects were also additive or even slightly more than additive. However, a different picture was obtained for subjective arousal, which was increased by both substances during rest but not during light physical activity.

A similarly puzzling result was already obtained earlier by Rose (1986). In his study, smoking and caffeine both increased subjective arousal but the combination showed a clear antagonistic effect rather than an additive one. Hasenfratz *et al.* (1991) observed additive cardiovascular effects of smoking and caffeine, but rapid information processing, which was facilitated similarly by either treatment alone showed no further improvement when both treatments were combined. Hasenfratz and Bättig (1992) obtained similar results with a numerical computerized version of the Stroop test. This task requires responses to conflicting and non conflicting information, whereby the Stroop effect is defined as the difference between the response times to the two types of information. Each substance given alone improved both types of response times, but the combination of both treatments was not more effective than either substance alone, whereas the physiological effects were again additive.

Besides an easy version of the task (1 second interstimulus intervals), the experiment also included a more difficult version (0 second interstimulus intervals). With this version, both treatments failed to improve the response times, but they did improve, that is shortened in contrast to the easy version the Stroop effect. However, this improvement seen with either treatment alone disappeared completely when both treatments were given together.

Taken together, these results suggest that the widely popular combination of coffee and cigarettes hardly results in any benefit of objective performance or subjective arousal, except that both provide distinct sensory/manipulatory pleasures, which may enhance each other.

THE TOBACCO WITHDRAWAL SYNDROME

The tobacco withdrawal syndrome is considered by many researchers as a critical factor for the maintenance of the habit as well as for relapse after cessation. By accepting this concept outlined by Siegel (1983) and Solomon (1977) for drug consumption, it would have to be assumed that the positive effects of smoking would rapidly be subject to complete tolerance and that smoking would be maintained mainly by the negative reinforcement of terminating or avoiding withdrawal symptoms. Although this view underlies a considerable part of actual smoking research as well as different smoking cessation therapies, it remains problematical in several respects.

The paradoxical profile of the withdrawal syndrome

Swanson *et al.* (1994) in their review on nicotine and caffeine withdrawal, list 42 studies on the effects of abrupt smoking cessation. With respect to the reported symptoms, these studies are quite consistent. A first group of symptoms includes an increased need to smoke and more thinking about cigarettes, commonly referred to as 'craving'. A second group refers to changes in mood, namely: anger, frustration and irritability; increased anxiety; difficulty in concentration and restlessness. The third group includes the two bodily symptoms of increased appetite with weight gain and reduced HRs. Whereas the first group of 'craving' symptoms is not specific for smoking and appears with the cessation of other habits as well and the third group of bodily symptoms does not create immediate discomfort, it is the second group consisting of mood changes which merits the primary interest when it comes to considering their role in maintaining the habit.

However, this profile of mood changes is hardly what one would expect when discontinuing the chronic consumption of a stimulant substance. Fatigue, reduced wakefulness and sluggishness would logically have to be expected, and this is in fact what happens after the cessation of coffee drinking – another widely consumed stimulant – as has been documented in many laboratory studies and recently also by Höfer and Bättig (1994) in an extended field study. On the other hand, the mood profile of tobacco *withdrawal* has striking similarities with the mood profile of a caffeine *overdose*, as outlined by Swanson *et al.* (1994). Since smoking accelerates caffeine metabolism, this could be a reason why smokers consume more coffee than non-smokers, as discussed above, and it is likely the reason why smoking abstinence goes along with rising plasma caffeine levels. If this were the reason for the paradoxical character of the tobacco withdrawal syndrome, one would also have to expect that quitting smokers would spontaneously reduce their coffee drinking. However, according to the review by Swanson *et al.* (1994), this is quite certainly not, or only modestly the case.

A second aspect of the withdrawal syndrome which merits consideration is its severity and duration. The severity varies across subjects and is probably not related to the intensity of smoking and nicotine exposure, as discussed below, but eventually rather to differences in the personality profile and socio-environmental conditions.

The duration of the withdrawal syndrome after quitting also seems to be of rather short duration. Hasenfratz and Bättig (1991) observed a nearly twofold decrease from the first to the third day and a threefold one to the ninth day of abstinence as already similarly reported by West *et al.* (1989). A follow-up of the 20 subjects who participated in the abstinence program of the Hasenfratz and Bättig (1991) study revealed, however, that despite the reduction of craving, nearly all of these subjects resumed smoking mostly within a few days after the termination of the experiment (unpublished post hoc analysis).

The impact of nicotine

If a nicotine deficit were the primary cause of the smoking withdrawal syndrome, nicotine-free cigarettes should elicit it. As already mentioned above, Hasenfratz *et al.* (1993a) found that smoking regular, ultra light and the nearly nicotine-free cigarette 'Next' all similarly reduced the subjective need to smoke and the intervals until the next cigarette was lit also did not differ between the three types of cigarette. This result was confirmed in a further laboratory study with the same cigarettes by Baldinger *et al.* (1995a). In the later field study by Baldinger *et al.* (1995b), smoking regular cigarettes, smoking 'Next' and complete smoking abstinence were required for two days each. Self-ratings of mood were included in order to detect possible changes. Impatience and irritability increased with abstinence, but not when smoking 'Next' cigarettes, and the same was the case for decreases of relaxation and subjective activity. There was also a tendency for nervousness and restlessness to increase under abstinence.

As a second postulate, one would expect compensatory smoking after a period of enforced smoking abstinence, such as smokers have to undergo when traveling in planes, or wherever smoking is prohibited. Jacober *et al.* (1994b) simulated the condition of a few hours of abstinence in a field study, for which the subjects were equipped with event recorders, combined activity and HR recorders and electronic diaries for repeated self-assessments of craving and smoking pleasure. The subjects were free to smoke for one complete day and were required to abstain from 1.00 to 5.00 p.m. on a second day, whereas on an additional day they had to double their number of cigarettes during the same 4 hours of the afternoon. Activity-adjusted HR decreased on the abstinence afternoon and increased on the over-smoking afternoon, indicating therefore a good compliance of the subjects with the instructions. The abstinence afternoon was preceded by an anticipatory doubling of the smoking rate, which started about 30 minutes before abstinence. However, abstinence was not followed by any compensatory increases of smoking in the evening. Craving increased considerably during abstinence and smoking pleasure did so for the first post-abstinence cigarettes. Over-smoking went along with considerably reduced craving and pleasure, but in the evening it was not followed by any compensatory savings in the number of cigarettes. Under all three conditions, the subjects were asked to select on the electronic diary before each lighting one among different reasons for smoking that particular cigarette. The answers differed in part between the times of day but not between the three different conditions. During the mornings and evenings the most frequent reason was 'to relax', whereas the second most frequent reason was 'to increase motivation' in the mornings and 'in company' during the evenings.

These results find confirmation in a study by Kolonen *et al.* (1992) with more drastic manipulations. Abstinence was required for 15 hours and over-smoking involved chain smoking 20 cigarettes (2 at a time!). These two manipulations did not affect any puffing parameters, a fact which according to the authors hardly agrees with the nicotine titration hypothesis. In the same experiment, ultra light cigarettes were smoked in addition to regular cigarettes and with these, the subjects compensated for the lower smoke yields by sufficiently intensifying puffing to reach plasma cotinine levels comparable to those reached with the regular cigarettes. This of course raises again the question discussed above as to whether intensified puffing of ultra light cigarettes should be seen as an adaptation to the reduced nicotine content or as an answer to the overall smoke dilution, as suggested by the Hasenfratz *et al.* study (1993a). It may be recalled at this point that the authors obtained intensified puffing with the ultra-low-nicotine cigarettes only when the low nicotine content was a consequence of smoke dilution by ventilated filters, but not when it was a consequence of smoking 'Next' which was produced with denicotinized tobacco, thus yielding a 'regular tar' content.

Tobacco substitutes

The most popular substitutes for cigarettes are now nicotine chewing gum and transdermal nicotine patches, both of which are modestly but significantly helpful in the cessation of smoking, as mentioned in the introduction.

Recently, nicotine nasal sprays were found in many studies to reduce the desire for smoking, perhaps even more efficiently than other forms of application. The rapid absorption of nicotine given in the form of a nasal spray, as demonstrated by Sutherland *et al.* (1992a), could be one reason for the appeal of nasal nicotine application. This is consistent with the arguments put forward by Henningfield and Keenan (1993) that abuse liability for psychotropic substances is directly related to the speed of absorption. As absorption is slow with transdermal application and chewing gum, they expect for these two delivery systems a low abuse liability. However, if this were the case one might have expected better consumer acceptance for the cigarette 'Premier', developed by the R.J. Reynolds Tobacco Company. This cigarette-like product delivers nicotine, although in small amounts only, without burning down, and according to Cone and Henningfield (1989), it most closely mimics the nicotine kinetics of normal cigarettes.

In contrast, it was also reported by Hatsukami *et al.* (1993), that up to 14% of chewing gum users may continue use beyond a year suggesting that they replace the smoking habit by a nicotine chewing habit. Even higher figures were presented by Hughes (1989; *et al.*, 1991a), and they suspect that nasal nicotine sprays might lead even more to continued use than chewing gum. Both chewing gum and nasal spray provide the smoker with stinging and burning sensations in the nose, the oropharynx or the throat in a way typical for nicotine. To the extent that this may be one of the important aspects of tobacco and nicotine consumption, it would have to be expected that transdermal patches would not induce any need for continued use, but no relevant reports are available so far on this question. In this context, the clinically well-known fact might be recalled that nasal decongestants are also often used on a regular basis beyond the recommendations of the physician.

The approach of facilitating smoking cessation by nicotine antagonists rather than by nicotine has gained less interest so far, although the results are in part not only fascinating but also not easy to understand. Rose *et al.* (1989) observed that mecamylamine, a centrally active nicotine antagonist, reduced dose dependently harshness and strength ratings of cigarette smoke, particularly for high nicotine cigarettes. Furthermore, with a smoke mixer allowing the regulation of the nicotine content of the smoke, the subjects increased nicotine intake with increasing doses of mecamylamine. However, the need to smoke was reduced with middle doses of mecamylamine but not with the lowest (2.5 mg) and the highest dose (20 mg). In a new study, Rose *et al.* (1994) compared cessation rates when treating the subjects either with transdermal patches alone or combined with mecamylamine. The rationale was that by combining an agonist with an antagonist, they might oppose each other, nullifying not only any therapeutic effect of either one but also any negative side effects resulting from overstimulation or under-stimulation of the nicotinic cholinergic system. Although the mecamylamine doses were low, starting with 2.5 mg and continuing with 5 mg, the combined mecamylamine and patch treatment was in fact considerably more effective than the patch treatment alone.

Predictors of relapse

Logically, one might expect that the difficulty of quitting or the probability of relapsing after cessation might be a direct function of the previous nicotine exposure. However, the outcomes of the relevant investigations reveal another picture. Kozlowksi *et al.* (1994) presented to nearly 3,000 participants of smoking cessation programs different questionnaires frequently used to assess the severity of dependence. Although statistical significance was reached in the expected direction, this relation explained less than 1% of the total variance. Coambs and Kozlowski (1992) reported from a larger sample which did not participate in a cessation program that among the younger smokers the heavier smokers were less likely to quit, whereas among the older smokers the heavy smokers were more likely to quit than the moderate smokers. Kenford *et al.* (1994) analyzed data from 200 smokers treated with transdermal patches. Pre-treatment assessments of smoking intensity including not only relevant questionnaires but also expired CO levels, plasma nicotine and cotinine showed no relation to successful quitting.

However, relapsing in the second week of treatment was in this study a powerful predictor of long-term failure and complete abstinence for two weeks a good predictor of long-term success. But even two weeks of enforced abstinence in an inpatient program with nicotine patches, group therapy, exercise, daily lectures, etc. resulted for a small sample of long-term smokers to a long-term success rate of no more than 30%, as reported from a Mayo Clinic study by Hurt *et al.* (1992). A larger study from the Mayo Clinic, reported by Rohren *et al.* (1994), included 650 subjects who contacted the clinic mostly for other reasons than seeking help for smoking cessation. The sample was divided into a pre-contemplation group made up of subjects who had not yet considered stopping smoking, a contemplation group made up of subjects who had considered giving up but had not yet taken into action toward this goal and finally an action group of subjects who already had started active attempts at quitting on their own initiative. The action group stood out against the other ones by a double success rate after 6 months, suggesting that personal attitude and will are perhaps the most potent predictors of successful quitting. A similar suggestion was already made by Hyman *et al.* (1986), who compared in a relatively small sample of 60 subjects the 3 and 6 month success rates with the four conditions of hypnosis, focused smoking (smoking while concentrating on smoking and health), attention placebo (being told they would give up without any help) and control (being put on a waiting list without any intervention). No significant differences appeared between the four groups and the authors suggested that active treatments are perhaps of minor help against the background of the manifold non-specific and unknown variables.

In this context it should also not be overlooked that the prescriptions for nicotine chewing gum and transdermal patches are mostly accompanied by insistence on the importance of the readiness to quit and having confidence in the effectiveness of the treatment. This is in contrast to the dispensation of most modern pharmaceutical preparations, as it is difficult to imagine getting the latest antibiotic from one's physician with the recommendation that it may help only if one believes in it and has a strong will to regain health.

DISCUSSION

At the end of a review on a complex and in many aspects controversial question one would be happy to be able to propose valid conclusions. However, the impact of the reports considered leads one to raise questions rather than to formulate any firm statements.

One serious question concerns the validity of the withdrawal and homeostatically based theories of the smoking habit. Such concepts assume that any possible positive effects of smoking are nullified by acute and chronic tolerance and that smokers light their cigarettes due to the negative reinforcement obtained through terminating or avoiding withdrawal distress.

Without going back into the details of the reports considered in this review, the main arguments against a purely pharmacological and negative reinforcement view are:

- Smokers fail to compensate for lowered nicotine levels of light cigarettes by additional smoking. Further, the partial compensation achieved by intensified puffing might be better explained as a response to smoke dilution than to a nicotine deficit.
- Not only the taste and smell of smoke constitute important elements of the smoking satisfaction, but nicotine *per se* is an important element of the chemosensory stimulation through tobacco smoke. Further, non-nicotinic elements of the chemosensory constituents of tobacco smoke, not only mediate in part the satisfaction of smoking, but also reduce the need to smoke as well as withdrawal symptoms.
- Smoking was seen to be continued with nicotine-free cigarettes at the same rate as with regular and habituated cigarettes across entire days without inducing withdrawal symptoms and without elevating the smoking need.
- Pre-lighting arousal of HR and activity as verified by all day physiological recordings was independent of the presence of nicotine. This suggests that smoking may be elicited more by learned habits than by an actual need of nicotine.
- Cognitive enhancement appears to be due specifically to the absorption of nicotine, and growing evidence strongly suggests that tolerance to this effect may at least not be complete.
- There are numerous occasions which elicit smoking without objective benefits. Stress smoking, post-meal smoking and smoking together with a coffee were discussed in more detail in this review, but they may hardly be the only possible examples. It remains therefore primarily the possibility that on such occasions smoking occurs for the sake of immediate behavioural pleasure and relaxation rather than for a pharmacological advantage. According to this view, it would be merely the chemosensory and manipulatory pleasure and relaxation that facilitates stress coping or enhances post-meal satisfaction and the pleasure of coffee drinking as well that of other positive experiences.
- Withdrawal symptoms are modest, transient and paradoxical in the resulting profile of mood. Withdrawal is not followed by compensatory over-smoking. Relapses occur in abstaining smokers after all withdrawal symptoms have dissipated. The best predictors of successful smoking cessation appear to be personal will and decision taking rather than any parameter of previous smoke exposure.

Regardless of these facts, the concept that smoking is a purely or predominantly nicotine regulated behaviour continues to underlie the rationale of many smoking cessation programs as well as to serve as a theoretical hypothesis in experimental investigations. This viewpoint culminated in the proposal suggested by Benowitz and Henningfield (1994) and Henningfield *et al.* (1994) to reduce or nearly eliminate nicotine in tobacco used to produce cigarettes.

Focusing primarily on nicotine delivery or bio-availability neglects the fact that other fractions of tobacco smoke are probably more relevant for health issues than nicotine as well as the fact that the increasing numbers of smokers who switch to light and ultra light cigarettes continue to smoke the same number of cigarettes regardless of the declining plasma cotinine levels. Therefore the reservations regarding this proposal formulated by the authors at the end of the first of these two articles certainly merit serious consideration.

A second question to be raised is whether concepts of positive rather than negative reinforcement might better explain the phenomenon of the smoking habit. The concept of positive reinforcement would not only offer the advantage of allowing the inclusion of the sensory and manipulative rewards of smoking into a theoretical framework of smoking motivation. It would also help to explain the different instances of 'paradoxical' smoking which are 'paradoxical' to the extent only that no pharmacological benefit can be demonstrated.

According to a positive reinforcement concept, particular brain systems mediating pleasure and reward would initiate smoking due to the heightened probability of pursuing operants previously experienced as rewarding in accordance with the theoretical model proposed by Tiffany (1990). An experimental verification of this view has been demonstrated by Zinser *et al.* (1992) of stress smoking.

The brain system of reward apparently organizes and facilitates all behaviours which previously have been of advantage to the organism in some way, as outlined by Wise (1988). This system is crucial not only for the organization of behaviours essential for the maintenance of life such as eating, drinking or sexual behaviour but also for the organization and repetition of the innumerable learned habits including the consumption of psychoactive substances. The reward system normally functions to preserve the motivational hierarchy of the different needs essential for the maintenance of the bodily, psychological and social equilibrium. It can, however, also decompensate. An example of the decompensation of a biological need may be seen with overeating resulting in pathological obesity. The use of psychoactive substances also can lead to an escalation of the need to the extent that the motivational hierarchy begins to deteriorate. This happens frequently with the use of opiates and cocaine and in a minority also with the use of alcohol. The need for smoking is also very strong in many smokers but the motivational hierarchy is maintained as in the case of coffee drinking.

Nevertheless it is still a controversial issue as to whether smoking should be considered as an addiction rather than as a (perhaps strong) habit. This issue was discussed by several authors among other occasions in a special issue of Psychopharmacology, Vol. 108 (4), 1992 and also at the occasion of an international symposium on nicotine (Clarke *et al.*, 1994). These discussions suffer not only from the facts that the differences between heavy smoking and decompensated needs for alcohol, opiates or cocaine are profound. Another more basic difficulty is that there is no generally accepted definition of addiction. DSM-IV, edited by the American Psychiatric association (1991) and ICD-10, edited by WHO (1992), the most widely accepted diagnostic manuals for behavioural disorders, do not define or use the term addiction, and this not even with respect to cocaine, the opiates or pathological alcohol consumption.

In the light of these many open questions it is not surprising that hardly any serious researcher would be able to state which of the particular actions of nicotine *per se*, of smoking or of both combined might be responsible for the formation and maintenance of the smoking habit. In the established smoker, non-nicotinic factors such as taste, smell and the manipulative habit certainly play important roles which should not be overlooked. It is more difficult to assess the importance of such factors for the formation of the habit in the beginner. It might be reasonable, however, to consider as candidates those effects of smoking which can be demonstrated reliably. The two most robust effects are represented by cognitive enhancement through nicotine and chemosensory stimulation through nicotine and other constituents of tobacco smoke. But how they work together and perhaps interact with other additional factors remains to be determined. Cognitive enhancement may arise, as indicated by electrocortical studies, through functional changes which facilitate the analysis, processing, evaluation and storing of information. However, the effect is weak and there is little evidence that smoking is used instrumentally to improve mental functioning. On the other hand, chemosensory stimulation through smoke is strong but predominantly unpleasant for a child at the pre-smoking state.

In order to gain a deeper insight into these questions it might be of help in future research not only to manipulate different constituents of smoke but also to consider more closely different populations of smokers. The results reported in this study from research with the nearly nicotine free but still smokable cigarettes 'Next' marketed for a short time by Philip Morris are cause for regret that they are not any more available. The same is also true for the counterpart of 'Next', the R.J. Reynolds cigarette 'Premier', a cigarette-like device which mainly delivers nicotine in small amounts without burning down. When, it comes to considering particular populations of smokers to be studied one would certainly be most interested in the young beginners in order to find out more about the reasons for the gradual development of the habit and the factors underlying this development. Among the adult and established smokers, the two most extreme minorities, the 'chippers' and the smokers who seek clinical help for smoking cessation, would attract the greatest interest. By far the great majority of smokers who have quit smoking in the last decades have done so as a consequence of a personal decision without outside help. The 'chippers', first described by Shiffman (1989) are smokers who for years regularly smoke only a few cigarettes per day, and it was estimated that they represent about 10% of the smoking population. They certainly stand out against the relatively few smokers who wear transdermal nicotine patches, chew nicotine gum and seek counseling. Shiffman *et al.* (1992) failed to detect any particularities of nicotine effects and kinetics in 'chippers' as opposed to the average smoker. The difference against the other extreme of the 'clinically dependent' smokers may therefore well be rooted to a considerable degree in the non-nicotinic aspects of smoking. Thereby not only the dynamics and kinetics of nicotine but also differences in personality, lifestyle and socio-environmental conditions remain candidates for a better understanding of the multi-faceted puzzle of the smoking habit.

SUMMARY

This review considers in 7 sections the relative importance of nicotinic and non-nicotinic factors involved in the maintenance of cigarette smoking behaviour:

- 'Nicotine titration' is commonly suspected as the reason for intensified smoking of light and ultra light cigarettes. However, the concept is questioned because the compensation is incomplete. It is also not complemented by smoking additional cigarettes in order to reach full compensation. Finally, puffing is not intensified with cigarettes with ultra-low nicotine but regular tar yields.
- Smoke taste is not only an attribute of the multiple substances of the tar and volatile fractions of smoke but most importantly of nicotine as well. Smokers like the stinging and burning taste of nicotine, but chemosensory substances of smoke other than nicotine also reduce the need to smoke. Even cigarettes with poor taste and nearly no nicotine are smoked under field conditions at similar frequencies and intervals as regular cigarettes.
- Activity and HR increase in parallel before a cigarette is lit and return to normal levels immediately upon lighting. This fluctuation is not subject to intraday tolerance and appears as well with nicotine-free cigarettes and imaginary smoking. After lighting, HR increases alone, but this effect is subject to intraday tolerance and does not appear with nicotine free-cigarettes or imaginary smoking. It is suggested that for the habituated smoker lighting up constitutes a reward *per se* which is triggered by learned and automated intraday programs rather than by the actual need for nicotine.
- Cognitive enhancement appears to be a specific effect of nicotine and there is growing evidence that it may not or only partially be subject to intraday tolerance. Whether it constitutes a primary reinforcer for smoking remains open.
- Paradoxical smoking can be assumed when no objective benefits can be demonstrated. Three examples are discussed, namely stress smoking, post-meal smoking and smoking with a cup of coffee, all of which do not seem to provide any objective advantage. It is assumed that the sensory manipulative enjoyment of smoking reduces distress when under stress and serves as a pleasure enhancer when drinking coffee, after a meal, or perhaps also at a multitude of other subjectively enjoyable occasions.
- The withdrawal syndrome includes mood changes which are opposite to those seen with the cessation of the chronic use of a stimulant and resembles more the effects of an overdose of caffeine. Whether this is due to the stop of the smoking induced acceleration of the metabolism of caffeine or to a nonspecific frustration when interrupting a learned habit remains open. Further, the withdrawal syndrome is moderate, transient and relapse occurs frequently also after its dissipation.
- The discussion questions the validity of the concept which sees smoking exclusively as a homeostatic nicotine regulation and withdrawal driven behaviour. Positive reinforcement through the pleasure of the repetition of an instrumental act appears to play an important role. It allows the explanation of non-pharmacological aspects of smoking which cannot be explained easily by the negative reinforcement concept of withdrawal alleviation.

ACKNOWLEDGMENTS

The author thanks Mrs. B. Strehler for her excellent editorial help in the preparation of the manuscript.

CHAPTER FOUR

Functional Utility of Nicotine: Arousal, Mood and Performance

Verner J. KNOTT, Anne HARR and Stacey LUSK-MIKKELSEN

INTRODUCTION

The psychobiological theories which have attempted to account for the wide-spread attraction of nicotine, particularly as inhaled from cigarette smoke, have for the most part been apportionable according to two fundamentally different and incompatible orientations, namely the addiction orientation and the functional orientation. The former theory holds that people use smoke-inhaled nicotine because it shares many features that cause compulsive use of prototypical drugs of abuse such as heroin and cocaine and as such fulfills the criteria of an addicting drug (U.S. Department of Health and Human Services, 1988; Robinson and Pritchard, 1992). The functional orientation on the other hand argues that people are not irrationally addicted but instead are rationally motivated to use nicotine as a 'tool' or 'resource' which offers psychological/behavioural benefits in the face of every day social/psychological needs and demands (e.g., Stepney, 1979; Warburton, 1988b; Warburton et al., 1988).

Popular accounts and self-reports by individuals who habitually self-administer nicotine throughout the day by repeated inhalation of cigarette smoke indicate that arousal control, mood regulation, enhanced concentration and stress reduction are frequently sought-after effects which motivate nicotine use (e.g., Gilbert, 1979; Wesnes and Warburton, 1983b). Objective identification of the purported psychobiological effects of smoke-inhaled nicotine have often proved, at least in the controlled environment of the research laboratory, to be subtle and elusive. Whereas empirical findings on the effects of nicotine on mood and stress regulation have been found to be highly variable (e.g., Gilbert, 1994; Gilbert and Gilbert, 1995, Ch. 5), nicotine-performance studies have reported consistent improvements in psychomotor vigilance-related task behaviours (e.g., response speed and accuracy) in cigarette-deprived smokers, non-deprived smokers and non-smokers (Warwick and Eysenck, 1963; Provost and Woodward, 1991).

The functional model, readily acknowledging the heterogeneous motives (see for example Bättig, Ch. 3) underlying nicotine use and conceding that nicotine effects may vary according to individual-situational interactions, postulates, as its basic tenet, that affective, cognitive and behavioural facilitation induced by nicotine are mediated via regulation of neural mechanisms modulating central arousal processes (e.g., Warburton, 1988b; Wesnes, 1987). Although electrophysiological studies have indeed marshalled abundant support for arousal shifts following nicotine administration (Clarke, 1990; Edwards and Warburton, 1983), there have been relatively few attempts to document nicotine-related arousal shifts and mood/performance changes together in the same study.

This chapter briefly reviews the results of three unpublished exploratory studies with particular emphasis on quantitative electroencephalography (qEEG) indices of arousal shifts, self-reported mood ratings, as well as reaction time (RT) and accuracy indices in psychomotor and cognitive (memory) tasks.

In each of these studies subjects were habitual cigarette smokers, between 20 and 40 years of age, with a smoking history of at least five years and smoking (with inhalation) 15 or more cigarettes/day. All were negative for psychiatric/neurological and alcohol-drug abuse histories. No subjects took CNS medications during the study period. Studies included an orientation session and separate test sessions (randomly ordered, counterbalanced and separated by 3–7 days) which were conducted in the mornings following 12 hours of smoking, alcohol, drug and caffeine abstinence. Monopolar, linked earlobe reference, EEG recordings were extracted from 3 or 16 scalp sites (10–20 system) and together with separate vertical and horizontal electro-oculographic (EOG) recordings, were amplified with bandpass filters of 0.1–40 Hz. For each analysis, a minimum of 30 two second artifact-free EEG epochs (sampled at 128 Hz) were processed through a high-pass auto regressive filter, weighted by a 5% cosine taper and a conventional fast Fourier transform algorithm which computed relative (%) power for six EEG frequency bands (delta, theta, alpha$_1$, alpha$_2$, beta$_1$, and beta$_2$). The band frequency ranges varied across studies, depending on software capabilities.

In the two studies with recordings from 16 scalp sites topographical power values were colour coded and mapped (using linear four point interpolation) with Nicolet Pathfinder II-T mapping software (version 7.1). Statistical analysis of EEG was carried out on log transformed data (Gasser *et al.*, 1982). EEG, mood, cognitive, and behavioural measures were subjected to repeated measures analysis of variance (ANOVA) procedures with significant effects being followed up by Neuman-Keuls multiple comparison tests.

STUDY I

Nicotine, EEG, and Mood: Comparative Effects with Caffeine and Alcohol

Rationale

Smoke-inhaled nicotine has been repeatedly shown to alter spontaneous electrical rhythms recorded from the intact scalp of cigarette smokers (Church, 1989; Conrin, 1980; Knott *et al.*, 1995). Quantitative spectral analysis of eyes-closed resting EEG has typically produced qEEG profile alterations characterized by reductions in slow-wave (delta and theta) activity as well as increases in fast (alpha and beta) activity (e.g., Knott, 1988, 1989a; Pritchard *et al.*, 1992a).

Although this neuroelectric response pattern parallels EEG activity changes observed with psychostimulants (Saletu, 1987), it is somewhat at odds with the frequently reported paradox that smokers report using nicotine for both its stimulant ('pick-me-up') and sedative ('calm-me-down') properties (Russel *et al.*, 1974). Elucidation of nicotine's putative dual actions may be best achieved by empirical studies which examine relationships between nicotine dose and smoker-situation interactions or, alternatively, by studies which compare nicotine with prototypical CNS stimulants and depressants. The present exploratory study adopted the latter approach and compared the qEEG and subjective mood-related effects of smoke-inhaled nicotine with a commonly used stimulant (caffeine) and a commonly used depressant (alcohol). Multiple site recordings were carried out to examine the topographical uniformity of electrical changes across the scalp.

Method

Sixteen, male smokers attended the lab for four separate, randomly ordered (incomplete counterbalanced) test sessions. During each session subjects consumed, over a 10 minute period, a 350 ml desugared and decaffeinated fruit beverage together with a single capsule. Sixty minutes later they inhaled, while blindfolded, a cigarette at a rate of one puff/min. In one session (placebo) mood assessments and five minutes of eyes-closed EEG recordings were taken before and after consuming a placebo beverage, placebo (dextrose) capsule and inhaling a placebo cigarette (0.06 mg nicotine, 4 mg tar; Robinson *et al.*, 1992). In a second session (nicotine) the placebo cigarette was replaced by a cigarette from each subjects' own brand (mean nicotine = 1.2 mg ± 0.1 mg; mean tar = 12.6 mg ± 2.0 mg). In a third session (caffeine) the placebo capsule was filled with 300 mg of caffeine (equivalent to 3–4 cups of coffee). In a fourth session (alcohol) the beverage included 2.2 ml/kg dose of 80 proof vodka which was previously shown to increase blood alcohol levels (BAL) up to 0.07–0.08% and to induce depressant-type EEG activity 45–120 minutes following ingestion (Schwarz *et al.*, 1981). Ten ml of vodka was floated on the surface of the beverages in the three non-alcohol sessions to provide an initial strong taste without any significant BAL increase.

Immediately before and after each EEG recording, two questionnaires were administered to evaluate mood changes. The first questionnaire assessed euphoria via 17 true-false items comparing the Morphine-Benzedrine Group (MBG) Scale of the Addiction Research Centre Inventory (Haertzen and Hickey, 1987). MBG effects have been shown to be marked with opiates, marijuana, and stimulants, only moderately evidenced with alcohol, barbiturates and hallucinogens, and are relatively absent with tranquilizers. The second questionnaire consisted of 17 bipolar analogue scales of various mood dimensions (e.g., attentive-dreamy; happy-sad; tense-relaxed) which, when analyzed result in three mood factors (contentedness, alertness, calmness) which have been shown to be sensitive to acute administration of psychotropic compounds (Bond and Lader, 1974). Subjects were instructed to rate their position on these scales according to how they felt during the respective eyes-closed EEG rest periods.

Statistical analysis included separate 4 (session) × 16 (site) repeated measure ANOVAs carried out on log transformed relative (%) power difference (post-smoking minus baseline activity) scores derived for delta (0.1–3.9 Hz), theta (4.0–7.9 Hz), $alpha_1$ (8.0–10.9 Hz), $alpha_2$ (11.0–14.9 Hz), $beta_1$ (15.0–17.9 Hz), and $beta_2$ (18.0–20.9 Hz) frequency bands. Similarly, separate one-way repeated measures ANOVAs were carried out on difference scores (post-smoking minus baseline rating) for MBG and each of the mood factors.

Results

Figure 4.1 depicts the topographical EEG changes induced in each of the four test sessions. Relative to placebo, none of the three substances significantly altered delta power. Follow-up analysis (Neuman-Keuls) of a significant session effect [$F(3/36) = 3.25$ $p \leq 0.033$] did show however that caffeine did result in delta power beyond that seen with alcohol ($p \leq 0.05$) and nicotine ($p \leq 0.05$). Additional follow-up analysis (Neuman-Keuls) of a significant session $\times$ site interaction [$F(45/540) = 2.38$, $p \leq 0.001$] showed that relative to placebo, caffeine and alcohol, nicotine decreased theta power at anterior (Fpz, FP1, FP2, Fz, F3, F4) recording sites ($p \leq 0.05$). Despite apparent power elevations shown in the topographic maps, $alpha_1$ was not affected by any of the active substances. Significant session $\times$ site interaction effects were observed for $alpha_2$ [$F(45/540) = 1.99$, $p \leq 0.001$], $beta_1$ [$F(45/540) = 1.43$, $p \leq 0.05$], and $beta_2$ [$F(45/540) = 1.76$, $p \leq 0.002$] power changes. Follow-up analysis (Neuman-Keuls, $p \leq 0.05$) showed only nicotine to have an effect on $alpha_2$ with increases being restricted to frontal-central-temporal sites (Fpz, FP1, FP2, F3, F4, Cz, T3, T4). For $beta_2$, only nicotine (at FZ) and alcohol (at Fpz, FP1, FP2, Fz, F3) increased power relative to placebo ($p \leq 0.05$). $Beta_2$ power alterations were restricted to nicotine with increases being observed primarily at frontal and midline recording sites (Fpz, Fz, F3, F4, Cz, Pz; $p \leq 0.05$).

Figure 4.2 displays the MBG and mood-related changes for each session. A significant effect was observed with MBG ratings of euphoria ([$F(5/78) = 2.41$, $p \leq 0.05$] and follow-up analysis indicated significantly higher MBG scores with alcohol compared to placebo or caffeine ($p \leq 0.05$). No significant changes were observed with the additional mood states.

Discussion

Smoke inhaled nicotine produced significant power alterations which paralleled previous reports describing fast wave power increments (Knott *et al.*, 1995). For the most part these electrical changes were limited to *anterior* recording regions, a topographic response profile which is similar to that observed with the smoke-inhalation of high nicotine-yield cigarettes (Knott, 1989b). Although alcohol elevated beta activity, a response pattern similar to that seen with some but not all studies utilizing similar or smaller 'social' doses of alcohol (Knott, 1990) and with minor tranquilizers (Saletu, 1987), caffeine failed to exert any of the previously reported subtle EEG changes observed with 100–200 mg dosages (e.g., Schwarz *et al.*, 1981). However as with Clubley *et al.*, (1979) who observed increases in delta with 100 mg of caffeine, this study observed significant increases in delta beyond that seen with alcohol.

Given that dosings of alcohol and caffeine were not self-regulated as would be the case during normal social occasions, the more robust neuroelectric

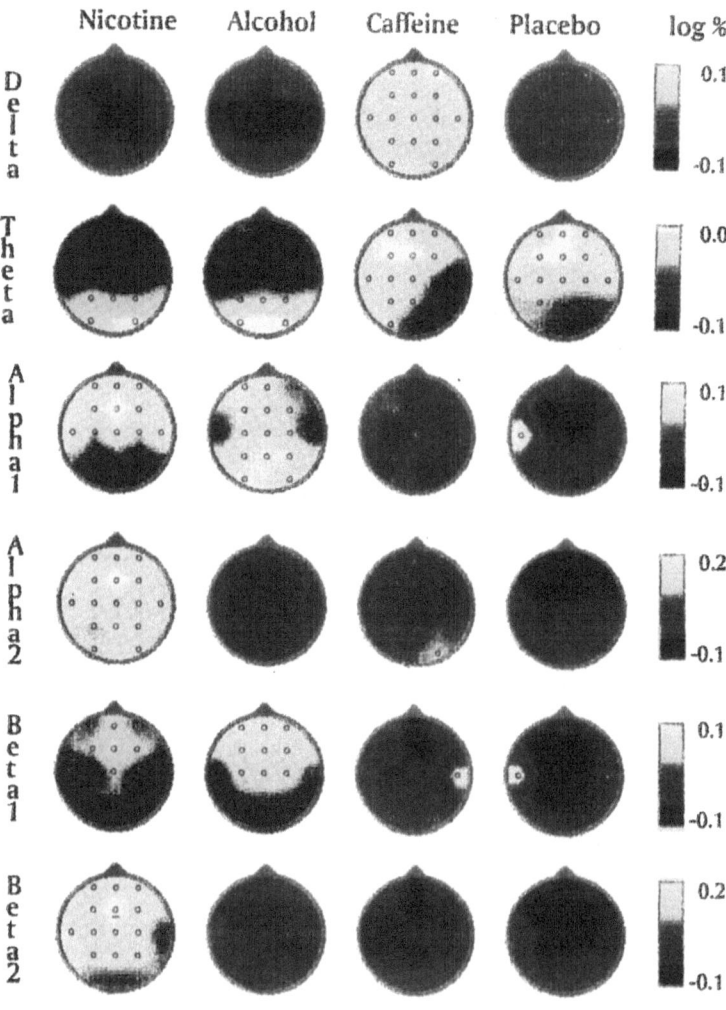

V.J. Knott

Figure 4.1 – Grand averaged topographic log relative (%) EEG power difference maps for delta, theta, alpha₁, alpha₂, beta₁ and beta₂ frequency bands following separate administration of placebo, caffeine, alcohol, and nicotine. For each condition, maps were calculated by subtracting the averaged pre-substance EEG values from the averaged post-substances. Decreases in relative power are evidenced by darker colours and increases in power by lighter colours.

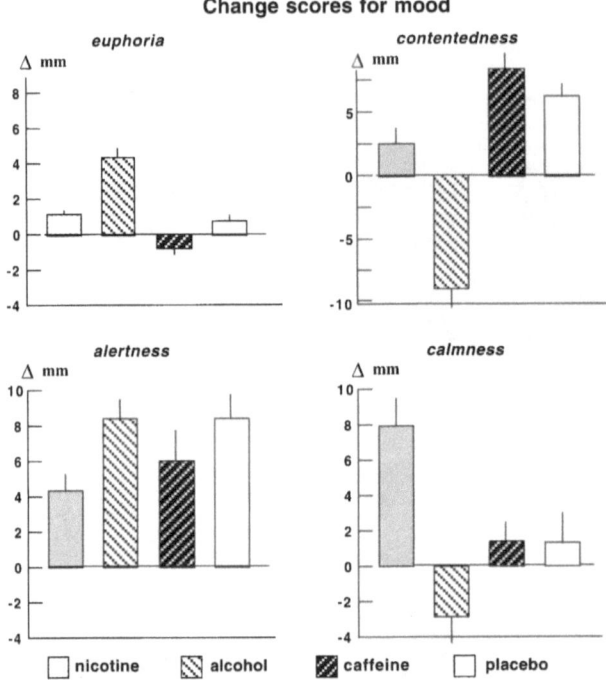

Figure 4.2 – Mean (and standard errors) euphoria and mood dimension changes (Δ) in each of the four substance conditions. Change scores (post-substance minus pre-substance ratings) are shown for nicotine, alcohol, caffeine and placebo.

alterations observed with nicotine may simply reflect a more 'optimal' dosing achieved via 'finger-tip' adjustment (e.g., puff volume variations) of inhaled nicotine for desired effects. Given that the laboratory environment and resting eyes-closed recording conditions may have induced a state of relaxed drowsiness, central stimulant effects may well have been the immediate desired goal of nicotine-inhalation in this present experimental condition. However this result was not evidenced in the MBG euphoria ratings or with the additional mood-related factors.

Although alcohol produced a typical euphoria response, the degree to which nicotine produces an 'addictive' euphoric effect, or for that matter, any noticeable mood-related changes, is being hotly debated and as yet there is no general consensus (e.g., Robinson and Pritchard, 1992) as to the degree or direction of laboratory assessed nicotine induced mood alterations.

Given that habituated nicotine users inhale most of their nicotine during the working day it would seem reasonable to conclude that any sought-after functional effects, be they psychological, behavioural or mood-related, would be more appropriately characterized in laboratory conditions which require work/performance assessments (Wesnes and Warburton, 1983b).

STUDY II

EEG, Mood and Psychomotor Alterations Following Morning-long Repeated Nicotine Use

Rationale

Although the central arousing effect of nicotine may function as a critical rein-forcing event motivating the repeated use of inhaled nicotine in a variety of social situations throughout the day, relatively few studies have attempted to track the central effects on successive occasions of self-administered nicotine. Acute toler-ance effects, appearing as increasingly diminished responses to repeated adminis-tration of a fixed dose of a substance over several hours or so (Kalant *et al.*, 1971) have been observed with some measures (e.g., heart rate, tremor, subjec-tive ratings; Benowitz *et al.*, 1982; West and Russell, 1986) but not with others (e.g., finger tapping; West and Jarvis, 1986). As vigilance performance incre-ments after repeated administration of nicotine (1×2 mg) tablets at 20 minutes intervals (Wesnes *et al.*, 1983) and arousal ratings (Sherwood *et al.*, 1991b) and psychomotor responsivity (Sherwood *et al.*, 1992) following repeated nicotine gum (2 mg) administration were not subject to tolerance effects, the present study examined the issue of acute tolerance with repeated smoke-inhaled doses of nico-tine and their effects on EEG, mood and psychomotor responsivity.

The study also examined the effects of inhaled nicotine in cigarette smokers in comparison to non-nicotine users (i.e. non-smokers) in order to determine whether nicotine enhances (i.e., induces states in users which are different from non-users) or normalizes (i.e., induces states in users which are similar to non-users) biobehavioural parameters.

Method

Twenty male cigarette smokers and ten male non-smokers were examined at five successive 30 minute intervals between 9.00–11.00 a.m. All of the smokers came to the laboratory following overnight smoking abstinence and ten ran-domly selected smokers were required to smoke (i.e., smoking smokers, SS) prior to each of the five EEG/mood/performance assessment periods (test blocks B_{1-5}) while the remaining smokers, designated as abstinent smokers (AS) and the ten non-smokers (NS) did not smoke throughout the testing period. SS sub-jects smoked, at a rate of one puff/min, one of their own preferred brand ciga-rettes immediately prior to the assessments at each test block.

EEG was recorded from the mid-frontal (Fz), mid-central (Cz), and mid-parietal (Pz) sites for a three minute period at each of the five time points and was sub-jected to spectral analysis (0.25 Hz resolution) for calculation of peak alpha fre-quency (PAF) and relative (%) power in delta (0.25–3.75 Hz), theta (4–7.25 Hz), alpha (8–12.75 Hz) and beta (13–20.25 Hz) bands. Mood was assessed by the 16 self-report analogue rating scales of Bond and Lader (1974) which allowed extrac-tion of three main mood factors: alertness (9 items), contentedness (5 items) and calmness (2 items).

Psychomotor responsivity was assessed by measuring choice reaction time in a modified Eriksen and Schultz (1979) paradigm in which subjects responded (by pressing, with left and right index fingers, a left or right RT button; direction counterbalanced) to the presentation (1 per 3 sec) of visual target letters (*A* or *H*) of 1000 ms duration (presented on computer screen positioned one m from subject, at eye level; letters subtending 0.25° in both width and height) flanked either by no letters (control), 4 stimulus-compatible letters, for example, *AAAAA* or stimulus-incompatible letters, such as *HHAHH*. Target letters were presented above a warning stimulus 'X' and each session consisted of 15 practice trials followed by 36 randomized trials, 12 per letter condition, at each of the five test blocks.

Separate 5 (test block) × 3 (recording site) × 3 (group) split-plot repeated measures ANOVAs were used to examine main and interactional effects. Neuman-Keuls tests were employed for follow-up analysis.

Results

Figure 4.3 shows the EEG variables which were affected by smoking, and as smoking effects were similar across the three electrode sites the graphs reflect the mean of the three recording regions.

No significant *group* or *group × block* effects were observed for delta or beta power but significant group differences were found with theta [$F(2/25) = 5.34$, $p \leq 0.02$], alpha [$F(2/25) = 3.53$, $p \leq 0.05$] and PAF [$F(2/25) = 4.31$, $p \leq 0.05$]. Follow-up analysis indicated that AS exhibited significantly more theta, less

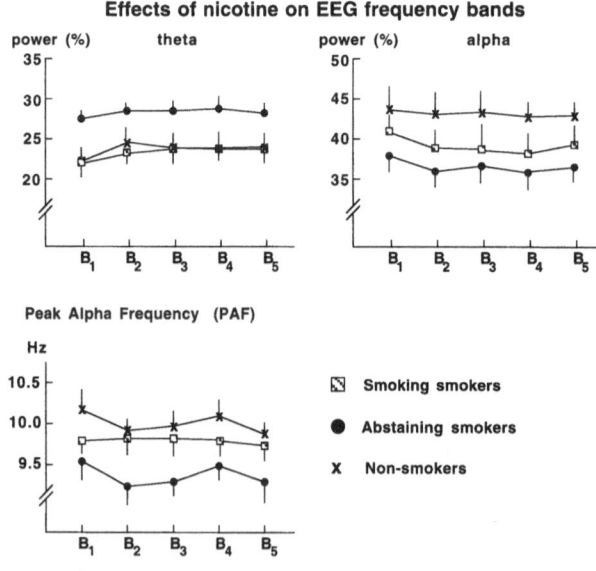

Figure 4.3 – Mean (and standard errors) peak alpha frequency (PAF) and power values of theta and alpha for Smoking smokers, Abstinent smokers, and Non-smokers at each of the five test blocks (B_1–B_5).

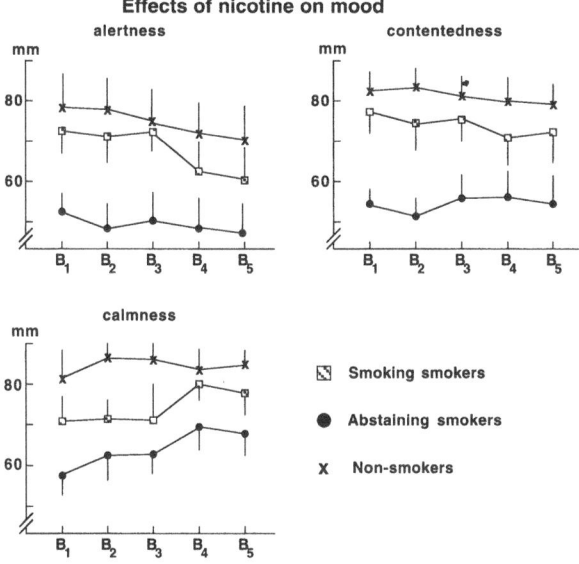

Figure 4.4 – Mean (and standard errors) subjective ratings for alertness, contentedness and calmness mood dimensions in Smoking smokers, Abstinent smokers, and Non-smokers at each of the five test blocks (B_1–B_5).

alpha and a slower PAF (all $p \leq 0.05$) than both NS and SS groups. No significant differences were observed between NS and SS groups on any of these three electrical measures. Figure 4.4 depicts the mood ratings for each of the three separate groups.

Significant main effects of group were observed for all three mood factors including alertness [$F(2/26) = 4.66$, $p \leq 0.02$], contentedness [$F(2/26) = 7.72$, $p \leq 0.003$) and calmness [$F(2/26) = 7.74$, $p \leq 0.003$]. For each of these factors NS were found to be similar to SS but AS exhibited lower scores than both NS and SS (both $p \leq 0.05$) on all three mood scales. Figure 4.5 shows the RT data for the 3 study groups. Subjects made few if any response errors in the three RT conditions. As with EEG and mood, significant group effects were observed for 'control' [$F(2/26) = 4.37$, $p \leq 0.05$], 'compatible' [$F(2/26) = 3.89$, $p \leq 0.05$] and 'incompatible' letter trials [$F(2/26) = 3.57$, $p \leq 0.05$].

Follow-up analysis showed the AS group to be markedly slower than NS and SS groups (both $p \leq 0.05$) in each of the three trial conditions. No significant differences were observed between NS and SS groups.

Discussion

Relative to both abstinent smokers and non-smokers repeated nicotine inhalation in smokers resulted in arousal increases as shown by the reduction in slow wave (theta) power and the augmentation of fast wave (alpha) power. Nicotine inhalation exerted similar relative effects with respect to mood and psychomotor

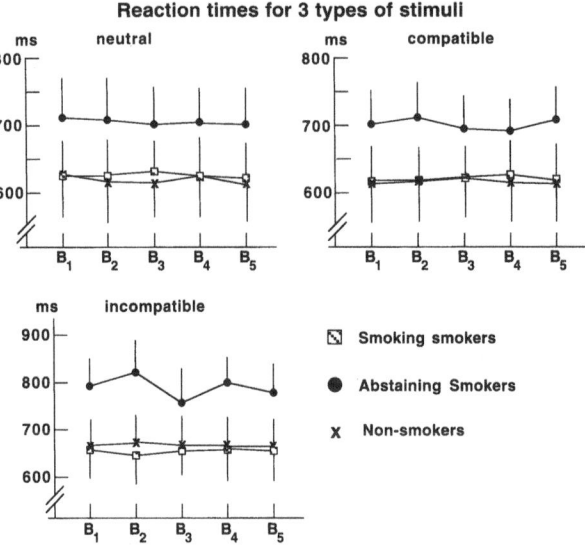

Figure 4.5 – Mean (and standard errors) reaction times for neutral, compatible and conflicting stimulus conditions in Smoking smokers, Abstinent smokers and Non-smokers at each of the five test blocks (B_1–B_5).

speed. The central arousing, mood and behavioural effects of inhaled nicotine were evident across all test blocks and as such they suggest that subjective affect ratings, physiological arousal indices and response speed are not subject to acute tolerance effects.

None of the response measures exhibited significant time-related effects and as such it is reasonable to argue that the EEG/mood/behavioural changes were due either to the first nicotine dosing, the effect of which was sustained on subsequent doses or to an alteration at each nicotine dosing. As nicotine-related changes in nicotine users resulted in arousal increments and mood shifts at levels comparable to those observed in non-smokers, the findings suggest that repeated nicotine inhalation by habitual nicotine users does not appear to enhance brain state arousal, positive affect or task performance. However it does place them on an equivalent functional level of non-users.

The reduced levels of arousal and positive affect observed in nicotine abstinent subjects relative to nicotine non-users would suggest that smoking restrictions in social and/or work environment may have significant impact on individuals who are habitually reliant on nicotine for interpersonal and/or performance related functions. The RT data is of particular interest in that the nicotine effects were evident in both the relatively simple and the more perceptually/cognitively demanding versions (i.e., the incompatible condition where RT was slower) of the tasks and as such suggest that nicotine may be a useful performance enhancer in both monotonous and stressful work environments.

STUDY III

EEG-Arousal and Nicotine-induced Memory Alterations

Rationale

Vigilance and sensorimotor functions are the most frequently examined parameters which have been shown to be positively influenced by nicotine (Wesnes and Warburton, 1983b), but additional evidence suggests that nicotine administration may also affect higher-order cognitive functions including abstract thinking and memory-related processes. Although heightened arousal has been generally found to enhance long-term retention (e.g., Eysenck, 1983), clinical and commonly used psychotropic substances influencing brain arousal states have been shown to be variable in their enhancement and distraction of memory processes (e.g., Wittenborn, 1988). Nicotine, either via smoke delivery or oral administration has been shown to have varied effects on memory, both negative and positive, but a number of reports have indicated that nicotine administered before a learning task conveys a long-term positive effect on delayed recall of verbal material (Mangan, 1983; Peeke and Peeke, 1984; Warburton *et al.*, 1986, 1992a,b; Rusted and Eaton-Williams, 1991) and at least two studies have shown memory improvements with post-learning smoking (Mangan and Golding, 1983; Colrain *et al.*, 1992).

Nicotine-induced increments in peripheral arousal indices (e.g., heart rate) have been shown to be related to improved retention and recall (Andersson and Post, 1974; Andersson, 1975) and a number of non-nicotine (Easterbrook, 1959; Hasher and Zacks, 1979) and nicotine (Rusted and Eaton-Williams, 1991; Warburton *et al.*, 1992a; Warburton *et al.*, 1992b; Rusted and Warburton, 1992) studies have suggested that arousal increments may alter memorial processes by optimizing attentional selectivity/focusing, resulting in more efficient encoding of test material.

As memory processes, both storage and retrieval aspects, are often carried out under less than optimal environmental conditions, this present study examined the effects of smoke-inhaled nicotine on EEG arousal and recall of verbal material presented under auditory 'noise' and 'non-noise' conditions (Peeke and Peeke, 1984). Previous studies have reported positive and negative effects of noise manipulations on cognition and performance and the numerous conflicting results have been interpreted in relation to several theories including theories which invoke arousal and attentional mechanisms (Jennings, 1986a,b).

Method

Twenty-four (12 females) cigarette smokers participated in two (sham vs. cigarette smoking) randomized morning test sessions following overnight smoking abstinence. During each session EEG (16 sites referred to linked earlobes) was recorded for three minutes before and after sham smoking (i.e. inhaling on a non-lighted cigarette) or real smoking of the subject's preferred cigarette with an inhalation rate of one puff/min.

The EEG was followed by two acquisition/learning trials of 60 item word lists (different lists for each session created from the 1959 Bättig and Montague norms) each ending with a two minutes written free recall. Words were presented through headphones every two seconds and half of the words (randomly selected) were embedded in a one second burst of binaural white noise (90 dB SPL). Each noise-free and noise-embedded word were clearly identified by all subjects. A subsequent 45 minutes interpolated task was performed to avoid list rehearsal. A second preferred cigarette (in the active smoking session) was smoked to maintain nicotine levels during middle and at the end of this interval. A four minute free recall test of the previously learned list followed the 45 minute task. Separate repeated measures ANOVAs (nicotine, sex, recording site) were carried out on log transformed relative difference scores (post-minus pre-smoking) for delta (1–4 Hz), theta (4–8 Hz), alpha$_1$ (8–10 Hz), alpha$_2$ (11–13 Hz) and beta (13–29 Hz). Repeated measures ANOVAs (nicotine, sex, noise) were also performed on the absolute number of total items remembered at each of the first two recall tests as well as on the number of items correctly remembered on the third (i.e. delayed) free recall which were expressed as a percentage of the items remembered (learned) on recall trial two.

Results

As shown in Figure 4.6, significant nicotine-EEG effects were observed for theta [$F(1/21) = 9.6$, $p \leq 0.006$], alpha$_1$ [$F(1/21) = 6.39$, $p \leq 0.02$], alpha$_2$ [$F(1/21) = 15.2$, $p \leq 0.0008$], and beta bands [$F(1/21) = 10.2$, $p \leq 0.005$] with smoke-inhaled nicotine reducing relative power in theta and alpha$_1$ while increasing power in the two higher frequency bands.

For alpha$_1$ nicotine interacted with recording regions [$F(1/21) = 3.58$, $p \leq 0.0001$] and follow-up analysis indicated nicotine-induced alpha$_1$ power reductions to be limited to central-posterior sites (Cz, C3, C4, Pz, P3, P4, O1, O2). No differences were observed between males and females in response to nicotine.

Analysis of data from the first recall trial failed to show any nicotine or gender effects on total items recalled but noise-embedded items were recalled less frequently than no-noise items [$F(1/21) = 9.06$, $p \leq 0.007$]. No significant nicotine, gender or noise effects were observed with recall on trial two, but a significant nicotine effect was found on trial three (Figure 4.7) with smoke-inhaled nicotine increasing the percentage recalled [$F(1/21) = 4.33$, $p \leq 0.05$].

Nicotine also interacted with noise and gender [$F(1/21) = 5.53$, $p \leq 0.05$] and follow-up analysis indicated that females exhibited superior recall of no-noise items ($p \leq 0.009$) during non-nicotine sessions. During nicotine sessions percentage recall of no-noise items was reduced in females ($p \leq 0.05$) while the percentage of no-noise items recalled by males increased ($p \leq 0.05$) as did the noise-related items in females ($p \leq 0.05$).

Discussion

Spectral analysis of background electrical rhythms showed that smoke-inhaled nicotine produced a pharmaco-EEG profile change similar to that seen in the first two experiments with power being shifted from slow to fast frequency bands in a manner similar to that observed with other psychostimulants such as amphetamine (Saletu, 1987). The present results, in addition to increasing our knowledge on the reliability of nicotine-electrocortical arousal relationships, also provide a

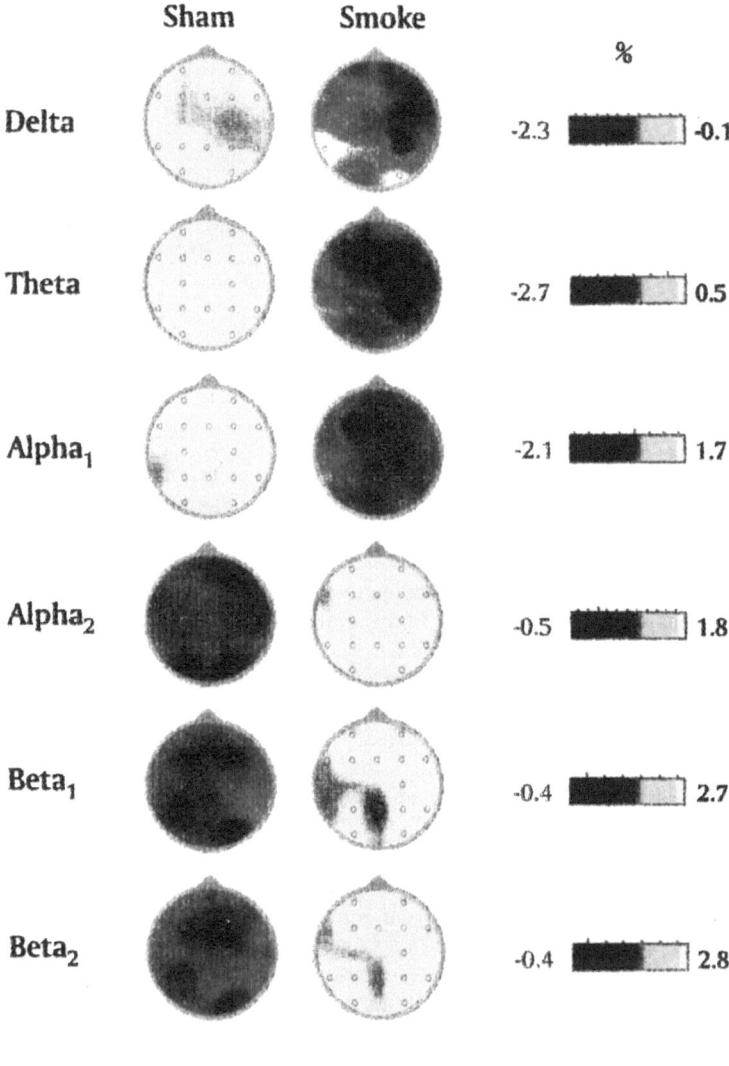

V.J. Knott

Figure 4.6 – Grand averaged topographic relative (%) EEG power difference maps (post-minus pre-values for sham and real smoking) for delta, theta, alpha₁, alpha₂, beta₁ and beta₂ frequency bands. Decreases in relative power are evidenced by darker colours and increases in power by lighter colours.

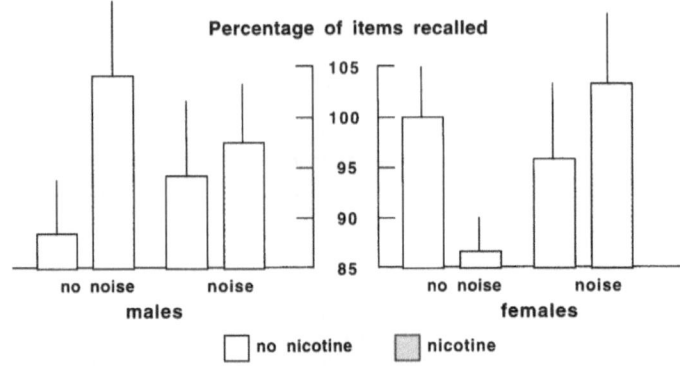

Figure 4.7 – Mean (and standard errors) trial three (delayed) recall values of no-noise and noise embedded items observed in males and females in non-nicotine (sham smoking) and nicotine (cigarette smoking) conditions. Values are expressed as a percentage of the items recalled (i.e., correctly learned) in trial two.

data base for generating hypotheses concerning the identification of arousal conditions which may modulate the effect of nicotine on memory processes.

A number of individuals (e.g., Kleinsmith and Kaplan, 1963) have related memory to arousal and have proposed that high arousal would result in more protected and intense memory traces leading to poorer access on immediate recall but better ultimate long-term memory. Noise-embedded items were recalled less frequently than no-noise items and this detrimental effect may be related to altered arousal levels which impact negatively on cognitive strategies (e.g., encoding) or attentional selectivity (Jennings, 1986b). Nicotine-induced increases in cortical arousal were not paralleled by enhanced performance on immediate recall tests and results on delayed recall suggest that nicotine-related arousal increments, despite being equivalent in males and females, may result in more performance-related advantages for certain individuals and situations, particularly males remembering no-noise items and females remembering noise-related items.

Although neither intellectual nor educational status was assessed in the two sexes, it is reasonable to suggest that the selective nicotine-induced improvements in males and females may be related to pre-nicotine arousal levels. Post-hoc analysis of pre-treatment baseline EEG did indicate that males, relative to females, exhibited reduced arousal in the form of greater amounts of slow alpha (alpha$_1$) in the posterior regions ($p \leq 0.05$) and less fast alpha (alpha$_2$) in frontal regions ($p \leq 0.05$). The idea that nicotine may be particularly performance enhancing in extreme arousal conditions receives additional support from the observation that nicotine-induced memory improvements were observed only in low-arousal subjects (i.e., males) recalling non-arousing (i.e., no noise items) and in high arousal subjects (i.e., females) recalling arousing (i.e., noise-embedded) items.

CONCLUSION

Habitual nicotine users frequently report that smoke-inhaled nicotine reduces feelings of anxiety and anger, provides hedonic pleasure and improves concentration and attention (e.g., Russell *et al.*, 1974). The neurobehavioural sciences have strived to objectify these subjective reports which motivate a significant proportion of the population to use nicotine in their daily working and leisure activities. The empirical approach of the functional modal has, for the most part, focused on nicotine's actions on the target organ – the human brain, and in so doing has argued that nicotine's attraction stems from its ability to modulate arousal systems underlying emotional and cognitive-behavioural processes (Warburton, 1988b).

Although it is not necessary to reiterate the many limitations of the arousal concept each of the present study results lend support to the arousal contention in that smoke-inhaled nicotine was repeatedly shown to shift power from slow to high frequency power bands, a neuroelectric profile pattern change frequently observed both with increased behavioural/emotional arousal (Lindsley, 1960) and with psychostimulant-induced arousal (Saletu, 1987). It should be cautioned of course that arousal-performance associations in themselves to not establish a cause-effect relationship between nicotine-induced arousal alterations and performance changes. Such cause-effect relationships may be examined via a number of design/analytical routes including the comparison of nicotine users displaying marked nicotine-induced activation with users who display no central responsivity to nicotine. Although the functional usefulness of nicotine-induced brain activation will be undoubtedly determined by the individual user and the immediate specific situational demands, the present results indicate that electrical activational increases can be associated with changes in mood, psychomotor function and memory.

As independent studies have observed similar improvements in minimally nicotine-deprived users and in non-smokers administered nicotine (Provost and Woodward, 1991) one may reasonably argue that these nicotine-related alterations reflect absolute (primary) effects and not secondary effects associated with relief of prolonged nicotine abstinence/deprivation. Given that nicotine dosing from smoke-inhaled nicotine is under the control of the cigarette user, that is to say via manipulation of puff volume, frequency, etc. and given that the effects appear repeatedly with each occurrence of nicotine use, it is perhaps understandable that the habitual nicotine user is faced with somewhat compelling and convincing evidence in deciding whether to utilize nicotine's assistance to cope with routine daily exigencies.

SUMMARY

Although many features of habitual nicotine use are believed to resemble those seen in people dependent on frequently abused substances heroin, alcohol and cocaine, some psychobiological properties of nicotine stand apart as potentially therapeutic in many different people, for many different reasons, and in many different situations. Functional models of nicotine use have argued that mood regulation and facilitation of cognitive-behavioural processes are critical factors in maintaining daily nicotine consumption. The putative beneficial effects of nicotine are frequently explained by alterations in central arousal processes. Although a plethora of infrahuman and human research has evidenced nicotine-induced neuroelectrophysiological response patterns congruent with this notion, there is a relative dearth of research attempting to assess arousal and mood/cognitive-behavioural changes together pre- and post- nicotine manipulation. Three studies which:

– compare the electrocortical and mood-related effects of nicotine, caffeine and alcohol;
– examine the effects of nicotine on electrocortical, mood and psychomotor activity and
– investigate nicotine effects on electrocortical activity and memory, are described which appear to support the contention that arousal increments may underlie the putative beneficial effects of nicotine.

CHAPTER FIVE

Nicotine and the Situation by Trait Adaptive Response (STAR) Model: Emotional States and Information Processing

David G. GILBERT and Brenda O. GILBERT

INTRODUCTION

Situational and individual-specific effects are the norm for nicotine (Gilbert, 1995), caffeine (Silverman et al., 1994), alcohol (Stewart and Pihl, 1993), and numerous other drugs (Janke, 1983). The STAR (Situation by Trait Adaptive Response) model of smoking and nicotine's effects (Gilbert, 1994, 1995) was recently proposed to make reactions to nicotine more predictable across persons and situations. The present chapter reviews and further articulates aspects of the STAR model that are directly related to nicotine's effects on emotion and information processing.

While it has frequently been observed that cognitive and emotional effects of many psychoactive drugs are a function of situational (Schachter and Singer, 1962), personality, and other individual different factors (Eysenck, 1980; Janke, 1983), until the STAR model was proposed, no well-articulated comprehensive model of how situational and trait factors interact to produce individual differences in drug response was available. The lack of such a model is related to the fact that historically, models and literature reviews have considered stimulus and environmental factors independently of individual differences in biological and psychological traits. Not surprisingly, such fractionated reviews resulted in limited and competing models with limited generalizability and predictive power. Some (e.g., Peele, 1985) have addressed the futility, inadequacy, and inaccuracy of comprehensive models that focus only on drug parameters to the exclusion of person by environment interactions. The STAR model was developed to provide a biopsychosocial framework from which to understand such interactions.

Changes in affect (emotion, feelings, mood) are the primary reason that smokers give for smoking. For example, Spielberger (1986) found that occasional smokers ranked tranquilizing effects (i.e., "Relaxes me when I'm upset/nervous") as the most important reasons for continuing to smoke among women and second most important among men. Since occasional smokers do not experience abstinence response symptoms (Shiffman, 1989, 1991), Spielberger's findings are contrary to the view that nicotine reduces negative affect only by reducing withdrawal. Among current smokers, tranquilization was rated by both males and females as the second most important reason for continuing to smoke, slightly behind "Because I enjoy it."

The subjective effects of smoking, alcohol, and caffeine/coffee were compared to each other and other substances and activities in a study by Warburton (1988a). Alcohol, amphetamine, amyl nitrite, cocaine, glue, heroin, marijuana and sex produced significantly more pleasurable stimulation than smoking, while sleeping tablets and tranquilizers were less stimulating. Coffee, chocolate, and smoking were seen as equally pleasurably stimulating. Relaxing effects of smoking were similar to those of chocolate, and significantly less than those of alcohol, heroin, tranquilizers, sleeping tablets, and sex.

Thus, the literature suggests that smokers see smoking as having modest, yet important tranquilizing and concentration-enhancing effects (Ikard *et al.*, 1969; Russell *et al.*, 1974). They report smoking more when in negative affective states and they believe that smoking helps them feel better when in such states (Spielberger, 1986). In some conditions some smokers believe that smoking increases positive affective states, but these situations have been less well established (Tomkins, 1966).

Some individuals report severe affective distress when abstaining from smoking for a few hours or days, while other equally heavy smokers experience minimal or no distress. The range of variability in response to nicotine abstinence is apparent in a sizable study by Hughes *et al.* (1991b), who found experimental abstinence-related incidences of self-reported increases in negative affect to be far from universal. Percentages of individuals experiencing anxiety increases was highest (59%), with increases in angry/irritable (52%), impatient (52%), restless (55%), and drowsy (40%) close behind. The percentages experiencing no negative affect of any kind was not reported and has received little attention in the literature where until recently the focus has been on group means, rather than individual differences. The nature of the biological and psychological mechanisms underlying these individual differences in affect-modulating and information processing effects of smoking and nicotine is reviewed within the context of the STAR model, the exposition of which we turn to now.

THE STAR MODEL

Situational specificity of affective responses to nicotine and tobacco abstinence

Compared to non-laboratory (self-report) studies, laboratory investigations of the acute effects of nicotine and nicotine abstinence have less reliably demonstrated affect modulation. One likely explanation for the relatively reliable increases in negative affect reported in non-laboratory studies of smoking cessation is that they assessed aggregated mood across the entire day in the natural environment

(Gilbert, 1995). Averaging across time and situations increases the probability of occurrence of situations in which nicotine can reduce negative affect. In contrast, in a controlled-environment, non-stressful, laboratory environments negative affect does not increase after nicotine deprivation of up to 12 hours or more (Arci and Grunberg, 1992; Gilbert *et al.*, 1992a,b; Meliska and Gilbert, 1991; Parrott and Craig, 1992; Parrott and Winder, 1989; Suedfeld and Ikard, 1974) and possibly not after any period of deprivation (D.G. Gilbert, unpublished data). The failure of negative affect to increase during relaxing and minimally stressful situations indicates that abstinence-related dysphoria is not an inherent nicotine 'withdrawal symptom' but occurs only in certain situations. The situation-specific nature of nicotine-related mood changes may reflect decreased coping capacity, psychophysiological over-reactivity, and/or negatively biased bioinformation processing, rather than a time-locked inherent syndrome.

Experimental studies show smoking/nicotine to reduce negative moods and emotions (e.g., anxiety, tension, irritability) when stressor stimuli are ambiguous and/or anticipatory/distal, but not when stimuli are proximal, unambiguous and without distractor stimuli (Gilbert and Welser, 1989; Gilbert *et al.*, 1989b; Jarvik *et al.*, 1989; reviewed by Gilbert, 1995). For example, during the anticipation of stressful tasks, but not during the tasks themselves, anxiety was reduced in a study by Jarvik *et al.* (1989). Similarly, Gilbert *et al.* (1989b) found smoking cigarettes with normal nicotine, relative to denicotinized tobacco cigarettes, to reduce movie-induced anticipatory anxiety.

Individual differences in nicotine, and tobacco abstinence responses

Evidence shows that a significant amount of the variability in response to smoking and abstinence reflects an individual's typical style of coping with stress, individual differences in trait adaptive responses (TARs) (Gilbert, 1994, 1995). The TAR hypothesis of the STAR model ignores situational parameters (the 'S' in 'STAR') and simply predicts an individual's dominant trait adaptive response to stress across situations and time. Thus, an individual for whom anger/irritability is a characteristic TAR will on the average experience anger more frequently when nicotine abstinent and will smoke more often to reduce anger/irritation. Similarly, a person whose dominant response to stress is depression will become more depressed more often when nicotine abstinent and smoke more often to reduce depressive affect. In a similar manner, individuals with attentional problems are likely to smoke to enhance attention and to respond to abstinence with increased difficulty coping and more frustration and negative affect in situations requiring high levels of sustained attention.

TARs reflect the major personality dispositional traits, habitual coping styles, and psychopathologies. The most important traits moderating nicotine-related adaptive responses are posited to be those associated with smoking initiation, prevalence, and relapse (neuroticism, depression, extraversion, psychoticism, impulsivity, unconventionality, antisocial behaviour, and low educational/long-term aspirations [reviewed by Gilbert, 1995]). Personality traits and other TARs related to stress predict abstinence response type, severity, relapse, and endurance to the degree that the individual's tobacco-associated learning history and biological sensitivities to nicotine make tobacco abstinence a stressor for the individual.

Evidence of TAR-influenced responses to nicotine and tobacco abstinence

The systematic nature and importance of individual differences in trait adaptive responses became clear as evidence mounted showing the high prevalence of depressive affect subsequent to abstinence in those with a clinical history of depression or scoring high in trait measures of depressed affect (Breslau *et al.*, 1992; Covey *et al.*, 1990; Glassman, 1993; Hall *et al.*, 1991; Kinnunen *et al.*, 1994; Russell *et al.*, 1990). For example, depressive symptoms occurred in 75% of those with a history of depression but only 31% of those without such a history in a study by Covey *et al.* (1990). Those with a history of depression reported more and more intense cessation-related effects. Kinnunen and associates (1994) found that smokers depressed immediately prior to quitting reported larger increases in dysphoric response to abstinence relative to those without depressed affect, but nicotine gum attenuated dysphoric response in these individuals.

Similarly, those high in anxiety report smoking more often to reduce anxiety (Pritchard and Kay, 1993; Spielberger, 1986). Furthermore, several recent studies indicate that anxiolytic drugs can reduce abstinence-related anxiety and negative affect increases in high anxiety individuals (Cinciripini *et al.*, 1995; Franks *et al.*, 1989; Glassman *et al.*, 1984; Prochazka *et al.*, 1992). Finally, Levin recently found nicotine patch to result in feelings of relaxation and attentional enhancement in individuals with attention deficit hyperactivity disorder (E.D. Levin, personal communication, September, 1994). The TAR proposition is thus supported by a convergence of findings.

Individual differences in vigilance and arousal responses to nicotine and abstinence

Few studies have assessed personality trait and other correlates of individual differences in information processing, attention, and arousal effects of nicotine and tobacco abstinence. Group mean scores have shown abstinence to increase reported difficulty in concentrating (Ward *et al.*, 1994; West *et al.*, 1987) and vigilance-task performance (Warburton, 1992; Warburton and Walters, 1989). Cessation generally is associated with lessened objectively measured ability to concentrate over the first week or two of abstinence (Gilbert *et al.*, 1992a; Hughes *et al.*, 1990). Decrements in self-reported alertness have been observed in a number of studies subsequent to cessation of habitual tobacco use (Hatsukami *et al.*, 1987; Hatsukami *et al.*, 1988; Hughes *et al.*, 1991b). Laboratory studies assessing effects of different nicotine content of cigarettes have fairly consistently found nicotine deprivation to increase drowsiness relative to smoking conditions (Gilbert *et al.*, 1994c; Meliska and Gilbert, 1991; O'Neill and Parrott, 1992). Nicotine substitution via gum also generally attenuates the effects of tobacco abstinence on concentration (Hughes *et al.*, 1990).

Neuroticism scores were positively associated with the concentration-enhancing effects of nicotine in two studies by Warburton and Wesnes (1978). Similarly, Gilbert *et al.* (1992a) found the severity of vigilance decrements during the first few days of abstinence to be a function of habitual nicotine intake and neuroticism. In addition, while the vigilance of most individuals had fully recovered within 31 days of cessation, those high in neuroticism had not returned to baseline levels. Finally, individuals scoring high in psychometrically assessed depression were found to exhibit more tobacco-abstinence-related EEG slowing, that is larger decreases in cortical arousal, especially in the left relative to the right frontal cortex (Gilbert *et al.*, 1994b).

Cognition and information modulation by nicotine

Nicotine appears to be especially effective in enhancing working memory performance (Levin, 1992; 1994; Sherwood, 1994). Evidence suggests that the cognitive and psychomotor enhancing effects of nicotine are not simply due to reversal of nicotine abstinence (Gilbert, 1995; Sherwood, 1994). Cognition and performance enhancing effects of nicotine include increased perceptual sensitivity, decreased reaction time and enhanced vigilance. Effects on short-term and long-term memory have been variable. However, memory tasks assessed to date have used affectively neutral stimuli. Effects of nicotine on memory for affectively laden words, during affective states, and in association with phasic increases in nicotine associated with each puff have not been determined.

Effects of nicotine and nicotine abstinence on cognitive tasks may vary according to the degree to which the task is mediated by functions relatively more based in one hemisphere of the brain than the other. Their review of the literature led Gilbert and Welser (1989) to suggest that the effects of nicotine on cognition and performance may in part be a function of the relative lateralization of the cognitive processing systems involved in the task and upon the stressfulness of the task. The findings of current research suggested that during moderately stressful tasks, predominantly left-hemisphere-based processes are enhanced, while right-hemisphere-based processes, including visual-spatial integration, are relatively attenuated. In contrast, during relaxing minimal-demand tasks nicotine appeared to stimulate right-hemisphere information processing.

For example, Schultze (1982), found that smoking enhanced anagram performance, a left-hemisphere verbal task, but interfered with performance on digit symbol substitution and Bender-gestalt form, relatively right-hemisphere tasks. Also, decrements in right-hemisphere processing efficiency attributed to acute smoking have been observed during several other somewhat stressful/arousing situations including facial emotional cue identification (Hertz, 1978) and jigsaw puzzle tasks (Schneider, 1978). During relaxing conditions, nicotine's ability to facilitate habituation to simple stimuli (Friedman *et al.*, 1974; Golding and Mangan, 1982) and its tendency to activate EEG in the right hemisphere more than in the left (see Figure 5.1) (Domino, 1995; Gilbert *et al.*, 1994c) are evidence of relatively greater right than left hemisphere activation effects of modest doses during low arousal states.

MECHANISMS UNDERLYING NICOTINE'S AFFECT- AND INFORMATION-MODULATING EFFECTS

The wide range of mechanisms hypothesized to mediate smoking's affect-modulating and reinforcing effects reflects the large number of effects produced by nicotine. Nicotinic-cholinergic receptors, in addition to constituting a primary component of cholinergic neurons, are also located on a variety of non-cholinergic neurons (Rosecrans and Karan, 1993). The pervasiveness of these cholinergic

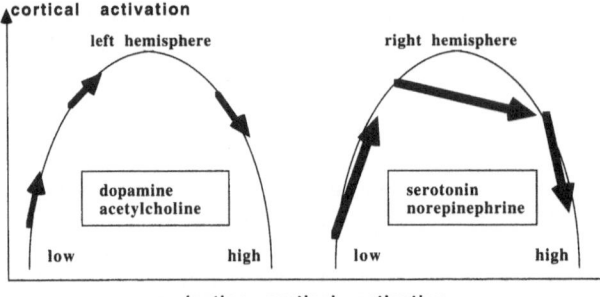

Figure 5.1 – Effects of a standard dose of nicotine as a function of hemisphere and pre-nicotine cortical activation (after Gilbert and Welser, 1989). Based on the assumptions of biphasic response curves and greater dose-response sensitivity of the right (longer response arrows) than left hemisphere, nicotine is predicted to exhibit state-dependent and differing effects on the two hemispheres. During low-activation state nicotine produces greater increases in right-hemisphere than left-hemisphere cortical activation. However, with higher levels of pre-nicotine cortical activation, a decrement in right-hemisphere activation can occur with simultaneous increases in left activation. Dopamine and acetylcholine are indicated as associated with the left hemisphere and serotonin and norepinephrine with the right hemisphere because of the hypothesized relative densities and influence of these neuromodulators. Serotonin and norepinephrine may play modulatory roles in promoting nicotine's inhibitory effects.

receptors allows nicotine to influence almost all biological systems throughout the body. Moreover, different systems have different sensitivities to nicotine's effects and thus the effects of nicotine are dose and biological state-dependent. Nicotine's multiple reinforcing effects appear to be in part a function of the ability of the smoker to modulate a number of different cholinergically influenced biological systems.

The ability of nicotine to act as a cholinergic agonist initially and in lower doses, while acting as cholinergic antagonist in other conditions may mediate a number of its reinforcing effects (Rosecrans and Karan, 1993).

Characterization of nicotine's affect-modulating, information processing, and reinforcing effects must take into account situation and trait dependency relationships noted previously. Nonetheless, heretofore, characterization of such mechanisms have been almost universally offered without addressing situational and trait dependencies. Psychological mechanisms most frequently hypothesized to mediate nicotine's reinforcing and mood modulating effects are listed in Table 5.1 along with their proposed biological substrates, which are discussed later in this chapter. A number of these mechanisms appear to be operative in most smokers, but some smokers are more likely to smoke for some of these reasons than others (Ashton and Golding, 1989; Eysenck, 1980; Gilbert, 1994; 1995; Pomerleau and Pomerleau, 1989; Warburton, 1990a). In spite of their conceptual advancement over unifactorial models, most multi-mechanism approaches to date are limited by their failure to specify *when* nicotine results in *what* effects, in *what* individuals, and by failures to note interactions among mechanisms. The question of in whom, after what dose history, in what circumstances, does what dose of nicotine result in what type of affect-modulating, information processing, and reinforcing effects is much more likely to produce useful answers than more simplistic questions (Gilbert, 1995). The mechanisms and their putative substrates are now addressed under separate headings.

Mood and cognitive enhancement via withdrawal alleviation

Increased irritability, anxiety, difficulty concentrating, and appetite are the most common responses to the cessation of habitual nicotine/smoking administration (Hughes *et al.*, 1990; Murray and Lawrence, 1984). These also are the psychological states most consistently reduced by acute doses of nicotine in nicotine-deprived smokers. Thus, some have argued that nicotine's negative-affect reducing effects are solely a result of its reducing internally driven, physiologically based, withdrawal symptoms (Perlick, 1977; Schachter, 1979).

The following findings (reviewed in detail by Gilbert, 1995) are contrary to the withdrawal-symptom-alleviation hypotheses of nicotine's affect and cognition modifying effects:

- many moderate and heavy tobacco users do not experience increased irritability, anxiety, difficulty concentrating, depression, and other withdrawal symptoms subsequent to abstinence;
- aversive tobacco abstinence responses are not time-locked and inevitable;
- withdrawal responses are only very slightly related to degree of physical dependence and nicotine exposure (Russell *et al.*, 1990; Sherwood *et al.*, 1990; Shiffman, 1989, 1991);
- individual differences in personality, temperament, and environment are as or more predictive of abstinence response and abstinence success than in nicotine exposure history (Gilbert *et al.*, 1992b; Gilbert *et al.*, 1994c; Masson and Gilbert, 1990);
- individuals not currently physically dependent on nicotine (occasional and very light smokers) experience negative-affect-reducing effects from nicotine (McNeill *et al.*, 1987; Shiffman, 1989; Shiffman *et al.*, 1993; Spielberger, 1986);
- situational factors determine both abstinence response and response to nicotine in habitual smokers (Gilbert *et al.*, 1989b; Gilbert, *et al.*, 1992b; Perkins and Grobe, 1992; Suedfeld and Ikard, 1974) and
- in certain conditions ex-smokers and non-smokers experience negative affect reducing and cognition-enhancing effects from nicotine (E.D. Levin, personal communication, September, 1994; Pomerleau *et al.*, 1984; Sherwood, 1994).

Mesolimbic dopaminergic activation system models

Many have argued that nicotine, as well as many other drugs, are self-administered because they activate a major dopamine (DA)-based reward system located in the base of the brain, known as the mesolimbic reinforcement system (Bozarth, 1989; Corrigall *et al.*, 1992; Willner and Scheel-Kruger, 1991). However, evidence does not support the view that the mesolimbic dopaminergic system is uniquely associated with positive reinforcement or is the primary biological mediator of 'reward' and 'euphoria' (Morgenson, 1987; Salamone, 1991; Scheel-Kruger and Willner, 1991). Aversive stimuli, like positive stimuli and self-administered drugs,

Table 5.1 – Mechanisms of nicotine reinforcement and affect modulation

Psychological mechanism	Negative affect reduction (Attenuation of anxiety, depression, irritability and related dysphoric states)
	Underlying biological mechanism(s)
Attenuation of (right-frontal-hemisphere dominant) negative cognitive-affective schema activation	Conditioned neurotransmitter/modulator profile states tending to elicit neutral or positive affect (proposition 8)
	Hippocampal memory indexing (proposition 8)
	Inhibition of right-hemisphere-dominant DA and ACh corpus callosal inhibition (proposition 6)
	Serotonergic and noradrenergic modulation by nicotine (proposition 6)
Attentional focus on proximal on anxiogenic stimuli (enhancement of)	DA, ACh modulation of frontal attention mechanisms (proposition 3 & 6)
Body weight reduction	Modulation of insulin, serotonin, and cortisol
Conflict reduction	DA and ACh-promoted psychomotor activation of approach behaviour (proposition 5)
	DA and ACh-activation priming of LH approach-oriented inhibition of RH avoidance schemas
Extinction of negative cognitive-affective shcema-elicited responses via attention, memory, and approach-mediated enhancement of exposure and learning	Cortisol/corticosteroid-based facilitation of homeostatic recovery from negative-affect-related bioinformational states thereby facilitating opportunity for extinction (Gilbert, 1995)
	DA and ACh-mediated approach facilitation (proposition 5 and 6)
	Vasopressin-, DA-, ACh-, epinephrine-facilitated focused attention and extinction enhancement (proposition 4)
Hedonically pleasing level of arousal	CNS arousal modulation (DA, ACh)
	Cortisol/corticosteroid-based facilitation of homeostatic recovery
Impulsivity control	Noradrenergic stimulation
	Serotonin modulation
Somatic relaxation	Muscle relaxation, Renshaw inhibitor cell stimulation

Table 5.1 – *Continued*

Psychological mechanism	Underlying biological mechanism(s)
Negative affect reduction (Attenuation of anxiety, depression, irritability and related dysphoric states)	
Performance-enhancement-related goal achievement leading to minimization of frustration-induced negative affect	DA, ACh-mediated increases in cortical activation and information processing capacity and reciprocal inhibition of right frontal cognitive-affective schema (proposition 2)
Positive affect enhancement	
Associative biasing towards positive cognitive affective schema, associations, interpretations, and active coping	Priming of left-hemisphere-dominant, DA- and ACh-modulated bioinformational networks via conditioned and unconditioned mechanisms (proposition 6 & 8)
Cognitive-performance-enhancement mediated goal achievement	DA, ACh-mediated increases in cortical activation and information processing capacity (proposition 2, 5, 6 & 7)
Self-efficacy enhancement via goal achievement (cognitive performance enhancement, weight control and affect modulation)	Indirect outcome of all the above positive and negative reinforcement mechanisms
Sensory gratification	Trigeminal throat, mouth and lung stimulation; taste, olfactory, and subjective stimulation

can increase DA activity in the accumbens and frontal cortex (Abercrombie *et al.*, 1989).

Mesolimbic DA is involved in the coping/engagement processes common to both positive and negative emotional states in that it "promote(s) interaction with and adaptation to the environment" (Salamone, 1991, p. 609). Thus, nicotine-induced increments in brain mesolimbic DA activity would be expected to increase the ability of the organism to avoid aversive stimuli and to obtain positive ones (Gilbert, 1994, 1995). Reductions in mesolimbic DA would be expected to have the opposite effects. While no direct test of this situation-specific hypothesis has been made for nicotine and smoking, there is evidence supporting the situation-specific view of caffeine. Whether or not caffeine is reinforcing in humans is a function of the situation-related goals of the user (Silverman *et al.*, 1994). Similarly, activation of the mesolimbic DA systems by nicotine (reviewed below) may simply facilitate reward and reinforcement obtainment, rather than directly causing such processes.

Cortical arousal modulation model

Eysenck (1973, 1980) proposed that nicotine is rewarding because it can increase cortical arousal when one is under aroused and can, in other situations, reduce cortical and subjective arousal when one is excessively aroused, as when in a state of emotional distress. Consistent with this bi-directional, inverted-U-hypothesis, a large number of studies does show that in low-arousal situations modest doses of nicotine increase cortical activation and alertness (reviewed by Knott, 1990). However, contrary to the view that nicotine always decreases cortical activation when activation is high, smoking increased cortical event-related potentials to electrical shock stimuli (Knott, 1990) and auditory stimuli during a stressful, shock-stress paradigm (Knott and De Lught, 1991). Also contrary to the arousal-reduction hypothesis, Hasenfratz and Bättig (1993b) found that during stressful tasks smoking decreased anxiety but simultaneously increased EEG activation. Thus, while some studies have observed nicotine-induced decreases in electrocortical arousal when pre-smoking arousal is high (Gilbert *et al.*, 1989b; Golding and Mangan, 1982), others have observed smoking-induced increases in both electrocortical and subjective arousal during stressful and high-arousal conditions even though subjective anxiety is decreased.

An alternative form of the arousal-modulation hypothesis is that cortical sedation may be obtained from smoking and other forms of nicotine intake because users increase their dose of nicotine during stress. Smoking has been found to attenuate cortical event-related potential amplitude in individuals with large nicotine intakes but to enhance amplitude in low-dose smokers (Norton and Howard, 1988; Norton *et al.*, 1991). The low nicotine group reported increases in subjective arousal subsequent to smoking, while high-dose smokers reported decreases. High-dose-dependent attenuations of electrocortical activity were also found by Ashton and associates (1974). While low and moderate doses of nicotine increased cortical arousal, very high doses decreased CNS arousal in animals (Armitage *et al.*, 1969; Guha and Pradhan, 1976). Evidence contrary to the view the cortical sedation model can explain all forms of nicotine-promoted negative affect reduction is provided by findings that nicotine can decrease negative affect while concomitantly increasing subjective arousal (O'Neill and Parrott, 1992; Parrott and Joyce, 1993; Perkins *et al.*, 1992c).

Thus, in contrast to the arousal-modulation model, in some cases smoking leads to simultaneous increases in subjective alertness and increased EEG arousal combined with decreased distress. Another problem with this model is the finding that during smoking abstinence, subjective and electrocortical indices of arousal decrease, but irritability and anxiety frequently increase (Gilbert, 1979; O'Neill and Parrot, 1992). Therefore, many of nicotine's tranquilizing and negative-affect-reducing effects cannot be attributed to its attenuating elevated cortical arousal.

Muscle relaxation

Another hypothesized mechanism to account for the stress-reducing effects of nicotine is that nicotine produces a state of tranquillity by reducing the level of tonic and/or phasic muscular activity (Gilbert, 1979). Nicotine has been found to reduce resting muscle tone in spastic patients (Webster, 1964), patellar reflexes, startle responses (Clark and Rand, 1968; Domino, 1973) and electromyographically measured tension in some muscle groups (Gilbert, 1995).

Appetite- and weight-related mechanisms

The tendency of smoking to lower body weight is perceived by many smokers as a beneficial effect of smoking that contributes to the decision of some individuals, especially women, to start smoking (Klesges and Klesges, 1988), to continue smoking (Pirie *et al.*, 1991), and to relapse subsequent to quitting (Pirie *et al.*, 1991). Nicotine reduces body weight in humans and rodents, and withdrawal from nicotine leads to increases in weight (Levin *et al.*, 1993). Control of one's weight may reduce negative affect and enhance positive affect and self-esteem by a number of psychological and social mechanisms. Biological substrates of nicotine's effects on appetite, weight, and affect appear to include serotonin (Spring *et al.*, 1993), possibly insulin (Grunberg and Raygada, 1991), and cortisol (Gilbert, 1995). In addition, negative affect promotes both smoking (Rose *et al.*, 1983; Pomerleau and Pomerleau, 1989) and eating (Cooper and Bowskill, 1986; Herman and Polivy, 1975; Morley *et al.*, 1983).

Additional mechanisms

There is reason to believe that a number of indirect and relatively more psychological mechanisms are required to provide a full account of nicotine's reinforcing effects. The STAR model builds on evidence that, while a number of the mechanisms proposed above are operative in some individuals in some circumstances, a complete, accurate, and maximally useful model must articulate *when* and *in whom what* processes occur. The propositions and supportive evidence outlined immediately below address bioinformational mechanisms by which nicotine exerts its situation- and trait-specific affect modulating effects.

STAR bioinformational propositions

The effects of nicotine and nicotine abstinence on affect-modulating and reinforcing effects vary as a function of individual traits, as well as stressor type and other situational parameters. Nicotine influences information processing that can modulate affect and facilitate both positive and negative reinforcement. Under certain circumstances nicotine appears to attenuate negative affect by altering internally driven associative and memory-based processes. In other circumstances, its stimulant effects enhance performance, helps the individual obtain a desired goal, and thereby increases positive and prevents negative affect. Weight and appetite suppression, muscle relaxation, and sensory pleasure are operative in many individuals at different times. Based on these situation, stress, and individual-dependent effects and associated review of the evidence, Gilbert (1995) generated the eight STAR-model-based propositions summarized below. These propositions are complimentary to and more specific and subtle than previously proposed mechanisms of affect modulation and information processing effects of nicotine.

PROPOSITION 1

Most of the affect-modulating and reinforcing effects of nicotine are indirect and mediated by brain bioinformation processing mechanisms preceding final emotional output mechanisms. Generally, internally driven (e.g., memory, associative and attention-based) processes are more influenced by nicotine than are externally driven processes. Externally driven process are associative processes resulting from potent, immediate-response-demanding external stimuli that are associated with rigid stimulus-response associations (e.g., direct exposure to a greatly feared object).

Evidence supporting Proposition 1

Empirical evidence indicates that the cognitive and affective effects of nicotine are situation- and trait-dependent (reviewed earlier in this chapter and in detail by Gilbert, 1995). Nicotine and tobacco abstinence have no direct and invariable affect-modulating and reinforcing effects that generalize across situations and individuals (Gilbert, 1995; Gilbert *et al.*, 1992b). In situations characterized by cues suggestive of future threat and those utilizing ambiguous cues requiring substantial associative elaboration to generate threat, nicotine attenuates negative affect by altering internally driven associative processes, while in others it promotes goal achievement by its attentional and performance-enhancing effects and thereby indirectly modulates mood states. Evidence in support of this general proposition is provided by propositions 2–8 and their supportive evidence.

PROPOSITION 2

Nicotine can enhance performance and other forms of goal attainment that prevent frustration-induced negative affect and facilitate positive affect.

Corollaries

a) Smoking and other forms of nicotine self-administration will increase positive and attenuate negative affect in situations where the smoker is motivated to attain a goal and smoking facilitates goal attainment.
b) Those individuals who obtain large acute performance enhancements from nicotine (e.g., heavy users, neurotics, depressives, and those suffering from attention-deficit disorders) will exhibit more frequent and larger performance-related smoking reinforcements and motivations than those who experience smaller effects.

Evidence supporting Proposition 2

Evidence reviewed earlier in this chapter indicates that nicotine can enhance performance and goal attainment and that these effects can not be solely attributed to relief of physical dependence (E.D. Levin, personal communication, September, 1994; Pomerleau *et al.*, 1984; Sherwood, 1994; Sherwood *et al.*, 1990). Smokers report smoking for improvement in cognitive performance (Peeke and Peeke, 1984; Russell *et al.*, 1974). Goal frustration and attainment are major determinants of affective states (Lazarus, 1991). Evidence suggests that acetylcholine, dopamine, cortisol, vasopressin, and ACTH mediate these effects but that the support is strongest for acetylcholine and dopamine (Gilbert, 1995). Roles of ACTH and vasopressin are more speculative, but have been implied by findings of Eysenck and Kelley (1987).

PROPOSITION 3

In situations where there are cues of a potential, non-extreme threat, nicotine will reduce negative emotional states to the degree that the situation also provides moderately distracting, non-threatening cues onto which nicotine facilitates attentional focus. The potent sensory cues (e.g., smell, taste, lung and throat irritation) associated with tobacco provides such distracting cues.

Evidence supporting Proposition 3

This proposition is based on evidence demonstrating reliable attention-enhancing effects of nicotine (see also Pritchard and Robinson, Ch. 2), on the limited capacity of attentional processes (M.W. Eysenck, 1982), and on the conclusion that nicotine's tranquilizing effects occur primarily, if not solely, when stressor stimuli are distal/anticipatory and/or ambiguous (Gilbert and Welser, 1989).

PROPOSITION 4

Nicotine reduces negative affect elicited by conditioned aversive stimuli by facilitating extinction of conditioned responses in conditions with repeated or protracted conditioned stimulus presentations to which nicotine enhances attention.

These attention-promoted, extinction-enhancing effects are in part mediated by nicotine-induced release of a variety of neuromodulators (ACTH, cortisol, vasopressin, dopamine, acetylcholine).

Evidence supporting Proposition 4

This speculative proposition is based on evidence showing that nicotine can elevate ACTH, cortisol, dopamine and vasopressin (reviewed by Gilbert, 1995) and on findings of Eysenck and Kelley (1987) showing that during conditions of prolonged conditioned stimulus exposure ACTH, vasopressin and other stress-related hormones facilitate extinction of the conditioned response to the conditioned stimulus. We suggest that the important roles of acetylcholine and dopamine in facilitating attention and approach behaviour will also promote extinction of aversively conditioned responses to conditioned stimuli.

PROPOSITION 5

Nicotine can reduce approach-avoidance-conflict-generated tensions and negative affect by raising the approach gradient and, in some cases, lowering the avoidance gradient. This relative elevation of the approach gradient also minimizes attention to external and internal (memory associated) cues of punishment and frustrating-non-reward, thereby reducing negative-affect-eliciting associations.

Corollary

Neurotic, introverted, attention-deficit-disordered, schizophrenic, and other individuals characterized by frequent approach-avoidance conflicts (e.g., ambivalence) will be especially attracted to smoking and will report conflict-reduction as a motivation for smoking.

Evidence supporting Proposition 5

The commonly reported experience of needing a cigarette before one can force oneself to undertake a task involving prolonged concentration or other effort is consistent with the proposal that nicotine has conflict-reducing effects. Additionally, Amsel (1990) has argued that the paradoxical calming and attention-enhancing effects of psychomotor stimulants on individuals with attention-deficit hyperactivity disorder is a function of conflict reduction resulting from dopaminergically based enhancement of approach tendencies. Gilbert (1995) proposed that nicotine has similar effects on such individuals, as well as on others in approach-avoidance conflict situations. Nicotine stimulates both dopamine and approach behaviour (Balfour, 1991).

PROPOSITION 6

Nicotine increases left-frontal (cholinergic and dopaminergic) cortical activation and thereby enhances controlled processing, approach mechanisms, and left-hemisphere (LH) dominant positive-affect-related associative networks. Thereby, feelings of self-control and well-being are increased, especially in conditions and in individuals predisposed to relatively low LH activation and/or relatively high right-frontal activation. Relative frontal right-hemisphere (RH) arousal and related negative-affect-based associative network activation (possibly modulated by and associatively connected to serotonin and norepinephrine states) are decreased by nicotine during stressful conditions. Such increases in LH and relative decreases in RH bioinformational activation reduces interruption sensitivity and decreases the probability of frustration-induced negative affect.

Corollary

Individuals characterized by decreased left-frontal and/or relatively increased right-frontal activation (e.g., depressed and neurotic individuals) will use smoking and other forms of nicotine as a form of self-medication and will experience larger increases in left-frontal activation from nicotine than those without such asymmetry.

Evidence supporting Proposition 6

There are a number of parallels between the effects of nicotine and lateralized CNS affect- and behaviour-related processes (Gilbert, 1995; Gilbert *et al.*, 1989; Gilbert and Welser, 1989). Thus, nicotine's reinforcing effects may stem directly or indirectly from its capacity to alter the relative activation and information-processing biases of relatively lateralized CNS neural networks. Several observations suggest that common genetically influenced, lateralized mechanisms mediate relationships between smoking, personality, psychopathology and stress (Gilbert and Gilbert, 1995). For example, clinical depression and the induction of depressive affect in non-clinical subjects is associated with an increase in the frontal right-relative-to-left-hemispheric EEG activation (Davidson, 1993), and smoking-delivered nicotine attenuates such asymmetry in smokers scoring high in psychometrically assessed depression (Gilbert *et al.*, 1994b; Gilbert *et al.*, 1994c). Moreover, the performance enhancing, frustration attenuating, and approach promoting effects of nicotine reduce state depression, anxiety, irritation and impulsive behaviour, all of which are related to personality, psychopathology, and different lateralized neural network setpoints (Gilbert, 1995).

Although it would be inaccurate to imply that one hemisphere or brain region mediates affect, it has been suggested that nicotine decreases negative affect by increasing the balance of activation in favor of the left relative to the right cerebral hemisphere (Gilbert and Hagen, 1985; Gilbert and Welser, 1989). Lateralized nicotine effects on electrocortical (Elbert and Birbaumer, 1987; Gilbert, 1985, 1987; Gilbert *et al.*, 1989, Gilbert *et al.*, 1994a,b; Norton *et al.*, 1992; Pritchard, 1991b; Stough *et al.*, 1995) and electrodermal activity (Boyd and Maltzman, 1984) have been reported. These electrophysiological investigations as well as task-performance studies (reviewed in Gilbert, 1995) suggest that during stressful conditions nicotine reduces right relatively more than left hemisphere processing/ activation, while during low-stress situations, it may activate right-hemisphere processing more than the left (Domino, 1995; Gilbert *et al.*, 1994c). Thus, the lateralized effects of nicotine appear to be situation-specific. Effects of nicotine on EEG laterality also vary as a function of neuroticism, depression, extraversion, and type A behaviour pattern (Eysenck and O'Connor, 1979; Gilbert, 1987; Gilbert *et al.*, 1994b,c; Golding, 1988).

A convergent body of electrocortical, neuropsychological, pharmacological, psychophysiological, and behavioural data demonstrates that the activation of the right-frontal cortex and/or relative deactivation of the left-frontal cortex is associated with the experience of and predisposition to depressed affect and avoidance behaviour (Davidson, 1984, 1993; Kinsbourne, 1989). More generally, the RH specializes in nonverbal, simultaneous/holistic, and spatial processes, while the left cortex is more associated with approach behaviour, sequential/logical, and verbal information processing (Davidson, 1993; Kinsbourne, 1989; Tucker and Williamson, 1984). Nicotine's ability to promote approach (Balfour, 1991) and attenuate avoidance (Pauly *et al.*, 1992) is consistent with its priming LH processes while attenuating those of the right frontal cortex. Evidence suggests that approach and active coping are largely mediated by relatively left-lateralized dopaminergic and cholinergic circuits, while withdrawal, phasic arousal is mediated by relatively right-lateralized noradrenergic systems (Kinsbourne, 1989; Tucker and Williamson, 1984).

Thus, nicotine-induced reductions of RH activation and associated information processing should reduce physiological and subjective responses to stress cues eliciting avoidance and RH dominant, negative-affect-related, associative networks. Nicotine-induced stimulation of left-frontal activation and/or reduction of right frontal activity during active coping tasks is thus expected to reduce negative affect. Enhancement of left-relative-to-right frontal activation is predicted to correlate with enhanced task performance and other forms of active coping. Decreased right posterior cortical activation/sensitivity by nicotine would decrease interruption sensitivity, distractibility, and arousal to emotion-related stimuli (Gilbert, 1995). Decreased interruption/frustration sensitivity during stressful conditions should lead to increased goal achievement and improved mood. On the other hand, during low-stimulation, nicotine's ability to enhance right-relative-to-left parietal activation (Gilbert *et al.*, 1989, 1994c) may mediate some of smoking's ability to enhance vigilance.

PROPOSITION 7

Low doses of nicotine administered in relaxing situations results in enhanced *posterior*, especially right posterior activation, and thereby enhances vigilance. Under such conditions a perceptual, rather than an overt motor orientation, predominates. This perceptual enhancement may be experienced as a satisfying and tranquil state of alertness. This perceptual enhancement may be relatively more mediated by cholinergic activation. Higher smoking-sized doses of nicotine will activate more frontally influenced dopaminergic projections involved in active motor engagement and active attention.

Evidence supporting Proposition 7

While there frequently appears to be some lateralization of nicotine's influence on cognitive and electrocortical processes, the strongest and most reliable effect is upon general arousal functions. During relaxing, low-arousal situations, Knott (1989c) found low doses of smoking-delivered nicotine to have relatively greater activating effects on posterior sites, while slightly higher doses activated anterior and posterior sites equally. Others (Domino, 1995; Gilbert *et al.*, 1994c) have found low and moderate nicotine doses to result in slightly greater right than left hemisphere activation, though Knott (1989c) found no such asymmetry.

PROPOSITION 8

The relatively localized and lateralized effects of nicotine alter the accessibility of hippocampal memory units and bias information in terms of left-frontal positive affect-related associations and schema and away from right-frontal negative-affect and withdrawal-related schema. Bioinformational states channel information flow not only by conditioned associative processes, but by lateralized neurotransmitter and neural-network-specific processes. Thus, nicotine primes memory cues and schema by interacting with and altering hippocampal memory indexing (Teyler and DiScenna, 1986). Negative-affect- and avoidance-related information is relatively more encoded in RH serotonergic and noradrenergic-related systems and cognitive-affective-behavioural propositional networks, while positive affect and approach-related memory/information processing is more integrated with left-hemisphere-dopaminergic neural networks. These lateralized and localized bioinformational networks (including schema) operate at both limbic and neocortical levels (Derryberry and Tucker, 1992) and are differentially primed by nicotine (Gilbert, 1995).

Corollary

The effects of nicotine on affective and cognitive processes are a function of the pre-existing state of various CNS bioinformational networks, which are affected by genetically influenced individual differences in biological structures, neuro-modulation, and biological tuning, as well as by relatively stable learning-influenced psychobiological structures.

Evidence supporting Proposition 8

This proposition integrates evidence indicating localized, lateralized, situation-, and trait-specific effects of nicotine with memory indexing theory (Teyler and DiScenna, 1986). Teyler and DiScenna (1986) proposed that the hippocampus retains memory by generating an index of neocortical regions aroused by stimulus events. Since nicotine is seen as producing situation- and trait-dependent localized and lateralized effects, smoking should be associated with unique memory indices. Reactivation of the hippocampal memory index is associated with a reactivation of the neocortical pattern and associated memorial experience. Such indexing is consistent with the situation- and trait-specific effects of nicotine on affective processes and can be seen as consistent with bioinformational models of emotion and affect (Davidson, 1993; Lang, 1979). Hippocampal indexing may also account substantially for the development and modification of complex cognitive-affective schemas found in smoking motivation, smoking abstinence, and psychopathology (Beck *et al.*, 1993; Tiffany, 1990, 1992).

CONCLUSIONS AND DIRECTIONS FOR RESEARCH

While the broad framework of the STAR model appears solidly established, it is in no way intended to be taken as a definitive theory of nicotine effects. It is intended to stand as a general theoretical framework for integrating findings showing important situational and trait influences on responses to nicotine and motivations for nicotine use. It is presented as a useful guide for future research, including the propositions outlined above and the additional ones detailed in the book devoted to the model (Gilbert, 1995). The value of the model will be evaluated largely in terms of the research it generates. Some of its specific proposals concerning mechanisms underlying nicotine's effects on cognitive and affective processes are likely to be wrong or in need of significant qualification and further articulation. Being a framework, it must be fleshed out and further articulated at various psychological and biological levels. The eight bioinformational propositions of the STAR model have not been adequately tested and, with the exception of proposition 1, must be considered speculative.

Studies assessing situation by trait interactions for various proximal and distal/ anticipatory stressors and cognitive tasks while assessing multichannel EEG will be especially useful tests of the model. Similarly, effects of nicotine on responses to lateralized visual field stimulus presentations, dichotic listening, and relatively lateralized cognitive tasks will provide important tests of the model. Characterization of the effects of nicotine on various cognitive and memory tasks as a function of affective state manipulations will be especially relevant to the model, as will dose-response analyses of nicotine's putative lateralized and localized effects in environments varying in stress level and performance demands. Given the trait-dependency of many of the effects of nicotine on affective processes, it is important that personality traits and habitual nicotine intake histories be characterized in future studies. Finally, careful quantification and/or standardization of nicotine dose administered is essential for research pertaining to the STAR model. Standard doses of nicotine cannot be obtained with standard smoking procedures, but must be obtained with quantified smoke delivery systems (Gilbert and Meliska, 1992; Gilbert *et al.*, 1989a, 1994c).

SUMMARY

This review and theoretical treatise highlights the effects of nicotine on emotion and information processing, and the underlying localized and lateralized brain systems mediating these effects. Situation- and personality-dependent cognitive, emotional, and biological effects of nicotine are addressed within the context of the recently developed STAR model. The biopsychosocial Situation by Trait Adaptive Response (STAR) model is presented as a means of predicting and explaining nicotine's biological, information processing, emotional, and reinforcing effects. The review of the STAR theory/model of nicotine's effects includes the model's propositions and empirical basis. The review includes conclusions and suggestions for further research. Nicotine is seen a having multiple bioinformation-processing-related effects that are situation and personality trait related. Nicotine and nicotine abstinence have no inherent effects on emotion and related affective processes. Instead, nicotine exerts its affect-modulating effects indirectly by means of localized and lateralized neuromodulators (e.g., acetylcholine, dopamine and serotonin) that are differentially associated with personality traits, coping styles, and cognitive and emotional processes. Emotion modulation is accomplished by two overall mechanisms:

- enhanced goal achievement via cognitive performance enhancement; and
- cognitive-affective information priming/biasing towards positive associations and away from negative schemas.

Part II — Coffee and Caffeine

CHAPTER SIX

Pharmacokinetics and Metabolism of Caffeine

Maurice J. ARNAUD

INTRODUCTION

Caffeine metabolism and pharmacokinetics have been reported in several reviews and monographs: Arnaud (1984, 1987, 1993a,b) and Anonymus (1991). This chapter reviews our present knowledge including the most recent published studies. Because there is a strong and significant relationship between coffee consumption and smoking, it was interesting to summarize the results obtained from metabolic studies looking at the interactions between caffeine and smoking habit (Swanson et al., 1994). It has also been shown that alcohol, nicotine and caffeine consumption are associated with chronotype, specially evening-types (Adan, 1994; see also Kole et al., Ch. 9). There are, however, only few studies looking at the interactions between caffeine kinetics and metabolism with alcohol consumption.

CAFFEINE ABSORPTION AND DISTRIBUTION

Absorption

Caffeine absorption from the gastrointestinal tract is rapid and complete and 99% of the administered dose is absorbed in man in about 45 minutes (Blanchard and Sawers, 1983), mainly from the small intestine but also 20% from the stomach. Complete absorption has also been demonstrated in animals using radio labeled caffeine (Arnaud, 1985).

Pharmacokinetics were independent of the route of administration and after oral or intravenous doses, plasma concentration curves were superimposable suggesting that there is no important hepatic first pass effect in man. In man, caffeine absorption does not seem to be dependent on age, sex, genetic and disease of the subject as well as drugs, alcohol and nicotine consumption. Only higher absorption rate constants were reported in a group of obese subjects (Kaminori et al., 1987). The efficacy of percutaneous caffeine absorption has been demonstrated in premature infants treated for neonatal apnea (Morisot et al., 1990). The outward transcutaneous caffeine migration is linearly correlated with the plasma area under the plasma caffeine Concentrations-time curve (AUC) and sweating did not play a significant role in the flux of caffeine (Conner *et al.*, 1991).

Distribution

Caffeine is sufficiently hydrophobic to pass through all biological membranes. In newborn infants, similar levels of caffeine concentration were found in plasma and cerebrospinal fluid (Anonymus, 1991). Both in the human and animal models no placental barrier prevents the passage of caffeine to the embryo or fetus and relatively high caffeine concentrations acquired transplacentally were reported in a premature infant (Khanna and Somani, 1984). Analysis of human and fetal gonads showed that their caffeine concentrations were the same as in plasma (Anonymus, 1991; Arnaud, 1993a). Caffeine has been detected in all body fluids. The concentration ratio between blood and semen is 1 and the decline after administration is similar (Beach *et al.*, 1984). In milk, values of the milk to serum concentration ratio of 0.52 and 0.81 were found. As the binding of caffeine to constituents of serum and whole breast milk was 25.1 and only 3.2% respectively, it was suggested that all the binding in breast milk was accounted for by the butter-fat content (Arnaud, 1993a). Caffeine can also be detected in umbilical cord blood, bile and saliva. The range of saliva concentrations was 65–85% of those in plasma.

After ingestion of caffeine, metabolites such as theophylline, theobromine and paraxanthine are also detected in body fluids. Theophylline and theobromine plasma concentrations increase to a small and similar extent in man. In contrast, the level of paraxanthine is 10-fold higher than those of theophylline and theobromine. Caffeine plasma concentrations decrease more rapidly than those of paraxanthine so that in spite of important interindividual differences, paraxanthine concentrations become higher than those of caffeine within 8–10 hours after administration. These metabolic plasma profiles characterize human from all animal species.

Because caffeine is submitted to a 98% renal tubular reabsorption, only a small percentage, 0.5–2%, of the ingested dose is recovered in human urine. In athletes participating in competitive sport, the presence of caffeine concentrations higher than 12 mg/l urine is considered as a disqualifying factor by the International Olympic Committee (IOC). A good correlation was found between urinary and plasma caffeine concentrations, but there was marked interindividual variations. Because of this large population variation in caffeine metabolism, the athletes should be advised to limit their intake (Birkett and Miners, 1991). Moreover, Dukhel *et al.* (1991) suggest that caffeine metabolism may be decreased during exercise to increase then during the post-exercise period. The reduction of caffeine elimination was also much greater in women than in man due to the excretion of smaller urinary volume in women than in men. At a caffeine intake of 1 g (about 10–12 cups of coffee), the recovery of caffeine in urine was 0.74–0.91% of the dose and urinary concentration of 14 mg/l can be reached. Thus, the relevance of this control can be questioned although it has been suggested that athletes would not attain these values by mere social intake of caffeine-containing beverages and foods.

The limited excretion of caffeine in urine indicates that its metabolism is the rate-limiting factor in its plasma clearance. No specific accumulation of caffeine and its metabolites in various organs or tissues can be observed, even after high doses.

CAFFEINE PHARMACOKINETICS

Dose-dependent Pharmacokinetics

Peak plasma caffeine concentration is reached from 15 to 120 minutes after oral ingestion in man and for doses of 5–8 mg/kg, the mean plasma values were 8–10 mg/l (Arnaud, 1993a). The delayed gastric emptying due to the presence of dietary constituents and pathologies such as gastric stasis, explains the pharmaco-kinetic variations observed (Brachtel and Richter, 1988). The fraction of caffeine reversibly bound to plasma proteins varies from 10–30% in man. Caffeine is elim-inated by apparently first order kinetics, described by a one-compartment open model system in man (Arnaud, 1993a). Non-linear kinetics was first observed in rats at doses higher than 10 mg/kg, indicating a limited capacity to absorb and/or metabolize caffeine (Latini *et al.*, 1978). From the results of several studies using a limited number of patients and a small range of doses: 2 to 10 mg/kg (Blanchard and Sawers, 1983; Newton *et al.*, 1981), the existence of dose-dependent kinetics in man at levels of normal consumption was not demonstrated. Dose-dependent kinetics was observed when caffeine plasma levels were higher than 30 mg/l, in the case of acute intoxication in an infant (Jarboe *et al.*, 1986) but other results obtained by different laboratories (Tang-Liu *et al.*, 1983; Cheng *et al.*, 1990; Denaro *et al.*, 1990) demonstrate that some metabolic transformations must be rapidly saturated in the lower dose range of 1–4 mg/kg. Such dose-dependent kinetics could be due to the formation of paraxanthine. It has been shown that both caffeine and paraxanthine plasma concentrations are dependent on the 3-demethylation of caffeine (Arnaud and Enslen, 1992), and a decrease of the clearance of paraxanthine was also reported (Tang-Liu *et al.*, 1983).

Experiments were performed in vitro to identify the different mechanisms involved. Microsomes prepared from human liver exhibit biphasic kinetics, indi-cating the potential participation of two different isozymes with different sub-strate affinities in the production of individual metabolites (Campbell *et al.*, 1987a). The demethylations of paraxanthine, theophylline and theobromine at high concentrations also show biphasic kinetics in the production of individual metabolites with human microsomes. At the concentrations reached *in vivo*, most demethylation activity is mediated by the high-affinity enzyme site with only a negligible contribution by the low-affinity site. Half-lives of 2.5–4.5 hours were observed in human volunteers receiving a 4 mg/kg caffeine dose (Anonymus, 1991; Arnaud, 1993a).

Effects of Subject Characteristics on Caffeine Pharmakinetics

Age

The study of the comparative pharmacokinetics in the young and elderly shows no significant differences in half-life suggesting that aging does *not* alter caffeine elimination in man, in contrast to the rat model where an age-dependent increase of caffeine half-life has been observed. In man, a slight decrease in plasma caffeine binding has been reported in the elderly who are characterized by a significantly lower plasma albumin concentration (Blanchard, 1982). Caffeine half-life is increased in the neonatal period due to the immaturity of the hepatic enzyme systems (Pons *et al.*, 1988) and similar results were reported in different animal species (Arnaud, 1993a). Half-lives of 50–103 hours were found in premature and newborn infants but these values decreased rapidly to 14.4 hours and 2.6 hours in 3–5 month and 5–6 month infants respectively. The clearance of 31 ml/kg/h in 1-month-old infants increases to a maximum of 331 ml/kg/h in 5–6 month infants with values of 155 ml/kg/h in adult subjects. A mean distribution volume of 0.7 l/kg (0.5–0.8 l/kg) was found either in newborn infants, adult or aged subjects.

Sex

No difference in the metabolic fate of caffeine was observed between men and women except a 25% longer caffeine elimination found in the luteal phase (half-life of 6.85 hours) of the menstrual cycle compared to the follicular phase (half-life of 5.54 hours) (Balogh *et al.*, 1987). The use of oral contraceptives has been shown to double the caffeine half-life. The half-life was also prolonged during the last trimester in pregnant women. The half-life returns to the pre-pregnant value a few weeks postpartum (Arnaud, 1993a).

Physical exercises and obesity

In adult volunteers, the effects of a one hour moderate exercise (30% $\overset{\circ}{V}o_2$ max) performed immediately after the ingestion of a capsule containing 250 mg caffeine caused a sharp rise in plasma caffeine concentrations, a decrease in the distribution volume and a reduction of the half-life (Collomp *et al.*, 1991). These results are however not in agreement with a previously published study showing no effect in lean, caffeine naive, untrained and non-smoking male volunteers.

In obese subjects with more than 30% body fat, a decrease in the maximal serum concentration and area under the curve was reported when exercising on a treadmill at 40% $\overset{\circ}{V}o_2$ max. In these subjects, a larger caffeine distribution volume was observed both at rest and when performing an exercise in comparison with lean subjects. Significantly higher absorption rate constants, lower elimination rate constants and a longer mean serum half-life (4.32 versus 2.59 hours) were also reported in obese subjects (Kaminori *et al.*, 1987). A previous study on caffeine disposition in obesity showed a pronounced increase in the apparent volume of distribution and a trend towards prolonged elimination half-life (Abernethy *et al.*, 1985). In contrast, Caraco *et al.* (1995), reported that caffeine elimination half-life and oral clearance rate were not altered significantly in severely obese subjects or following a substantial weight reduction. They confirmed, however, that obese individuals exhibited an increased volume of distribution and this volume was decreased with weight reduction. The effect was more important in females. They concluded that caffeine distribution was incomplete into excess of body weight.

Disease

A recent study performed in patients with decompensated Type I and Type II diabetes mellitus showed that caffeine half-life, apparent clearance and distribution volume were similar to controls (Zysset and Wietholtz, 1991).

Smoking

Caffeine clearance is stimulated by smoking (Hart *et al.*, 1976; Parsons and Neims, 1978; Wietholtz *et al.*, 1981; May *et al.*, 1982; Arnaud *et al.*, 1982; Kotake *et al.*, 1982; Fraser *et al.*, 1983; Joeres *et al.*, 1988; Busto *et al.*, 1989; Caraco *et al.*, 1995), while cessation of cigarette smoking significantly reduced it (Murphy *et al.*, 1988) and changed the pattern of caffeine metabolism (Brown *et al.*, 1988). More recent results obtained on regular caffeine users showed either no effect (Oliveto *et al.*, 1991) or significant higher caffeine plasma levels during non-smoking conditions (Brown and Benowitz, 1989; Benowitz *et al.*, 1989). These contradictory results could be explained by the different doses of caffeine used as well as different smoking habits and lengths of the protocol.

Alcohol

Chronic consumption of alcohol leads to cirrhosis and in these patients an increase of caffeine half-life was observed up to 50–160 hours (Statland *et al.*, 1976; Statland and Demas, 1980; Desmond *et al.*, 1980; Renner *et al.*, 1984; Wang *et al.*, 1985; Jost *et al.*, 1987; Scott *et al.*, 1988; Varagnolo *et al.*, 1989; Sánchez-Alcaraz *et al.*, 1991).

CAFFEINE METABOLISM

Liver Metabolism

Hepatic microsomal enzyme systems and caffeine metabolism

Caffeine is metabolized by hepatic microsomal enzyme systems and a significant contribution by any other organ cannot be demonstrated. The major role of the liver is assessed by the impaired elimination of caffeine observed in subjects with liver diseases and particularly in alcoholics (for references see 'Alcohol' Statland *et al.*, 1976; Statland and Demas, 1980; Desmond *et al.*, 1980; Renner *et al.*, 1984; Wang *et al.*, 1985; Jost *et al.*, 1987; Scott *et al.*, 1988; Varagnolo *et al.*, 1989; Sánchez-Alcaraz *et al.*, 1991). Caffeine clearance in patients following orthotopic liver transplantation was also used in the differential diagnosis of early post operative complications (Nagel *et al.*, 1990).

Several animal studies have suggested that caffeine at doses as high as 100–150 mg/kg may induce some forms of cytochrome P-450 and thus induce its own metabolism (Berthou *et al.*, 1995). There was a discrepancy for the results obtained with lower doses showing either an induction, inhibition or absence of effect. In healthy volunteers, the daily ingestion of 480 mg caffeine for one week failed to alter caffeine pharmacokinetics and metabolism (George *et al.*, 1986). In healthy volunteers, resuming coffee drinking after 21 days restriction from caffeine-containing foods and beverages, showed a significant but relatively small decrease of the elimination rate constant of caffeine with large and opposite interindividual variations (Caraco *et al.*, 1990). Their study does not provide a clear answer about the time period required for deinduction to occur.

Induction of caffeine metabolism has been repeatedly demonstrated in animals pretreated with 3-methylcholanthrene and b-naphthoflavone. Similar effects of 3-methylcholanthrene and phenobarbital were demonstrated in cultured human hepatocytes. The induction by 3-methylcholanthrene of the demethylation giving paraxanthine and theophylline was significantly correlated but not for the formation of theobromine, suggesting that at least two isozymes of the P-4501A family are involved in the demethylation of caffeine. Caffeine was proposed as a safe probe for measuring the relative P-4501A2 activity in human populations and it was demonstrated that this enzyme does not support the N-7 and N-1 demethylation of caffeine which is mediated, at least partly, by other P-450 enzymes (Berthou *et al.*, 1991).

Metabolites

Extensive reviews on the discovery of the products of caffeine metabolism have been published (Arnaud, 1984; Arnaud, 1987; Anonymus, 1991). Caffeine is eliminated through liver biotransformation to dimethyl and monomethyl xanthines, dimethyl and monomethyl uric acids and various uracil derivatives. The quantitative profile of urinary caffeine metabolites will depend on the rate of their formation, their body distribution, their plasma concentration and their renal excretion. The metabolic pathways (Figure 6.1) show the multiple and separate demethylation, C-8 oxidation and uracil formation observed in man.

These transformations occur in liver microsomes except the C-8 oxidation of 1-methylxanthine into 1-methyluric acid which is mediated by xanthine oxidase. The reverse biotransformation of theophylline into caffeine was first reported in premature infants treated in the management of apnea (Bory *et al.*, 1979). The accumulation of caffeine observed in theophylline-treated babies is due to the immaturity of the hepatic microsomal enzymes. In adult subjects, caffeine produced from the conversion of theophylline to caffeine is extensively metabolized and has been evaluated at 6% of the theophylline dose (Arnaud, 1984).

The major metabolic difference between rodents and man is that in the rat, 40% of the caffeine metabolites are trimethyl derivatives but in man they are less than 6% (Arnaud, 1985). Because of the importance of demethylation in man, breath tests were developed using caffeine labeled on the methyl group with radioactive or stable isotopes to detect impaired liver function. Man is characterized by the quantitative importance of 3-methyl demethylation leading to the formation of paraxanthine. This first metabolic step represents 72–80% of caffeine metabolism (Arnaud and Welsch, 1982). The quantitative urinary excretion of caffeine metabolites in man, expressed as the percentage of the administered dose is shown in Table 6.1.

The analysis of urinary caffeine metabolites in man shows the presence of uracil derivatives produced from caffeine, 6-amino-5-[N-formylmethylamino]1, 3-dimethyluracil (1,3,7-DAU), from theobromine, 6-amino-5-[N-formylmethylamino]-1-methyluracil (3,7-DAU) and from paraxanthine, 6-amino-5-[N-formylmethyl-amino]-3-methyluracil (1,7-DAU) (Arnaud and Welsch, 1979; Latini *et al.*, 1986). The uracil metabolite of caffeine represents about 1% of the

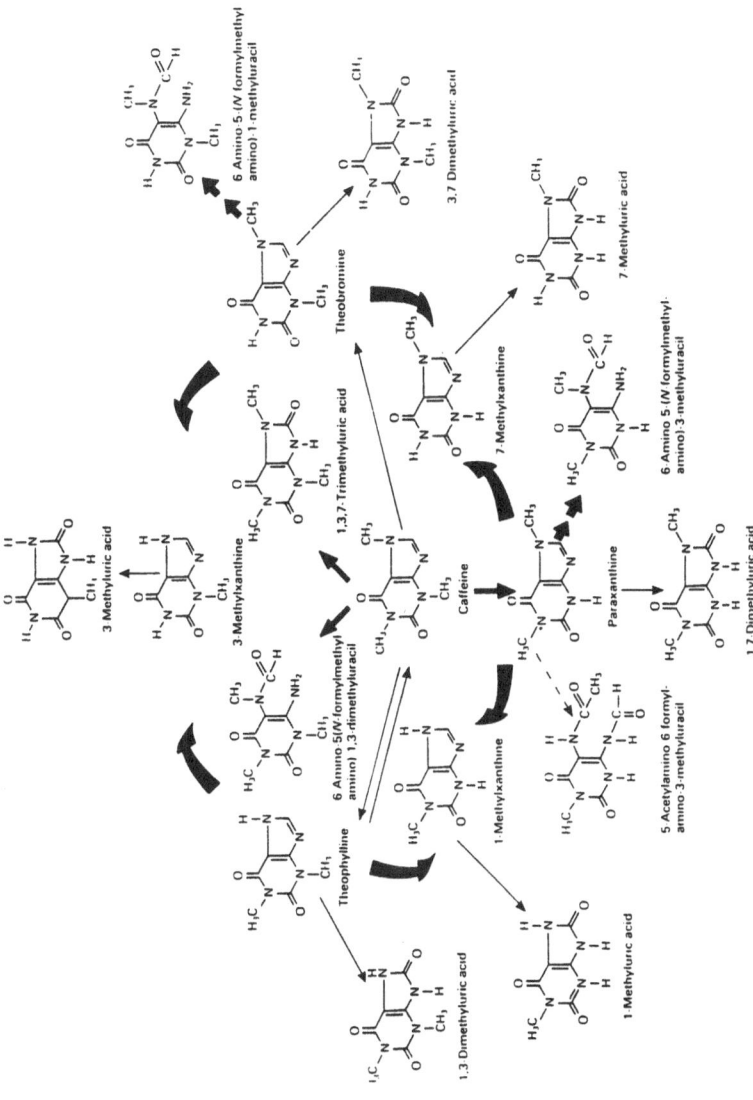

Figure 6.1 – Metabolic pathways of caffeine.

Table 6.1 – Urinary excretion of caffeine and its metabolites in humans

Recoveries expressed as the percentage of the administered dose

caffeine	1.2	6-amino-5-[N-formylmethylamino] 3-methyluracil (1,7-DAU)	2.4
trimethyluric acid	1.3	6-amino-5-[N-formylmethylamino] 1-methyluracil (3,7-DAU)	2.0
6-amino-5- [N-formylmethylamino] 1,3-dimethyluracil (1,3,7-DAU)	1.1	1-methylxanthine	18.0
paraxanthine	6.0	7-methylxanthine	7.0
theophylline	1.0	3-methylxanthine	3.0
theobromine	2.0	1-methyluric acid	25.0
1,7-dimethyluric acid	6.0	7-methyluric acid	—
1,3-dimethyluric acid	2.5	3-methyluric acid	0.1
3,7-dimethyluric acid	0.8	5-acetylamino-6-formylamino- 3-methyluracil (AFMU)	15.0

administered dose in the urine of adult subjects while its excretion increased in the urine of a premature infant in the case of caffeine overload (Gorodischer *et al.*, 1986). An acetylated uracil derivative, 5-acetylamino-6-formylamino-3-methyluracil (AFMU), the major caffeine metabolites in man has not been identified in animal species. After identification of the acetylated urinary metabolites, 5-acetylamino-6-amino-3-methyluracil (AAMU) and AFMU, it was shown that their production and excretion rates were related to the acetylation polymorphism (Grant *et al.*, 1983a) with a bimodal distribution of the general population into fast and slow acetylators. Paraxanthine is the precursor of AFMU which accounts for 67% of paraxanthine clearance (Grant *et al.*, 1983b). The rate of AFMU production and clearance approximates and changes according to the rates for the production of 1-methylxanthine and 1-methyluric acid (Yesair *et al.*, 1984), suggesting that its formation occurs through a common precursor of AFMU and 1-methylxanthine. This intermediate is not yet identified.

The ratios of urinary concentrations of AFMU/1MX or AFMU/1MX+1MU (Grant *et al.*, 1984; Hardy *et al.*, 1988; Evans *et al.*, 1989; McQuilkin *et al.*, 1995) or with the complete conversion of AFMU into AAMU, the ratios of AAMU/1MX or AAMU/1MX+1MU (Tang *et al.*, 1986; Tang *et al.*, 1987; Kilbane *et al.*, 1990) give markers of acetylator status in man. In addition, the ratio of 1MU/1MX represent an index of xanthine oxidase, 1,7DMU/1,7DMX of microsomal 8-hydroxylation, AFMU+1MX+1MU/1,7DMX of microsomal 7-demethylation (Kalow, 1984) and the caffeine metabolic ratio (CMR), AFMU + 1MX + 1MU/1,7DMU reflects microsomal 3-demethylation and also systemic caffeine clearance as well as polycyclic aromatic hydrocarbon-inducible cytochrome P-450 activity (Arnaud and Enslen, 1992; Campbell *et al.*, 1987a,b). The molar ratio of paraxanthine/caffeine in urine taken 3-4 hours after caffeine administration was proposed as an alternative to evaluate hepatic cytochrome P-4501A2 activity (Kadlubar *et al.*, 1990).

Effects of Subject Characteristics on Caffeine Metabolism

Age

In the rat, an age-related decline in the hepatic demethylation and caffeine elimination half-life was demonstrated but this effect was not confirmed in man.

During neonatal development, a progressive increase in the activity of the hepatic microsomal system is demonstrated. The maturation of caffeine elimination in infancy has been evaluated from the patterns of some urinary caffeine metabolites (Carrier *et al.*, 1988). Total demethylation, 3 and 7-demethylation increases exponentially with postnatal age to reach a plateau 120 days after birth while 8-hydroxylation is mature as early as one month of age. The ratio of AFMU/1MX did not change during infancy and was below 0.4 (Pons *et al.*, 1989). This ratio appears to increase in fast acetylators between 6 and 12 months of age. Using the urinary metabolite ratio 1MU/1MX, the total xanthine oxidase activity was shown to increase in premature infants with severe respiratory distress syndrome and in infants in severe acute clinical states (Boda and Németh, 1989).

The urinary metabolite profile of caffeine given either orally or intravenously was compared at a dose of 5 mg/kg in young and elderly males and showed no influence of the route of administration on the excretion of urinary metabolites (Blanchard *et al.*, 1985). Significantly greater amounts of 1-methyluric acid, 7-methyluric acid and 1,7-dimethyluric acid were excreted in the urine of the elderly (69 ± 2 years). However, other studies did not find any significant age or gender-related changes (Grant *et al.*, 1983b; Campbell *et al.*, 1987b).

Sex and hormones

No sex differences in caffeine metabolism were observed from urinary metabolite patterns or metabolite ratios and no correlation was found between age or weight of subjects (Grant *et al.*, 1983b).

Following one month growth hormone therapy in growth-deficient children, the administration of [3-Me-^{13}C]caffeine showed a decreased caffeine 3-demethylation (Levitsky *et al.*, 1989).

Upon administration of oral contraceptives, the urinary excretion of caffeine, paraxanthine and 1,7-dimethyluric acid was increased at the expense of 1-methylxanthine, 1-methyluric acid and the acetylated metabolites: AFMU and AAMU. A 33% decrease of the caffeine metabolic ratio was reported in women using oral contraceptives (Kalow and Tang, 1991a).

During pregnancy, the excretion of 1-methylxanthine and 1-methyluric acid was also increased (Scott *et al.*, 1986). These results are in agreement with a caffeine study showing a significantly increased hydroxylation activity during pregnancy. Late pregnancy was also characterized by a decrease in P-4501A2, xanthine oxidase and acetyltransferase activities (Bologa *et al.*, 1991).

In non-smoking pregnant women and in smoking and non-smoking women using oral contraceptives, the caffeine metabolic ratio was reduced by 29 and 20% respectively compared to a control group, so demonstrating an inhibition of P-4501A2 (Vistisen *et al.*, 1990).

Phenotypes

The caffeine metabolic ratio and the xanthine oxidase index were not different between Chinese and European populations although individual metabolites differed, due to a higher proportion of rapid acetylators in the Oriental compared with the European population. Within a log normal distribution, the metabolic ratio showed a 6.3-fold range variation and only a 1.7-fold range for the xanthine oxidase ratio (Kalow and Tang, 1991a).

Wide human interindividual differences appear for caffeine 3-demethylation which was shown to vary 57 fold in a study using more than twenty human liver microsomal preparations. Another study using microsomal preparations confirmed the high degree of inter-liver variation in metabolic rate and showed a 20-fold range in paraxanthine demethylation rates (Campbell *et al.*, 1987a). Genetic and/or environmental factors are suggested as an explanation for the larger variability of caffeine clearance (Nagel *et al.*, 1990). In man, the acetylated uracil derivative (AFMU) is produced according to the polymorphism of acetylator phenotype but this pathway does not seem to play a role in the interindividual variations of caffeine pharmacokinetics (Arnaud and Enslen, 1992).

Intraindividual variations could also be explained by chronovariation in caffeine elimination all along the day, although the effect appears to be small (−25 to 16%) in most of the subjects (Levy *et al.*, 1984).

Obesity

Recent results demonstrated that obese women excreted more theobromine, theophylline and paraxanthine than lean women after the administration of 5 cups corresponding to a total of 20 mg/kg caffeine. Also significantly higher paraxanthine and lower theophylline concentrations were reported in saliva of obese women. These results confirm that obesity alters caffeine metabolism and modify the urinary metabolite concentration ratios used as indexes of enzyme activities (Bracco *et al.*, 1995).

Drugs and disease

Many drugs taken simultaneously with caffeine competitively inhibit its metabolism through the first most important metabolic transformation of caffeine in man, the 3-demethylation (Tarrús *et al.*, 1987). These inhibitory effects on caffeine metabolism have also been demonstrated in vitro using human liver microsomes (Sesardic *et al.*, 1990). Furafylline, a highly selective inhibitor of P-4501A2, is a potent inhibitor of 3-demethylation in man and to a lower extent of 1- and 7-demethylation. Quinolones were also shown to inhibit caffeine 3-demethylation (Fuhr *et al.*, 1990). Some drugs such as allopurinol, cause a dose-dependent inhibition of the conversion of 1-methylxanthine to 1-methyluric acid (Lelo *et al.*, 1989).

The urinary ratio of 1MU/1MX decreases from 0.8–1.0 in the control subjects to 0.15–0.3 and 0.07–0.1 after a 300 and 600 mg/day allopurinol treatment, respectively.

Clinical studies report increasingly frequent drug interactions leading to impaired caffeine elimination. Decreased clearance is observed both for caffeine and its main metabolite, paraxanthine (Lelo *et al.*, 1989). These effects are explained by competitive inhibition at the hepatic microsomal level (Joeres *et al.*, 1987). The excretion of caffeine metabolites and the different urinary ratios were measured in insulin-dependent diabetes receiving 200 mg caffeine (Bechtel *et al.*, 1988). The results showed a variation of caffeine metabolism in the patients with an uncontrolled diabetic state. These conclusions have been questioned (Denaro and Benowitz, 1989) because they are based on large 59% and 89% differences, in the recoveries of urinary metabolites in the two tests on the diabetic patients. Since the clearance of caffeine is not modified in diabetic patients, it is suggested that cytochrome P-4501A2 activity is not impaired (Zysset and Wietholtz, 1991).

The recovery of caffeine metabolites in the urine collected 48 hours after oral administration of 400 mg caffeine was studied in patients with compensated and decompensated cirrhosis (Scott *et al.*, 1988). No significant differences were observed in the overall pattern of metabolite excretion between the control group and patients with decompensated cirrhosis except for a higher caffeine excretion (5% of the dose vs. 2%). Further work is necessary to know whether or not cirrhosis changes the pattern of caffeine metabolite as only very low dose recoveries of 46–57% were reported.

Smoking

Changes in the pattern of caffeine metabolites excreted in urine have been evaluated in cigarette smokers after 3 or 4 days cigarette abstinence. During abstinence, 24-hour urine ratios of dimethylxanthines to caffeine and monomethylxanthines to dimethylxanthines were reduced, suggesting that cigarette smoking accelerates both demethylation steps. Less than 50% of the ingested dose was recovered in 24-hour urine. A study on the enzyme-inducing effect of smoking reports changes on caffeine urinary metabolites. In a population of 178 students including 19 smokers, analyses of the caffeine metabolic ratio demonstrated a dose-effect relationship between this cytochrome index and the number of cigarettes smoked per day as well as the urinary cotinine levels (Kalow and Tang, 1991b). However, the highest enzyme indexes were observed more frequently in non-smokers suggesting that other unknown factors determine P-4501A2 activities. The ratio of urinary metabolites reflecting xanthine oxidase activity was increased in subjects smoking 1–9 and 10 or more cigarettes per day to 1.26 and 1.29, respectively, compared to 1.04 for control subjects. These results suggest that even light smoking increases xanthine oxidase activity (Vistisen *et al.*, 1990).

After a one month training period with eight hours vigorous exercise per day, the caffeine metabolic ratio and the ratio for xanthine oxidase activity increased by 58% and 110%, respectively, while the ratio reflecting acetylation was unchanged. These results confirmed an inducing effect of physical exercise on cytochrome P-450 activity (Vistisen *et al.*, 1990).

Diet and alcohol

Dietary constituents may have an effect on drug metabolism and an impairment or an increased elimination of caffeine was reported with alcohol (George *et al.*, 1986; Mitchell *et al.*, 1983) and broccoli (Vistisen *et al.*, 1990) respectively. Cruciferous vegetables are known to induce cytochromes P-450. The collection and analyses of urine samples excreted by subjects fed a diet supplemented with either 500 g green beans or broccoli for ten days showed a 19% increase of the caffeine metabolic ratio corresponding to an induction of P-4501A2 activity by broccoli (Vistisen *et al.*, 1990). Using urinary ratios to estimate P-4501A2 activity, modifications were reported when dietary habits were changed but these effects were not significant because of the large inter individual variations (Kall and Clausen, 1995).

An induction of hepatic microsomal Cytochrome P-4501A2 in mice treated chronically with ethanol suggests that caffeine metabolism could be affected by ethanol intake (Lucas *et al.*, 1992). In a longitudinal study of 11 healthy male volunteers over several months, lifestyle factors, dose of caffeine, multivitamin and ethanol intake were shown to have modest effects on the caffeine metabolic ratio and no effect on xanthine oxidase index (Kalow and Tang, 1991a).

A recent study shows that the AUC for caffeine administered at a dose of 400mg was significantly higher when administered with 0.8 g/kg alcohol (Azcona *et al.*, 1995). It has also been shown that the administration of 400 mg caffeine one hour before 1 g/kg ethanol did not influence the blood elimination kinetics of alcohol in male volunteers. However, the ingestion of two cups of coffee 30 minutes after 0.5 g/kg ethanol produced a significant increase of blood ethanol levels one hour later (Strubelt *et al.*, 1976). The same group showed earlier in the rat a dose-dependent depression of blood alcohol concentrations when caffeine oral doses of 10, 20, 60 or 100 mg/kg was given 15 minutes prior to 4.8 g/kg ethanol (Siegers *et al.*, 1974).

Environment

A field biochemical study showed the effect of polybrominated biphenyls (PBB) on caffeine 3-methyl demethylation and on the caffeine urinary metabolic ratio (Lambert *et al.*, 1990). The exposure of a rural population to this enviromental factor increased their values in comparison to urban subjects not exposed to PBB. This induction was relatively small and the median value of the smokers of the urban group was even higher. The authors observed an unexplainedly lower induction in female than in male subjects of the PBB-exposed group.

Fecal excretion of metabolites

After oral administration, the use of radiolabelled caffeine has showed that fecal excretion amounts to 8–10% of the dose in rats and 2–5% in man. The products identified in the feces of human volunteers were 1,7-dimethyluric acid, 1-methyluric acid, 1,3-dimethyluric acid, trimethyluric acid and caffeine which amounted to 44, 38, 14, 6 and 2% of fecal radioactivity, respectively (Callahan *et al.*, 1982).

Summary

Dose-dependent pharmacokinetics of caffeine reported in man is explained by a saturation of metabolic transformations. This saturation was observed for a long time in the neonatal period, leading to an impairment of caffeine elimination due to the immaturity of the hepatic enzyme systems. An age-dependent impairment of caffeine metabolism has been observed in animal models, but not in man.

Caffeine can be detected in all body fluids and it passes through all biological membranes, including the blood-brain and placental barriers. No accumulation of caffeine or its metabolites has been observed in various organs or tissues. The metabolic pathways is characterized by an extensive metabolism of caffeine and the predominance of paraxanthine formation in man. Only 0.5 to 2% of an ingested caffeine dose is excreted unchanged in the urine and more than twenty metabolites have been identified in the urine of man.

No differences have been observed in the metabolic rate of caffeine between men and women except a weak effect of the menstrual cycle and a decreased elimination of caffeine as well as changes of metabolic pattern in women using oral contraceptives and during the last trimester of pregnancy.

Obesity and physical exercise were also shown to modify caffeine pharmacokinetics and metabolite formation. Drugs taken simultaneously may impair caffeine elimination due to a competitive hepatic microsomal inhibition. Changes in caffeine elimination and in the pattern of metabolites excreted in urine are observed in cigarette smokers while patients with liver disease exhibit impaired elimination with no apparent change in the metabolic profile.

In man, chronic ingestion or restriction of caffeine intake does not modify its metabolism. The effects of some dietary constituents such as broccoli, alcohol and vitamins have been studied. Alcohol intake was shown to modify plasma pharmacokinetics but no effect on caffeine metabolism was yet reported.

Genetic expression of Cytochrome P-4501A2 plays a role in caffeine elimination and metabolism in man but the polymorphism of acetylator phenotype evaluated from urinary metabolite concentration ratios has no effect on caffeine clearance. Several other urinary concentration ratios have been shown to constitute useful indexes of hepatic enzyme activities for clinical and epidemiological studies.

CHAPTER SEVEN

Caffeine and Human Performance

Odin van der STELT and Jan SNEL

INTRODUCTION

One of the earliest laboratory studies of the effects of caffeine on mental performance carried out under carefully controlled conditions was published more than 80 years ago. The study (Hollingworth, 1912) employed a placebo-controlled and double-blind design in which the effects of caffeine were assessed on sleep and several tests of human performance. The main conclusion was that caffeine possesses clearly properties for the facilitation of performance. Further study was found necessary to elucidate the mechanisms underlying the observed caffeine effects. Since then a number of studies have confirmed these early observations (Weiss and Laties, 1962), but the literature also includes many conflicting and ambiguous reports of caffeine's effects. In fact, despite a large number of papers, some of the more recent reviews (Bättig, 1985; Dews, 1984) found it difficult to arrive at a coherent account of the principal effects of caffeine on human performance.

There are several possible reasons for the inconclusive results reported in the caffeine literature. One main reason probably relates to the great variability in experimental approach that have been employed to examine performance effects of caffeine (Lieberman, 1992). Another reason is that caffeine's effects on performance may be moderated by a wide variety of variables, which include dose, the time since ingestion, and personal, situational and task variables. The main thrust of this chapter is that only if the experimental design and method as well as such 'moderator' variables are carefully chosen, caffeine's effects on aspects of performance may be detected more consistently and, eventually, may be better understood.

This chapter aims at providing relevant information with respect to two basic issues in the study of caffeine's effects on performance. The first is the question of *when* effects of caffeine occur. This issue is an empirical one and concerns the identification of the aspects of human task performance that might be sensitive to caffeine. It also concerns the identification of variables that moderate the caffeine-performance relationship. These moderator variables may be seen (see Baron and Kenny, 1986) as the factors that specify the conditions under which effects of caffeine most likely occur. The second, more theoretical, issue involves the question of *how* or *why* certain effects of caffeine occur. Within an information-processing theoretical framework, this *how* or *why* issue is concerned with the processing mechanisms and processes internal to the subject that are hypothesized to underlie or mediate the observed caffeine-performance relationship (see Lorist, Ch. 8). While both issues are important, this chapter is more concerned with the *when* question than the *how* or *why* one.

STUDIES ON CAFFEINE AND PERFORMANCE 1984–1995

Study selection

Eighty-five studies concerned with the effects of caffeine on mental performance that were published between 1984 and 1995 were identified by conducting computer literature searches (MedLine and PsycLIT, American Psychological Association), using the combined key words 'caffeine' and 'performance'. The reference citations of the identified articles also were reviewed, in order to ensure that other relevant studies were not missed. The criteria for inclusion were that the study: (a) involved the acute effects of caffeine on mental performance, (b) included a placebo condition, and (c) utilized healthy adult human subjects. 'Mental performance' was meant broadly, ranging from the measurement of subject's psychophysical thresholds, like the critical flicker fusion threshold, to higher-order cognitive performances, such as problem solving, and to simple speeded motor performances, as finger tapping. Only studies published in the form of journal articles were included in this review.

Excluded were studies of the effects of caffeine on physical performance, on mental performance-induced stress, on the combined effects of caffeine with other substances or experimental manipulations, and also those studies on the efficacy of caffeine to counteract the effects of benzodiazepines on next-day sleepiness and performance.

General features

About half of the studies used a within-subjects (cross-over) design to assess the effects of caffeine. The other half employed a between-subjects (independent groups) design. The potential advantage of the former over the latter design is increased statistical power to detect true caffeine effects by preventing inter-individual differences from contributing to the error variance. A disadvantage is the presence of a potential differential carryover effect of caffeine, which confounds estimates of its effects (Maxwell and Delaney, 1990).

Several studies have examined caffeine's effects by means of a test battery purporting to sample a diversity of mental functions. The main disadvantage of this approach is that most tests in these batteries have no history of reliability and validity, which seriously hampers the interpretation of the results (Gaillard, 1988; Hindmarch, 1980; Parrott, 1991a,b,c). By comparison, other studies have evaluated caffeine's effects within an information-processing framework.

This approach typically has employed a single task with a solid theoretical basis, and by that could evaluate caffeine's effects as a function of experimental task manipulations (see Lorist, Ch. 8). This approach usually allows more robust and specific interpretations of results (Sanders, 1983; Sherwood and Kerr, 1993; Warburton and Rusted, 1989). Problems may arise, however, in the generalization of results beyond the particular task paradigm used.

Most studies had at least 10 subjects, mainly young male adults, mostly students. The self-reported habitual level of daily caffeine consumption usually is low-to-moderate (about 200–300 mg), and in nearly all studies they are instructed to abstain from caffeine-containing substances for a period of time (usually 10 hours or more) prior to testing. In some studies, the subjects are also asked to abstain from alcohol and/or to fast for some hours before the experiment. Experimental control is often, but certainly not always, exerted over a number of well-known factors associated with inter individual differences in caffeine metabolism, such as smoking, liver disease, and for females the use of oral contraceptives and pregnancy. We noticed that many studies did not assess caffeine saliva or plasma levels. As a consequence, uncontrolled variations in baseline and/or achieved plasma caffeine concentrations, due to the subjects' failure to comply with the abstinence instructions or to uncontrolled individual differences in caffeine metabolism, may have confounded some of the reported results.

The caffeine given to the subjects were taken orally, as a fixed dose or a dose of grams per kilogram body weight, in anhydrous form (e.g., in gelatine capsules) or dissolved in a drink (e.g., fruit juice or decaffeinated coffee). In general, the amounts of caffeine (often 200–300 mg or more) were substantial larger than those typically consumed in beverages, foods, or over-the-counter drugs on a single occasion (Lieberman, 1992). In addition to the fact that the subjects were typically instructed to abstain from caffeine and sometimes also from specific foods for some period of time prior to the experiment, apparently caffeine's effects generally have not been studied under ecologically relevant consumption conditions.

Results

A wide diversity of tasks has been used in the study of caffeine's effects on performance. Unfortunately, there exists no generally accepted taxonomy of human task performance (Fleishman and Quaintance, 1984; Hockey, 1986). This notwithstanding, in an attempt to organize and facilitate generalizations of research, we have chosen to order the tasks, if possible, on the basis of their nature or structure. This resulted in the distinction of five broad but related areas of mental function: (I) sensation and perception, (II) cognition, (III) memory and learning, (IV) motor co-ordination, and (V) attention. Principally, two kinds of behavioural measures have been employed, namely those based on judgement and those based on time (Sanders, 1988). Succeeding sections present the major caffeine results with regard to these measures.

SENSATION AND PERCEPTION

There have been few studies of caffeine's effects on sensory-perceptual tasks. Böhme and Böhme (1985) have reported that caffeine can facilitate the ability to discriminate colours, for both male and female subjects. But for females, caffeine was found to impair colour discrimination when also oral contraceptives were administrated.

Several studies (Fagan *et al.*, 1988; Kerr *et al.*, 1991; King and Henry, 1992; Mattila *et al.*, 1988; Nicholson *et al.*, 1984; Swift and Tiplady, 1988; Yu *et al.*, 1991) have evaluated caffeine's effects on the subject's threshold to the fusion of a flickering light source, referred to as the critical flicker fusion threshold (CFFT). The CFFT often is used as an index of the state of CNS arousal. While the measurement of this threshold seems to be generally a reliable, valid, and pharmacosensitive psychophysical technique (Curran, 1990), five (Fagan *et al.*, 1988; Kerr *et al.*, 1991; Mattila *et al.*, 1988; Swift and Tiplady, 1988; Yu *et al.*, 1991) out of seven studies failed to find effects of caffeine on the CFFT.

There are few data available with regard to caffeine's effects on the sensory and perceptual processing of auditory information. Bullock and Gilliland (1993) recorded the human brainstem auditory evoked potential (BAEP), and observed that caffeine decreased the latency of Wave V of the BAEP. Schicatano and Blumenthal (1994) reported that caffeine (4 mg/kg) delays the habituation of the human acoustic startle reflex. These findings indicate that caffeine alters the integrity of the auditory sensory pathways, presumably at the level of the brainstem.

Gupta and co-workers (Gupta *et al.*, 1994; Gupta and Gupta, 1990, 1994) have shown that various amounts of caffeine facilitate perceptual-tactile judgements, as indexed by an improved estimation of the felt width of blocks employing haptic presentation, although personality modified these effects.

Caffeine when applied directly to the surface of the tongue has also been found to enhance the perceived taste intensity of some sensory stimuli (Schiffman *et al.*, 1985, 1986) This taste potentiation of caffeine, however, is probably mediated by a peripheral rather than a central physiological mechanism, since caffeine does not appear to affect perceived taste intensity under natural conditions of ingestion (Mela, 1989). Similarly, as implicated by clinical and animal studies, caffeine may have the potential to affect the processing of nociceptive (pain) information (Snel and Lorist, 1996). This caffeine effect may well represent the net outcome of parallel and, possibly, physiologically opposing peripheral and central actions (Sawynok and Yaksh, 1993).

COGNITION

Simple and choice reaction time

A relatively large number of caffeine studies have employed a simple reaction time (SRT) and/or a choice reaction time (CRT) type of task. In SRT tasks, the subject is required as rapidly as possible to make a fixed response to a single stimulus. CRT-tasks are similar to SRT-tasks, except that the subject is exposed to different stimuli, each of which requires a different response. In about half of the experiments, utilizing either visual or auditory stimuli, caffeine was found to reduce SRT or CRT, while response accuracy was either improved or not affected by caffeine (Jacobson and Edgley, 1987; Kerr *et al.*, 1991; King and Henry, 1992; Lieberman *et al.*, 1987a,b; Lorist *et al.*, 1994a; Roache and Griffiths, 1987;

Smith, A.P., 1994; Smith, A.P. *et al.*, 1993a, 1994a,b; Swift and Tiplady, 1988). These caffeine effects were observed utilizing a wide range of doses, although Jacobson and Edgley (1987) reported that a large dose as 600 mg caffeine exhibited no effect. In the other half of the experiments, however, reaction time (RT) was not affected by caffeine (Bruce *et al.*, 1986; Bullock and Gilliland, 1993; Fagan *et al.*, 1988; Kuznicki and Turner, 1986; Landrum *et al.*, 1988; Lieberman *et al.*, 1987a,b; Münte *et al.*, 1984; Smith, A.P. *et al.*, 1991; Smith, B.D. *et al.*, 1991; Swift and Tiplady, 1988; Zahn and Rapoport, 1987).

Possible reasons for the difference between the positive and null findings seen with caffeine may relate to differences between experiments in dose, protocol, task, and/or subject variables (Lieberman, 1992). In addition, most studies fail to distinguish between the '*decision*' and '*motor*' component of the task. The former component is believed to index central processes underlying task discriminatory and decisional processes, while the latter is assumed to index only peripheral, motor output or execution processes (e.g., Frowein *et al.*, 1981; Welford, 1968). Accordingly, it may be the case that caffeine's effects on RT are subtle and selective in that only the relatively minor, more peripheral aspects of the reaction process are affected by caffeine. Indeed, some of the studies have shown (Kerr *et al.*, 1991; King and Henry, 1992; Swift and Tiplady, 1988), although not all studies concur (Jacobson and Edgley, 1987), that caffeine is more likely to facilitate the *motor* than the decision component of the task.

Recent studies by Lorist *et al.* (1994a), applying the additive factors method (Sternberg, 1969; Sanders, 1983, 1990) to assess main and interaction effects of task variables and caffeine on visual CRT, confirmed and extended these findings. Caffeine was observed to facilitate both input-perceptual (*feature extraction*) and output-motor (*response preparatory*) aspects of the reaction process, while central or cognitive (*response choice*) aspects were not affected.

Speeded decision-making

In studies on caffeine's effects on the performance of a verbal reasoning task (Bonnet and Arand, 1994a; Borland *et al.*, 1986; Mitchell and Redman, 1992; Rogers *et al.*, 1989; Smith, A.P., 1994; Smith, A.P. *et al.*, 1991, 1992b, 1993a, 1994a,b) subjects typically are shown statements about the order of the letters A and B, each sentence being followed by a pair of letters *AB* or *BA* (e.g., *A* follows *B: BA*). The subject's task was to read the statement, look at the pair of letters, and then decide, as fast as possible, whether the statement was a true or false description of the order of the letters. In 9 out of 12 experiments, caffeine was found to improve the speed or accuracy of logical reasoning (Bonnet and Arand, 1994a; Borland *et al.*, 1986; Mitchell and Redman, 1992; Rogers *et al.*, 1989; Smith, A.P., 1994; Smith, A.P. *et al.*, 1991, 1992b, 1993a, 1994a,b).

In 3 out of 6 experiments, caffeine has been shown also to improve the performance of a speeded semantic processing task in which the subjects were shown a series of sentences (e.g., canaries have wings) and had to decide, on the basis of general knowledge, whether the sentence was true or false (Smith, A.P., 1994; Smith, A.P. *et al.*, 1993a, 1994a).

Cancellation

In examining caffeine's effects on the performance of a cancellation task (Anderson, 1994; Bättig *et al.*, 1984; Borland *et al.*, 1986; Bruce *et al.*, 1986; Frewer and Lader, 1991; Loke, 1988, 1990; Loke *et al.*, 1985; Rogers *et al.*, 1989) subjects usually are presented sheets with digits, letters or symbols. The task is to cancel as many specified target items as fast as possible. In some studies caffeine improved task performance (Anderson, 1994; Bättig *et al.*, 1984; Borland *et al.*, 1986; Frewer and Lader, 1991; Loke, 1988), other studies found no effects (Bruce *et al.*, 1986; Loke, 1990; Loke *et al.*, 1985; Rogers *et al.*, 1989). The benefits seen with caffeine, however, were modified by dose (Frewer and Lader, 1991), time on task (Bättig *et al.*, 1984), and the memory load of the task (Borland *et al.*, 1986; Frewer and Lader, 1991). The latter interaction signifies that caffeine improves cancellation performance only when relatively few target items have to be retained in memory.

Substitution

Digit or symbol substitution tasks, which require subjects as rapidly as possible to replace symbols by digits (or vice versa), do not seem to be sensitive to caffeine's effects (Bruce *et al.*, 1986; Lieberman *et al.*, 1987b; Loke, 1989; Mattila *et al.*, 1988; Roache and Griffiths, 1987; Rush *et al.*, 1994a,b; Yu *et al.*, 1991), except perhaps when performed under suboptimal conditions, such as during the night (Borland *et al.*, 1986; Nicholson *et al.*, 1984; Rogers *et al.*, 1989) or in the later part of the test session (Rush *et al.*, 1993).

Other cognitive tasks

In two studies (Borland *et al.*, 1986; Nicholson *et al.*, 1984) caffeine was observed to improve the copying of symbols during the night-time, but did not so during the day-time (Bruce *et al.*, 1986). Only Anderson (1994) reported that caffeine dose-dependently improved the performance of a verbal abilities task, but only in high impulsive subjects. For low impulsives, performance first improved and then declined with increasing dose. As for intelligence caffeine has been found to improve some measures of intelligence (Gupta, 1988a,b), but again these effects were dependent on dose and personality. Additionally, one study has shown that a 200 mg dose caffeine tended to improve the number of solved problems and the number of correct solutions on a concentration performance test (Dimpfel *et al.*, 1993), whereas a 400 mg dose tended to impair performance.

Other cognitive tasks that have been used in caffeine research include arithmetic (Loke, 1990; Loke *et al.*, 1985; Mitchell and Redman, 1992; Roache and Griffiths, 1987) writing speed (Landrum *et al.*, 1988), reading comprehension (Landrum *et al.*, 1988; Mitchell and Redman, 1992), sentence completion (Loke, 1990), solving anagrams (Smith, B.D. *et al.*, 1991), classification of pictures (Swift and Tiplady, 1988), and card sorting (Loke, 1990; Loke *et al.*, 1985). None of these activities were found to be affected significantly by caffeine, except that one study found an improvement of addition performance (Uematsu *et al.*, 1987), although a 600 mg dose reduced the number of completed additions (Loke, 1990).

MEMORY AND LEARNING

Free recall

In an immediate free recall task with a supraspan word list, subjects are presented a list of unrelated words, exceeding their memory span, and then asked to recall as many words as possible in any order. The basic finding is (Arnold *et al.*, 1987; Barraclough and Foreman, 1994; Erikson *et al.*, 1985; Foreman *et al.*, 1989; Loke, 1988, 1993; Loke *et al.*, 1985; Smith, A.P., 1994; Smith, A.P. *et al.*, 1991, 1994a,b; Terry and Phifer, 1986) that words occurring at the beginning and end of the list are recalled better than middle-list words, producing a so-called primacy and recency effect, respectively. This recall pattern is referred to as the serial position effect. It is usually hypothesized (e.g., Atkinson and Shiffrin, 1968) that the more recent words are retained in primary, short-term, or working memory, while the earlier items are retained in secondary or long-term memory.

Caffeine was found either to exert no effects on recall performance (Foreman *et al.*, 1989; Loke, 1988, 1993; Loke *et al.*, 1985; Smith, A.P., 1994; Smith, A.P. *et al.*, 1991) to improve recall (Arnold *et al.*, 1987; Barraclough and Foreman, 1994; Smith, A.P. *et al.*, 1994b) or to impair recall performance (Erikson *et al.*, 1985; Loke, 1993; Smith, A.P. *et al.*, 1994b; Terry and Phifer, 1986). In one study (Arnold *et al.*, 1987) improvements with caffeine were seen only in females at the third level of practice, while decrements were seen in males at the second and third levels of practice. In Erikson's *et al.* study (1985), detrimental caffeine's effects were seen only for females under a slow rate of stimulus presentation, while no effects were observed for males. Similar effects were found in high impulsives (detrimental effects) but not in low impulsive subjects (Smith, A.P. *et al.*, 1994b).

A few studies have evaluated caffeine's effects on recall more precisely by taking into account the serial position effect (Arnold *et al.*, 1987; Barraclough and Foreman, 1994; Erikson *et al.*, 1985; Smith, A.P. *et al.*, 1994b; Terry and Phifer, 1986). Some of the results indicated that neither the primacy nor the recency effect is affected specifically by caffeine (Arnold *et al.*, 1987; Erikson *et al.*, 1985; Smith, A.P. *et al.*, 1994b). Instead, it was observed (Barraclough and Foreman, 1994) that small to moderate amounts of caffeine, at least for males, exerted its largest effect on the middle portion of the list. According to the authors, this selective caffeine effect may be the consequence of a general increase in CNS activity "leading to increased salience of stimuli that are normally of low recall priority".

Other immediate free recall tasks used, include the recall of 8-digit numbers (Nicholson *et al.*, 1984), and the recall of five- and six-letter nouns (Mitchell and Redman, 1992). In the former study, caffeine tended to impair recall, although not significantly, while caffeine showed no effects in the latter study. In addition, caffeine has been observed to impair the immediate reproduction of numbers during the night (Nicholson *et al.*, 1984), but in an other study (Rogers *et al.*, 1989), also measuring during the night, caffeine exerted no effect on number reproduction. Finally, caffeine has been reported (Linde, 1994) to impair the immediate reproduction of spatial relationships of verbal information after normal sleep, but to improve it after a vigil, indicating a compensation of fatigue.

Delayed free recall and recognition

Caffeine appears not to affect the *delayed* free recall of word lists (Loke, 1988; Loke *et al.*, 1985), nor the *delayed* recognition of words (Loke, 1988; Loke *et al.*, 1985; Smith, A.P., 1994; Smith, A.P. *et al.*, 1994a). In one study (Smith, A.P. *et al.*, 1994b), however, caffeine was seen to impair the delayed recognition of words, but this was observed only for high impulsive subjects; no caffeine effects were seen for low impulsives. Also, caffeine has been reported to impair, although nonsignificantly, the delayed recognition of pictures (Roache and Griffiths, 1987).

Memory-search

Based on Sternberg's memory-search paradigm (Sternberg, 1966, 1969) to evalu-ate caffeine's effects on the retrieval of information from primary memory, three studies (Anderson *et al.*, 1989; Kerr *et al.*, 1991; Lorist *et al.*, 1994b) have employed RT-tasks. In this task paradigm, the subject is presented a sequence of stimuli, and on each trial has to decide, as fast as possible, whether the stimulus is a member of a small memorized set of stimuli. Task memory load is manipulated by varying the size of the memory-set. The slope of the linear regression of RT on memory-set size, then, is assumed to index the rate of scanning in short-term memory, while the intercept is believed to index peripheral perceptual-motor (and other unknown) processes. It appeared that caffeine lowered the intercept, while the slope was either not affected (Lorist *et al.*, 1994b) or increased (Anderson *et al.*, 1989) by caffeine. The latter finding implied that caffeine slowed memory scanning. The remaining study (Kerr *et al.*, 1991), using a constant memory-set size of four items, reported only that caffeine speeded RT.

Memory span

In particular the digit span task has been used to assess caffeine's effects on the capacity of primary memory (Borland *et al.*, 1986; Davidson and Smith, 1989, 1991; Rogers *et al.*, 1989; Smith, B.D. *et al.*, 1993). Span tasks involve the eval-uation of the maximum number of unrelated items that can be recalled in the correct order immediately after presentation. In four out of five studies, caffeine did not affect memory span performance (Borland *et al.*, 1986; Davidson and Smith, 1989; Roache and Griffiths, 1987; Smith, B.D. *et al.*, 1993). In the remaining study (Davidson and Smith, 1991), caffeine was seen to impair back-wards digit recall under a noise condition, but to improve it under a no-noise condition.

Paired-associate learning

In a paired-association learning paradigm, subjects typically are exposed to word pairs of a high and a low degree of semantic association. They are instructed to learn the word pairs. Subjects are then given the first word of each pair as a stim-ulus for recall of the second. Thus, paired-association learning tasks involve cued recall, where the cue is provided by the experimenter rather than 'created' by the subject self as in free recall. Caffeine does not seem to affect paired-association learning performance, neither when recall is assessed immediately (Mattila *et al.*, 1988; Smith, B.D. *et al.*, 1991; Yu *et al.*, 1991), nor when assessed after a delay of 30 minutes (Smith, B.D. *et al.*, 1991).

Serial learning

The effect of caffeine on learning has been examined by assessing changes in recall performance as a function of repeated presentation of information. Caffeine appears neither to exert influence on the learning of words across two or six presentations (Landrum *et al.*, 1988; Terry and Phifer, 1986) nor on the learning of numbers across six presentations (Loke *et al.*, 1985). Also, caffeine does not seem to affect the learning of mental mazes (Bättig *et al.*, 1984). No caffeine effects have been observed on the learning of 10-response sequences using three buttons on a response panel (Rush *et al.*, 1993, 1994a,b), although caffeine did attenuate decrements in learning performance that occurred across the test session (Rush *et al.*, 1993), that were induced by alcohol (Rush *et al.*, 1993) and benzodiazepines (Rush *et al.*, 1994a,b).

Incidental learning

In contrast to the tasks discussed before, in which subjects are told that their memory is tested later (*intentional* learning), a few caffeine studies have used memory tasks where subjects are not told before that there will be a memory test (*incidental* learning). The basic idea behind the use of the incidental learning paradigm is that more experimental control is gained over the subjects' processing activities on the information at the time of learning. In intentional learning, subjects may be inclined to perform additional processing activities in order to improve their performance (Craik and Tulving, 1975).

Gupta (1991) used an incidental learning paradigm to assess caffeine's effects on different encoding processes involved in long-term memory. The subjects first performed an acoustic and a semantic word categorization task, and then were tested, unexpectedly, for the free recall of the words. Within the levels of the processing-framework of memory (e.g., Craik and Tulving, 1975), it was assumed that the acoustic categorization task required shallow or non-semantic processing of the verbal material, while the semantic task demanded deep or semantic processing. The results showed that caffeine facilitated recall for high impulsive subjects after rhyming acquisition, but impaired it after semantic acquisition. Caffeine had no effect on the recall performance of low impulsives. In a subsequent study (Gupta, 1993), similar results were obtained with respect to recognition performance. These findings seem to indicate for high impulsives that caffeine facilitates the encoding of the physical properties of verbal material, but impairs the utilization of semantic information.

Only one other study (Loke, 1993) examined the effect of caffeine on incidental learning. The results show that caffeine did not affect the free recall of words following an incidental learning condition in which the subjects had to repeat the words.

MOTOR CO-ORDINATION

Studies of caffeine's effects on motor performance as measured by RT-tasks have been discussed before. The scanty caffeine studies on motor performance are mainly concerned with fine muscular co-ordination such as hand steadiness (Bonnet and Arand, 1994a; Gemmell and Jacobson, 1990; Jacobson and Thurman-Lacey, 1992; Jacobson *et al.*, 1991; Kuznicki and Turner, 1986; Loke *et al.*, 1985). Caffeine seems to impair hand steadiness (Jacobson and Thurman-Lacey, 1992; Jacobson *et al.*, 1991; Kuznicki and Turner, 1986; Loke *et al.*, 1985), although this effect appeared to depend on dose (Kuznicki and Turner, 1986; Loke *et al.*, 1985) and the habitual level of caffeine consumption (Jacobson and Thurman-Lacey, 1992; Kuznicki and Turner, 1986). Caffeine had no effects on hand steadiness in two other studies (Gemmell and Jacobson, 1990; Kerr *et al.*, 1991), although in the one of Gemmell and Jacobson (Gemmell and Jacobson, 1990) power analysis indicated that the sample size (n = 23), utilizing a between subjects design, might not have been sufficient to detect effects of caffeine. Similarly, two studies (Jacobson and Thurman-Lacey, 1992; Jacobson *et al.*, 1991) reported that caffeine impaired manual dexterity performance, at least for low but not high habitual caffeine users (Jacobson and Thurman-Lacey, 1992). Fine motor co-ordination as assessed by a pursuit rotor tracking task seems to improve by caffeine (Fillmore and Vogel-Sprott, 1994) but other studies (Kuznicki and Turner, 1986; Smith, A.P. *et al.*, 1991) did not find caffeine effects on similar tasks. Other measures of fine motor control, involving tracing (Jacobson and Thurman-Lacey, 1992; Jacobson *et al.*, 1991; Kuznicki and Turner, 1986), trail making (King and Henry, 1992), and peg-board tests (Lieberman *et al.*, 1987b; Smith, A.P. *et al.*, 1991), have not been found to be affected by caffeine. Several studies have included a tapping test to assess caffeine's effects on the speed and endurance of finger movements. Although it was found that caffeine increased the rate of finger tapping (Fagan *et al.*, 1988; Swift and Tiplady, 1988), in most research caffeine exhibited no effects on this measure (Bruce *et al.*, 1986; Fagan *et al.*, 1988; King and Henry, 1992; Landrum *et al.*, 1988; Lieberman *et al.*, 1987a,b; Loke *et al.*, 1985; Mattila *et al.*, 1988).

ATTENTION

In a general sense, attention is concerned with the *control* of information processing (Posner and Peterson, 1990; Shiffrin, 1988). Tasks that have been used to evaluate caffeine's effects on attention may be distinguished in focused-attention, divided attention, and sustained attention tasks (Van der Stelt, 1994).

Focused attention

There have been done only a few behavioural studies of caffeine's effects on focused attention. Focused or selective attention tasks generally are used to assess the subject's ability to select information. Three caffeine studies (Borland *et al.*, 1986; Foreman *et al.*, 1989; Hasenfratz and Bättig, 1992) have used variants of the well-known Stroop-Colour Word Test. This test requires subjects to focus their attention on the colour of printed words, while ignoring their meaning. The results, however, are inconsistent, with one report of positive caffeine's effects (Hasenfratz and Bättig, 1992), a null finding (Borland *et al.*, 1986), and a negative report (Foreman *et al.*, 1989).

Smith (Smith, A.P, 1994; Smith, A.P. *et al.*, 1991, 1992b) used a visual focused attention ('filter') task that required subjects to respond, as fast as possible, to a centrally presented letter that was sometimes surrounded by other, irrelevant letters. Also an other attention ('search') task was used, similar to the former, except that the subject did not know at which of two possible locations the letter would appear (see for a discussion of these two types of attention tasks Kahneman and Treisman, 1984). Basically, neither of these tasks were found to be affected by caffeine.

Loke (1992) used two versions of a visual search task, utilizing stimulus frames of four items. One version required subjects to search for target digits among consonant distractors. The other version demanded search for target consonants among consonant distractors. Following Schneider and Shiffrin (1977), the assumption was that the former task involved automatic target detection, and the latter subject-controlled search. Processing (memory) load was manipulated by varying the number of targets (either 2 or 4 targets). The results showed that caffeine exhibited neither main nor interaction effects on performance.

Divided attention

These studies examining caffeine's effects on divided attention have all used the dual task paradigm. Mostly subjects were required to perform a compensatory tracking task simultaneously with a visual or auditory vigilance task. In three studies (Borland *et al.*, 1986; Kerr *et al.*, 1991; Rogers *et al.*, 1989) caffeine was seen to improve dual task performance, but in three other studies (Croxton *et al.*, 1985; Rosenthal *et al.*, 1991; Zwyghuizen-Doorenbos *et al.*, 1990) caffeine exhibited no effects. Although the basis for the difference in outcomes is not clear, it should be noted that in two of the studies reporting benefits, the caffeine was given shortly before midnight, and the dual task was carried out during the night.

Sustained attention

Considerable research efforts have been devoted to effects of caffeine on sustained attention as assessed by vigilance tasks. Vigilance tasks are designed to assess the subject's ability to sustain attention and performance over time. The important aspects of vigilance performance are the overall level of performance and the decrement in performance over time. These aspects of vigilance performance may be differentially affected by experimental manipulations (Davies and Parasuraman, 1982; Mackworth, 1969). Various vigilance tasks have been utilized to study caffeine's effects on sustained attention, most often these were versions of the auditory vigilance task (AVT, Wilkinson, 1968) and the Bakan task (Bakan, 1959).

In auditory vigilance tasks (Fagan *et al.*, 1988; Lieberman *et al.*, 1987a,b; Rosenthal *et al.*, 1991; Zwyghuizen-Doorenbos *et al.*, 1990), lasting 60 minutes or sometimes 40 minutes, subjects were asked to detect the occurrence of slightly deviant (e.g., longer duration) tones occurring infrequently and randomly within a continuous series of standard tones. The tones were presented against a background of white noise at a rate of one every 2 sec. Caffeine was found to improve the overall number and/or speed of correct detections, while the number of false alarms was not significantly affected (Fagan *et al.*, 1988; Lieberman *et al.*, 1987a,b; Rosenthal *et al.*, 1991; Zwyghuizen-Doorenbos *et al.*, 1990). Two studies (Fagan *et al.*, 1988; Rosenthal *et al.*, 1991) also provided information regarding changes in AVT performance over time. It appeared that caffeine reduced the vigilance decrement seen with placebo.

Several studies have used, in one or another form, versions of the Bakan task (Bakan, 1959; sometimes also referred to as the Rapid Information Processing task (RIP), after Wesnes and Warburton, 1978). In this type of vigilance tasks, the subjects usually are presented single digits on a visual screen, and instructed to detect the occurrence of three successive odd or even digits. Stimulus presentation rate typically is fast (100 digits per minute), and task duration across studies ranges from 1 min to 30 min (usually 20–30 min). In comparison with the versions of the AVT which may be identified as 'sensory' types of vigilance task, the Bakan task versions may be better characterized as 'cognitive' vigilance tasks (Davies and Parasuraman, 1982).

Six studies used visual versions of the Bakan task where caffeine was found to improve the overall number and/or speed of correct detections, whereas the number of false alarms (when reported) was reduced or not affected by caffeine (Borland *et al.*, 1986; Eaton-Williams and Rusted [in Rusted, 1994]; Frewer and Lader, 1991; Nicholson *et al.*, 1984; Rogers *et al.*, 1989; Smith, A.P. *et al.*, 1990, 1992b). The later authors found in 1991 that caffeine showed no effects on vigilance performance, but in this study the task used lasted only 1 minute and had a lower presentation rate. Additional information on performance changes over time, indicated (Frewer and Lader, 1991; Smith, A.P. *et al.*, 1990) that vigilance decrement in this type of task was not affected by caffeine. These results have been confirmed in several studies of Bättig and colleagues (Bättig and Buzzi, 1986; Hasenfratz and Bättig, 1994; Hasenfratz *et al.*, 1991, 1993, 1994) using a visual version of the Bakan RIP-task in which the digits were presented in a subject-paced manner rather than at a fixed rate. The benefits induced by caffeine for the level of vigilance performance were observed for both males and females and did not appear to depend on subjects' personality or habitual level of caffeine consumption.

In sum, these findings indicate that caffeine regularly improves the level of vigilance in AVT and Bakan types of vigilance tasks. Caffeine may also have the potential to reduce vigilance decrements in the AVT. Interestingly, however, caffeine may not compensate for decrements in Bakan type of tasks (but, see Hasenfratz, *et al.*, Exp. 3, 1994). Although differences in the information-processing demands between tasks, for example sensory versus cognitive, could play a role in the failure of caffeine to affect vigilance decrements in the Bakan task, other differences in task parameters remain to be studied such as sensory modality, event rate and task duration.

Most other studies examining caffeine's effects on vigilance performance while utilizing a diversity of visual and auditory vigilance tasks, found that caffeine, to a more or lesser extent, improved the overall level of vigilance

(Bonnet and Arand, 1994; Borland *et al.*, 1986; Fine *et al.*, 1994; Frewer and Lader, 1991; Kozena *et al.*, 1986; Rogers *et al.*, 1989; Smith, A.P. *et al.*, 1994a,b). Caffeine exerted some effect on the decrement in vigilance over time (Frewer and Lader, 1991), although there are findings (Fine *et al.*, 1994) that caffeine failed to affect the vigilance decrement or vigilance performance (Croxton *et al.*, 1985; Loke and Meliska, 1984; Smith, A.P., 1994; Smith, A.P. *et al.*, 1992b; Yu *et al.*, 1991). Finally, Swift and Tiplady (1988) reported that caffeine increased the ratio of false alarms to hits, indicating a shift in the subject's response criterion rather than a change in detection efficiency with caffeine.

MODERATOR AND MEDIATOR VARIABLES

A striking feature in research on effects of caffeine on mental performance is the role of moderator variables, i.e., variables that moderate the strength and/or direction of caffeine-performance relationships. Those variables for which there is experimental or hypothetical evidence of such influence are presented in Table 7.1. In general, these variables implicate that effects of caffeine on performance depend on dose, the time since ingestion, and on both personal, environmental, and task factors, which represent the internal and external conditions under which caffeine is used. For detailed information regarding the effects of these moderator variables, the reader is referred to the particular articles. It is of particular interest to consider the moderator variables for their association with distinct aspects of the mental task given to the subject. Some of these task variables may provide clues about the mechanisms underlying observed caffeine effects. For instance, the hypothesis that caffeine's effects can only be found under conditions of 'reduced alertness' and/or 'fatigue' (e.g., Bachrach, 1966; Dews, 1984) was based originally on the findings that caffeine interacted with time on task to improve performance only toward the end of a work period and not at the beginning (Barmack, 1940).

In other words, some of the task variables may not be considered, conceptually nor statistically, as moderator variables, but rather as variables that underlie observed caffeine-performance relationships. For this reason these variables may be seen as potential '*mediator*' variables, that is to say as variables that are related to the tentative generative mechanism through which caffeine is thought to affect performance. Unfortunately, the conceptual, design, and statistical implications of the moderator-mediator variable distinction (Baron and Kenny, 1986) have not yet been addressed explicitly in caffeine research.

CONCLUSION AND DISCUSSION

This chapter considered the behavioural results from recent studies on caffeine and mental performance. The results from different studies often were at variance with each other. A major potential source for these varying results appears to relate to differences in caffeine dose, experimental design, protocol, and methods and procedures of testing. This diversity across studies seems to reflect a lack of consensus on the appropriate methods to employ in human psychopharmacology in

Table 7.1 – Variables potentially moderating caffeine-performance relationships*

Variables	References
CAFFEINE	
dose	(1), (4), (15), (21), (22), (32), (36), (42), (47-49), (52)
time after treatment	(16), (18), (40), (48), (74), (75)
SUBJECT	
age	(80)
sex	(3), (70), (40)
personality	(1), (24-30), (75), (77)
expectancy	(19)
genetic (?)	
habitual caffeine use	(20), (37), (42), (49)
subject's actual condition:	
diet	(35), (74), (76)
drugs	(31), (33), (34), (39), (41), (52), (62), (65-67)
oral contraceptives	(7)
common cold	(71)
wakefulness	(8)
SITUATION	
time of day	(9), (46), (54), (60), (63), (70), (71), (77)
time of session	(22), (67)
noise	(14), (79)
TASK	
time on task	(6), (18), (22), (64)
stimulus rate	(17), (31)
practice	(3), (14)
task structure:	
sensory (?)	
perceptual	(54)
memory	(2), (8), (22)
response	(54)

* Numbers between brackets correspond with the references in the reference list provided with numbers in italics.

general (Gaillard, 1988; Lieberman, 1992). Another potential source for the variable caffeine results lies in the nature of caffeine's effects itself. It seems that the influence of caffeine on performance typically is (a) of a modest size, (b) selective, in that some features of performance are more sensitive than others, (c) complex, perhaps representing patterns of behavioural facilitation and interference, and (d) not constant, in that it can be moderated by a wide variety of variables.

General conclusions

Taking into account these considerations, however, and emphasizing some consistency in results, several conclusions can be drawn.

1 Caffeine can apparently improve the performance of a wide variety of mental tasks indirectly by reducing decrements in performance under suboptimal condi-

tions of alertness. This conclusion is based on the findings that caffeine can improve the performance across various cognitive, learning, memory, and attention tasks by counteracting performance decrements associated with, for instance, time of day (e.g., Nicholson *et al.*, 1984; Schweitzer *et al.*, 1992; Smith, A.P., 1994; Smith, A.P. *et al.*, 1993a), time on task (e.g., Bättig *et al.*, 1984; Fagan *et al.*, 1988; Frewer and Lader, 1991), length of test session (Frewer and Lader, 1991; Rush *et al.*, 1993), prolonged work without sleep (Bonnet and Arand, 1994), lunch (Smith, A.P. *et al.*, 1990; see Smith, Ch. 10), upper respiratory tract illness (Smith, A.P., 1994), or drugs (Hasenfratz *et al.*, 1993, 1994; Kozena *et al.*, 1986; Loke *et al.*, 1985; Roache and Griffiths, 1987; Rush *et al.*, 1993, 1994a,b). The efficacy of caffeine to reduce impairments in mental efficiency under states of reduced alertness is one of the most consistent findings in caffeine research (Dews, 1984; Van der Stelt and Snel, 1993; Weiss and Laties, 1962). The results, however, are not in accordance with the view that caffeine's effect can only be observed under suboptimal conditions (e.g., Bachrach, 1962; Barmack, 1940; Dews, 1984), since benefits of caffeine have been observed under both suboptimal and optimal alertness conditions (e.g., Lorist *et al.*, 1994a,b), and sometimes even to a comparable extent across alertness conditions (Bättig and Buzzi, 1986; Hasenfratz *et al.*, 1994; Lorist *et al.*, 1994b; Smith, A.P. *et al.*, 1993a; Rosenthal *et al.*, 1991). In fact, the further conclusions are based on evidence presumably obtained under more optimal alertness conditions.

- Another conclusion is that caffeine may affect performances of sensory-perceptual tasks. These caffeine effects take the form of either performance facilitation or inhibition, possibly depending on dose, subject variables, and other yet to be specified variables. The available evidence is very limited, however, and much more research is clearly needed here. Earlier studies concur in showing that caffeine decreases the subjects' visual threshold for colour (Kravkov, 1939) and luminance (Diamond and Cole, 1970) or sensory threshold in general (Snel and Lorist, 1996).

- A more robust result is that caffeine decreases hand steadiness. This caffeine effect has clearly been demonstrated (Hollingworth, 1912; Dews, 1984; Nash, 1962; Weiss and Laties, 1962). Hand unsteadiness may occur particularly if relatively large amounts of caffeine are ingested by subjects whose habitual use of caffeine is low (Jacobson and Thurman-Lacey, 1992; Kuznicki and Turner, 1986). The extent to which this caffeine effect represents a generalized effect on fine motor control remains to be determined.

- A fourth conclusion is that caffeine often does not affect performance on cognitive tasks, as noticed also by Weiss and Laties (1962). This conclusion, however, should be qualified in that caffeine does seem to have the potential to improve cognitive performances that are timed, as assessed by RT, decision-making, or cancellation tasks. Lienert and Huber (1966) and Nash (1962) also indicated that cognitive tests requiring 'speed' might be more sensitive to caffeine's beneficial effects than tests involving intellectual 'power'. The results from an additive factor study (Lorist *et al.*, 1994a) suggested, however, that caffeine's effects on speeded cognitive performances may be slight and selective in that only the speed of more peripheral perceptual and motor processes, and not of central cognitive processes, is increased by caffeine.

Accordingly, the perceptual-motor task demands may be the principal determinants of caffeine's effects on speeded cognitive performance, with effects
most likely observed if these task demands are relatively high, for instance if
stimulus quality is low. Similarly, caffeine usually does not affect performances
of learning and memory tasks. Although some studies occasionally have
found caffeine effects on memory and learning performance, either facilatory
or inhibitory, these effects typically emerged as complex interactions with
dose, subject, and task variables. These caffeine effects may represent effects
on the encoding, or the attention devoted to, the information, rather than
direct and specific effects on the storage or retrieval of information in
memory. However, additional research is demanded here, especially in light
of possible detrimental caffeine effects. Similar as cognitive performance,
additive factor studies (Lorist *et al.*, 1994b) have implicated that caffeine can
improve speeded memory-performances through its action on the rate of perceiving and responding.

• Finally, the ingestion of caffeine is likely to improve general levels of performance in vigilance tasks. That is, its efficacy is evident throughout a period of
vigilance, resulting in a steady overall higher level of performance. This
caffeine effect is rather robust and does not appear to depend on the type of
vigilance task. There is limited evidence available regarding the influence of
caffeine on decrements in vigilance that indicates that the influence of caffeine on vigilance decrements is more task-specific.

Theoretical implications

Theoretical interpretations of caffeine's effects on performance, have mainly
been based on arousal and information-processing theories (e.g., Bachrach,
1966; Barmack, 1940; Humphreys and Revelle, 1984; Smith, B.D., 1994; Nash,
1962; Van der Stelt, 1994).

Two tentative general mechanisms or factors that may account for most of the
observed caffeine effects emerge as particularly salient (Van der Stelt and Snel,
1993):

– an indirect, nonspecific 'alertness', 'arousal', or 'processing resources' factor,
presumably accounting for why effects of caffeine generally are most pronounced when task performance is sustained or degraded under suboptimal
conditions,
– a direct, and more specific, cognitive, or 'perceptual-motor' speed or efficiency factor, that may explain why under optimal conditions certain aspects
of human performance and information processing (e.g., those related to sensation, perception, motor preparation and execution) are more sensitive to caffeine's effects than other aspects (e.g., related to cognition, memory, and
learning).

On the basis of the evidence currently available, more definite and detailed conclusions and theoretical claims can hardly be drawn.

FUTURE DIRECTIONS

If progress is to be made in this field of caffeine research, standarization of procedures and techniques, replication of results, and extremely well-designed, controlled, and fine-grained studies are necessary. Studies on effects of caffeine on basic sensory and motor function are demanded. In addition, theory-based and process-oriented studies are needed concerning caffeine's effects on, for instance, perception, motor skill, focused and divided attention, and vigilance. There is a need for further study of effects of moderator variables to determine the conditions or domains of maximal effectiveness of caffeine with regard to a particular performance parameter. The ecological validity of results should be addressed, and the possible role of caffeine withdrawal effects in performance enhancements (James, 1994). Finally, recent developments in cognitive neuroscience (Wikswo *et al.*, 1993) are challenging enough to allow the simultaneous applications of functional brain imaging techniques (e.g., EEG, MEG, PET) while at the same time performing tasks. Application of these advanced techniques will give a better insight into the brain structures involved in the effects of caffeine on mental performance. We have the feeling that the research has just begun.

SUMMARY

Over the last decade, many studies have been conducted to evaluate effects of caffeine on mental performance. The results of these studies, however, are not always consistent, presumably because of the large differences in experimental approach and the nature of caffeine's effects self. Nevertheless, certain consistencies in results do emerge. Caffeine reliably reduces decrements in performances of a variety of mental tasks in states of reduced alertness. Under more optimal alertness conditions, caffeine impairs hand steadiness, and may improve the performance on sensory-perceptual tasks. Caffeine typically does not affect performances on cognitive, memory, and learning tasks, although the rate of perceiving and responding, as assessed by time measures, can be improved by caffeine. Limited evidence is available with respect to caffeine's effect on focused- and divided-attention, although caffeine has been found to increase rather consistently levels of vigilance performance.

It may be hypothesized that both an indirect, nonspecific 'arousal' or 'attention' factor and a direct, more specific 'perceptual-motor' factor account for various effects of caffeine on performance. Further well-controlled, theory-based, and detailed studies are necessary, however, to substantiate and specify the general findings and the invoked explanatory concepts.

CHAPTER EIGHT

Caffeine and Information Processing in Man

Monicque M. LORIST

INTRODUCTION

The increasing emphasis on cognitive work and mental efficiency in daily life, justifies research trying to answer the question what the effects of caffeine are on human information processing activities. Does caffeine have an influence on task performance, and more important, if so, which specific cognitive processes are affected?

Caffeine is generally accepted to be a mild stimulant acting on the central nervous system, producing diverse and complex effects, even when consumed in small quantities (for reviews, see Dews, 1984a; Garattini, 1993). Measures of cortical brain activity might serve as an indication of these stimulating effects. With increasing energy, the electroencephalogram (EEG) shows more activation and changes towards faster frequency and lower-amplitude activity. After caffeine a reduction in alpha and delta power has indeed been found (Kenemans and Lorist, 1995; Bruce et al., 1986; Etevenon et al., 1989; Newman et al., 1992; Pollock et al., 1981; Saletu et al., 1987). Moreover, these studies have revealed some task-specific patterning in EEG activity (Gale and Edwards, 1983), although these measures seem less suitable to uncover more detailed effects of caffeine on cognitive processing.

This chapter tries to shed more light on the specific effects of a single dose of caffeine, comparable to two average cups of coffee, on cognitive activities. After some general remarks, the effects of caffeine on specific operations underlying human information processing are discussed. Knowledge of the involvement of different brain areas and neurotransmitter systems in the regulation of performance served to create more specific hypotheses about the actions of caffeine.

GENERAL REMARKS

Information processing is generally thought of as a sequence of information transformations that take place between stimulus and response. These information transformations or structural processes are always necessary for task performance (Gaillard, 1988; Posner and Rothbart, 1986). The involvement of different structural processes depends on both the task to be performed and the strategy used by subjects. In Figure 8.1 a model is shown of human information processing, based on Pribram and McGuinness (1975), Sanders (1983), Tucker and Williamson (1984) and Mulder (1986). Generally three stages can be distinguished, which correspond to sets of functionally different structural processes (mid part of Figure 8.1).[1]

The first stage consists of perceptual or input processes, which are information transformations related to the encoding and identification of incoming information (e.g., feature extraction, stimulus identification). More specific, input from the retina produces activity in different brain areas analysing either spatial or nonspatial stimulus features. The output of these analyses is compared to stored memories in order to identify stimuli. The second stage includes controlled processes necessary for further processing of information (e.g., serial comparisons, binary decisions and response selection). Different memory systems play an important role in the adequate functioning of these processes. Finally, one can discern the response or output stage. Processes in this stage are involved in the organisation of motor related activities, that is, preparation, activation, and execution of responses, resulting in, for example, a button press.

In addition to the necessary presence of different structural processes, the quality of human information processing is supposed to be regulated by energetical mechanisms (Heemstra, 1988; Hockey *et al.*, 1986; Sanders, 1983; Wickens, 1991). These processes are related to higher level control over information processing and focus on explaining mechanisms responsible for changes in the efficiency of performance. These mechanisms determine performance in those cases in which the structural demands of the tasks remain unchanged but the intensity of performance is manipulated (e.g., well-rested versus fatigued subjects).

Various neurotransmitter systems are assumed to be involved in the regulation of these energetical aspects of the information processing system, acting through processes of facilitation and/or inhibition (Posner and Rothbart, 1986). The behavioural effects of caffeine are thought to be mainly due to its antagonism of adenosine, that is, through blockade of adenosine receptors (Daly, 1993; Phillis, 1991). Adenosine is known to have depressant effects on behaviour by inhibition of, predominantly, excitatory neurotransmitters, and these depressant effects can be reversed by caffeine. The indirect effects of caffeine through antagonism of adenosine receptors on noradrenaline, acetylcholine and dopamine neurotransmitter systems, suggest that caffeine might be able to affect a broad range of information processing activities.

In order to be able to examine effects of caffeine on the timing and organisation of cognitive processes occurring in the brain during task performance, event-related brain potentials (ERP) can be used (Renault *et al.*, 1982; Ritter *et al.*, 1982, 1983). ERPs are sequences of voltage deflections in the spontaneous

---

[1] Although a distinction is made between different stages, the distinction is not as clear as indicated. In fact, experimental evidence suggests that information might undergo continuous transformations, with an overlap in time between different processing stages. Although this should be noted, this issue is not the scope of this chapter and the interested reader is referred to some extended reviews (e.g., Miller, 1988; Smid, 1993).

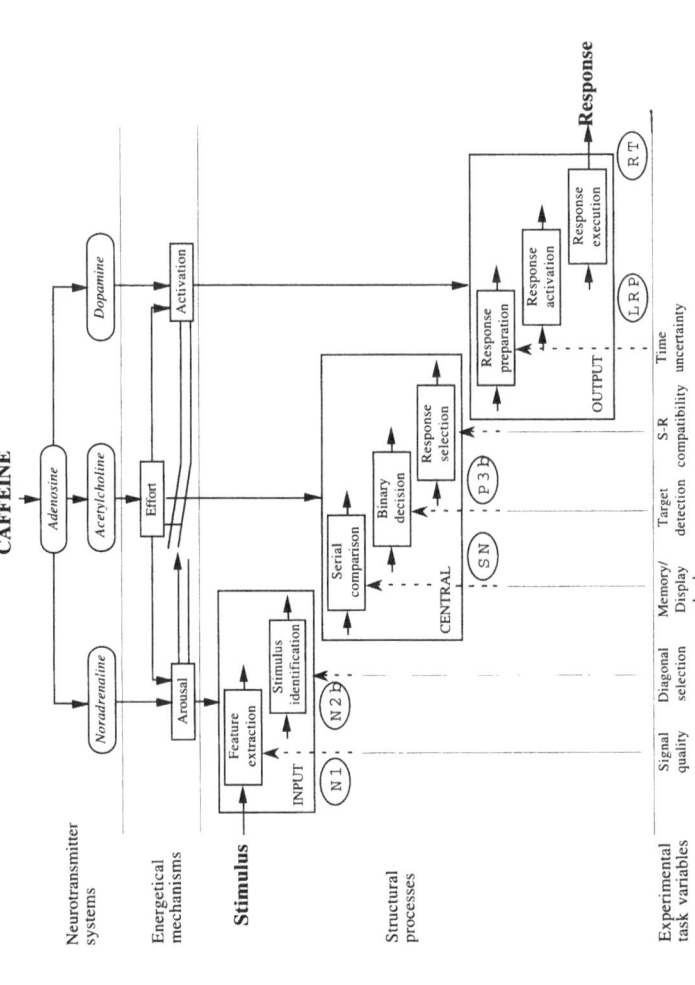

Figure 8.1 – A model of the human information processing system, in which different energetical mechanisms are related to specific structural processes. The energetical mechanisms are in turn controlled by specific neurotransmitter systems. Physiological and behavioural indices related to different processes are depicted below the concerning processes. Experimental task variables related to different information processing stages are mentioned in the lower part of the model (SN = search negativity, LRP = lateralized readiness potential, RT = reaction time). The model is based on those presented by Mulder (1986), Pribram and McGuinness (1975), Sanders (1983), and Tucker and Williamson (1984).

electrical activity of the brain, which are time-locked to particular cognitive events such as the onset of a stimulus. They are revealed by averaging brain activity recorded during many trials (Cooper *et al.*, 1989). ERPs can be to recorded on trials in which stimuli are presented to which a response is or should be given and stimuli which should be ignored, all within the same experimental task.

The positive or negative deflections or components in the ERP components are very sensitive to changes in structural and energetical task demands. Moreover, these components have been found to be differentially sensitive to input, central, and output processing dimensions of information processing (see Figure 8.1; Donchin *et al.*, 1986; Kok, 1990; Kramer and Spinks, 1991; Mulder, 1986). More specifically, it has been argued that the *latencies* of different ERP components reflect the timing of information processing, while the *amplitude* of ERP components have been found to be very sensitive to changes in the mobilisation of energetical mechanisms involved in task performance (Gopher and Donchin 1986; Kok, 1990; Mulder, 1986). It should be noted that excitatory and/or inhibitory influences of different neurotransmitter systems on energetical mechanisms might modify the speed of structural operations, reflected in latencies of ERP components, in addition to the above mentioned amplitude changes (Posner and Rothbart, 1986).

The components, elicited by visual stimuli used in this chapter, are:

- *N1*: This early occipital negative ERP deflection, usually peaking at about 100 ms from stimulus onset, is an exogenous component (Donchin *et al.*, 1978), that is, a more or less obligatory response to the onset of the stimulus (see Okita *et al.*, 1985). It reflects the receptivity of subjects to external stimuli (Hillyard and Kutas, 1983). In addition, the amplitude of this component might be used as an index of phasic cortical arousal.
- *N2b*: The selective processing of the relevant stimuli has been associated with a centro-frontal negativity, 200 ms after stimulus onset, called the N2b component. This component is interpreted as being a phasic alerting reaction that facilitates further controlled processing in the limited capacity system of relevant information (Näätänen and Gaillard, 1983). Moreover, this component is sensitive to the state of the subject and to resource allocation (Gunter *et al.*, 1987).
- *Search negativity* (SN): Search negativity is regarded as a neurophysiological manifestation of subject-controlled memory search processes, involving serial comparisons between stimulus and memory set items (Okita *et al.*, 1985; Shiffrin and Schneider, 1977; Wijers *et al.*, 1987). Larger memory sets resulted in a more extensive search process of longer duration and a more pronounced negativity, reflecting increased demands on serial comparison processes. The amplitude of the search negativity is regarded as a reflection the amount of energetical resources invested in task performance.
- *P3b*: The latency of this component is determined by stimulus evaluation time and is relatively independent of the time required for response selection and execution. The P3b amplitude is sensitive to energetical mechanisms related to perceptual and central processes (Isreal *et al.*, 1980). Moreover, this component has been found to be affected by fluctuations in noradrenergic activity in the locus coeruleus (Pineda *et al.*, 1989).

- *Lateralized readiness potential* (LRP): LRPs provided us with a measure of the time at which, at the level of central motor processes, preparation of the overt response has begun (Coles, 1989; De Jong *et al.*, 1988; Smid *et al.*, 1987). More precisely, LRPs are measures of brain activity, indicating activity of the motor system several hundred milliseconds before overt movement occurs.

CAFFEINE AND INFORMATION PROCESSING

Evidence of Caffeine Effects on Input-related Processes

Information transformations related to the encoding of incoming information, that is, to the identification of incoming information, are supposed to be input-related processes (see Figure 8.1). A method of disentangling the precise action of caffeine on these processes is based on the additive factor method (Sanders, 1980, 1983; Sternberg, 1969). The idea underlying this method is that by systematically manipulating specific task variables, which are known to be related to certain processes, differential effects of caffeine can be traced. More specific, if caffeine effects interact with the effects of a change in a certain task variable, both factors are thought to affect a common process. If the effects are additive, it is assumed that the drug does not influence the process affected by the manipulated task variable (see also Frowein, 1981; Gaillard, 1988). The effects of caffeine on input-related processes are examined by manipulating stimulus quality, spatial frequency and orientation and diagonal selection.

Experimental Manipulations

Stimulus quality

The perceptual or encoding stage was manipulated, using stimulus quality as task variable (Lorist *et al.*, 1994a). Stimulus quality is supposed to affect feature extraction processes (Sanders, 1983). The stimuli (the digits 2, 3, 4 and 5) in the intact condition consisted of a dot pattern surrounded by a rectangular frame of dots. In the degraded condition dots were replaced from the frame into random positions within the frame, which were not occupied by dots making up the stimuli.

The spatial arrangement of the dot patterns used in the degradation task impaired the identification of the stimulus, as reflected in longer RTs and increased error rates in this condition as compared with the intact condition. The effects of caffeine were found to be more pronounced when the quality of visual information was impaired, that is, in the degraded condition in which a greater demand was placed upon feature extraction processes. Concerning ERP measures, a more negative going N1 in combination with a shorter latency of this component was observed after caffeine. The influence of caffeine on this component indicates that caffeine increases the receptivity of subjects to external stimuli and accelerates input-related information processing (Hillyard and Kutas, 1983). Both the behavioural and ERP results, therefore, point to a positive effect of caffeine on the input stage, or more specific, on feature extraction processes.

Spatial frequency and orientation

Stimuli used by Kenemans and Lorist (1995) consisted of gratings of a particular spatial frequency (3.2 and 0.8 cpd) and orientation (vertical and horizontal). Subjects were required to respond to a specific grating on the base of a combination of these stimulus features and to ignore the other gratings. The stimuli were presented at a fixed position in the centre of the visual field.

The results of this experiment indicated that caffeine improved the speed and efficiency of information processing as reflected in faster RTs and a decrease in error rate. ERP components were found to be differentially sensitive to the processing of spatial frequency and orientation of visual stimuli (Harter and Previc, 1978; Kenemans *et al.*, 1993, 1995), however, were not affected by caffeine. In addition, neither an effect of caffeine was found on selective attention mechanisms involved in the processing of relevant stimuli defined by a conjunction of spatial frequency and orientation. Selective attention was indexed by the difference between brain potentials elicited in gratings possessing the combination of relevant feature dimensions and gratings having irrelevant feature dimensions. Thus, contrary to the observed effects of caffeine in the stimulus degradation task the results in the grating task suggest that caffeine has no effect on input-related information processing.

Diagonal selection

A third manipulation of input-related information processing relied upon the manipulation of selective attention mechanisms involved in the processing of stimuli presented on specific spatial locations. Subjects had to direct their attention to a restricted number of display positions (left-up or right-up diagonal positions) on which stimulus items could be presented. They had to ignore stimuli presented on other display positions. Thus subjects had to make a selection between relevant and irrelevant information on the base of spatial features. The task was to press a button every time a memory set letter was presented on one of the relevant diagonal positions (see for details Lorist *et al.*, 1994b, 1995). The number of items presented simultaneously on the display (i.e., display load) was held constant at two letters, and they were randomly presented on either the left-up (50%) or right-up (50%) diagonal positions.

The difference in ERP signature associated with selection on the base of spatial frequency and orientation, and the ERP pattern elicited by diagonal as a stimulus-selection criterion indicates that both types of selection are distinctive attention mechanisms, requiring different brain processes (Kenemans and Lorist, 1995; Harter and Previc, 1978; Kenemans *et al.*, 1993; Wijers, 1989). The selective attention mechanism involved in the second case, the feature unspecific attention mechanism, is indexed by the N2b component (Wijers, 1989).

This selection process was found to be influenced by caffeine. The amplitude of the N2b component increased after caffeine as compared with placebo. The conclusion was that the amount of caffeine comparable to two cups of regular coffee makes people less easily distracted by irrelevant information and improves the sensory perceptual processing of stimuli presented at attended locations (see Näätänen and Gaillard, 1983). In other words, caffeine appears to improve the selective processing of information presented on specific locations in the visual field.

Theoretical Interpretations

Evidence for an effect of caffeine on input-related processes was found in tasks in which stimulus quality and diagonal selection were used as task variables (Lorist *et al.*, 1994a,b; 1996). In contrast, however, effects of manipulating spatial frequency and orientation were not changed by caffeine and indicated that these input-related processes were not affected by caffeine. Apparently, the influence of caffeine on input-related processes depends on the used experimental manipulations. To be able to understand the effects of caffeine found in the different experimental conditions, a more precise description of these processes and their interrelations seems necessary.

The perception and identification of visual information encompass several levels of analysis performed in various parts of the brain. Input from the retina produces activity in different visual areas in the occipital lobe. Kosslyn and Koenig (1992) call these areas the 'visual buffer'. The spatial relations of images that fall on the retina, that is, information about edges and regions of colour and texture are preserved in this visual buffer; the 'master map of locations' postulated in the feature-integration theory of Treisman (1988, 1993). Treisman states that this map is formed automatically in an early stage of stimulus identification, and shows *where* all feature boundaries are located. It does not indicate *which* features are located where, it cannot identify stimuli.

Next, information from the visual buffer is sent to two different projection pathways where specific aspects of the stimuli are encoded. In the 'ventral system' stimuli features such as colour (hue, brightness, and saturation) and form (frequency, orientation, shape and size) are analysed. The brain areas in this system go from the occipital lobe out into the inferior temporal lobes. The 'dorsal system' is engaged in the analysis of information about spatial properties of visual stimuli. The brain areas involved in these encoding processes go from the occipital lobe out into the posterior parietal lobes. These ventral and dorsal projection systems are also referred to as the *'what'* and the *'where'* systems, respectively (Ungeleider and Mishkin, 1982). The output of analyses performed in both systems is matched to information stored in memory and if a match is obtained, the stimulus is recognised.

The visual buffer contains more information than can be handled in subsequent stages of information processing (Luck and Hillyard, 1994; Mangun, 1995). For this reason a function of attention mechanisms in visual information processing is the selection of information from the visual buffer. This selection process alters the signal-to-noise ratio of inputs, thereby producing a relative facilitation of selected information to be transported to higher stages of information processing (Mangun, 1995). Attention can be directed selectively to spatial and nonspatial stimulus features. Spatial selective attention is said to depend on the fast, retino-tecto-pulvinar pathway, while nonspatial selective attention is assumed to be mediated by the retino-geniculo-striate pathway.

After this closer look at the different levels of processing involved in the perception and identification of visual information, we now return to the initial question: How can the observed effects of caffeine be interpreted in terms of an influence of caffeine on specific processes in the perceptual processing stage?

The spatial arrangement of the dot patterns used in the degradation task, described by Lorist *et al.* (1994a) determines the ease with which figure-ground segregations can be made. The computations of these segregations are assigned to early levels of spatial analysis (Van der Heijden, 1993). Treisman (1993) stated that once visual information is organised in objects against a background, selective attention can be directed at objects isolated in this way, a process which, as has been argued before, might involve the tecto-pulvinar projection system. The implication for the interpretation of the effects of caffeine in the degradation task is that an effect of caffeine on the spatial selective attention mechanism consists of improving the direction of attention towards objects isolated from the background.

Additional evidence for an effect of caffeine on the processing of spatial stimulus features comes from the effects on the N1 ERP-component. The occipital N1 peak is assumed to reflect activity in the dorsal projection pathway encoding spatial stimulus features (Mangun, 1995; Mangun *et al.*, 1993). The effects of caffeine on the N1 component might therefore be interpreted as an effect on the dorsal system. This is in accordance with the idea of Kosslyn and Koenig (1992), which implies that the identification of objects from the background is performed in a later stage than in the early tecto-pulvinar projection system. So far, experimental evidence indicates that caffeine affects the processing of spatial stimulus features either in the early tecto-pulvinar projection system or in the dorsal system.

The stimuli used in the experiment, reported in Kenemans and Lorist (1995) differed with regard to spatial frequency and orientation, features which are analysed in the *what* pathway. The absence of an interaction between caffeine and these task variables indicates that the processing of what features in the ventral projection system is not influenced by caffeine. Also caffeine had no effect on the selective attention mechanisms involved in the processing of non-spatial stimulus information. Putting the results together it seems that caffeine affects the input stage in a specific way. Apparently, the effects of caffeine are restricted to the processing of spatial stimulus features.

The selective processing of specific locations in the visual field improves after caffeine, in case diagonal is used as the selection cue. Diagonal, however, is no simple elementary feature and becomes available at a rather late stage in information processing. Attention to visual stimulus features on the base of diagonal consequently seems to rely upon a later feature-unspecific attention mechanism, reflected in the N2b component (Mulder *et al.*, 1994). Brain scans have demonstrated that the N2b component might be generated in the anterior cingulate cortex (Wijers *et al.*, in press), an area which is known to be active in the detection of relevant stimuli (Posner and Raichle, 1994).

In summary, the effects of caffeine on input-related processes appear to be selective rather than general. It can be concluded that caffeine affects the input stage of the human information processing system by affecting the analysis of spatial features. Moreover, caffeine appears to improve later, feature-unspecific selection mechanisms. No effects of caffeine were observed on the perception and identification of nonspatial stimulus features.

Pharmacological Interpretations

The normal functioning of the information processing system depends both on structural processes and on the modulation of these processes by energetical mechanisms, which in turn are influenced by neurochemicals (Figure 8.1). The next question which was summoned in the interpretation of the observed results, is whether pharmacological evidence supports the conclusion drawn above about the effects of caffeine on input-related processes.

The spatial analysis of a visual stimulus seems to concern the retino-tecto-pulvinar pathway that goes into the posterior parietal lobes. Animal research indicated that each of these areas receives a heavy input of noradrenaline from the locus coeruleus. Noradrenaline influences the speed and efficiency of information processing operations by inhibition of background cell activity and augmentation of cell-responding to sensory stimulation, which leads to more favourable signal-to-noise ratios (Posner and Raichle, 1994; Robbins and Everitt, 1995; Tucker and Williamson, 1984). Moreover, noradrenaline appears to facilitate a link between attention and its object, making subjects less easily distracted by unimportant events.

The analyses of spatial stimulus features are under the influence of noradrenaline to a greater extent than analyses of nonspatial features. If indeed caffeine acts through the modulation of the noradrenergic system, it might be hypothesised that the analysis of spatial information will be affected by caffeine induced changes in noradrenaline to a greater extent than the analysis of nonspatial features. The presence of caffeine effects on the analysis of spatial features, in addition to the absence of an effect on the analysis of nonspatial features supports the hypothesis that caffeine acts through a modulation of the noradrenergic system.

EVIDENCE OF CAFFEINE EFFECTS ON CENTRAL PROCESSES

The second stage discerned in the human information processing system is the central stage, which includes controlled processes necessary for further processing of information. These processes are highly flexible and are serial in nature, that is, they do not operate in parallel with other processes. In addition, these processes, in general demand attention and are strongly capacity limited. The influence of caffeine on three different kinds of processes in this stage were examined: Serial comparisons, binary decisions and response selection processes.

Experimental Manipulations

Memory load and display load

In both the focused and the divided versions of a selective attention task, used by Lorist and associates (Lorist *et al.*, 1994b; Lorist *et al.*, 1995, 1996), subjects were instructed to detect the possible appearance of a memory set item on relevant positions of the display. Targets and distracters were mixed from task block to task block, therefore no automatic target detection could develop (see Schneider and Shiffrin, 1977; Shiffrin and Schneider, 1977).

The increase in RT with increasing task load, defined by the product of memory set size and display set size (Shiffrin, 1988), suggests in accordance with Schneider and Shiffrin (1977) and Shiffrin and Schneider (1977), that subjects used a serial, limited-capacity search process in which each memory set item was compared to all display items sequentially. ERPs evoked by stimuli presented on relevant display positions showed a long latency negative shift, starting around 250 ms after stimulus onset (Okita *et al.*, 1985), which was regarded as a neurophysiological manifestation of these serial comparisons made between stimuli and memory set items (Okita *et al.*, 1985; see also Schneider and Shiffrin, 1977; Shiffrin and Schneider, 1977). An increase in the number of targets to be memorised or items presented on the display resulted in a more extensive search process of longer duration and a more pronounced negativity.

No effects of caffeine were observed on the electrophysiological indices of serial comparison processes performed by the subjects. In those task conditions in which task load was varied by varying the *memory set size*, no caffeine effects were observed on RTs as well. But, concerning the RTs in those tasks in which task load was manipulated by varying *display load*, an interaction between caffeine and task load was observed (Lorist *et al.*, 1996). In these task conditions the influence of caffeine was restricted to the low display load condition, showing that subjects reacted faster after caffeine than after placebo. Thus, memory load effects were not affected by caffeine, but display load effects were.

Target detection

In the selective attention task, after the selection of relevant information and comparisons of stimuli to internal representations of information in memory, subjects had to decide about whether a memory set item was present on the display and whether a response should be made (Lorist *et al.*, 1994b, 1995, 1996). The ERPs elicited by correctly detected targets on the relevant diagonal contained a large positive deflection (P3b) with a maximum amplitude at Pz and Oz. The P3b peak latency was found to depend on the duration of stimulus evaluation processes (i.e., input and central information processing activities) and to be relatively independent of the time required for response selection and execution (Donchin and Coles, 1988; Mulder, 1986). The peak latency of this component was on average shorter after caffeine than after placebo, and this effect was dependent on display load. The absence of an effect of caffeine on search related negativity in the selective attention task (see above), which is regarded as a reflection of central processes, indicates that the observed effects of caffeine on P3b peak latency originate from an effect on input related processes and not from an effect on central processes.

Stimulus-response compatibility

The next step in the chain of processes necessary to transform a stimulus into a response is the decision which response should be made. Lorist, Snel and Kok (1994a) manipulated stimulus-response compatibility in order to affect the complexity of this response decision process. Stimulus-response compatibility refers to the degree of association or compatibility between members of a stimulus-response pair in a choice reaction task. In the task used, the side of the display on which the stimulus was presented determined the hand with which the response had to be made.

The results indicated that this response decision process is insensitive to caffeine. Neither the behavioural results, nor the electrophysiological results showed an interaction of task manipulation with caffeine. Apparently, there is no effect of caffeine on the central information processing stage.

Theoretical Interpretations

Memory plays an important role in the processing of information, it enables us to store and retrieve information. Following the perception of different objects in the input stage, formed by spatial and nonspatial feature analyses, these objects are matched to stored descriptions in memory to allow recognition in the central information processing stage. The information received in this way might in turn be used as the basis for target detection and the selection of an adequate response.

Human memory can be divided into short-term memory (STM), long-term memory (LTM) and working memory (e.g., Baddeley, 1983). STM stores information over brief intervals of time during which further processing can be performed (e.g., recognition). Only limited amounts of information can be stored in this STM of which we are aware. The visual buffer mentioned in the description of input-related processes serves as a STM structure (Kosslyn and Koenig, 1992). LTM contains large amounts of information stored for considerable periods of time. We are not aware of this information until it is activated and becomes part of working memory. Working memory, thus, corresponds to activated information from the LTM-system needed to perform a task. This information is temporarily kept in the STM-system to enable matching of information from both STM- and LTM-systems.

STM, LTM and working memory can be deliberately accessed during task performance, they are then *explicit* memories. On the other hand, there are memory representations which cannot be directly accessed, these memories are called *implicit* memories. An example of implicit memory is the association between a specific stimulus and a response. This association is formed after practice and can only be evoked by specific stimuli. These connections are implemented in the striatum, and strengthened by the influence of the substantia nigra (Kosslyn and Koenig, 1992).

In the used selective attention task (Lorist *et al.*, 1994b, 1995, 1996) an association between memory set items and a specific context is formed in the memory system, that is, the memory set letters are associated with being targets. When subsequently stimulus letters are presented and perceived by the subjects these associations get activated and identify a stimulus letter as being a target to which a response should be made. These activation processes seem to be unaffected by caffeine as indicated by the absence of an interaction between caffeine and memory load or display load on electrophysiological indices of visual- and memory-search processes.

A positive effect of caffeine on RT in the low display load condition, which was absent in the high display load condition, however, indicated that display load manipulations were influenced by caffeine. The decrease of caffeine effects with increasing display load corresponds to results indicating that caffeine facilitates performance only in simple or moderately complex tasks. In more complex tasks caffeine may have either no effect or even to impair performance (Humphreys and Revelle, 1984; Weiss and Laties, 1962). However, in simple tasks as well as in complex tasks different processes are involved which determine the quality of performance and therefore merely this conclusion adds nothing to the understanding of the specific effects of caffeine on information processing. The question is whether these effects can be explained more precisely in terms of effects on specific processes involved in the processing of visual information.

The information presented on a display, as in the selective attention task, will be temporarily held in STM. The high display load condition makes a greater appeal to the STM system than the low display load condition. The decreasing effect of caffeine with increasing display load implies that a moderate dose of caffeine affects the STM system. This conclusion supports the Humphreys and Revelle's notion that caffeine facilitates performance on tasks presumed to appeal to STM to a small degree, but hinders performance on tasks presumed to appeal to STM heavily.

It is also known that performance improves as long as energetical supplies increase up to a certain peak, beyond which it deteriorates (Yerkes and Dodson, 1908). Because caffeine influences the availability of energetical resources, the positive effects of caffeine in the *low* display load condition, and the absence of effects of caffeine in the *high* display load condition, might be interpreted as reflecting the turning point from an optimal level to a too high level of energetical supplies involved in STM processes. It might be speculated that a further increase of display load in combination with the used dose of caffeine may lead to an impairment of performance.

So far, it can be concluded that memory scanning seems to be unaffected by caffeine, while caffeine positively affects or has no effect on STM, dependent on the demands placed upon this process.

Stimulus evaluation processes, as reflected in the P3b component of the ERP were on average faster after caffeine than after placebo. One should be cautious, however, to ascribe this caffeine effect simply to an effect on central processes, because stimulus evaluation involves both input and central processes. The absence of an influence of caffeine on indices of visual- and memory search processes, in addition to the presence of caffeine effects found on the encoding of spatial stimulus features indicate that the effects of caffeine on P3b latency might come as well from an effect on early input-related processes.

Memory structures involved in the processing of visual information discussed so far, were explicit memories. Memory processes addressed in the stimulus-response compatibility task (Lorist *et al.*, 1994a) were implicit memory systems. The absence of an interaction between caffeine and the difficulty of the response decision process indicates that this part of memory is insensitive to caffeine. The question might be raised, however, whether the stimulus-response compatibility manipulation results in a direct connection between a representation of the stimulus and a motor output program. An alternative interpretation is that stimulus-response compatibility affects cognitive processes related to decision making (Sanders, 1983). Subjects had to decide on the base of specific aspects of the stimuli presented on the display which response would be correct. No evidence in favour of one of these interpretations is available, but it is clear that caffeine did not affect the concerning processes.

In summary, the central processing stage consists of a broad range of functionally different structural processes. In the described series of experiments the effects of caffeine were examined on a subset of these processes: Serial comparisons, binary decisions and response selection processes. Evidence for caffeine effects on central-related processes is limited to the observed effects on STM. The absence of an effect of caffeine on the performance of serial comparisons, that is, on visual- and memory search and on response selection processes supports experimental evidence, showing that caffeine has no effects on so-called higher mental processes performed in the central information processing stage. Furthermore, effects found on indices of binary decision processes cannot unequivocal be interpreted as an effect on central processes.

Pharmacological Interpretations

When new information or associations between stimuli and a new context (i.e., memory set items being targets in specific task conditions) enter the information processing system, the basal forebrain issues a signal that this new information should be stored in memory (see Kosslyn and Koenig, 1992). The signal involves the release of acetylcholine. However, there is neither much evidence in the literature (Daly, 1993) nor from experiments (see above) available for an effect on performance regulated by acetylcholine due to the influence of caffeine on this neurotransmitter system.

The association between a specific stimulus and a response depends on the striatum, which consists of the caudate nucleus and the putamen. These brain structures receive dopamine from the substantia nigra. Dopamine is mainly associated with activation, and thus indirectly to motor organisation, but it has also a role in cognitive functions such as the organisation of responses (Robbins and Everitt, 1995). The absence of interactions between caffeine and stimulus-response compatibility (Lorist *et al.*, 1994a) suggested that this latter role of dopamine is not influenced by caffeine.

EVIDENCE OF CAFFEINE EFFECTS ON OUTPUT-RELATED PROCESSES

Processes in the output stage are those involved in the organisation of motor related activities. These processes are preparation, activation, and execution of responses. We manipulated response preparation processes in order to evaluate the effects of caffeine on this third stage of the information processing system.

Experimental Manipulations

Time uncertainty

The efficiency of a motor response can be increased by a more optimal preparation of this response. But, preparation can only be maintained for a short period of time. In the time uncertainty task (Lorist *et al.*, 1994a) the degree of uncertainty about the moment of stimulus presentation and thus the possibility to maintain an optimal level of motor pre-setting, was manipulated by varying the inter-stimulus interval. The effects of caffeine were found to be more pronounced in the task condition in which the foreperiod was relatively long. In this condition in which it was relatively difficult to maintain an optimal level of motor pre-setting caffeine improved performance, indicating that caffeine has an effect on processes in the output stage.

Response preparation the lateralized readiness potential (LRP)

Rather than examining the interactions between caffeine and time uncertainty on RT data (Lorist *et al.*, 1994a), Kenemans and Lorist (1995) used the LRP as a more direct index of the onset of hand-specific processes that prepare the motor system for action (Coles, 1989; De Jong *et al.*, 1988; Smid *et al.*, 1987). LRPs are measures of brain activity in the primary motor cortex indicating activity of the motor system several hundreds of milliseconds before overt movements occur. Using LRPs in addition to RTs might indicate to what extent response related processes are central motor processes or more peripheral motor processes. Although the results showed that caffeine shortened the RTs, caffeine had no effects on the onset latency of the LRP. This finding indicates that, contrary to the observed effects in the time uncertainty task, caffeine has no effects on output-related processes involved in the preparation of movements.

Theoretical Interpretations

The organisation of motor related activities includes several levels of processing (Kosslyn and Koenig, 1992). Before the actual response can be executed subjects have to prepare an adequate set of movements, which after execution fulfil the task requirements. The computation of these movements can be performed either by using stored motor representations in motor memory or by computing novel movement combinations. The supplementary motor area is involved in these movement planning activities at a coarse level. The pre-setting of motor actions at a more refined level is carried out by the premotor cortex (area 6), which receives also sensory input from the posterior parietal lobe where spatial stimulus features are processed. The premotor area uses this spatial information to adjust movements to a specific environment. A third structure concerned in the planning of movements is the caudate nucleus, which forms a part of the basal ganglia. After the preparation of movements, the appropriate muscles have to be activated and the response has to be executed. Brain areas which play a role in this process are the putamen and area M1 (area 4). A third neural structure involved in these processes is the cerebellum, which serves two output-related activities, first guidance of motor activity and secondly taking care of the storage of motor memories.

The output-related processes involved in a simple movement such as pressing a button depend upon different brain areas, each performing different parts of the organisation of motor related activities, that is, preparation, activation, and execution of responses. Using time uncertainty as task variable we (Lorist *et al.*, 1994a) tried to manipulate the motor preparation processes involved in the generation of a correct button press. The interaction found between caffeine and time uncertainty indicated that caffeine has an influence on these processes, although the validity of using time uncertainty as an output related task variable is questioned (Smulders, 1993). Smulders' study suggested that the process manipulated by varying inter-stimulus interval is probably nonmotoric and might even precede response decision processes performed in the central stage.

An electrophysiological index of response preparation is the LRP, measuring the differential activation of the left and right hand representations in the motor cortex. Although caffeine did not influence the LRP onset latency, caffeine did shorten reaction times. The absence of caffeine effects on LRP onset, in addition to an effect of caffeine on RT, led to the conclusion that the faster reactions were due to an effect of caffeine on processes taking place *after* response preparation, either at the central or peripheral motor execution level (Kenemans and Lorist, 1995; Lorist *et al.*, 1994b, 1995, 1996). Previous studies examining the effects of caffeine on these 'late' motor processes, indeed, showed stimulating effects of caffeine, although these effects seem to be largely dose-dependent (Bättig and Welzl, 1993; Daly, 1993).

In summary, the maintenance of an optimal level of motor pre-setting over a period of time seems to be positively influenced by a moderate dose of caffeine (3 mg/kg or 200 mg). But it might be questioned whether the manipulation of time uncertainty affects the output stage. It was further concluded that caffeine affects output-related processes taking place after the start of response preparation.

Pharmacological Interpretations

Brain structures involved in the preparation and execution of motor activities depend mainly on dopamine. This neurotransmitter is known to affect different aspects of output-related processes (Robbins and Everitt, 1995; Tucker and Williamson, 1984). Effects of caffeine on striatal dopamine systems and other brain structures have been demonstrated in several paradigms. Moreover, caffeine effects are partly mediated by an effect on these dopamine systems (Daly, 1993), which supports the conclusion that caffeine affects output-related processes.

DOSE

In the present chapter the effects on the human information processing system of a single dose of caffeine (3 mg/kg or 200 mg), comparable to two average cups of coffee, were discussed. This moderate dose of caffeine appeared to have beneficial effects on specific processes in the input and output stages of the information processing system. This is in accordance with studies examining the effects of caffeine on neurotransmitter systems; noradrenergic areas and dopaminergic areas involved in the control of input and output processes, respectively, are very sensitive to caffeine (Nehlig and Debry, 1993).

It is known, however, that effects of caffeine are dose-dependent. Low and medium doses might produce improvements and high doses impairments or no effects on performance. In addition, it is important to realize that the 'optimal' dose depends on habitual caffeine use. Moreover, different brain areas have been found to be differentially sensitive to caffeine. For example, low doses of caffeine (0.1–1.0 mg/kg) mainly affect dopaminergic areas such as the caudate nucleus and ventral tegmental area (Nehlig and Debry, 1993), while higher doses of caffeine (>10 mg/kg) might affect different mechanisms (e.g., inhibition of phosphodiesterase). The involvement of different brain systems with different doses of caffeine might underlie the dose dependent effects of caffeine. Studies, using various doses of caffeine should be done to achieve a better understanding of the effects of caffeine on human performance.

SUMMARY

An attempt was made to gain a better insight in the specific effects of caffeine on cognitive activities. To attain this goal ERPs and behavioural indices of performance were used. In addition, knowledge about the involvement of different brain areas and neurotransmitter systems in the regulation of performance was used to create specific hypotheses about the actions of caffeine.

It was found that the effects of caffeine on input-related processes are selective, since they appear to affect the analysis of spatial features. It also appeared that caffeine improves later, feature-unspecific selection mechanisms, but that caffeine has no effects on the processing of nonspatial stimulus features. Support for caffeine effects on central processes, such as serial comparisons, binary decisions and response selection was found only for short term memory. No effects of caffeine on output-related processes involved in movement preparation were found. Nevertheless the maintenance of an optimal level of motor pre-setting over a period of time seems to be positively influenced by an amount of caffeine similar to the contents of two cups of coffee. A further conclusion was that caffeine affects output-related processes that take place after the start of response preparation.

ACKNOWLEDGEMENTS

This research was supported by the Physiological Effects of Caffeine Research Fund of the Institute for Scientific Information on Coffee, Paris.

CHAPTER NINE

Caffeine, Morning-Evening Type and Coffee Odour: Attention, Memory Search and Visual Event Related Potentials

Adriaan KOLE, Jan SNEL and Monicque M. LORIST

INTRODUCTION

Caffeine is said to have arousing effects (e.g., Zwyghuizen-Doorenbos *et al.*, 1990), although its effects on perceptual and cognitive functions are not always consistent. For example, the influence of caffeine on short-term memory search processes is occasionally positive (Anderson and Revelle, 1983), sometimes negative (Anderson *et al.*, 1989; Erikson *et al.*, 1985), but mostly no effect has been obtained at all (Clubley *et al.*, 1979; Mitchell *et al.*, 1974; Erikson *et al.*, 1985; Loke *et al.*, 1985, 1988). One reason for this lack of consistent results is the influence of person bound factors such as boredom, fatigue, motivation and morningness-eveningness and factors in the environment like noise and odours. In general, effects of caffeine appear to be most pronounced in subjects who are in suboptimal conditions as for example when tired (Dews, 1984; Weiss and Laties, 1962; Lorist *et al.*, 1994a). Since many experimental designs in drug research usually do not take account of these factors, it is impossible to disentangle their influences from the desired experimental effects (Gaillard, 1988).

Diurnal type and coffee odour might be such arousal affecting factors. Compared to people who have no outspoken tendency to be morning or evening types, the former are reported to have a relatively advanced and the latter a relatively delayed circadian phase position (Kerkhof, 1981, 1991; Clodoré *et al.*, 1987; Horne *et al.*, 1980). Consequently, morning types are usually more active and perform better than evening types in the morning and evening types better than morning types in the evening (Horne *et al.*, 1980; Kerkhof, 1981, 1985; Pátkai, 1970, 1985). For this reason we used diurnal type as a suitable factor to study the effects of caffeine on information processing. The assumption we tested was that when measured in the evening, the morning types benefit more from caffeine than evening types.

The other factor we were interested in was coffee smell. Since it is not very probable that caffeine is transported through the air, a possible arousing effect of the smell of coffee could be caused by a semantic and social association with the psychoactive effect of caffeine (Knasko, 1995) and its most well known beverage: coffee. The possibility of odour conditioning has been demonstrated, for example by Kirk-Smith *et al.* (1983). Evidence that odours affect arousal-related mechanisms is both anatomical and psychophysiological. Different odours at both sub- as well as supra-threshold concentrations resulted in decreases of alpha-frequencies and increases of higher beta-frequencies (Freeman and Viana Di Prisco, 1986; Freeman and Baird, 1987; Freeman, 1987, 1991; Schwartz *et al.*, 1992; Lorig *et al.*, 1988, 1990, 1991; Lehmann and Knauss, 1976; Van Toller *et al.*, 1993). Such changes are usually associated with increased (attentional) activity in the brain (e.g. Cacioppo and Tassinary, 1990). On the other hand, some findings of increasing alpha-, together with decreasing beta-frequencies do suggest a relaxing effect of odours (e.g. Sawada *et al.*, 1992; Torii *et al.*, 1991). These contradicting results may pertain to specific odour qualities.

Also, there is some anatomical evidence for the involvement of the olfactory system in arousal. In 1973, Motokizawa and Furuya showed that either odorous or electrical stimulation of the olfactory bulb in the nose of human subjects could elicit EEG-desynchronization (indicating increased arousal) in neocortical and hippocampal areas. This increase of arousal was produced essentially by the mesencephalic reticular formation (MRF), innervated by olfactory impulses from the olfactory bulbs which pass through the medial forebrain bundle (MFB). Warm and colleagues (1991) showed that periodic 30 second whiffs of a hedonically positive fragrance (Muguet) was able to enhance signal detectability in a sustained attention task. In stress-research, several authors report that high noradrenaline levels enhance the signal-to-noise ratio in perceptual processes (Geen, 1986; Gray, 1982; Hebener *et al.*, 1989). Following Hebener *et al.* (1989) and others, it suggests that the connections of the olfactory system to noradrenergic and serotonergic locations in the brain stem prepares a system for the processing of incoming stimuli (Foote *et al.*, 1980). We tested the assumption that when drinking coffee, some of its effects could stem from its smell. We did so by offering the subjects a continuous olfactory stimulation from freshly milled coffee. The influence of this odour on human information processing was assessed from (visual-) task performance and electrophysiological measures of brain activity.

In Experiment 1 the results of the diurnal type factor are presented; in experiment 2 those of coffee odour.

EXPERIMENT 1: DIURNAL TYPE

Method

Thirty healthy and non-smoking first-year students (age 18–30 years, m = 21.41 ± 3.13) with normal or corrected-to-normal vision participated in this study. Their participation was part of a study-requirement. They were regular coffee drinkers (range 2–6 cups a day, m = 3.77 ± 1.47) and were either moderate to extreme morning or evening types. Their selection was based on the criteria of the validated 7 item morning type-evening type questionnaire, range of scores: 7 to 31 (Kerkhof, 1984). The morning types scored from 23 to 31 (m = 24.93 ± 2.15), the evening types between 7 and 13 (m = 12.20 ± 1.61). Subjects were requested to refrain from caffeine containing products and alcohol for twelve hours prior to the experimental sessions, not to work night shifts and not to engage in activities which might interfere with their habitual sleep-wake pattern the day prior to the experimental sessions.

Thirty minutes before testing, the subjects received one cup of decaffeinated coffee in which either 200 mg lactose or 200 mg caffeine was dissolved. The procedure was deceptive (Kirsch and Weixel, 1988), double blind and in random order. In a weakly lit room, subjects were sitting before a VGA screen at a distance of 80 cm. Each task started with the presentation of a cue frame, indicating which diagonal was relevant (left up or right up), followed by a memory set of 1, 2 or 4 letters. Each frame lasted for 5 seconds. One task consisted of 192 display frames of 300 ms, inter-stimulus-interval (ISI) 1380 ms briefly shown on an Apple Macintosh computer-screen. Display set size consisted of two letters presented on a diagonal. A small fixation cross remained in the center of the screen throughout the task (Figure 9.1).

Fifty percent of the display frames belonged to the right up category, the other half to the left up category. In each category, fifty percent of the frames contained a target letter. The frames were presented in random order. Targets and non-targets were randomly chosen from all consonants except *Q* and *X*, with the limitation that a memory set item was never used in successive tasks in one session. Stimuli were black on a white background. The visual angle between the center of fixation to each of the elements of the display frame was 2.2°. The stimulus letters had a height of 1 cm and a width of 0.4 cm. An AT computer was used to control the timing and duration of the stimuli and to measure reaction times, errors and to record the ERPs.

Recording and data analysis

The EEG was recorded from the midline positions Fz, Cz, Pz and Oz and referred to the linked earlobes. Signals were digitized at a rate of 100 Hz and amplified with a time constant of 5 seconds and a high frequency filter of 35 Hz.

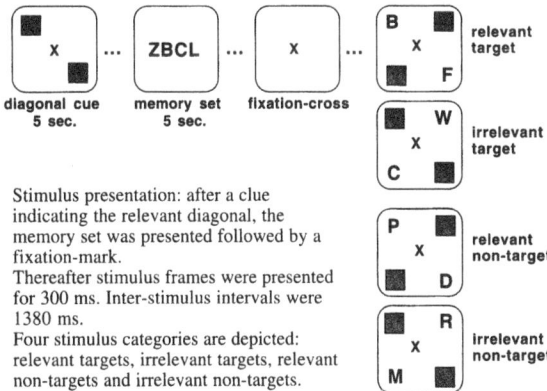

Figure 9.1 – Visual focused attention and memory search task.

All sessions started at 19.30 p.m. with a two week interval; one with caffeine and another with placebo. The order of the three experimental tasks was randomized for each subject. The instruction was to attend to the diagonal which was designated relevant by the cue frame, and to press a button as quickly and accurately as possible whenever a target letter was detected. The other diagonal had to be ignored. Before each task the instruction was repeated.

Being in a less aroused state at that time of the day, morning types were expected to benefit more from the consumption of caffeine than evening types.

The data of the first ten trials within each task were excluded from further analysis. For all subjects, average Event-Related-Potentials (ERPs) were computed for each drug and memory load condition separately. In addition, separate ERPs were made for the four stimulus categories, that is relevant targets, relevant nontargets, irrelevant targets and irrelevant nontargets, and electrode positions. A 200 ms prestimulus period was used as baseline. The epoch to be averaged lasted from 200 ms prior to stimulus onset to 1080 ms post-stimulus.

Selection processes were assessed by comparing ERPs of relevant and irrelevant nontargets. In order to evaluate memory search processes, the effects of memory load differences were compared for relevant stimuli. Effects of stimulus category were assessed on the relevant diagonal. Caffeine effects and effects of morningness-eveningness were studied using MANOVA for repeated measurements in a Drug × Type × Memory Load × Stimulus Category × Attention × Location design. When the main design showed interactions, additional analyses were performed with an adjusted alpha. Factors were Drug: caffeine and placebo; (Diurnal) Type: morning and evening type; Memory load: 1, 2 and 4 letters; Stimulus: targets and nontargets; Attention: relevant and irrelevant diagonal; Location: scalp sites, Fz, Cz, Pz and Oz.

Results

Performance

In higher memory load conditions reaction times were slowed [$F_{(2/27)} = 343.8$, $p < 0.001$]. Memory load also had an effect on the number of false alarms [$F_{(2/27)} = 5.4$, $p < 0.01$] and the number of misses [$F_{(2/27)} = 20.7$, $p < 0.001$] (Table 9.1).

In the caffeine condition, RTs were significantly shorter than in the placebo condition [$F_{(1/28)} = 5.8$, $p < 0.023$]. There were no effects of caffeine on the amount of false alarms or misses, and no interactions between diurnal type, caffeine and memory load.

Event Related Potentials

The ERPs in response to relevant nontarget stimuli showed a positive peak at anterior locations around 150–250 ms (P2) post-stimulus, followed by a negative peak at about 250–325 ms (N2) and a large positive peak between 325–550 ms (P3). At the posterior scalp locations appeared, a large negativity (N1) with a maximum around 200 ms (Oz). ERPs in response to irrelevant targets and non-

Table 9.1 – Effects of caffeine on task performance as function of type, drug and memory load

memory load	morning types (n = 15)		evening types (n = 15)	
	placebo	caffeine	placebo	caffeine
	reaction times in ms			
1	477 (51.6)	467 (60.3)	477 (55.4)	466 (56.8)
2	549 (63.8)	530 (66.5)	541 (56.7)	518 (61.7)
4	650 (52.4)	615 (73.4)	626 (56.3)	615 (51.3)
	% of false alarms			
1	0.44 (0.52)	0.26 (0.35)	0.37 (0.40)	0.11 (0.23)
2	0.44 (0.52)	0.44 (0.43)	0.33 (0.54)	0.18 (0.33)
4	0.77 (0.78)	0.62 (0.65)	0.33 (0.40)	0.55 (0.59)
	% of misses			
1	0.15 (0.33)	0.22 (0.46)	0.29 (0.72)	0.00 (0.00)
2	1.06 (1.18)	0.48 (0.62)	0.55 (0.75)	0.33 (0.65)
4	2.71 (1.99)	3.15 (2.41)	2.34 (2.37)	1.80 (1.53)

targets were almost identical. Compared to the relevant nontargets they show a more prominent N2 and a larger, earlier and sharper P3 at the anterior scalp positions. ERPs for relevant targets contained similar peaks as those of the relevant nontargets except for the more prominent positivity of the P3 component, and of a longer duration.

Selective attention effects (relevant or irrelevant diagonal) were mainly concentrated at the anterior positions and predominantly consisted of a more negative going N2 in relevant nontarget ERPs beginning at approximately 200 ms, followed by a less pronounced P3 in the 200–900 ms region: (*Attention × Location*; $F(3/26) = 7.5$–45.5, all $p \leq 0.001$). The findings also reflected a selective attention effect in the 100 to 200 ms area, shown as a small positivity, which was maximal at the central and parietal locations [*Attention × Location*: $F(3/26) = 4.4$–7.3, both $p \leq 0.01$]. Especially at Pz, irrelevant nontarget ERPs showed an earlier onset and an enhancement of the amplitude of the N1 than relevant nontargets. At the Cz location, P2 onset was later for irrelevant compared to relevant nontarget ERPs.

In the 350–600 ms range, caffeine produced an enlargement of the N2b [Drug × Attention: $F(2/27) = 5.5$–4.4, $p < 0.05$]. As will be argued below, this does not only represent an effect of caffeine on selection but also on memory search processes. For a more detailed account of the effects of caffeine see the caffeine section.

The effects of *memory* load started at 100 ms [Memory load × Location in the 100–250 ms area [$F(6,23) = 3.3$–6.3, all epochs $p < 0.02$] but were independent of attention. After 250 ms there was a prolonged negative shift which was most pronounced for relevant stimuli and for higher memory load conditions. This negativity is commonly referred to as the 'search negativity' or SN (Okita *et al.*, 1985). Search negativity started around 250 ms at posterior and around 300 ms at anterior sites, lasted for approximately 450 ms and was broadly distributed over the scalp in the 200–600 ms time domain [$F_{(6/23)}$ = 6.3–22.9, all p ≤ 0.005]. After around 300 ms, memory load effects were larger for relevant targets [$F_{(6/23)}$ = 3.9–9.8, all p ≤ 0.006].

For the *morning* types search processes started at approximately 325–350 ms in the placebo condition. For the *evening* types in the placebo condition SN onset was after around 250 ms, and this effect was mainly produced by the lowest memory load. In the caffeine condition SN appeared to start around 250 ms in both groups.

Caffeine also enhanced the SN amplitude [overall: Drug × Memory Load × Attention between 350 and 400 ms: $F_{(2/27)}$ = 4.3, p = 0.024, and Drug × Memory Load × Location between 400 and 450 ms: $F_{(6/23)}$ = 2.6, p = 0.043]. There were large, although nonsignificant, effects of diurnal type on the amplitude of SN.

Effects of caffeine

Between 350 and 450 ms, caffeine had a time effect on the descending flank of the P3b; it produced a shorter peak latency [Drug × Stimulus Category: $F_{(2/27)}$ = 8.8–8.2, both p ≤ 0.01 and Drug × Stimulus Category × Location: $F_{(3/26)}$ = 3.2, p ≤ 0.04]. This effect was mainly generated by the morning types in the highest memory load condition in the 400–450 ms area (p ≤ 0.028). Between 100 and 200 ms after stimulus onset (Figure 9.2) caffeine induced an increase of the negativity of the N1, which was most prominent at the centro-posterior sites.

Thereafter, caffeine produced a negative shift of the N2 which was most pronounced for *relevant* stimuli in the 200–300 ms range [$F_{(3/26)}$ = 3.6–4.1, p < 0.03]. In this epoch caffeine affected SN onset, in particular by the relevant nontarget ERPs and with its maximum effect at the central location [$F_{(6/23)}$ = 3.1, p < 0.03]. A closer inspection of the data revealed that this effect was not only due to caffeine, but was influenced also by morningness-eveningness (Figure 9.2).

Especially for morning types in the lower memory load conditions, there was a trend to an earlier onset of search processes in the caffeine than in the placebo condition. Furthermore, the SN amplitude was larger in the caffeine condition between 350 and 400 ms [$F_{(2/27)}$ = 5.4, p ≤ 0.01]. *After* around 500–550 ms, caffeine effects on memory processes were dominant for *irrelevant* stimuli; in the caffeine condition memory load effects were larger than in the placebo condition (p < 0.05); most obvious for evening types on posterior locations (p < 0.04).

As expected, caffeine effects on stimulus category prevailed on relevant stimuli consisting of a shorter P3b latency in the 350–450 ms area, especially in the lowest memory load condition [$F_{(3/26)}$ = 4.7–9.0, p ≤ 0.04].

This effect was induced by morningness-eveningness in the 400 to 450 ms area. However, the effects of diurnal type, a positive shift of the P2, N2 and P3, were hardly confirmed by the statistical analyses. There was one effect of diurnal type on memory search processes; morning types had longer onset latencies of the SN than evening types. The differences between the groups disappeared with caffeine, as was confirmed by the statistical analyses for the 250 and 300 ms epoch (p < 0.027). In the placebo condition, the P3b was smaller and somewhat later for the morning types. This effect was counteracted by caffeine in the 400 and 450 ms area (p = 0.01).

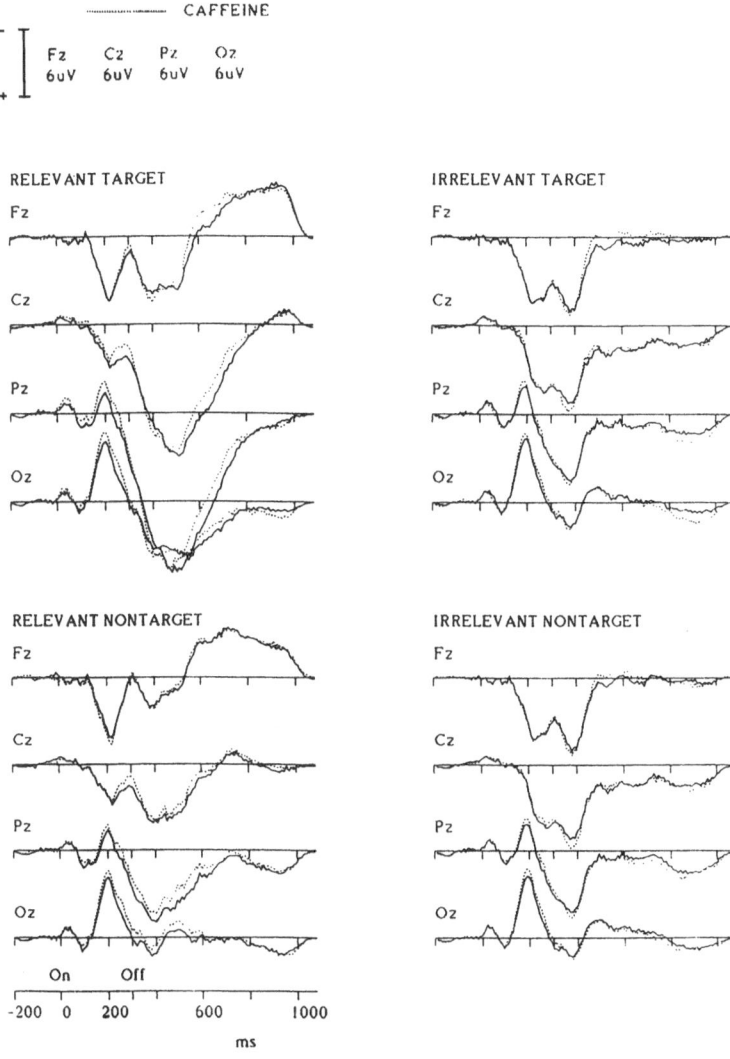

Figure 9.2 – Grand average ERPs of the 4 stimulus categories of the placebo versus the caffeine condition.

Discussion Experiment 1

The central question was whether diurnal type plays a role in the effects of caffeine in task performance. Caffeine speeded *reaction times* in morning types, but also in evening types. This finding supports the view that beneficial effects of caffeine can be found irrespective of whether subjects are in a suboptimal state or not. The reaction time data showed no interactions of diurnal type and caffeine with *memory* load, which implies that the arousal manipulations had *no* effect on *central* processes.

When looking at the impact of caffeine on the ERPs, the earliest effect of caffeine was an increase of the negativity of the N1 component (cf. Lorist *et al.*, 1994a,b) which indicates that caffeine affects early processes such as feature analysis or encoding. Thereafter, it enhanced the amplitude of the later part of the N2b, which differs from earlier findings. First, the enlargement of the N2b did not consist of a positive shift of the ERPs in response to *irrelevant* stimuli as was found by Lorist (1995), but mainly of a negative shift of the N2 elicited by *relevant* stimuli which have to be further processed. Second, there were also large effects of caffeine on memory load on the relevant diagonal in this area. We suggest that this enlargement may be mainly due to an influence of caffeine on memory search and selection processes. For this suggestion there are two arguments. Caffeine reduced the latency and enlarged the amplitude of the search negativity. These results point to an effect of caffeine on the central stage of memory search. Memory search starts sooner, is more intensive and consequently stops earlier. These effects are in contrast to those of Lorist *et al.* (1994b).

A possible explanation for these contradicting results could be the fact that in the present experiment the inter-stimulus intervals were shorter and also with a shorter stimulus duration. Moreover, the smallest memory set, consisting of 1 letter, was again used in the other memory sets and put forward the onset of the search processes. Finally, the enhancement of the P3 amplitude by caffeine supports the notion that caffeine facilitates the effectiveness of transmission and processing of information, although this was partly due to a latency shift. In relevant target ERPs however, the most prominent effect of caffeine was an earlier peak latency of the P3b. We regard this effect as belonging to an advanced onset of memory search processes.

State was manipulated by having all morning and evening types tested at the same time of day at 19.30 p.m. Evening types are reported to perform better than morning types in the evening. This finding was not supported by the reaction times, although the evening types tended to react somewhat faster.

If it is true that the influence of caffeine is most pronounced in suboptimal conditions, morning types should have benefitted more from caffeine than evening types. Caffeine had an unexpectedly large effect on the onset of memory search processes, which was mainly found in the morning types. While in the placebo condition morning types showed memory load effects starting at around 350 ms post stimulus, which confirms largely the results of Okita and co-workers (1985), under the influence of caffeine memory search processes started already around 250 ms after stimulus onset. In both cases memory processing was largely confined to the relevant stimuli. For evening types, the onset of memory search processes did not change very much after caffeine.

The very early onset of memory search processes is remarkable. This could constitute a genuine effect of the manipulations on memory processes. Since search related processes however, have been reported to start around the peak latency of the N2b (Mulder and Wijers, 1991), while in the present experiment memory search already started close to the onset of this negativity, it is difficult to separate selection and search processes and the effects of caffeine and morning- and evening-type on these processes. Nevertheless, it is plausible to adhere the view that in the caffeine condition, less time is needed to end the selection process, thereby affecting the onset of search processes. This is conceivable since the N2b is considered to reflect the post selective processing of relevant stimuli rather than reflecting the selection process directly. Posed in this way, caffeine improved the 'quality' of selection processes, a point of view which is compatible with the reaction time data.

Concerning the results of the drug and state manipulation it can be said that caffeine in addition to a beneficial effect on perceptual processes, had a specific influence on memory search processes. Experimental evidence suggests that an influence of caffeine on other information processes is concealed or indirect and weak (Lorist, 1995). The classification of caffeine effects as direct or indirect is difficult. However, in one way, caffeine effects can regarded to be indirect or energetic since they were most pronounced for the morning types who are supposed to be in a suboptimal arousal condition in the evening. However, since caffeine and type jointly affected the onset but not the amplitude of search processes and thereby affected the timing of the information processing sequence, it can also be argued that caffeine effects were direct. Unfortunately, there are no generally agreed methods to distinguish between direct effects on structural processes and indirect effects on energetic mechanisms (Gaillard, 1988). In spite of this lack of clarity, the preliminary conclusion is that caffeine may counteract suboptimal conditions of diurnal type, and that its positive effects may go beyond these conditions.

Summary

Morning and evening types are found to differ in the phase of their circadian arousal rhythm. In the evening, when morning types have lower arousal levels than evening types, stimulating effects of caffeine are expected to be more pronounced in morning-compared to evening types. In a caffeine and a placebo session, subjects performed a visual focused attention task with a display load of 2 letters and a memory load of 1, 2 or 4 letters, during which EEG and RTs were recorded.

Beside faster reactions, caffeine produced an increase in the amplitude of the N1, N2 and P3. Furthermore, caffeine advanced the onset and increased the negativity of search related processes and the amplitude of the P3b, which was due to a latency shift. The ERPs of evening types showed a positive shift of the amplitude of the P2, N2 and P3 components. Caffeine and diurnal type interacted to produce a latency shift of the search negativity. Apparently, caffeine can counteract suboptimal arousal states, although its positive effects are not restricted to these situations.

EXPERIMENT 2: COFFEE ODOUR

The question we liked to answer was whether olfactory stimulation can be seen as a general arousing factor on cognitive information processing. In order to attain the answer odour-induced brain activity was recorded to be better able to relate the brain activity to specific information-processing stages as reflected in ERP-components. The reason for addressing the olfactory system continuously was the ecological validity of the odorous stimulation. In most real life situations smells come to us in a more or less continuous flow, lasting from seconds to hours. To illustrate this point: Warm *et al.* (1991) found an amelioration in a detection task performance when subjects were presented with odour-*trials*. In fact this is not a very realistic task setup.

The questions we liked to study were:

- Is it possible to measure an effect of odour on information processing in an ERP-paradigm.
- Is it possible to relate a hypothetical influence of a continuous odour to its quality. If the effect of coffee odour is due to its semantic connection to caffeine, then would the effect be comparable to the effect of caffeine?

Since olfactory stimulation has been found to change cortical arousal, it is legitimate to expect that coffee smell will decrease alpha levels and increase beta-EEG-activity levels, which might interact with specific components of task-related information processing, such as feature extraction, stimulus identification, controlled central processing and response-execution (Sanders, 1983). A possible influence of this odour on human information processing should reveal itself in (visual-) task performance and electrophysiological measures of brain activity. Since odour ERP effects in visual or auditory tasks until now have mainly been found 100–250 ms after stimulus onset (Kobal and Hummel, 1991; Lorig, 1989; Lorig *et al.*, 1990; Torii *et al.*, 1991), we expect to find coffee-odour effects on reaction time in this 100–250 ms epoch as well.

Method

Subjects were 13 male and 14 female psychology freshmen, 18 to 25 years of age. Inclusion criteria were: healthy, age under 25, normal olfaction, right-handed, non-smoking, average coffee consumption (2–4 cups a day), no extreme morning-evening type (Kerkhof, 1984) and without medication. Subjects were asked to abstain from caffeine-containing products, alcohol, other psychoactive substances, drugs and perfumes for twelve hours prior to the experimental sessions.

They all engaged in two sessions, in which they performed a twenty minute visual decision-reaction task (480 trials). During this task they were required to push as fast as possible the appropriate button out of four pertaining to each of four different digits. Digits were briefly (400 ms, ISI 450–800 ms) shown on an Apple Macintosh computer-screen. Each of four digits could possibly in two qualities: intact or in a degraded quality. This task was designed so as to manipulate the required arousal-level for the task; degraded digits can be considered to invoke a higher processing intensity at the feature extraction information processing stage, relative to intact digits (Frowein, 1980; Frowein *et al.*, 1981; see also Lorist: Ch. 8). When an effect of the coffee odour interacts with the degradation-effect of the digits, these manipulations can be considered to affect the same processing stage (additive factors method; Frowein, 1980; Van Duren and Sanders, 1988; Sternberg, 1969). Therefore half of the digits shown would be intact, the other half degraded. Before the two experimental sessions, subjects got one training session to get used to the experimental setting and to the task. Recorded were reaction time to each trial, error-scores (misses, false positives), and brain potentials at Fz, Cz, Pz, F3, F4, C3 and C4 (electrocap with Nihon Kohden Polygraph, 100 Hz sample frequency, corrected for eye movements and extreme shifts).

In a 2 by 2 within-subject design, the independent variables were odour (present or absent), presented in counterbalanced order, and stimulus-condition (intact/degraded). Coffee odour was present continuously and supraliminally. Coffee, 125 g freshly milled, was spread on a 30×25 cm cardboard and placed outside the subjects' sight at a distance of about 250 cm in the air-conditioned experimental room. Any association with brewing or consumption of coffee was carefully avoided. Intensity-level of the coffee-odour appeared to be strong enough to be smelled by anyone with normal smelling-capacity. No subject mentioned to have noticed anything special either seen, heard, smelled, tasted or felt, except incidentally other details, like brightness of the screen and sounds. It indicates there was no explicit notice of the drug condition (coffee odour present or absent).

Results and Discussion

Consistent with expectations, the intact digits were scored considerably faster than the degraded digits (average 521.6 ms versus 575.1 ms, $p < 0.01$) and more accurate (7.8% error versus 13.1%, $p < 0.001$). These behavioral measures did not show any significant differences for coffee odour being present or absent. Apparently, coffee odour did not affect the overt performance for this decision-reaction task.

To a certain degree this corresponds to the averaged ERPs: visual inspection of the wave forms shows that peak-latency of the P3 component (about between 300–550 ms) occurs later for the degraded stimuli. Also this component extends longer (amplitudes being more positive) for degraded stimuli. If we consider the P3 component to be related to memory-search processes, these differences might reflect an increase in difficulty with respect to the identification of the stimulus-digit. This corresponds with the degradation of the digits. However, on this component, no effect of the coffee odour has been found. Apparently, coffee odour did not influence the identification or controlled evaluative processing, if any, of the stimuli (Mulder, 1986; Sanders, 1983).

This influence did actually occur somewhat earlier, between 100–300 ms. Apart from a pronounced odour-main effect for a very pronounced P1 component at Pz, an interaction is seen frontally between coffee-odour being present or absent and the potential being evoked by intact or degraded stimuli. Frontally, in the control condition (no coffee odour) the most pronounced P2, N2 components show consistent lower amplitudes for the intact stimulus-digits than the degraded digits. Latencies are equal. These differences in amplitudes are clearly

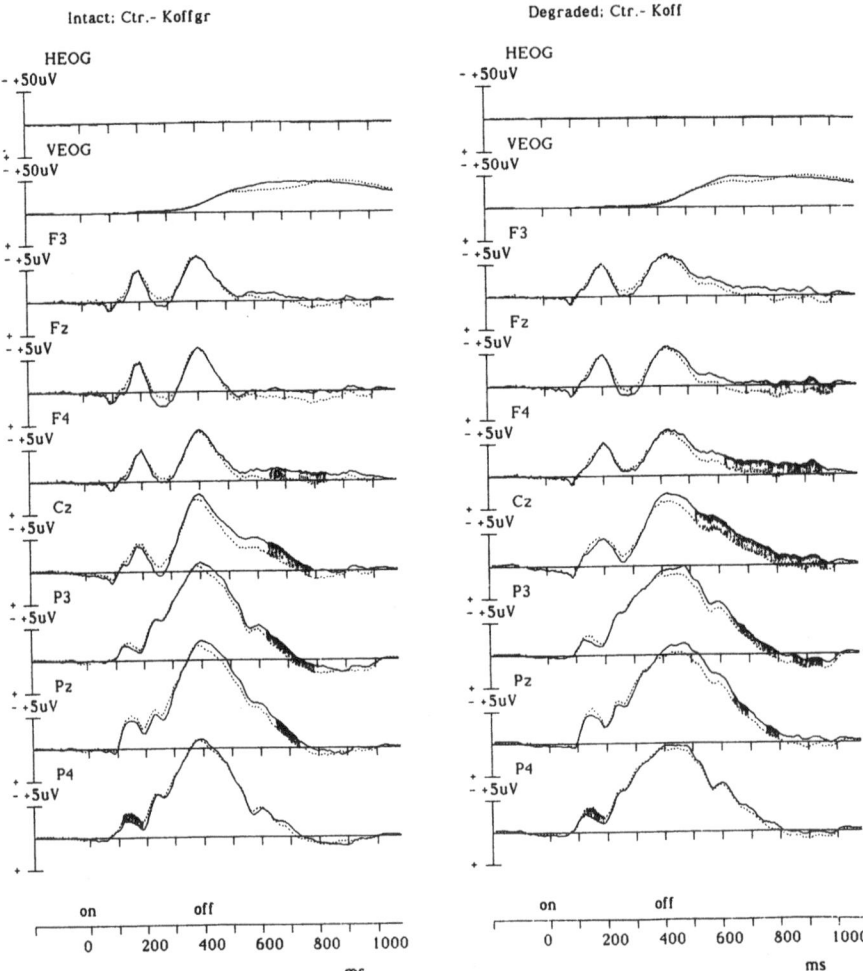

Figure 9.3 – Effects of coffee odour (dotted line) vs. no-odour (control, solid line) on the amplitude of event related potentials of intact (left half of figure) and of degraded stimuli (right half). Black areas depict significant amplitude effect of coffee odour (more positive). HEOG and VEOG are horizontal and vertical electrooculograms respectively.

absent in the odour present condition during the task. These effects are significant (sample point comparisons; F-values range from 5.658 to 12.412, p < 0.05). Additional analyses showed the same interaction-effects at frontal locations F3 and F4.

More positive amplitudes on the early, stimulus-related, components for degraded stimulus-quality have been found before (e.g. Frowein *et al.*, 1981; Wijers *et al.*, 1987) and seem to be related to an early selection based on the physical characteristics of stimuli (Frowein, 1981; Mulder, 1986; Mulder and Wijers, 1991). Wijers and co-workers found more positive P100 amplitudes when the discernment of the relevant features of the stimulus was difficult, and less negative N2a components with less clear stimulus contours. The same occurs in our study with degraded digits, when pixels of the digit are placed outside the original digit-contours. This effect disappears in the presence of coffee odour. Intact stimuli appear to yield higher amplitudes with coffee odour than without coffee odour. For the degraded stimuli there is no difference between odour-conditions. So, when coffee odour is present in the experimental room, even unnoticed as reported afterwards by the subject, early frontal ERP-amplitudes are elevated for the easier (intact digits), but not for the more difficult stimuli (degraded digits).

In other words, selection-type processes can be referred to as the orientation of attention (e.g., Desimone and Duncan, 1995; Posner, 1994). The allocation of more attentional resources would, within limits, be reflected in an increase of arousal. According to some cognitive information processing models, it is especially the early processings that are affected by the level of arousal (Pribram and McGuinness, 1976; Sanders, 1983; Shiffrin and Schneider, 1977). As has been said before, there are indications that olfactory stimulation might affect the level of arousal. Effects of olfactory stimulation similar to ours, have been associated with arousal as cited by Frowein *et al.* (1981), and as studied by Kobal and Hummel (1991) and by Lorig (1991).

Intact stimuli profit more from the assumed alleviation of arousal by olfactory stimulation with coffee odour than degraded stimuli. Apparently, the allocation of additional attentional resources to more complex stimuli (degraded digits) does not add to the extra arousal that seems to stem from the olfactory stimulation per se as appears from the parietal main odour effect as well. This might suggest that the increase of arousal affects the intensity of relatively simple processing. Said in more general terms, intact stimuli appear to be processed through an easier or different routine (Shiffrin and Schneider, 1977; Wijers *et al.*, 1987); or stated otherwise in simpler, more automatic processing evaluation and response become less dissociated than with more complex stimuli. Apparently, coffee odour is able to enhance arousal, but without alleviating dissociation substantially, since the behavioral measures did not differ between odour conditions.

This still leaves the possibility open that this apparent increase in arousal mediated by the smell of coffee, contributes to the presumed stimulating effect of coffee as a beverage. However, the results of Lorist *et al.* (1993), using a similar stimulus-degradation task and a 200 mg doses of caffeine put to decaffeinated coffee, are in contrast to the odour effects. Caffeine induced a more negative N100 and P200, and more positivity between 400–500 ms. Apparently, since we are dealing with an olfactory effect only, it might mean a deterioration of performance, due to the extra claim it puts on the attentional resources. Indeed, Lorig *et al.* (1991) found some evidence for this "divided attention" hypothesis.

The cautious conclusion is that the beneficial effect of the smell of coffee, if any, on early information processing stages in the brain is caused by its widely accepted positive hedonic characteristics and its specific association with caffeinated coffee.

Summary

Odours produce arousal effects of a physical-sensory or cognitive nature or both. To many coffee drinkers the smell of coffee is cognitively conditioned to some sort of arousing effect. Our goal was to compare visual information processing in a situation with or without coffee odour as a background stimulation. During a visual stimulus degradation task, brain potentials and performance measures were recorded. Coffee odour did not affect behavioral performance. Of the ERPs components, the parietal P1 and frontal N1-P2 amplitudes were more positive going in the coffee odour condition. Early frontal amplitudes in the non-odour condition were significantly smaller for intact than for degraded stimuli, such a difference was absent in the odour condition. This result suggests an influence of coffee odour especially on input related processing.

The general conclusion is that the smell of coffee increases the intensity of early stages of information processing. Whether this effect is specific to coffee-odour or whether it is a general odour-effect remains unclear.

ACKNOWLEDGEMENTS

The authors thank Dr. Ingmar Spreeuw for her enthusiastic participation in collecting the data and writing an internal report on the first experiment.

CHAPTER TEN

Effects of Caffeine on Attention: Low Levels of Arousal*

Andrew P. SMITH

INTRODUCTION

Like most factors which change a person's state, the effects of caffeine are selective and influenced by contextual factors such as the nature of the activity being performed, the personality of the subject, and so on (e.g., Clubley *et al.*, 1979; File *et al.*, 1982; Leathwood and Pollet, 1983; Lieberman *et al.*, 1987a,b; Lieberman, 1992; Bättig, 1994). The two major issues considered in this paper are the dose of caffeine and state of arousal of the person. The major question under consideration is whether there are positive effects of caffeine or does it just restore function. This has recently been re-phrased to consider whether there are positive effects of caffeine or whether the claimed improvements in performance reflect withdrawal effects in decaffeinated conditions. This issue will be considered at the start of the paper as it has important methodological implications. The present paper will be mainly concerned with discussion of performance of attention tasks, for example, cognitive vigilance tasks, sustained self-paced responding, or a variable fore period simple reaction time task where the subject does not know exactly when to respond). This is not because the effects of caffeine are restricted to such tasks.

Indeed, our own research (e.g., Smith *et al.*, 1994a) has shown that caffeine improves performance of a wide variety of memory tasks.

However, as the present article is concerned with both the effects of caffeine and those of low arousal states it is advisable to focus on tasks which have already been used in both contexts, hence the choice of the attention tasks. A number of studies have demonstrated effects of caffeine on these tasks and data from one of our studies (Smith, 1995) shows the dose-response relationship between caffeine and performance of the tasks (Figure 10.1).

James (1994) has questioned whether the superior performance and increased alertness found in caffeine conditions in a recent series of studies (Smith *et al.*, 1990, 1991, 1992a,b, 1993a,b) were due to actual enhancement by caffeine or

---

*The present chapter is a part from an earlier published paper: Smith, A.P. (1994). Caffeine, performance, mood and states of reduced arousal: Pharmacopsychoecologia, 7(2):75–86.

Effects of caffeine on vigilance, serial response and reaction time

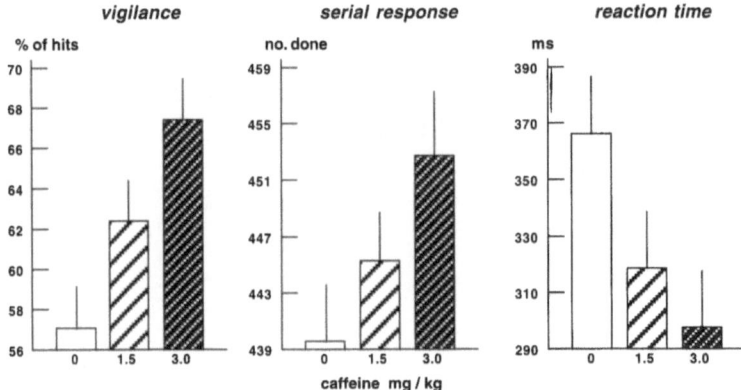

Figure 10.1 – Effects of different doses of caffeine on performance of a repeated digits vigilance task hits), five-choice serial response task (number done) and variable fore period simple reaction time task (RT ms). Scores are the adjusted means from analyses of covariance.

merely reflected performance and alertness being degraded by caffeine withdrawal in the two caffeine-free conditions. The present section of the present article addresses this issue and considers a number of pieces of evidence which produce problems for the 'caffeine withdrawal' explanation of beneficial behavioural effects of caffeine.

The view that beneficial effects of caffeine reflect degraded performance and alertness in the caffeine-free conditions crucially depends on the strength of the evidence for withdrawal effects. James (1994) states that "There is an extensive literature showing that caffeine withdrawal has significant adverse effects on human performance and well-being".

If one examines the details of the studies James cites to support this view (e.g., Hughes *et al.*, 1991; Silverman *et al.*, 1992; Van Dusseldorp and Katan, 1990; Hughes *et al.*, 1992; Griffiths *et al.*, 1990) one finds that effects of withdrawal were selective, influencing some functions only, and were apparent in a small percentage of the subjects. Indeed, these studies provide no evidence that the functions examined in our research, performance of the repeated digits task, are degraded by caffeine withdrawal.

The other attention tasks known to be sensitive to caffeine also showed no effect of caffeine withdrawal. Furthermore, even most recent studies of the effects of caffeine withdrawal on performance and well-being (e.g., Strain *et al.*, 1994) can be criticized on methodological grounds for example, failing to consider the importance of order of caffeine/placebo conditions.

A second point made by James himself in an earlier article (James, 1991) is that the effects of caffeine on performance and mood are variable and influenced by contextual factors. If withdrawal was the major factor in these studies one should find that, provided enough caffeine was given to prevent withdrawal, further increases in dose should have no effect. This is clearly not the case (Roache and

Griffiths, 1987). Another problem for the 'caffeine withdrawal' explanation is that it cannot account for effects in naive users or in animals that never had caffeine before. Yet behavioural effects clearly occur in these groups and one might expect them to be larger than those found with regular users, where some habituation may take place. James rejects such data as reflecting only 'a small minority of the population'.

Three results from our 1994 and 1995 studies also cause problems for the 'caffeine withdrawal' explanation. First, it is possible to demonstrate the same effects of caffeine on performance when a person has abstained for only 1 hour (Smith *et al.*, 1994b) as when they have had no caffeine for over 12 hours (Smith *et al.*, 1994a). Secondly, level of regular caffeine usage is not correlated with changes in alertness and psychomotor performance following 24 hour abstinence (Smith, 1995). Finally, beneficial effects of a single low dose of caffeine (e.g., 100 mg) have, in studies with a relatively small number of subjects (e.g., n = 48) only been obtained when a person's state of arousal is low (Smith and Phillips, 1993). This last result is of major relevance in that subjects in caffeine-free conditions would, according to the 'caffeine withdrawal' view, be impaired whether their alertness level was low or normal. In order to explain why the effects of caffeine vary with a person's arousal level the 'caffeine withdrawal' view would have to show that caffeine withdrawal effects depend on a person's alertness. No evidence for this is presented by James.

The aim of the above account is not to argue that caffeine withdrawal effects do not exist at all but rather to suggest that they are unlikely to be the major factor involved in the performance and alertness effects observed following the administration of caffeine. It is also important to point out that the current debate should not distract from the practical benefits of caffeine in low arousal situations. One of the most important points of James' 1994 article was to identify new experimental paradigms for investigating the effects of caffeine on performance. For example, James argues that one should withdraw caffeine for a week before testing for effects of caffeine. After this period of withdrawal any effects of caffeine should reflect direct effects rather than changes in the decaffeinated condition. This approach has recently been used (Smith, 1995) and the results show that caffeine improves performance of the attention tasks even in subjects who have had no caffeine for the previous week (see Figure 10.2).

It appears, therefore, that caffeine does produce improvements that cannot be attributed to performance decrements in the decaffeinated conditions. One must now consider its effects in low alertness situations.

EFFECTS OF 1.5 AND 3 MG/KG OF CAFFEINE

Caffeine and the post lunch dip

Smith *et al.* (1990) demonstrated that caffeine removed the post-lunch dip in sustained attention (Figure 10.3).

Close inspection of the data shows that caffeine had a beneficial effect both pre- and post-lunch. The bigger effect of caffeine after lunch reflects reduced performance by the decaffeinated group at this time.

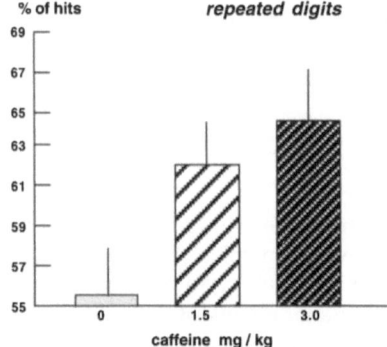

Figure 10.2 – Effects of caffeine on performance of the repeated digits task by subjects who have had caffeine withdrawn for a week. Scores are the adjusted means from analyses of covariance.

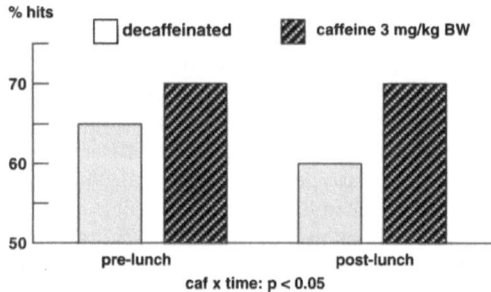

Figure 10.3 – Effects of caffeine on sustained attention in the late morning and after lunch.

Caffeine and performance during the day and night

An other experiment was conducted to examine the effects of caffeine on performance and mood over the day and night (Smith *et al.*, 1992a,b). Interest lay in whether the effect of caffeine would be constant, or whether it would be much greater when circadian arousal was decreased (at night).

Method

Design of the study

An experimental study of nightwork was carried out. Each subject was tested in all of the conditions found by combining day/night shifts with different drink

Table 10.1 – Schedule of test sessions and drinks

Day	Night	
09.00–09.30	22.00–22.30	drinks
09.30–11.00	22.30–24.00	first session of tests
11.00–11.30	24.00–00.30	drinks
11.30–13.00	00.30–02.00	second session of test
13.00–14.00	02.00–03.00	lunch, more drinks
14.00–15.30	03.00–04.30	third session of tests
15.30–16.00	04.30–05.00	drinks
16.00–17.30	05.00–06.30	fourth session of tests

Orders of tests

(1) alertness rating
(2) verbal reasoning
(3) semantic processing
(4) variable fore-period simple RT
(5) five choice serial response task

conditions. The order of the conditions was counter-balanced across subjects. The experiment was based on the study carried out by Smith and Miles (see Smith and Miles, 1985, 1986a,b, 1987a,b; Smith, 1988, 1990, 1991). Each subject attended for day shifts, from 09.00 to 17.30, and night shifts from 22.00 to 06.30. Before the first test session subjects were familiarized with the testing procedure and also completed personality questionnaires measuring dimensions known to be related to the effects of caffeine (e.g., impulsivity) and nightwork (e.g., morningness). The subjects were 24 students (12 male, 12 female) who were all regular, but moderate coffee drinkers (2–4 cups a day). Half of the subjects was tested in the order day/night and the others in the order night/day. In the caffeinated coffee condition the subjects were given decaffeinated coffee with 1.5 or 3 mg/kg body weight of caffeine tablets added. In another condition subjects were given decaffeinated coffee and in the third they were given fruit juice. The order of the three beverage conditions was counter-balanced across the subjects. The caffeine manipulation was double-blind with neither the experimenter nor subjects knowing whether they were given caffeinated or decaffeinated coffee. Subjects were given a drink at the start of the shift, after two hours' work, at the mid-shift meal and two hours after the meal. The performance tests, which were done on a PC, alertness ratings, and physiological measures were organized as shown in Table 10.1.

Only the results from the variable fore preiod simple reaction time task and the five-choice serial response task are reported here and details of these tasks are shown below (see Smith *et al.*, 1992b, for other details).

Variable fore period simple reaction time task

In the variable fore preiod task a box was displayed on the screen and at varying intervals (from 1–8 seconds) a square appeared in the box. The subject had to press a key as soon as the square was detected. This task lasted for 10 minutes.

Five choice serial response task

In the five choice serial response task, five boxes were displayed on the screen and a square appeared in one of the boxes. The subject pressed the corresponding key and the square then appeared in another box and the subject had to press the next key. This task lasted for 10 minutes and the number of responses made, number of errors and number of gaps (occasional long responses) were recorded.

Results

Since one subject failed to complete the study, the following analyses were carried out on the complete data of the remaining 23 subjects. Analyses of variance were carried out on the data and these distinguished the following factors: *Beverage*: caffeine, decaffeinated and juice; *Day* versus *night*; *Sessions* within a shift (4); *Order* of beverages; *Dose* of caffeine.

In none of the analyses was there a significant effect of order of beverages. Similarly, comparable effects were obtained for both doses of caffeine so the combined data of the two groups are presented here.

Serial response task

There was a significant effect of time of testing on the number correct (*day/night* × *sessions*: [$F(3/51) = 5.50$; $p < 0.01$]), with performance being impaired at 03.00 and 05.00. Significantly more correct responses were made after drinking caffeinated coffee than after the other beverages [$F(2/34) = 9.76$; $p < 0.0005$]. Performance declined over the night shift following consumption of juice and decaffeinated coffee but this was not observed after drinking caffeinated coffee. Indeed, the only time when there was little difference between the drinks was at 22.30 (Figure 10.4) (for more details see Smith *et al.*, 1993a).

Simple reaction time task

Reaction times were slower at night ([$F(1/17) = 7.62$; $p < 0.05$], especially at 03.00 and 05.00 (day/night × sessions: [$F(3/51) = 8.55$; $p = 0.0001$). Consumption of caffeinated coffee improved reaction times relative to the other drinks [$F(2/34) = 9.13$; $p < 0.005$] (Figure 10.4). Improvement was especially large at 03.00, 05.00, and 09.30. Indeed, after caffeinated coffee reactions at 05.00 were faster than the day time reaction for the other groups. Furthermore, at this time caffeinated coffee was associated with 20% faster responses than those found after the other beverages.

Discussion

This experiment has demonstrated quite clearly that caffeinated coffee improves performance of attention tasks. These effects were observed across both the day and night and the caffeinated coffee greatly reduced the night-time decline observed following consumption of the other drinks. In contrast to the effects of

Effects of caffeine during the day and night

Five choice serial response task

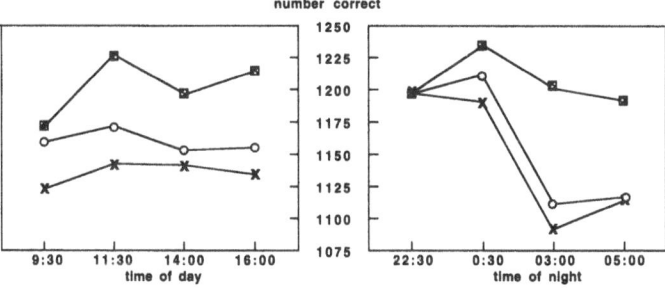

Variable fore-period simple reaction time

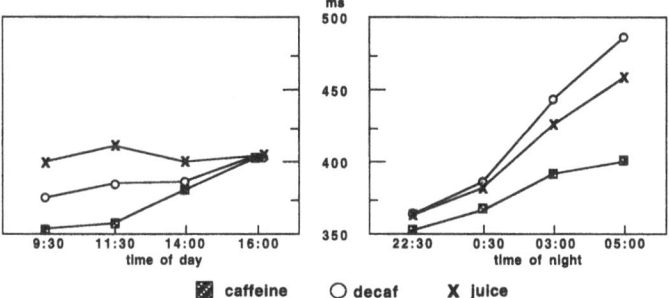

▨ caffeine   ○ decaf   ✕ juice

Figure 10.4 – Effects of caffeine on performance of the variable fore period simple reaction time and five-choice serial response tasks during the day and night.

the caffeinated coffee, the differences between the decaffeinated coffee and juice were small and variable. It would appear, therefore, that caffeine may be of great benefit in reducing the acute impairments associated with working at night. Further research is now required to examine whether it can remove some of the longer term effects, and whether it is also beneficial in other low arousal situations. The next experiment examined the effects of caffeine on the performance of individuals whose alertness had been reduced by prolonged work.

CAFFEINE AND PERFORMANCE FOLLOWING PROLONGED WORK

The main aim was to examine the effects of caffeine on impairments in performance and changes in mood induced by prolonged work (see Craig and Cooper, 1992, for a review of the effects of fatigue on performance). This paradigm has great potential for examining the effects of stimulants on performance and mood for it is possible to see how much recovery is produced by the drug. The literature on caffeine suggests that it has selective effects on performance and mood with some functions being more sensitive than others. The present study examined performance of a range of tasks to determine whether these would show similar impairments over a 12 hour period and whether caffeine would produce different levels of recovery in the different tasks.

Experimental Procedure

Twenty four undergraduate students (12 male, 12 female) took part in the study. They were paid for participating in the study. Each subject carried out three 12.5 hour sessions (from 09.00–21.30), with a rest day between each. The first eight hours of each day was constant with regard to beverages consumed, with all subjects being given fruit juice at the breaks. On one of the days subjects were given caffeinated coffee at the end of 8.5 hours, with further caffeinated coffee after 10.5 hours. On the other days subjects were given decaffeinated coffee or fruit juice instead of caffeinated coffee. The order of the different drinks was counterbalanced across subjects. The schedule for each day and the order of the tests within each session was identical to the previous study. After the first four sessions were completed the subjects had a drink break where either caffeinated coffee, decaffeinated coffee or juice was given. The subjects then carried out the tests again, had another drink break and then repeated the tests for a sixth and final time.

Prior to the start of the experiment subjects were given a familiarization session during which they practiced the performance tasks. Subjects refrained from drinking alcohol for 12 hours before the start of each session and got no caffeinated drinks just prior to the first tests. In between test sessions subjects remained in the laboratory and were not allowed to smoke or eat snacks.

The mood rating and performance tests were identical to the previous study. In the caffeine condition 3 mg/kg body weight of caffeine tablets were added to decaffeinated coffee. The caffeine/decaffeinated manipulation was double blind.

Results

Analyses of variance were carried out distinguishing the following factors: Drinks; Sessions 5 and 6 (the two test sessions after administration of the caffeine); Order of drinks.

Simple reaction time task

Again, caffeinated coffee improved speed of reactions ($p < 0.05$). This effect was greatest at session 6 but was apparent at session 5 as well (Figure 10.5).

Serial response task

As in the nightwork experiment, subjects made more correct responses following the caffeinated coffee ($p < 0.05$) and again, this effect was observed at both sessions 5 and 6. (Figure 10.5).

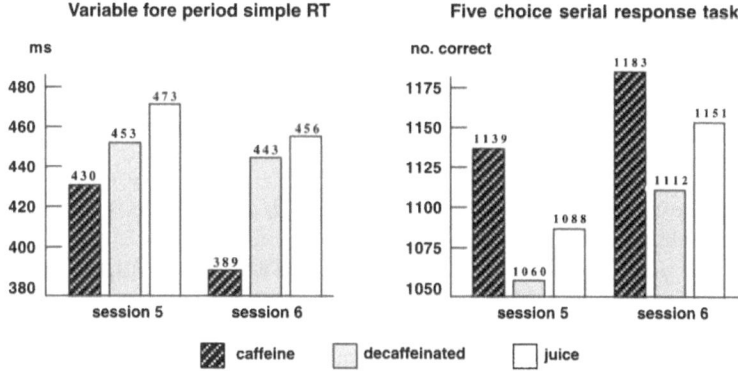

Figure 10.5 – Effects of caffeine on performance of the variable fore period simple reaction time task and five-choice serial response task after 8 hours prolonged work (session 5) and 10 hours work (session 6).

Discussion

This experiment confirmed that caffeinated coffee has beneficial effects on performance of attention tasks. These results confirmed that caffeinated coffee can reduce the low levels of subjective alertness induced by prolonged work, and that it can also remove the performance impairments associated with such situations. This has important implications for operational efficiency and safety and further research is now required to determine what doses of caffeine are successful in achieving these effects. The next set of experiments examined whether a single realistic dose of caffeine (equivalent to two cups of instant coffee) produced similar beneficial effects in low arousal states. In this experiment was examined whether a low dose of caffeine influenced mood and performance when circadian alertness was reduced after lunch.

The final study examined whether a 1.5 mg/kg dose of caffeine produced different effects on mood and performance in individuals suffering from an upper respiratory tract illness and those who were healthy.

1.5 MG/KG OF CAFFEINE AND PERFORMANCE BEFORE AND AFTER LUNCH

Method and Procedure

Forty six female subjects took part in the study; they were paid £36 for participating. Subjects were tested on 3 consecutive days. On the first day subjects were practised at the performance tests and completed questionnaires assessing personality (introversion/extraversion; impulsivity; sociability; neuroticism (Eysenck Personality Inventory); trait anxiety (Middlesex Hospital Questionnaire);

morningness (Horne-Ostberg morningness/eveningness scale), recent mood (anxiety, depression, perceived stress, positive and negative mood (Bipolar visual analogue scales), and food and caffeine intake. The weight and height of the subjects were also recorded at this time. On the second and third days each subject was tested 3 times; once in the late morning, once in the early afternoon (90 minutes after the start of lunch) and then finally in the late afternoon (3 hours after the start of lunch). On one of the days subjects were given lunch and on the other they abstained from eating. All subjects were given a standard lunch consisting of orange juice, a cheese salad roll, packet of crisps and a chocolate bar. The order of lunch conditions was counterbalanced across subjects. At the end of the lunch (and at the corresponding time on the no lunch day) the subjects were given a 150 ml cup of decaffeinated coffee. Half the subjects had caffeine (1.5 mg/kg body weight) added to this. The caffeine manipulation was carried out double blind.

At the start of each session blood pressure and pulse were recorded. Subjects then completed a number of performance tasks, and results from one of them, the five-choice serial response task, are described here.

Results

Analyses of covariance, with the pre-lunch scores as covariates, were performed on the data. The between subject factors were caffeine condition, and time of testing (half the subjects were tested at: 11.00, 13.30 and 15.30; the others at: 12.00, 14.30 and 16.30). The within subject factors were lunch/no lunch days and time of day (early vs. late afternoon). Task parameters (e.g., minutes doing the task) were also included as within subject factors. The effects of caffeine on mood after performing the tests was especially beneficial in the early afternoon testing session. Similarly, the effect of caffeine on the number completed in the five choice serial response task was greatest at this time, whereas this time of day was associated with the lowest level of performance in the decaffeinated group. This effects is shown in Figure 10.6. These effects of caffeine were apparent on both days when lunch was consumed and when the subjects abstained from eating.

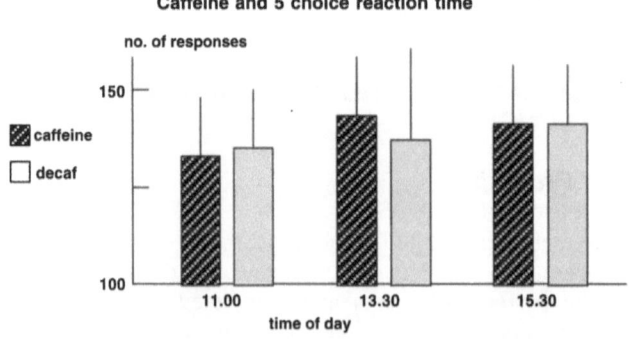

Figure 10.6 – Effects of a low dose of caffeine (1.5 mg/kg) on performance of the five-choice serial response task in the morning, after lunch (13.30) and later in the afternoon.

Discussion

The present results showed that the effects of low doses of caffeine were especially pronounced in the early afternoon testing session. This confirms previous findings obtained with higher doses of caffeine. The effects of caffeine were similar on days when lunch was consumed and when the subjects abstained from eating, which suggests that either caffeine influences the endogenous component of the post-lunch dip rather than the meal-related part, or that it has a more global effect which is independent of these but more easily detected when alertness is reduced.

The next experiment examined the effects of low doses of caffeine on performance and mood of subjects with upper respiratory tract illnesses and healthy controls.

CAFFEINE AND PERFORMANCE OF HEALTHY SUBJECTS AND THOSE WITH A COMMON COLD

This study investigated whether caffeine produced different changes in performance and mood in subjects with an upper respiratory tract illness (common colds) and healthy controls. The effects of having a cold on mood and performance are now well-documented (see Smith, 1992, for a review) and show that subjective alertness is reduced and the speed and accuracy of performing psychomotor tasks is impaired. The present account describes the effects of caffeine on the variable fore period simple reaction time task.

Method and Procedure

Ninety-nine students participated in the experiment (46 female, 53 male). All subjects were tested when healthy and then returned to the laboratory when they developed a cold. When approximately half of the sample had returned, the others were recalled to act as healthy controls. On the second visit subjects completed another test session and were then given a drink (either caffeinated coffee, decaffeinated coffee or fruit juice). They were then re-tested about an hour after they had finished their drinks. Fifteen subjects in the caffeine condition had colds and 20 were healthy. Fifteen subjects in the decaffeinated condition had colds and 16 were healthy; 16 subjects in the juice condition had colds and 17 were healthy. The caffeine manipulation 1.5 mg of caffeine/kg body weight was double blind.

Results

The pre-drink data confirmed that subjects with a cold felt less alert and had slower response times than the healthy subjects. The main interest here is whether caffeine produced different effects in the two groups of subjects. Figure 10.7 shows that in the variable fore period simple reaction time task, the difference between subjects with colds and the healthy subjects was modified by the nature of the drink. Subjects with colds who drank juice were slower (381 ms on simple

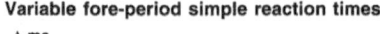

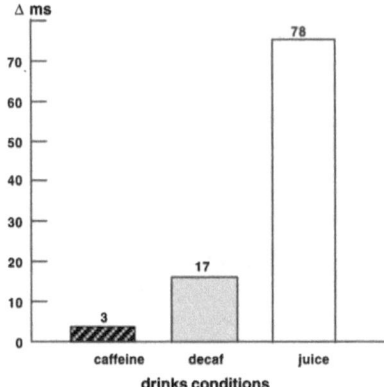

Figure 10.7 – Effects of a low dose of caffeine on the *difference* in speed of performing the variable fore period simple reaction time task by subjects with colds and healthy volunteers. Scores are the adjusted means from the analysis of covariance.

RT) than healthy subjects (303 ms) and decaffeinated coffee (339 ms and 322 ms respectively) reduced this effect. In the caffeinated coffee condition there was no difference between those with colds and those who were healthy (309 vs. 306 ms).

Discussion

The results from this last study confirm that caffeine may be especially beneficial in people with reduced arousal (see also Smith, 1994; Lorist, 1995). This conclusion applies not only to high doses but also to amounts of caffeine (1.5 mg/kg) that one might reasonably expect a person to consume in a real-life setting. Such results have obvious practical implications for real-life safety, performance efficiency and well-being. Further research is needed to determine which mechanisms underlie such effects but it is possible to speculate about this and this is done in the final section.

Mechanisms to account for the beneficial effects of caffeine in low arousal states

It is clearly the case that high doses of caffeine improve performance éven when arousal is at normal levels. Lower doses produce smaller effects and it is, perhaps, necessary to compare them with reduced performance (as is observed in low arousal states) for the effects of caffeine to be significant. This view suggests that low and high doses of caffeine produce identical changes in brain function (see for a review Lorist, Ch. 8). An alternative view is that certain neurotransmitter systems are important at high doses and others at low doses. For example, cholinergic or dopaminergic mechanisms may be the key ones when a high dose of caffeine is given, whereas changes in central noradrenaline may be most relevant when the dose is lower. Further studies examining the effects of caffeine on drug-induced changes in the neurotransmitter systems are necessary to test which of the above views is or are correct.

CONCLUSIONS

The research reviewed in this paper shows that high doses of caffeine can improve performance of attention even when alertness is at normal levels. Such changes cannot be accounted for in terms of impaired performance due to withdrawal of caffeine in the decaffeinated conditions. Effects of low doses of caffeine are more easily demonstrated in low arousal situations for example after lunch, at night, after prolonged work, or when the person has an illness like the common cold. These beneficial effects of caffeine are of great practical importance and they suggest that consumption of even low doses of caffeine may reduce the problems of safety and loss of performance efficiency which are often associated with low arousal states. At the moment it is unclear whether the effects of low and high doses reflect identical mechanisms or whether different neurotransmitter systems have primary roles depending on the dose of caffeine.

ACKNOWLEDGEMENT

Research described in this paper was supported by the Physiological Effects of Caffeine Research Fund of the Institute for Scientific Information on Coffee, Paris.

CHAPTER ELEVEN

Coffee Consumption and Subjective Health: Interrelations with Tobacco and Alcohol

Ronald A. KNIBBE and Ypie T. De HAAN

INTRODUCTION

Around the middle of the 17th century West European countries like England, the Netherlands and France started to import coffee at a large scale. Before that time it was mainly distributed by apothecaries and used for medical purposes. At the introduction of coffee on a larger scale in Europe, coffee houses played an important role in the diffusion of coffee drinking habits. The appreciation of coffee was not only determined by the intrinsic qualities of coffee, but also by the fact that it did not lead to drunkenness as did the beverages most frequently consumed in that period: beer and wine (Jansen, 1976; Dekker *et al.*, 1993).

Nevertheless, coffee consumption has also always been viewed with some suspicion as being not really healthy. In 1674, at the start of the popularity of coffee houses in England, an (in)famous pamphlet "The women's petition against coffee" was distributed. As the subtitle indicated, this pamphlet expressed worries about the "... grand inconveniences accruing to their sex from the excessive use of that drying, enfeebling liquor" (cited in: Jansen, 1976). The complaints focus very much on impotency because of coffee consumption "... they are become as impotent, as age and as unfruitful as those desserts whence that unhappy berry is said to be found", but include also an increased talkativeness "For here like so many frogs in a puddle, they sup muddy water, and murmur insignificant notes till half a dozen of them out-babble an equal number of us at gossiping ... You may as soon fill a quart pot of syllogismes, as profit by their discourses".

It is very unlikely that negative opinions about coffee were shared by large parts of the population in Europe. The fast diffusion of coffee drinking habits in Europe points almost to the opposite: an overwhelmingly positive appreciation. This appreciation has led to the present situation in which the large majority of the adult population in Europe (and most likely in the largest part of the rest of the world) regularly consumes coffee.

This does not mean that coffee consumption is viewed as completely without any risks. The fact that in most societies the habitual consumption of coffee is discouraged among the youngest age groups already indicates that some maturity is deemed necessary to handle such a 'strong' drink appropriately. As we will see in further detail, a considerable part of the general population reports that their coffee consumption is negatively commented upon by others and report health complaints attributed to coffee. Because of this, the public's attitude towards coffee consumption is likely to be somewhat ambivalent.

The idea that coffee consumption might have unwanted effects is, of course, also found in scientific research. To give but one example, Bak (1990) cites no less than 25 epidemiological studies, published between 1963 and 1989 in international journals, about the relation between coffee consumption and cardiovascular disease. However, studies which focus not on the consequences but on possible behavioural and social determinants of coffee consumption very rarely restrict their subject to coffee consumption alone. Coffee consumption is mostly studied in combination with tobacco consumption and/or alcohol consumption. The common element in these behaviours is either conceptualized as 'substance use', or more specifically the potential addictive properties of these substances (e.g., Carmelli *et al.*, 1990) or it is conceptualized in terms of a 'healthy life style' (e.g., Schwarz *et al.*, 1994). For both conceptualizations there is some support. Cessation of coffee consumption has been found to lead to a withdrawal syndrome in many studies (e.g., Silverman *et al.*, 1992) and considering the extent to which smoking as well as alcohol consumption and coffee consumption are associated with negative health consequences, it seems plausible to assume that a healthy life style is associated with an avoidance of all three substances. However, one can wonder to what extent in the public at large the health risks of coffee, tobacco and alcoholic beverages are perceived as similar. A robust indicator that the health risks associated with these behaviours are not perceived as equally dangerous to one's health is that the prevalence of smoking, coffee and alcohol consumption is different in the general population. Of course, subcategories of the population might find the common element of health risks more important than the differences in health risks, but it remains to be seen how large a proportion of the population views coffee consumption in that respect as similar to tobacco and alcohol consumption.

The major aim of this chapter is to describe some aspects of coffee consumption in the general population. The literature on this subject is less abundant than, for example, tobacco or alcohol consumption. Because of this situation we will discuss in this chapter a variety of themes rather than concentrating on one specific aspect of coffee consumption in the general population. The more specific problems we will focus upon are:

- To what extent do diary and questionnaire estimates of coffee consumption in the general population differ?
- What is the prevalence of coffee consumption and of coffee consumption in combination with alcohol consumption and/or tobacco use?
- What is the prevalence of criticism on coffee consumption and of perceived health related consequences attributed to coffee consumption?
- Which factors are associated with differences in coffee consumption?

The rationale for the first problem is the frequently reported under-estimation of alcohol and tobacco consumption when self reports are used. Coffee consumption shares one important characteristic with tobacco and alcohol consumption. All three behaviours are mostly habitual. This means among other things that people are not likely to monitor exactly how many units they have consumed over a period of time. From a theoretical point of view this may lead both to underreporting and overreporting. For alcohol and tobacco consumption it has been ascertained that most types of self reports tend to systematically underestimate real consumption. For coffee consumption this has not been studied as far as we know.

Because the appreciation of coffee consumption is probably somewhat ambivalent, subpopulations might differ in their norms about what constitutes a 'proper way' to integrate coffee consumption in their life style. This differences might lead to systematic differences in coffee consumption. From a theoretical level, status factors like gender, social class, age, ethnicity are most likely to be associated with differences in norms how to properly integrate coffee consumption in one's lifestyle (see Knibbe *et al.*, 1987 for a more elaborate discussion of the function of status roles). We will analyze the relation of coffee consumption with social class, age and gender. Employment and marital status might also be related to coffee consumption. In as far as the attraction of coffee drinking is also located in the opportunity it offers for some ritualized sharing of time, work and living with a partner may contribute to a higher level of consumption.

DATA AND METHODS

Sample

The data used in this chapter are from a Dutch population survey of which the fieldwork was done in 1985. One of the major aims of this survey was to answer some persistent questions concerning the reliability and validity of survey estimates of alcohol consumption estimates in a general population. With respect to the measurement of coffee consumption this survey has the following advantages:

- The interviewing was spread over the whole year to account for seasonal fluctuations. For alcohol consumption this has been shown to be of influence on survey-estimates (Lemmens and Knibbe, 1993).
- The respondents were not only interviewed, but also asked to keep a diary for 7 consecutive days. This allows a comparison between questionnaire and diary estimates of coffee consumption which can give some insight in measurement errors.
- It is the only Dutch general population survey with data about harmful consequences attributed to coffee consumption and criticism of others on one's own coffee consumption.

The fact that the survey data is from 1985 might mean that the consumption of coffee, alcohol and tobacco has changed. However, there are no indications that coffee, alcohol and tobacco consumption have changed strongly in the general population since 1985. According to the Central Bureau of Statistics (1995) the estimated per capita consumption of coffee increased somewhat between 1985 and 1994 from 7.9 kilo to 8.4 kilo per year, the per capita consumption of cigarettes was estimated to be about stable in this period (1985: 1124 per year; 1994:

1090 per year). According to the Productschap (1993), the per capita consumption of alcohol did not change much between 1985 and 1991 (respectively 8.5 and 8.2 liter 100% alcohol per year).

In the first stage of the field work a stratified (according to region and urbanization) random sample of all municipalities was drawn. In the next step a random sample of persons between 16-70 years (n = 1807) was drawn from the municipal population registers. The selected persons had to cooperate both to the interview and diary to be considered as response. The response was 69% (n = 1240). Deviations from the overall percentage of non-respondents according to gender, age, marital state, urbanization and region were small (maximum 5.2%). After completion of the fieldwork, a sample of 116 non-respondents were approached to gain more insight in possible selective drop-out, especially with respect to alcohol consumption. There were no indications for selective drop-out of heavy drinkers of alcoholic beverages, but among the non-respondents a higher proportion of female abstainers of alcoholic beverages (defined as no consumption in the preceding 6 months) was found (see Lemmens *et al.*, 1988 for more details).

It can be concluded that, with the exception of female abstainers of alcoholic beverages, there are no indications that the interviewed sample is biased when compared with socio-demographic characteristics of the general population or with respect to habits like smoking, drinking and coffee consumption. In one of the following sections it will be reported that male and female abstainers of alcoholic beverages are more likely not to drink coffee. Because of the relation between the non-use of alcoholic beverages and non-use of coffee it might be that the interviewed sample underestimates the percentage of women not drinking coffee.

Measurement of the variables

Coffee and alcohol consumption have been measured both by questionnaire and diary. Tobacco consumption was only included in the diary. The type of question used in the questionnaire to measure coffee and alcohol consumption was 'How many cups of coffee (glasses of alcoholic beverages) do you usually drink per day?' The diary was filled in for seven consecutive days. At the introduction the interviewers explained how the diary should be filled in and, as an exercise, showed the respondent how the day before the interview should be noted in the diary. The respondents were asked, among other things, to note every day at which times they had consumed the following 7 types of beverages: (1) beer, (2) wine, (3) fortified wines, (4) strong alcoholic beverages, (5) soft drinks (including water, fruit juice), (6) milk and similar types of drinks and (7) coffee or tea. The diary did not distinguish between coffee and tea. Nevertheless, it was possible to estimate coffee consumption during the diary week because in the questionnaire coffee and tea consumption were asked for separately. By computing the percentage of coffee of the total coffee and tea consumption according to the questionnaire and applying this on the diary data we obtained an estimate of coffee consumption during the diary-week. In individual cases the share of coffee and tea in the total consumption of these beverages might have changed. However it seems very unlikely that within the period of one week in the sample as a whole the change was systematically in the direction of a higher percentage of coffee or to a higher percentage of tea in the total consumption of these beverages. Therefore we assume that aggregate estimates of coffee consumption with the diary data are not invalidated by this procedure.

Worries about, criticism of others upon and consequences attributed to coffee consumption in the last half year were measured by direct questions in the interview. An example is: 'Did you experience stomach ache due to coffee consumption in the last half year?'. Also the socio-demographic variables have been measured by direct questions in the interview.

DIARY AND QUESTIONNAIRE ESTIMATES OF COFFEE AND ALCOHOL

Introduction

For the Netherlands it has been shown that since 1958 survey estimates cover varying 41% to 56% of the sales estimate of alcohol consumption (Neve *et al.*, 1993). For other countries with both survey and sales data about alcohol consumption the same applies: survey estimates are consistently lower than the sales estimates. Although there may be some error in the sales estimate there is no doubt about what the discrepancy between these two estimates mean: survey estimates of alcohol consumption underestimate 'real' alcohol consumption. Considering that one of the major aims of these types of studies is to estimate the prevalence of heavy drinkers, the fact that about 50% of the real consumption is not covered is a very severe drawback.

For tobacco consumption it has also been shown that surveys underestimate the number of cigarettes (Van Reek, 1986; Kozlowski and Heatherton, 1990). However most studies on smoking use a dichotomous variable: smoking or not. For this variable there are no major indications that the measurement is unreliable (Kozlowski and Heatherton, 1990).

About the measurement of coffee there are, as far as we know, no studies which concentrate on the reliability of measurement in general population surveys. We will do so by comparing diary estimates of coffee with questionnaire estimates of coffee of the same respondents. The main differences between diary and questionnaire estimates are that with the diary method the time between consumption and reporting is much shorter and that it is much more sensitive for temporal variations in consumption.

Lemmens *et al.* (1992) showed with respect to alcohol consumption that one of the main differences between diary and questionnaire estimates is that especially those respondents who drink 3-5 times a week, tend to report less drinking days when interviewed compared with a diary. However, whereas only a minority of 25% of the men and 15% of the women drink daily alcoholic beverages, the large majority of coffee drinkers consumes daily. Therefore it is unlikely that forgetting of drinking days is an important source of underreporting of coffee consumption. However, the number of cups of coffee consumed per day is likely to vary over the days. In the interview respondents were asked 'how many cups of coffee do you usually drink per day?'. It is unlikely that in reaction to such a question, respondents will do the arithmetics necessary for an accurate answer, for example remembering and adding up the number of cups on the preceding seven days and correcting for the variation in coffee consumption over weeks). More likely the number of cups consumed on most days will be reported in reaction to such a question. Depending on whether the other days he/she drinks less or more coffee the questionnaire estimate will overestimate or underestimate his real consumption and will be higher or lower than a diary estimate of coffee consumption.

Results

We collected questionnaire and diary estimates of average consumption of coffee and alcohol; for cigarette consumption we do not have a questionnaire estimate. The most surprising outcome is that questionnaire estimates of coffee consumption at 37.1 cups of coffee per week are higher than diary estimates of 31.5. Provided that diary estimates are more accurate, it seems that people tend to over-report their coffee consumption when asked 'how much coffee do you usually drink'. The questionnaire estimate of 9.4 glasses of alcohol beverages per week is lower than the diary estimate, which is 10.9.

To inspect to which extent the overreporting of coffee in interviews is about similar at all levels of coffee consumption, we plotted the differences (questionnaire minus diary estimate) in consumption against the diary estimate of consumption and did a regression analysis.

It appears that there is no systematic increase or decrease in difference scores at higher levels of coffee consumption. Both the correlation between different score and diary estimate of coffee consumption and the slope of the regression line are 0.00, meaning that at all levels of coffee consumption the questionnaire estimate is about 0.80 cup per day higher than the diary estimate.

In the last paragraph we will more fully discuss these outcomes. Because the outcomes do not mean, in our opinion, that the diary estimates are less precise than questionnaire estimates, we will use the diary estimates of coffee and alcohol consumption in the following paragraphs.

THE PREVALENCE OF COFFEE, ALCOHOL AND TOBACCO CONSUMPTION

In principle, the decision to use, for example, alcoholic beverages does not have to influence the decision to use coffee or tobacco. Similarly, for those who decided for example to drink both alcoholic beverages and coffee, the decision about how much alcoholic beverages one drinks does not necessarily influence how much coffee one drinks. Yet, as indicated in the introduction theoretical concepts like 'substance use' or 'life style' suggest that the use of these substances is interrelated in such a way that the use of one of these substances is positively correlated with the use of the other substances. In this paragraph we will analyze to which extent in the general population the use of coffee, alcohol and tobacco is combined and the extent to which the amounts consumed of these substances are related. More specifically the following questions will guide the analyses:

- What is the prevalence of the use of coffee, alcoholic beverages and tobacco and combinations of these substances?
- To which extent is the level of coffee consumption influenced by the level of alcohol consumption and intensity of tobacco consumption?

Respondents were classified as abstainers of alcohol if they answered on a question about frequency of alcohol consumption that they did not consume any alcohol in the preceding 6 months. Respondents who answered on the question 'how many cups of coffee do you usually drink per day?' that they never drank coffee were classified as non-users of coffee. Non-smoking was operationalized as not having smoked cigarettes, cigars or pipe during the diary week. The level of consumption of coffee, alcohol and tobacco is operationalised with the diary data.

The percentage of non-coffee drinkers appears to be very small: 4.4%. For alcohol consumption (18.7%) and smoking (57%) higher percentages of non-use are found.

Men are more likely to drink alcoholic beverages (87.6%) and to smoke (48.3%) than women (alcohol: 73.6%; smoking: 36.7%). The gender difference in coffee use (man: 96.1%; women: 94.9%) is not significant.

Coffee and alcohol drinking is most often combined (81.9%), followed by coffee and smoking (44.1%). A considerable minority of 38.3% uses more or less regularly all three substances.

The percentages above do not allow conclusions about the extent non-users of a substance are less likely to use the other substances. It appears that abstainers of alcoholic beverages are also more likely to abstain from coffee (8.6% versus 3.5%; $p < 0.05$) and tobacco (71% versus 54%; $p < 0.05$). Among male non-smokers the prevalence of coffee use is lower (93.6% versus 98.8%; $p < 0.05$)

Smoking and drinking status are also related to how much coffee is consumed. Compared with respectively non-smokers and no-drinkers, the average coffee consumption is higher among smokers (35 versus 28 cups per week; $p < 0.001$) and among drinkers (32 versus 29 cups per week; $p < 0.05$).

In the analyses of the relations between level of coffee consumption with respectively intensity of smoking and alcohol consumption, the non-smokers respectively non-drinkers have been excluded. An inspection of plots and regression analysis shows that coffee consumption tends to increase with the number of cigarettes, cigars or pipes ($r = 0.31$; $p < 0.001$) and number of glasses of alcoholic beverages ($r = 0.28$; $p < 0.001$). No systematic deviation from linearity could be observed.

To sum up the main results, it appears that in most cases the non-use of one of the substances predisposes to the non-use of other substances and that those non-smokers and abstainers from alcoholic beverages who do drink coffee, drink less coffee than respectively smokers and drinkers of alcoholic beverages. Among drinkers of alcoholic beverages and smokers the relation between the intensity of use of respectively alcohol and tobacco with coffee consumption is rather high. All in all, these outcomes support the notion of an interrelatedness of the use of these three substances. This interrelatedness might be conceptualized as health-conscious life style or substance use.

COMMENTS ON AND CONSEQUENCES ATTRIBUTED TO COFFEE

The survey contains two categories of questions indicating perceived negative consequences of coffee consumption. One category consists of 9 questions asking whether the respondent him/herself or 9 referent persons (partner, colleagues, parents, friends, physician, family or partner, others, brothers or sisters, sons or daughters) thought that he/she consumed too much coffee. The other category of questions concerned 4 physical and one behavioural complaints (stomach pain, failed attempts to reduce coffee, not able to sleep, feeling sick, trembling hands) attributed to too much coffee consumption. Both categories of questions specified that the respondent should restrict himself to comments on coffee consumption or consequences attributed to too much coffee in the half year preceding the interview.

We will use these two sets of questions for two purposes. Firstly, to describe the prevalence of worries about, social comments on and consequences of coffee consumption in the general population. Secondly we will analyze whether there is a relation between the reported consequence of coffee and how much coffee is consumed. Respondents who in the interview reported not to have consumed coffee (4.5%) were excluded.

The first thing to be noted is the high percentage of respondents who did not answer these questions. In total 15.9% of the 1.178 respondents systematically skipped all items about worries about, comments upon and consequences of coffee consumption. The average coffee consumption of those who systematically failed to fill in these items was somewhat lower compared with those who answered these questions but reported not to have experienced these consequences (29 versus 31 cups per week). With some caution this can be taken to mean that if these respondents had filled in the questionnaire, the prevalence would be about the same.

With respect to worries about and comments of others on coffee consumption it is clear that coffee consumption is much more frequently a topic of personal worry (13%) than of comments of others. Of the others, the partners of respondents most frequently (6.3%) comment that the respondent drinks too much coffee. For all other referent persons, the reported prevalence of comments is about 2% or less. Almost 20% of the coffee drinkers report that, in the last half year, 1 or more persons (including the respondent) thought they consumed too much coffee.

The outcome that about 20% of the coffee drinkers report that they themselves and/or others think they drink too much coffee can be interpreted as indicating an ambivalent attitude towards coffee consumption, may be based on health problems associated with coffee.

It appears that 4 of the five consequences of too much coffee consumption are reported by 11% or more of the coffee drinkers. The sum score indicates that almost 1/3 of the coffee drinkers report 1 or more consequences and 5.4% report 3 or more consequences of coffee consumption.

For a large minority of the coffee drinking population the pleasure of drinking coffee is mixed with worries about or negative consequences attributed to coffee consumption. In the last paragraph we will discuss somewhat further the possible interpretations of these outcomes.

Relation between level of coffee consumption and reported consequences

Implicit in the evaluation of the above presented data is that those who drink more coffee are more likely to worry about their consumption, to get comments from others and to experience consequences because of coffee consumption. Because in particular for health consequences of coffee a dose-response relation can be expected, we restrict the analysis to this type of consequences.

Before presenting data about this relation it is important to note the difference in how consumption level on the one hand and consequences of coffee on the other hand, are measured. The time frame used in the questions about consequences of worries about and comments upon coffee consumption was the last half year. A report about stomach ache because of too much coffee, might be due to a relatively isolated and restricted period in which the respondent consumed more coffee than usually. Such periods of increased coffee consumption are not reflected in the diary estimate of coffee consumption. It will be clear that in such cases the relation between reported consequences of coffee and the indicator of consumption level will be weak or absent.

To inspect the relation between how much coffee is consumed and the chance on 1 or more consequences, we divided the frequency distribution of consumption level of coffee in 10 deciles of increasing consumption. For these 10 categories the reported prevalence of 1 or more consequences due to too much coffee was computed. Figure 11.1 shows the results.

The prevalence of coffee related consequences increases in the categories with higher consumption levels. However, except for the two highest percentiles, the differences in prevalence are not very large. With one exception, the coffee consumption is higher among respondents reporting more complaints. The exception is formed by the 26 respondents who report three complaints but have a lower

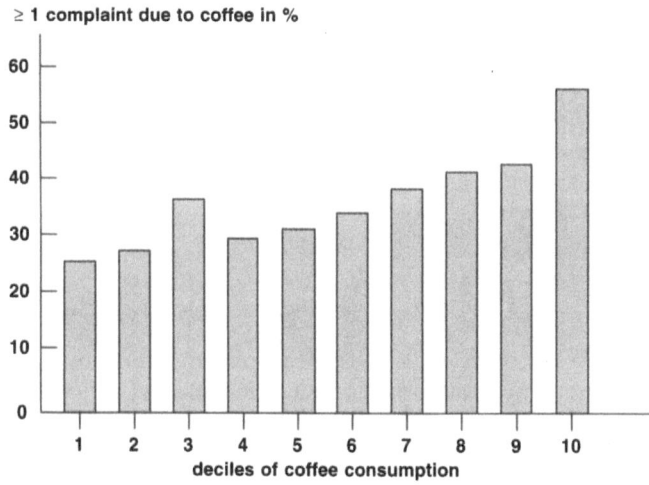

Figure 11.1 – Percentage of respondents reporting 1 or more coffee related complaints at increasing deciles of level of coffee consumption.

coffee consumption (33.1 cups/week) than those (n = 103) reporting two complaints (37.6 cups/week). The difference in coffee consumption between those (n = 193) reporting 1 and those (n = 37) with 4 or more complaints is about 8 cups of coffee per week (33.0 cups/weeks and 41.2 cups/week respectively). This is, in our opinion, a rather modest difference.

Our conclusion is that the proportion of coffee drinkers reporting consequences due to too much coffee increases at higher levels of coffee consumption and that the number of consequences reported increases with level of consumption. However, the relationship is not completely linear and not very strong.

FACTORS RELATED WITH LEVEL OF COFFEE CONSUMPTION

Considering that at the social level the appreciation of coffee is to some extent ambivalent, one can expect that cultures (or nations) and within cultures, specific subpopulations have developed their own norm as to how to integrate coffee into one's lifestyle. In principle these norms may concern a variety of aspects of coffee consumption such as at which times to drink coffee, with whom, how to prepare the coffee, etc. Dekker and his team (1993) describe for the Netherlands some of these aspects. We will restrict ourselves to one aspect of coffee consumption: how many cups of coffee are consumed. It is an exploratory analysis guided by rather general ideas about which social categories might differ in coffee consumption. As said in the introduction, we expect that status groups possibly differ in how much coffee is consumed. Considering that the attitude towards coffee consumption is somewhat ambivalent, people might orient themselves on those socially nearest to them as to how to drink coffee. It is known that status factors like gender, religion, social class, age define in most cases the social categories people identify with. Therefore these social categories might differ in coffee consumption. We will analyze whether there are differences in coffee consumption according to gender, age and level of education (as an indicator of social class).

Another general notion concerns structural differences in the number of situations in which the social aspect of coffee drinking (as a more or less ritualized form of time sharing) is optimized. Those who are in paid employment and/or are married might find themselves more often in situations in which the function of coffee consumption is (also) to articulate the social aspect. Therefore we will also analyze whether those who are in paid employment and those who are married differ in how much coffee is consumed.

Results

In Table 11.1 the outcomes of an ANOVA-analysis of differences in coffee consumption according to gender, level of education and age are presented. A low educational level has been operationalised as primary school and lower general and vocational schooling. The highest level of education has been operationalised as either university or higher occupational training.

All status characteristics are associated with differences in coffee consumption. The largest difference is between the youngest age group of 16-25 years

Table 11.1 – Average weekly coffee consumption according to gender, age
and level of education

factor	cups of coffee per week
Gender*	
men	32.7
women	29.6
Age**	
16–25 year	23.0
26–45 year	32.5
> 45 year	33.7
Level of education***	
lower	35.6
middle	30.7
higher	28.2

* $p < 0.05$; ** $p < 0.001$; *** $p < 0.001$ controlling for age

and the older age groups. Younger people drink about 10 cups per week less than people older than 25 years. Because of the large difference between the younger and the older, we controlled for age when analyzing the relation with level of education. The difference according to level of education is largest between the lower educated with more than 35 cups per week and the middle and higher educated with respectively about 30 and 28 cups per week. The gender difference, although significant, is small. Women drink on the average per week 3 cups of coffee less than men.

With respect to marital state it appears that those who live with a partner consume about 9 cups of coffee per week more than those who live alone (35.2 versus 26.2 cups per week). When controlling for age, the difference decreases somewhat but is still significant ($p < 0.001$). Because of the difference in paid employment between men and women we have analyzed separately for men and women and again we have controlled for age.

Unemployed men drink significantly ($p < 0.05$) less cups of coffee per week (30.2) than paid employed men (33.6 cups/week). Also unemployed women to drink less coffee, 29.3 cups of coffee per week than women in paid employment (31.0 cups/week), although the difference is not significant.

DISCUSSION

Four aspects of coffee consumption in the general population have been analyzed: the measurement of coffee consumption in general population surveys, the interrelation between the consumption of coffee, alcoholic beverages and tobacco, the prevalence of negative consequences ascribed to coffee consumption and differences in coffee consumption according to sex, age, level of education, marital status and work situation. In this paragraph we will summarize the main findings and discuss possible interpretations of these findings.

Questionnaire and diary estimates of coffee consumption

It was found that the questionnaire estimate of coffee consumption is higher than the (more precise) diary estimate. For alcohol consumption the reverse is found: the diary estimate is higher than the questionnaire estimate. There is, in our opinion, no reason to doubt that, compared with a questionnaire estimate, a diary offers a more precise estimate of habitual behaviours among which coffee consumption.

The difference between coffee and alcohol consumption is probably partly due to the difference in frequency of consumption of respectively coffee and alcoholic beverages. Almost all coffee drinkers consume daily while only a minority of the alcohol drinkers in the general population consume daily. Lemmens *et al.* (1992) have shown that, when interviewed, especially alcohol drinkers with an intermediate frequency of 3 to 5 times per week tend to forget drinking days. In the case of coffee consumption, forgetting of drinking days when asked about 'usual' consumption in an interview is far less likely to be a source of error. When this interpretation is followed, one would expect the diary and questionnaire estimate of coffee consumption to be about equal and not, as we found, a questionnaire estimate which is higher than the diary estimate. Inspection of the difference-scores of diary and questionnaire estimates with levels of coffee consumption according to the diary showed that at all levels of coffee consumption the proportion drinking less than 'usual' according to the diary was somewhat larger than the proportion drinking more. Apparently, questionnaire estimates of coffee consumption suffer most from the error of forgetting that one or two days per week less coffee than 'usual' is consumed. On the aggregate level questions about usual coffee consumption are likely to overestimate real coffee consumption.

The interrelation between coffee, alcohol and tobacco consumption

The proportion of non-coffee drinkers (4.4%) is much smaller than the proportion of non-alcohol drinkers (18.7%) and non-smokers (57%). It seems plausible to assume that the differences in non-use of these substances is to a large extent due to differences in the extent coffee, alcohol and tobacco are associated with health risks. Yet our analyses show that there are interrelations between the use of coffee, alcohol and tobacco. Because of the small proportion non-coffee drinkers, the interrelations with non-smoking and non-alcohol drinking are not very informative. Interrelations for two other aspects were found. Non-alcohol drinkers and non-smokers appear to drink significantly less coffee and there appears to be a fairly strong correlation between coffee consumption and number of cigarettes, cigars and pipes (0.31) and number of glasses of alcoholic beverages (0.28). These interrelations point to associations between the use of coffee, alcohol and tobacco. As mentioned in the introduction these associations can be conceptualized in terms of a health-conscious life-style or in terms of substance use.

The prevalence of consequences due to coffee

The two main outcomes are the high prevalence of subjectively felt consequences due to coffee and the mostly moderate relation between these consequences and the level of coffee consumption.

It appeared that 13% of the coffee drinkers report to worry about their consumption and 6.3% reports that their partner criticizes the coffee consumption of the respondent. All in all 35% of the coffee drinking population reports worries about or criticism of others on coffee consumption. The prevalence of health complaints attributed to coffee consumption is even higher. Almost half of the coffee drinkers in the general population (46%) reports one or more health complaints because of coffee. Stomach ache (13.8%), not able to sleep (11.8%) and feeling sick (11.5%) are the most frequently reported complaints.

When evaluating these figures one should bear in mind the time frame used in the questions about worries, criticism and health complaints. This was specified as the last half year and means, for example, that of the 13.8% reporting stomach ache because of coffee, a larger or smaller proportion may have experienced this only once in the last half year. Still, even taken into account the time frame of the questions, the prevalence is rather high.

The essential question is, of course whether in all cases coffee consumption is the sole or true cause of worries, criticism or health complaints. The questions to the respondents left no doubt that worries, criticism or health complaints should only be reported in as far as they were due to coffee. However, it might be found attractive to blame coffee for a wider range of worries and health complaints than those solely due to coffee. Drinking too much coffee is not associated with a social stigma and probably almost all coffee drinkers are confident that they can control their coffee consumption. To attribute worries or health complaints to coffee has the advantage that it reduces uncertainty over the cause and that it is a cause of which one is in control. In that way more worries, complaints and criticism may be attributed to coffee than those in which coffee is clearly the sole cause. When this line of reasoning is valid, the prevalence figures may be somewhat inflated due to a tendency to ascribe too many worries and health complaints to coffee.

One could interpret the mostly fairly small differences in coffee consumption between those who do and do not report health complaints due to coffee and those who differ in the number of health complaints due to coffee as support for the above interpretation. If coffee consumption functions as a focal point for more general worries and health complaints, this is likely to occur at all levels of coffee consumption and it will weaken the relation between level of consumption and consequences due to consumption. However, two other interpretations for the mostly modest relation between consumption and consequences should also be mentioned.

- The difference in temporal context between the questions measuring consequences (the last half year) and the questions measuring coffee consumption (one week) should be mentioned. A report of, for example, stomach ache might have as its context a single and restricted episode of increased coffee consumption. It will be clear that this period of increased coffee consumption will not be reflected in the diary estimates of coffee consumption and thereby weaken the relation between consumption level and consequences.
- Drinkers trying to restrict their coffee consumption because of health reasons may more fully recall and report the extent they experience consequences because of, in their eyes, too much coffee. In this category of drinkers a relatively low coffee consumption would be combined with a relatively high

chance of reports about consequences. Both last mentioned interpretations might also partly explain the deviations from linearity in the relation between coffee consumption and consequences.

To conclude, the prevalence of consequences due to coffee is high, however, the relation of coffee consumption with consequences is not very strong and deviates from linearity. The modest relation between coffee consumption and consequences might be due to variations in coffee consumption not covered in our measurement of coffee consumption. However, it might be that there is a tendency to ascribe worries and health complaints to coffee which do not have coffee as their sole cause. If that is the case the prevalence figures might be somewhat inflated. Further research is needed for a more precise evaluation of the relevance of the reported prevalence figures.

Differences in coffee consumption

Because of a possibly somewhat ambivalent attitude towards coffee, we expected that status groups defined by sex, age and level of education might differ in how they integrated coffee consumption into their life style. This might also be expressed in differences in how much coffee is consumed. Moreover, we expected that those who do not live with a partner and those who do not have a paid job might find themselves less often in situations in which coffee consumption functions as a ritualized form of time sharing. Therefore we expected these categories to drink less coffee.

Generally speaking, the outcomes support the notion that status groups differ in coffee consumption. However, it should be added that, with the exception of the difference according to age and marital state, the differences are small.

The outcome that, compared to older age groups, the youngest (<25 year) age group drinks about 10 cups of coffee per week less can be interpreted in three ways. The difference might be due a new trend of reduced coffee consumption, started among the youngest age group and possibly inspired by health considerations. In that case the age difference is likely to be temporary both because later generations, when starting to drink regularly coffee, are likely to follow this trend and because such trends have a tendency to diffuse throughout the population. The age difference can also be attributed to a more structural difference in living situation between the generations. Young people's life is probably to a lesser extent organized around regularly occurring routines concerning work and obligations at home (e.g., care for children). In as far as coffee consumption also functions as a break in the routines of daily life, this function of coffee consumption is likely less important in the younger age group. In this case the age difference in coffee consumption will be stable. Finally, young people may just have picked up the habit of coffee consumption and therefore drink less than older people who drink coffee for a longer time. Time trend studies, or even better, longitudinal studies are needed to decide which of the interpretations account for most of the age difference in coffee consumption.

The gender difference might have as its background the difference in paid employment between men and women. However, considering that the difference between working men and women is only marginally lower than the overall difference, there is no support for this interpretation. A plausible interpretation is that of a gender difference in concerns about health (Verbrugge, 1989), leading to a more limited coffee consumption among women. The difference according to level of education could have as a background that among the lower educated coffee, is more often consumed with meals. As Dekker *et al.* (1993) point out, a meal with coffee is perceived as a sort of substitute for a warm meal by a large part of the blue collar workers. The differences in coffee consumption according to marital state and employment status were as expected and point to the social function of coffee in daily life.

SUMMARY

Questionnaires estimate coffee consumption at all levels about 0.80 cup per day higher than diary estimates. Non-smokers and abstainers from alcoholic beverages who do drink coffee, drink less coffee than respectively smokers and drinkers of alcoholic beverages. Drinkers of alcoholic beverages and smokers have a related coffee consumption, supporting the idea of an interrelatedness of the use of these three substances. The joy of drinking coffee is for a large proportion of the coffee drinking population mixed with worries about or negative consequences attributed to coffee consumption. The higher coffee consumption is, the more negative health consequences are reported. Younger people drink less than people older than 25 years. Lower educated people drink on average 5 cups more per week more than the middle educated and approximately 7 cups more per week compared to higher educated people. Women drink per week 3 cups of coffee less than men. Living with a partner increases the weekly coffee consumption with about 9 cups. Paid employed men drink about 3 cups more per week than unemployed men. Unemployed women tend to drink 2 cups of coffee less per week than women in paid employment.

CHAPTER TWELVE

Caffeine, Nicotine and Attentional Performance

David M. WARBURTON

INTRODUCTION

In this chapter, I will present evidence on the effects of moderate amounts of caffeine and nicotine on attentional performance. By moderate amounts of caffeine, I mean the amount which produces plasma levels which are close to the levels which result from normal coffee, tea or cola drink consumption, that is to say four cups of coffee or less. Assuming a cup contains 75 mg of caffeine, this maximum dose of 300 mg would give caffeine levels of less than 45 μg/ml of plasma (interpolated from Passmore *et al.*, 1987). For nicotine, a moderate dose would be one which produces no more than 50 μg/ml of plasma, that is a smoking dose from a single cigarette (Feyerabend *et al.*, 1975; Warburton, 1989b).

A second aspect of this review will consider the possibility that caffeine and nicotine only minimize a performance decrement, such as that resulting from fatigue or boredom, but can never enhance performance above the best level which was achieved without caffeine or nicotine (James, 1991).

A related question is whether caffeine and nicotine improve performance absolutely (*absolute facilitation* hypothesis), or are merely restoring performance degraded by caffeine or nicotine withdrawal (*withdrawal relief* hypothesis) (West, 1993; James, 1994). Certainly, most of the early studies used the conventional, psychopharmacological paradigm of testing people after a wash-out period when they were drug free, that is to say after caffeine and nicotine abstinence (usually 10 or more hours) and so this explanation may have some validity.

At a common sense level, the belief of coffee drinkers and smokers that their habit does improve processing throughout the day argues against the idea that there is tolerance to the performance effects. However, some recent studies which have explored this issue with respect to attention will be presented.

ATTENTION

Attention is intimately involved in thought and action and, undeniably, the normal person is continuously engaged in these activities throughout their waking day. It is an aspect of consciousness, in the sense that consciousness of something is enhanced by focussing on it. Consciousness uses attention in order to exert its control functions (Umilta, 1988).

In this chapter, the research will be described within the framework of the Supervisory Attentional System model. This conception of attention has evolved in a series of papers by Norman and Shallice (Norman and Shallice, 1980; Shallice, 1988; Shallice and Burgess, 1993). This system does not control behaviour, but modulates systems by activation or inhibition of thought or action schemata (routine cognitive and motor programs). Recently, it has been argued that attention and Working Memory have properties in common and Baddeley (1993) has argued that Working Memory is the same as the Supervisory Attentional System.

The Supervisory Attentional System is involved in effortful (less automatic), more controlled tasks. Thus, it comes into play when we are coping with novelty, when routine selection of actions is unsuitable and when a habitual response competes with the appropriate response. It is essential for the genesis of willed actions, for planning and decision making. It is certainly important for coping with technical difficulty, overcoming temptation and dealing with danger. Overcoming temptation and dealing with danger are difficult to study in the laboratory, but the function of the Supervisory Attentional System can be illustrated in attentional tests.

ATTENTIONAL PROCESSING

Kinchla (1980) suggested three classes of studies which are relevant for the issue of attentional processing; namely *sustained attention* tasks, *attentional switching* tasks and *selective attention* tasks.

It is important to note that the psychopharmacological effects of the amounts of caffeine equivalent to two or three cups of coffee and nicotine at smoking doses are very subtle, in comparison with drugs, like amphetamine. Consequently, sensitive attentional tests have been required in order to evaluate their psychological properties. The majority of these studies have been experimenter paced, that is a fixed rate of presentation of information by the equipment.

CAFFEINE AND SUSTAINED ATTENTION

In an encyclopaedic review by Koelega (1993) of the effects of caffeine on sustained attention performance up to the start of 1993, there were 14 studies with at least one dose in the less than 300 mg range, which reported an improvement in detections or a reduction of errors, while three found no improvement. Of these two were subject-paced (Rapoport *et al.*, 1981; Lieberman *et al.*, 1987). Rather than repeat the review, this section will give some selected examples of the paradigms.

Vigilance

One classical type of sustained attention test is the vigilance task. Vigilance tasks are highly monotonous and attention has to be directed to an input for long periods of time and the subject is required to detect and respond to brief, infrequent changes in the input. During a typical vigilance session, the detection rate decreases – a change called the vigilance decrement, resulting from boredom and fatigue impairing performance.

Vigilance tests represent that particular category of mental performance for which most often improvements with caffeine have been reported (Dews, 1984). It could be argued that the effects during these vigilance tests may be more representative of the beneficial effects of caffeine in daily life than are effects in tests of short duration. Consequently, it is worth examining the effects of moderate doses on vigilance performance.

One experimenter-paced version of a vigilance test is the Wilkinson Auditory Vigilance Task (Wilkinson, 1970). A tone was presented to the subjects every two seconds, for one hour. The majority of the tones were 400 ms, but 40 were approximately 70 ms shorter. The number of correct detections and false alarms were recorded. It was found that 32 mg, 64 mg, 128 mg and 256 mg doses of caffeine, administered after 12 hours abstinence, significantly improved correct detections (Lieberman *et al.*, 1987). With an almost identical test, faster reaction times were obtained with 75 mg of caffeine (Clubley *et al.*, 1979).

In a second study with this test, caffeine improved performance and an analysis of the time course of the enhancement revealed that these improvements are usually in the later part of the test (Fagan *et al.*, 1988). This finding is in agreement with the suggestion that caffeine only minimizes a performance decrement, such as that resulting from boredom or fatigue.

The original vigilance task was the Mackworth Clock (Mackworth, 1950), in which subjects watched a clock face with the clock hand rotating once a minute. Once every 30 seconds on average the hand stopped briefly. Performance was assessed in terms of the detection rate and the proportion of signals correctly detected showed the typical vigilance decrement. In an unpublished study in our laboratory by Derek Turner-Smith, a dose of 250 mg of caffeine produced a significantly reduced the vigilance decrement (as did the high dose of 500 mg) in subjects who abstained for 10 hours, but there was no improvement over the initial 20 minutes of performance in the placebo condition. Once again, this finding is consistent with the notion that caffeine only improves performance which has been degraded by boredom or fatigue.

The problem with vigilance tasks is that they involve detection of infrequent targets, by definition. Consequently, estimates of detection probabilities must be done over relatively long time periods (e.g., 20 minutes) in order to obtain reliable values, and any absolute improvement could be masked in the averaging process. Tasks with a higher density of signals are more suitable for determining absolute improvements. Tasks of this sort will be considered next.

Continuous performance tests

A frequently used version of the Continuous Performance Test is one in which a random sequence of single digits is presented on a computer screen in rapid succession. Subjects press a button if they can see a target digit or sequence of digits. A difficult, experimenter-paced version of this test requires subjects to detect three odd or three even numbers in succession and has been called the rapid Visual Information Processing Test (Wesnes and Warburton, 1978). Koelega (1993) described this test as one of the most sensitive for detecting the effects of drugs.

A number of studies have revealed improvements by caffeine with this test. Doses of 3 mg per kg body weight (around 200 mg of caffeine for men and less for women) have produced enhanced performance (Smith *et al.*, 1990; Frewer and Lader, 1991; Smith *et al.*, 1993a,b). Of particular importance was the fact that the enhancement was similar across the entire 20 or 30 minutes of testing, showing that caffeine improves attentional processing directly rather than only counteracting fatigue or boredom and restoring performance to baseline levels.

However, in a subject-paced version of this test, several experiments (Bättig and Buzzi, 1986; Hasenfratz *et al.*, 1991) obtained positive effects with 250 mg of caffeine. Of particular significance was the fact that these effects were obtained from the beginning of testing and maintained throughout the test session. Once again, this would argue in favour of caffeine doing more than merely neutralizing fatigue or boredom.

Some studies have used another continuous performance test: one in which subjects are presented with a series of geometric patterns at 2 to 4 seconds intervals. The subjects had to respond when two consecutive patterns were the same. Caffeine reduced the error rate in the elderly (Swift and Tiplady, 1988; Yu *et al.*, 1991), but not in young volunteers (Fagan *et al.*, 1988; Swift and Tiplady, 1988). Although the differences between the groups (each of only six subjects) were not statistically significant, Swift and Tiplady (1988) stated that the elderly may be more responsive to caffeine and argued in a later paper that this may be because the elderly show attentional deficits (Yu *et al.*, 1991). However, it is clear that performance was close to ceiling in the young (mean error rate on the Continuous Attention Task was 2.67, s.d. = 1.41) in comparison with the elderly (mean error rate 17.4; s.d. = 27.1) and so there was more scope for improvement in the latter case.

These studies were performed with prolonged caffeine abstinence, but more recently some research has shown that with minimal abstinence of one hour attentional performance was improved (Warburton, 1995). In this study, the participants were given a dose of caffeine mixed with decaffeinated instant coffee to give a total dose of 75 mg caffeine. This dose was drunk in the morning, one hour prior to coming to the laboratory. This was explained to the volunteers as a method of ensuring that everyone was in exactly the same state before testing.

At the laboratory, they received either 75 mg or 150 mg of caffeine (giving a maximum possible dose of 225 mg) and were tested on the rapid visual information processing task. Performance was improved for both correct detections and reaction times, that is there was no speed-accuracy trade-off. Thus, complete abstinence was not necessary for improvements in performance.

Of course, it could be argued that there were still some residual withdrawal effects which had not been reversed by the single dose of caffeine. Convincing evidence that this is not the explanation comes from two further studies. Caffeine effects were assessed in the evening with one hour abstinence and ad libitum coffee consumption throughout the day. Improved performance on the rapid visual information processing task was found with this regimen (Smith *et al.*, 1994b).

In the second study, Jarvis (1993) examined the relation between habitual tea and coffee consumption and cognitive performance using data from a 9003 person Health and Lifestyle survey (Cox *et al.*, 1987). Jarvis (1993) found that there was a dose-related improvement in cognitive performance as a function of daily caffeine consumption.

These findings and ours suggest that tolerance to the performance enhancing effects of caffeine, if it occurs at all, is incomplete and contradict claims that caffeine only produces improvements because it is reversing the effects of caffeine withdrawal.

ATTENTIONAL SWITCHING

In an attentional switching test, volunteers must attend to more than one source of information and process material from all sources. One test of attentional switching required subjects to monitor a series of visual stimuli at one of four locations on a computer screen and press one of four buttons to indicate the location. During the 10 minutes test, reaction times were reduced by 32 mg, 64 mg, 128 mg and 256 mg doses of caffeine, administered after 12 hours of abstinence, while commissions errors were unchanged (Lieberman *et al.*, 1987).

CAFFEINE AND SELECTIVE ATTENTION

A task which demonstrates selective attention and perceptual intrusions from unattended material is the Stroop test (Stroop, 1935). This is a complex information processing task in which volunteers are required to process information under conditions of distraction. The Stroop effect is the name given to the distracting effect of the to-be-ignored distractor stimuli on the processing of the attended material.

In the prototypical Stroop test, a list of colour words are presented, with the words written in different-coloured inks. The ink colours are incongruent with the written words, for example, the word *red* may be written in *blue* ink. The task is to name the colour of the ink in which each word is written, ignoring the actual printed word. Ink colour naming of incongruently printed colour words takes much longer than ink colour naming of non-colour words written in different inks. The difference in time required for these conditions provides a measure of the volunteer's capacity to selectively attend to the relevant dimension (the ink colour) while ignoring the irrelevant one (the printed word). This test is a case of a habitual response (colour reading) competing with the appropriate response (colour naming).

To my knowledge, the traditional, colour version has not been used to test caffeine. However, equivocal results have been obtained with a numerical version of the test in which the subjects had to respond with the answer 'three' for the display of '1 1 1'. While one study found that there was a dose-dependent impairment of performance with 125 mg and 250 mg (Foreman *et al.*, 1989), a computer version of the same test revealed improvements with 250 mg of caffeine (Hasenfratz and Bättig, 1992).

SUMMARY OF CAFFEINE EFFECTS ON ATTENTIONAL PROCESSING

The bulk of the evidence indicates that moderate amounts (maximum dose of 300 mg) of caffeine improve attentional performance. The strongest evidence is for an enhancement of sustained attention.

A second aspect of this section was to consider the possibility that caffeine only minimizes a performance decrement, such as that resulting from boredom or fatigue, but could never enhance performance above the best level which was achieved without caffeine. The evidence suggests that when the test is sensitive to improvements, *absolute* improvements in attentional performance are possible. The analogue question is whether caffeine does not improve performance absolutely, but only restores performance degraded by caffeine withdrawal, that is after caffeine abstinence of 10 or more hours. However, attentional studies with minimal abstinence show that performance can be improved without caffeine abstinence.

This is compatible with the conclusions from a study by Jarvis (1993), which found that there was a dose-related improvement in cognitive performance as a function of daily caffeine consumption. It suggests that tolerance to the performance-enhancing effects of caffeine, if it occurs at all, is incomplete and contradicts the claim (e.g., James, 1994) that caffeine only produces improvements because it is reversing the effects of caffeine withdrawal.

NICOTINE AND SUSTAINED ATTENTION

Similar tests to those used for testing caffeine have been used to test nicotine, either in pure form, in tobacco smoke or smokeless tobacco. Reviews of the literature on nicotine and attention have been made by Warburton (1990b) and more recently by Sherwood (1993). Once again the research on nicotine and sustained attention performance has been reviewed exhaustively in Koelega (1993). He concludes that 11 of 17 studies show increased detections and in most cases reaction times were improved as well (9 out of 11 studies). In view of these exhaustive reviews, only selected studies will be presented here.

Vigilance

In a study of smoking using the Mackworth Clock (Wesnes and Warburton, 1978), there was no significant difference between abstaining smokers and non-smokers in their performance, i.e. no evidence of a withdrawal effect. This suggests that smoking was not just reversing degraded attentional processing due to abstinence but was enhancing performance absolutely.

In order to investigate the possibility of withdrawal reversal, we studied the effects of nicotine tablets (0, 1 or 2 mg) on nonsmokers, light smokers (less than 5 per day) and heavy smokers (more than 15 per day). Subjects took nicotine or placebo tablets at 20, 40 and 60 minutes. Nicotine tablets reduced the vigilance decrement to the same extent in non-smokers, light smokers (<10 per day) and heavy smokers (>15 per day) (Wesnes et al., 1983).

These two studies give no evidence for tolerance to the effects of nicotine. In addition, the improvement with non-smokers indicates that improvements in performance can be obtained without nicotine abstinence and that the effects were not simply due to reversing the effects of nicotine withdrawal.

Continuous performance test

The other attention task that has been used in nine studies of nicotine and sustained attention is the rapid visual information processing test which was described earlier. The initial studies (Wesnes and Warburton, 1983a, experiment I and II; 1984; Revell, 1988) found that smoking after 10 hours of abstinence produced improved performance in terms of both speed and accuracy, and that the nicotine performance was better than the pre-smoking baseline.

A critique of these studies by West (1990, 1993) has misinterpreted the study. He states that "... the fact that the results were expressed as a difference from baseline (the first 10 minutes on the task) makes it impossible to know whether the result was due to a nicotine-induced temporary decrement in performance in the early part of the session." (West, 1990; p. 215). However, the improved performance was expressed as a difference from baseline, which was *prior* to nicotine being given and so the improvements could not have been due "to a nicotine induced temporary decrement in performance".

Of course, these results could have been only a reversal of a withdrawal effect in smokers. Accordingly, we tested the same smokers after minimal abstinence of one hour and after 10 hours deprivation (Warburton and Arnall, 1994). Smoking increased detections and decreased reaction times (the original paper had a misprint on this last point) for both groups. Importantly, there was no difference in the improved performance when the subjects were smoking after one hour or 10 hours of abstinence, in other words there was no evidence of desensitization to the effects.

In order to investigate the importance of tolerance, we tested doses of 0.5 mg, 1.0 mg and 1.5 mg of nicotine with non-smokers. Performance on the Rapid Visual Information Processing task was improved in terms of both correct detections and reaction time (Wesnes and Warburton, 1984c). The highest dose of nicotine (1.5 mg) produced a performance improvement in non-smokers which closely resembled that produced by smoking in smokers.

These findings provide strong evidence that nicotine plays the major role in the improvements in focused attention tasks that are produced by smoking. The improvement in the performance of non-smokers argues against hypothesis that the improvement is only a reversal of withdrawal.

The effects of subcutaneous doses of nicotine on the performance of patients with senile dementia of the Alzheimer type have been examined (Sahakian *et al.*, 1989). Doses of nicotine produced a dose-related improvement in performance in the detection of signals in the rapid visual information processing task, and subjects under these conditions approached the performance of the normal elderly. The equivalent data for reaction time show that nicotine doses produce improvements in reaction times in the patients with senile dementia of the Alzheimer type in comparison with the baseline and placebo conditions. A much larger study by the same group (Jones *et al.*, 1992) has replicated this finding and a comparison of smokers with non-smokers found no differences in the effect of nicotine, which was in accord with the results with healthy volunteers.

The evidence in this section is consistent. Nicotine improves attentional processing and the results do not depend on abstinence from nicotine. Consequently, the effects cannot be attributed to the reversal of nicotine withdrawal.

ATTENTIONAL SWITCHING

As we described earlier, volunteers must attend to more than one source of information and process inputs from all sources in an attentional switching task. We have used a more complex version of the rapid information processing task which required subjects to divide their attention between two sources of information (described in Warburton and Walters, 1988).

Subjects were simultaneously presented with digits at a rate of 50 per minute in both the visual and auditory modality; a different sequence for each modality. The detection of sequences in both modalities improved significantly after a cigarette. Smoking a cigarette also prevented the increase in reaction times that occurred in the non-smoking condition. We have not carried out a study with pure nicotine, but only with nicotine presented in cigarette smoke.

SELECTIVE ATTENTION

A task which demonstrates selective attention and perceptual intrusions from unattended material is the Stroop Test which was described earlier (Stroop, 1935). In the colour version, administration of nicotine tablets reduces the size of the Stroop effect in both abstaining smokers and non-smokers (Wesnes and Warburton, 1978), indicating enhanced selective attention for relevant information. The magnitude of the effects on smokers and non-smokers was similar, and the fact that the effects were found with non-smokers indicates that they do not depend on nicotine abstinence and that they are simply due to a reversal of nicotine withdrawal.

SUMMARY OF NICOTINE EFFECTS ON ATTENTIONAL PROCESSING

Thirty-five studies (of which twenty-nine are listed by Sherwood 1993) indicate that pure nicotine and nicotine presented in cigarette smoke improve attentional performance. The largest amount of evidence is for an improvement on sustained attention tests.

The second question which was addressed in this section was whether nicotine only minimizes a performance decrement, such as that resulting from boredom or fatigue, but could never enhance performance absolutely. The evidence indicates that when the test is sensitive to improvements, absolute improvements in attentional performance are possible, such as those observed in the Rapid Visual Information Processing test.

The third question was whether nicotine only improves performance after nicotine abstinence of 10 or more hours, in other words whether it merely reestablishes performance degraded by nicotine withdrawal. The previously described attentional studies with minimal abstinence show that performance can be improved without abstinence. These data are in agreement with a sizeable body of evidence involving testing the effects of pure nicotine or nicotine presented in cigarette smoke with non-abstaining smokers. Smoking by non-abstaining smokers reduced reaction times in the "odd man out" task (Frearson *et al.*, 1988), a continuous performance task (Pritchard *et al.*, 1992a) and choice reaction time task (Frearson *et al.*, 1988). Smokeless tobacco in non-abstaining users had a similar effect on choice reaction times (Landers *et al.*, 1990).

Nicotine gum decreased reaction times in non-abstaining smokers (Hindmarch *et al.*, 1990a; Sherwood *et al.*, 1990, 1991c; Kerr *et al.*, 1991) as well as improving tracking in a simulated driving task (Hindmarch *et al.*, 1990a; Sherwood *et al.*, 1992; Kerr *et al.*, 1991). It has also been reported that nicotine gum speeded memory scanning in the Sternberg paradigm in non-abstaining smokers (Sherwood *et al.*, 1991c; Kerr *et al.*, 1991).

The studies of non-smokers given nicotine tablets demonstrate that there can be an improvement in performance without nicotine withdrawal (Wesnes and Warburton, 1984c). No differences were found between the baselines on the Stroop Task of the abstaining smokers and non-smokers, or in the amount of improvement produced by nicotine (Wesnes and Warburton, 1978). Similarly, subcutaneous nicotine produced identical attentional improvements in both smoking and non-smoking Alzheimer patients (Sahakian *et al.*, 1989; Jones *et al.*, 1992).

COMBINED EFFECTS OF CAFFEINE AND NICOTINE ON ATTENTIONAL PROCESSING

Few studies have investigated the effects of a combination of caffeine and nicotine on any sort of performance, in order to discover if both drugs affect the same underlying behavioural systems.

One study which examined the combined effects of caffeine and smoking on sustained attention was carried out by Bättig's group (Hasenfratz *et al.*, 1993). While they found improvements in the detection rate with both 250 mg caffeine and smoking (nicotine) after overnight abstinence from both caffeinated beverages and smoking, there was no significant additive effect, providing no evidence for a common mechanism of action. However, as the authors point out, a performance ceiling may have been reached for performance improvements with either of the two treatments alone. Certainly, each of the treatments was capable of producing improvements in detection rate of over 16 percent. The results of this study are consistent with a psychophysiological study which failed to find interactive effects on any EEG measures (Pritchard *et al.*, 1995).

To my knowledge, only one study has found an additive effect on any kind of performance test and that was with a three-choice reaction time task (Smith *et al.*, 1977). This finding contrasts with the absence of synergistic effects on six choice reaction time, a tracking task and critical flicker fusion (Kerr *et al.*, 1991).

A COMMON MECHANISM OF ACTION?

One important neurochemical system for attentional processing is the ascending *cholinergic* pathway to the cortex (Warburton, 1991). Warburton (1981) proposed that the release of acetylcholine at the cortex increases electrocortical arousal, and hence the size of the evoked potentials, and that this improves the probability that they are distinguished from background cortical activity. The cholinergic pathways do not cause a cognitive operation, but modulate the cortical state to make the operation more accurate and more efficient. This idea would fit neatly with the view of the Supervisory Attentional System as a system which is not controlling behaviour, but is modulating behavioural programmes. Warburton (1991) has argued that improved information processing can be achieved by nicotine-induced stimulation of the cholinergic system, because nicotine acts to reduce fluctuations in cortical arousal as a result of sustained acetylcholine release at the cortex. This may be a consequence either of enhanced activity in the ascending cholinergic pathway, or by increased presynaptic action at the cortex.

Recent evidence has indicated that caffeine may act on the same neurochemical system. It is known that caffeine blocks adenosine receptors throughout the body (James, 1991) and recently, a possible link between the adenosine receptors and the ascending cholinergic pathways to the cortex has emerged.

It has been found that mesopontine cholinergic neurons are under the inhibitory control of endogenous adenosine (Rainnie *et al.*, 1994). As caffeine blocks the adenosine receptors, it would release the ascending cholinergic neurons from inhibitory control and so increase electrocortical arousal the same effect as nicotine (Warburton and Rusted, 1993). Thus, there is no incompatibility between the qualitative similarity in the effects of caffeine and nicotine or, indeed, the quantitative equivalence which has been found in studies with moderate doses of caffeine and those with 'smoking doses' of nicotine.

CONJOINT USE OF CAFFEINE AND NICOTINE

From the biochemical data showing a common neurochemical mechanism of action, and the behavioural data indicating a common behavioural mode of action for the two compounds, it might be expected that there would be a negative association between the use of caffeine and nicotine use.

One study in support of this idea showed that smokers inhaled more nicotine when given decaffeinated coffee than when they drank coffee containing caffeine (Kozlowski, 1976). In contrast, several experimental studies have failed to show any correlation between the dose of caffeine and cigarette consumption (Ossip *et al.*, 1980; Ossip and Epstein, 1981; Chait and Griffiths, 1983). However, none of the latter studies measured plasma levels of nicotine and it is known that cigarette consumption is a poor measure of nicotine intake.

In fact, instead of a negative correlation, there is an overwhelming body of epidemiological evidence which shows a clear positive association between coffee use and smoking. A recent analysis of six epidemiological studies (from 1972 to 1993) revealed that smokers were significantly more likely to drink coffee than non-smokers in every study (Swanson *et al.*, 1994). In the Health and Lifestyle Survey of 9003 people (Cox *et al.*, 1987), there was a significant positive association between the number of cups of coffee drunk per day and the percentage of people smoking (Jarvis, 1993).

One explanation for the higher levels of coffee consumption among cigarette smokers involves the shorter half-life and increase in the metabolism of caffeine when nicotine is also present (Kalow, 1985). The scientific data for this hypothesis are impressive.

In a study of the rate and pattern of caffeine metabolism after smoking abstinence, it was found that the peak plasma caffeine levels were greater during abstinence from smoking than during smoking (Brown *et al.*, 1988). After three days of smoking abstinence, the rate of caffeine metabolism was substantially slower. Benowitz and his colleagues proposed that smokers tend to ingest more caffeinated products because of their higher rate of caffeine metabolism and their desire to maintain precise plasma levels of caffeine.

In a companion study, Benowitz *et al.* (1989) investigated the plasma caffeine levels and caffeine consumption patterns of participants in a smoking cessation programme prior to cessation (baseline), at 12 weeks and at 26 weeks. For quitters, they reported a 269% increase in plasma caffeine levels from baseline to 12 weeks, and a 266% increase from baseline to 26 weeks. Caffeine consumption levels in the cessation group were approximately the same at baseline and at 12 weeks, but had decreased from baseline at 26 weeks.

Thus, there is some evidence from smoking abstinence studies to support the hypothesis that there is a faster rate of caffeine metabolism in smokers.

1 However, the evidence against the hypothesis is the fact that there seemed to be only poor down regulation of consumption, despite a doubling of plasma caffeine levels, which argues against precise maintenance of plasma caffeine levels.

1 Another argument against the hypothesis is from the coffee consumption of ex-smokers. Four of the studies in the review cited above (Swanson *et al.*, 1994) provided information on the use of coffee by ex-smokers and so allowed a comparison of coffee drinking habits of smokers, ex-smokers, and never smokers (see Table 12.1). In these studies, smokers were more likely to be consumers of coffee (weighted mean 94.87%) than never smokers (weighted mean 91.53%), but only slightly more likely than ex-smokers (weighted mean 83.56%). The difference in percentage between smokers and ex-smokers using coffee was only significant in the two larger studies (Hrubec, 1973; Murray *et al.*, 1981), but not in the two smaller studies of less than 2000 people (Swanson *et al.*, 1994). This suggests that the caffeine consumption of ex-smokers is more like that of smokers than non-smokers.

1 A third explanation is that some third variable might influence both coffee and cigarette use and so result; in a positive association. In their study on smoking and caffeine and EEG measures, Pritchard *et al.* (1995) speculated that the epidemiological link between cigarettes smoking and coffee use may be non-pharmacological.

   – They suggest two possibilities taste and ritual. Thus, users find that coffee tastes better with cigarettes, or cigarettes taste better with coffee or they both taste better with each other. Alternatively, the two behaviours are ritually compatible with a sip of coffee fitting neatly between puffs on a cigarette.

   – Another possibility is the association being mediated by alcohol use (Swanson *et al.*, 1994). Several studies found that both coffee and alcohol

Table 12.1 – Summaries of studies showing percentage of smokers and non-smokers (never smoker and ex-smoker) consuming coffee (adapted from Swanson et al., 1994)

Reference	Sample	Smokers	Ex-Smokers	Never smokers
Boston Collaborative Drug Surveillance Program (1972)	1.104 men	80.2%	77.9%	71.3%
Hrubec (1973)	10.615 men	95.2%	90.1%	79.5%
Murray et al. (1981)	16.800 men	97.0%	95.0%	88.6%
Rosenberg et al. (1980)	980 women	71.5%	63.1%	55.0%

use were correlated with cigarette smoking and with each other (Ayers et al., 1976; Conway et al., 1981; Friedman et al., 1974). It was also clear from the analyses of the Health and Lifestyle Survey that there is a positive association between caffeine use and alcohol consumption (Cox et al., 1987; Jarvis, 1993; Warburton and Thompson, 1994).

– A third explanation is that the person who is predisposed to use coffee is also predisposed to take nicotine. The data on ex-smokers which were described earlier are consistent with this suggestion. Our own unpublished analysis of the Health and Lifestyle Survey (Cox et al., 1987) revealed a positive association between coffee drinking and extraversion as have other studies (Ayers et al., 1976). Extraversion is also positively associated with smoking and alcohol consumption (Ayers et al., 1976) and so this relationship would explain the association of caffeine use with both alcohol use and nicotine use. Extraversion is characterized by low arousal in long and boring tasks, and so this would explain the use of caffeine in these situations.

FUNCTIONAL MODEL

Previously, I have outlined a functional view of nicotine to explain nicotine use (Warburton, 1987). The functional approach considers that a person uses nicotine to control his psychological state. Caffeine (in low doses) and nicotine are unique in having both calming and stimulating effects (Swift and Tiplady, 1988; Lieberman et al., 1987; Warburton, 1995; Warburton et al., 1988). These mood effects occur in individuals who are minimally abstinent (Warburton, 1995) and so the compounds would produce their effects throughout the day.

James (1991) caustically observed that the effects of caffeine were typically only 10 percent or less and the improvement found with nicotine has been about the same amount (Warburton, 1990b). However, it must be remembered that the majority of studies have used healthy, young students who have been selected for a university on their ability to concentrate and so would have higher than average baselines.

In any case, which of us would sneer at a 10 percent improvement in our own performance?

Part III — Social Drinking

CHAPTER THIRTEEN

Pharmacokinetic and Pharmacogenetic Aspects of Alcohol

Dharam P. AGARWAL

INTRODUCTION

Most people who choose to drink alcohol (ethanol, ethyl alcohol) have little or no problems limiting their consumption to amounts that does not generally cause serious health or social consequences. However, there are significant health affects associated with excessive alcohol consumption. Moreover, a given dose of alcohol may affect different people differently. Many of the toxic effects of alcohol are due to disturbances of a wide variety of metabolic functions. The elucidation of pharmacokinetic and pharmacogenetic factors that control and influence elimination and metabolism of alcohol in humans is thus fundamental to understand the biochemical basis of alcohol toxicity and alcohol abuse-related pharmacological and addictive consequences.

This chapter deals with biochemical, epidemiological, and genetic findings regarding the pharmacokinetic and pharmacogenetic aspects of ethanol in humans – with particular emphasis on the role of genetic factors in alcohol-drinking habits and differences in initial sensitivity to alcohol.

ABSORPTION, BODY DISTRIBUTION AND ELIMINATION OF ALCOHOL

Ethanol enters rapidly into the circulation by diffusion mainly across the lining or membrane of the duodenum and jejunum and to a lesser extent from the stomach and large intestine (Batt, 1989). The absorption of ethanol is normally over in two hours and overlaps with the diffusion phase during which it is distributed throughout the body water. Variation in hormonal status, that is stage of the menstrual cycle affects ethanol absorption. During the absorption phase, the ethanol concentration is found to be higher in the arterial blood than in the venous blood (Martin *et al.*, 1984).

Only 5–15% is excreted directly through the lungs, sweat, and urine, the remainder being metabolized by the liver, via oxidation to carbon dioxide and water. The total alcohol eliminated by the human body per hour is usually in the range of 100–300 mg/litre/hour which is equivalent to about 6–9 g alcohol per hour for a healthy subject with an average body weight.

Pharmacokinetics of alcohol

Interplay between the kinetics of absorption, distribution, and elimination of the ingested alcohol is an important determinant of the blood alcohol concentration. Historically, Widmark (1930) first introduced mathematical equations to explain the disposition and fate of ethanol in the body. Accordingly, after the absorption and distribution processes were completed, the blood alcohol concentration time course decreased at a constant rate following a zero order elimination kinetics. However, in the meantime, many complex pharmacokinetic models have been suggested describing alcohol concentration-time curves in blood, breath, and urine. The models vary in the number of pharmacokinetically relevant compartments, in the kinetics of absorption, and the kinetics of the elimination processes.

The rate constants for absorption, distribution, and elimination of alcohol have been estimated on one hand by experimental results, and on the other by simulating whole blood curves with computers. Models involving two or more compartments have been proposed to explain differences in alcohol concentrations over time between venous and arterial blood or breath, and to account for differing absorption rate constants for the absorption from stomach and small intestine and the effect of food thereupon (Von Wartburg, 1989).

Various studies on the pharmacokinetics of alcohol reported in recent years indicate that a non-linear course of ethanol elimination may best explain the observed differences in the metabolic rate of alcohol in humans. Changes in the gastro-intestinal absorption rate have an influence on the so called first-path pharmacokinetic effect. In most pharmacokinetic models, the first-pass effect is neglected while calculating the alcohol elimination rates.

Alcohol Elimination Rate (ER)

Alcohol elimination rate (ER) is normally taken as a measure for alcohol metabolism, since only a 5–15% of ingested alcohol is eliminated through breath, urine, or sweat (Jones, 1984). The slope (β) of the post absorption pseudo linear blood curve is used to estimate alcohol elimination rate. For convenience β^{60} was introduced which expresses the grams of alcohol eliminated per litre of blood per hour. The slope β^{60} multiplied with the volume of distribution (V) yields the total elimination rate (ER) in the whole body in grams per hour. As an average, rates of 100 mg/kg/hour (range: 70–130) have been suggested. Thus maximal ER for normal human subjects would not exceed 200–240 g alcohol per day for a 70 kg body weight individual.

The pharmacokinetic data for ethanol at different times of the day are conflicting, although the slope of the ethanol elimination curve was found to vary according to the time of the day, suggesting a circadian rhythm (Sturtevant and Sturtevant, 1979).

Factors affecting ER

Many factors appear to influence the course of BAC. As the orally ingested alcohol is absorbed in the stomach and intestine by diffusion, any change in the absorption rate will affect the ER. Factors that retard gastric emptying or lead to dilution will cause slower absorption and thus effect systemic availability of alcohol. Consumption of a 8 concentrated solution of ethanol results in lower blood alcohol levels than does a dilute solution when subjects are tested in a fed state (Roine *et al.*, 1991). This is probably due to the fact that in fasted condition ingested alcohol passes rapidly from the stomach into the duodenum with minimal exposure to the gastric mucosa. High alcohol concentration can delay gastric emptying; a concentrated solution of alcohol will remain in the stomach much longer than a dilute solution. Fasting has also been shown to decrease gastric first-pass metabolism of ethanol in humans. A swift increase in the rate of ethanol metabolism also occurs after intake of food intravenously (Hahn *et al.*, 1994). Possibly, eating a meal increases the activity the alcohol metabolizing enzymes, and may also lead to increased blood flow through the liver post-meal. Recently, a three compartment model for BAC was applied to study the effect of sex, alcohol dose, and concentration, physical exercise and several aspects of meal consumption on BAC (Wedel *et al.*, 1991). The ingestion of a meal prior to the intake of alcohol reduced both the gastric emptying rate and absorption efficiency of alcohol, increased the gastric emptying delay and reduced the rate of elimination. Ethanol elimination rate is also found to be higher in heavy drinkers when they consume higher amounts of alcohol (Keiding *et al.*, 1983). Perhaps, higher ethanol concentrations stimulate hepatic ethanol metabolism via ADH and also due to an increased activity of the sympathetic nerve system. An increase in the rate of alcohol metabolism after fructose has been demonstrated after either oral or intravenous administration in both normal volunteers and acutely intoxicated alcoholics. Such an unique effect of fructose probably occurs because fructose metabolism consumes NADH and allows faster disassociation of the ADH-NADH complex. However, more recent studies have shown that the alcohol elimination rate is reported to be generally enhanced by high-sucrose and high carbohydrate meals (Mascord *et al.*, 1991).

The blood alcohol concentration is roughly equal to the concentration of alcohol in the tissue fluid throughout the 'lean body mass', which includes most of the body except those tissues such as fat, which do not freely exchange water with the blood. Thus, volume of distribution of alcohol is said to reflect the amount of water present in various tissues of the body. This would suggest that an obese person, with proportionately increased adipose tissue, should have a higher BAC when consuming an alcohol dose similar to that taken by a lean person. A relatively increased amount of body water, when there is less adipose tissue, would be capable of holding larger amounts of alcohol. This may explain the sex related differences in blood alcohol concentrations: females show a higher blood alcohol concentration than males after administration of an equal dose of alcohol per kg body weight. The relatively higher body fat in females leads to a higher alcohol concentration in tissue fluids.

Alcohol-elimination rates can increase significantly with chronic alcohol ingestion as the result of induction of the microsomal alcohol-oxidizing system (MAOS). Female sex hormones, particularly estrogen, inhibits ethanol consumption as well as interact with ethanol pharmacokinetics by lowering peak blood alcohol concentration (Mishra *et al.*, 1989).

Effects of habitual moderate and heavy alcohol use

Habitual heavy alcohol consumption significantly alters the pharmacokinetics of alcohol. Chronic alcohol ingestion may increase the amount of the cytochrome P450 IIE1 isoenzyme, thereby increasing the hepatic oxidation of alcohol. Low to moderate alcohol use too effects the alcohol pharmacokinetics. Recently, Whitfield *et al.* (1994) observed that both peak BAC and rate of metabolism were strongly associated with habitual alcohol consumption level in both men and women. The rate of decline of BAC in these subjects increased with increasing habitual intake over the range from no alcohol to 30 standard drinks/week. The authors further observed that the threshold for the effects was much lower than previously assumed, and that there are factors that reduce pre-absorptive or first-pass metabolism but increase the post-absorptive metabolism.

Environmental and genetic factors

Both environmental and genetic factors influence the rate of alcohol degradation. Twin studies indicate that interindividual variability in the rate of ethanol metabolism is under genetic control. A striking similarity in ethanol metabolic rate was observed in identical twins, with much greater variability between fraternal twins (Vesell *et al.*, 1971; Kopun and Propping, 1977; Martin *et al.*, 1985). Ethnic differences in the metabolism of alcohol have been also known for years (Agarwal and Goedde, 1987; Mizoi *et al.* 1994) on the rate of disappearance of ethanol from the blood (Reed, 1978). However, due to significant differences in body mass and dietary habits of the subjects examined, results of such studies are ambiguous in terms of genetics.

Gastric alcohol metabolism

Comparison of the blood alcohol concentration (BAC) after oral ingestion versus intravenous administration of alcohol indicates that a fraction of ingested alcohol never reaches the peripheral circulation. Recent studies have shown that indeed a significant fraction of ingested ethanol is metabolized in the gastrointestinal tract by the so-called first-pass metabolism. First-pass metabolism has been shown to disappear when stomach is bypassed via intra-duodenal or portal vein infusion of ethanol.

The enzyme responsible for ethanol oxidation in human stomach is the alcohol dehydrogenase (ADH). Two major isozymes of ADH, σ (Sigma) and γ (Gamma) forms have been characterized from gastric mucosa. Differences in blood alcohol levels are seen among individuals who have different liver ADH isozymes (Meier-Tackmann *et al.*, 1990). Gastric ADH activity in women was found to correlate with first-pass metabolism of ethanol which is significantly lower in women, and may contribute to the higher blood ethanol concentrations seen in females (Frezza *et al.*, 1990).

ALCOHOL METABOLISM: BIOCHEMISTRY AND GENETIC POLYMORPHISMS

Liver is the principal organ responsible for the metabolic degradation of alcohol and may account for about 75% of the total elimination. Organs such as kidney, stomach, intestine, lung, heart, brain, blood cells and skeletal muscle also exhibit low but significant alcohol-oxidizing capacity. Ethanol is oxidized in the liver mainly by ADH to hydrogen and acetaldehyde – to which many of the well known toxic effects of ethanol can be attributed. Acetaldehyde is further oxidized to acetate which is then converted to carbon dioxide via the citric acid cycle. ADH catalyzes the reversible oxidation of many primary and secondary alcohols to aldehydes and ketones, and the slower, irreversible oxidation of aldehydes to the corresponding alcohols and acids. Some alcohol may also be oxidized within the endoplasmic reticulum via the complex microsomal ethanol-oxidizing system containing the cytochrome P450 IIE1 isoenzyme. Catalase also has the potential to oxidize ethanol, albeit only at very high concentrations of alcohol. However, its precise role in alcohol metabolism is unclear. Alcohol may also get converted to fatty acid ethyl esters via a non-oxidative pathway mediated by fatty acid ethyl ester synthases. Although ADH has a major rate-limiting role in the pathway of alcohol metabolism, other factors such as an increased NADH/NAD$^+$ ratio may also be important.

Human ADH is a dimeric protein of molecular weight of 80,000 daltons consisting of two subunits of molecular weight of 40,000 daltons each. At least seven different genetic loci code for human ADH arising from the association of different types of subunits. The different molecular forms of ADH can be divided into four major classes or distinct groups (I, II, III, and IV) according to their subunit and isozyme composition as well as the electrophoretic and kinetic properties (Table 13.1).

The ADH isozymes vary in their pharmacokinetic properties:

- the types of alcohols they preferentially oxidize;
- the amount of alcohol that must accumulate before appreciable degradation occurs; and
- the maximal rate at which they oxidize alcohol.

Class I ADH is composed of isozymes with α, β and γ subunits coded by ADH1, ADH2, and ADH3 loci, respectively. These isozymes belong to the low Km (<4 mM) forms.

Class II and III are composed of π and X (chi) subunits respectively, encoded by ADH4 and ADH5 loci. X-ADH has an extremely high Km (>3 M), whereas π-ADH has an intermediate Km (18–34 mM). A recently cloned ADH6 gene would encode for liver and stomach ADH. Finally, class IV ADH, represented by a stomach specific isozyme (σ subunits), is encoded by ADH7 locus. This isozyme also exhibits an intermediate Km.

Table 13.1 – Human ADH: classes, alleles and subunit composition of various isozymes

Class	Allele	Subunit	Subunit combination
Class I ADH			
ADH_1 locus:			
	ADH_1	α	$\alpha\alpha,\ \alpha\beta_1,\ \alpha\beta_2,\ \alpha\gamma_1,\ \alpha\gamma_2$
ADH_2 locus:			
	$ADH_2{}^1$	β_1	$\beta_1\beta_1,\ \beta_1\gamma_1,\ \beta_1\gamma_2$
	$ADH_2{}^{2-1}$	$\beta_1,\ \beta_2$	$\beta_1\beta_2$
	$ADH_2{}^2$	β_2	$\beta_2\beta_2,\ \beta_2\gamma_1,\ \beta_2\gamma_2$
	$ADH_2{}^3$	β_3	$\beta_3\beta_3$
ADH_3 locus:			
	$ADH_3{}^1$	γ_1	$\gamma_1\gamma_1$
	$ADH_3{}^{2-1}$	$\gamma_1,\ \gamma_2$	$\gamma_1\gamma_2$
	$ADH_3{}^2$	γ_2	$\gamma_2\gamma_2$
Class II ADH			
ADH_4 locus:			
	ADH_4	π	$\pi\pi$
Class III ADH			
ADH_5 locus:			
	ADH_5	X	XX
ADH_6 locus:			
	ADH_6	?*	??*
Class IV ADH			
ADH_7 locus:			
	ADH_7	σ	$\sigma\sigma$

* Identified at the gene level only.

The electrophoretic pattern of human liver and stomach class I isozymes varies within and among different racial groups because of allelic polymorphism. Any one particular isozyme may be composed of homodimeric subunits, consisting of two identical polypeptides (e.g., $\alpha\alpha$, $\beta\beta$, $\gamma\gamma$) coded by a specific allele at one of the loci or heterodimeric subunits, consisting of two nonidentical polypeptides (e.g., $\alpha\beta$, $\beta\gamma$) coded by alleles at separate loci, or heterodimeric subunits, but coded by different alleles at the same locus (e.g., $\beta_1\beta_2$, $\gamma_1\gamma_2$)

A variant enzyme form produced at the polymorphic ADH2 locus is commonly known as the 'atypical ADH' (Von Wartburg and Schurch, 1968). The atypical enzyme contains a variant B2 subunit instead of the usual B1 subunit, and exhibits much higher catalytic activity than the normal enzyme at a relatively physiological pH (pH 8.8). On electrophoresis or isoelectric focusing, the various ADH2 and ADH3 phenotypes show a typical but complex isozyme pattern (Yin *et al.*, 1984; Goedde and Agarwal, 1992). The homodimeric $\beta_2\beta_2$ variant form has significantly higher Km values for ethanol and NAD$^+$ and a higher Vmax for ethanol than the $\beta_1\beta_1$ and $\beta_1\beta_2$ forms (Bosron and Li, 1987). Many allelic variants encoded at the ADH2 and ADH3 locus have been identified in different individuals. Livers containing the normal ADH2 pheno-type ($\beta_1\beta_1$ subunits), exhibit a pH optimum at about 10.0–11.0, whereas livers with either an atypical heterozygote ($\beta_1\beta_2$) or an atypical homozygote ($\beta_2\beta_2$) show a pH optimum at 8.8.

The distribution of ADH alleles is quite different in Caucasians, Japanese, Chinese, native Americans, black Americans and Brazilians. While about 5–10% of the English, 9–14% of the German, and 20% of the Swiss population have been found to possess the 'atypical' phenotype of ADH (ADH2 locus) this variant form occurs in at least 85% of the Japanese, Chinese, and other Oriental popula-tions. The frequency of the variant forms of ADH3 locus is relatively higher in Caucasians than in Oriental and African populations (Yin and Li, 1989).

Alcohol metabolism rate

Differences in the kinetic properties of the polymorphic forms of ADH isozymes may contribute to differences in *in vivo* elimination rate of ethanol (Bosron and Li, 1987). Since the activity of atypical ADH is several times higher than the usual enzyme at relatively physiological pH (pH 8.8), it is possible that individu-als with the atypical isozyme form (ADH2 locus) metabolize ethanol differently compared with the normal ADH2 phenotype (Kassam *et al.*, 1989). Thus, the pharmacokinetic curves for ethanol elimination at higher alcohol concentrations should differ significantly for individuals with the atypical ADH form. Indeed, differences in kinetic constants among the four classes of ADH isozymes have been reported (Bosron and Li, 1987) and could offer an explanation for the large inter individual and inter ethnic variation in alcohol elimination rates (Farris and Jones, 1978).

Metabolism of acetaldehyde

Acetaldehyde is the first metabolic product of enzymatic ethanol oxidation in human liver, and is far more toxic than the parent compound ethanol. Therefore, many of the pharmacological and toxic effects of alcohol and alcohol related physical alterations have been attributed to acetaldehyde rather than to ethanol itself. The major oxidation of acetaldehyde in the liver and other organs is cat-alyzed by the NAD^+-dependent aldehyde dehydrogenase (ALDH). The ALDH catalyzed reaction is irreversible and a wide range of straight chain and branched chain aliphatic and aromatic aldehydes serve as substrates producing the corre-sponding keto acids.

At least four isozymes of ALDH coded by different gene loci have been detected in human organs and tissues and differ in their electrophoretic mobility, kinetic properties, as well as in their cellular and tissue distribution (Goedde and Agarwal, 1992). The various ALDH isozymes differ in their molecular size, subunit structure and isoelectric point as well as in their chromosomal assign-ment. Besides, some of the ALDH isozymes show genetically determined struc-tural variation. Genetically controlled polymorphic forms of ALDH have been described (Goedde and Agarwal, 1992). About 50% of the autopsy liver samples from Japanese show a lack of the ALDH2 isozyme activity band. Recent studies on the enzyme structure at the protein and DNA level have revealed that a point

Table 13.2 – Distribution of ALDH2 genotypes in different populations

Population	Allele frequency	
	ALDH2[1]	ALDH2[2]
Caucasians		
Germans	1	0
Swedes	1	0
Finns	1	0
Turks	1	0
Asian Indians	0.980	0.020
Hungarians	0.987	0.013
Orientals		
Chinese	0.841	0.159
Japanese	0.764	0.236
Koreans	0.849	0.151
Filipinos	0.994	0.006
Thais	0.950	0.034
Malayans	0.966	0.034
American Indians		
Caboclos (Brazil)	0.826	0.174
Aboriginals		
Papuans New Guinea	0.996	0.004

Data source: Goedde *et al.*, 1992

mutation is responsible for inactivation of the ALDH2 isozyme in Orientals. Glutamic acid at the 14th position from the C-terminus (487th position from the amino-terminus) is substituted with lysine in the catalytically inactive ALDH2 isozyme form (Hempel *et al.*, 1985).

Oriental populations of Mongoloid origin show a varying frequency of ALDH2 isozyme abnormality, whereas none of the Caucasian or Negroid populations have this isozyme variation (Table 13.2). Among native Indians, the ALDH2 isozyme deficiency genotype was detected only in a very small percentage (Goedde *et al.*, 1992; Voevoda *et al.*, 1994).

ACUTE REACTIONS TO ALCOHOL AND ITS METABOLITES

Alcohol is a known vasodilator, and this property is not the direct effect of alcohol on the blood vessel but is a consequence of its actions on the central nervous system, The effects of ethanol *per se* are influenced by its sympathomimetic activity and also by its metabolites, acetaldehyde and acetate. Acetaldehyde shows stronger sympathomimetic action than alcohol, and facilitates the release of catecholamines from the chromaffin cells of the adrenal medulla and from the sympathetic nerve endings. Increase of plasma cate-cholamines apparently leads to an elevation of heart rate, dilation of peripheral vessels accompanied with the rise of blood flow in carotid arteries and increased cardiac output.

In some individuals, ingestion of moderate amounts of alcohol exerts the so-called alcohol sensitivity symptoms (facial flushing, increase in heart rate, enhancement of left ventricular function, hot feeling in stomach, palpitation, tachycardia, muscle weakness, etc.). Wolff (1973) reported significant differ-ences among the Caucasian group on one hand with a very low percentage (5%) of subjects showing a flush response to alcohol and the Mongoloids and American Indians on the other hand with over 80% who showed flushing reac-tions. The apparent individual and racial differences in euphoric and dysphoric response to alcohol were replicated and extended in various ethnic and racial groups (Agarwal and Goedde, 1990).

American Indians are also sensitive to alcohol and exhibit facial flushing associated with various subjective and objective vasomotor symptoms after drinking moderate amounts of alcohol. Wolff (1973) reported that Eastern Cree Indians who consumed no alcohol or less than 5 bottles of beer per week responded more intensely than those who reported drinking more than 10 bottles of beer per week or an equivalent amount of alcohol in other forms.

Mechanism of biological sensitivity to alcohol

As stated above, acetaldehyde and not ethanol *per se* seems to be mainly respon-sible for most of the severe symptoms of alcohol related cardiovascular sensitiv-ity. Indeed, higher steady-state blood and breath acetaldehyde levels have been noted post-drink in those Japanese and Chinese subjects who show flushing after drinking mild doses of alcohol (Agarwal and Goedde, 1990). The maximum alcohol absorption takes place in the small intestine. Anatomic variations in the internal organ size may be important in this respect; Orientals and American Indians have longer intestines. Since alcohol diffuses through the lining of the stomach and the small intestine, any variation in the surface areas will lead to a more rapid absorption rate (Hanna, 1976). Individual and ethnic differences in the alcohol metabolism rate (mg ethanol/kg BW/hour) and alcohol clearance rate (mg ethanol/100 ml blood/min) also vary considerably between and within various racial and ethnic groups (Reed, 1978). Thus, any genetically determined variation in the ethanol metabolism rate could also influence the steady-state blood acetaldehyde levels.

The atypical ADH which is quite frequent in the Japanese, was initially thought to be responsible for a rapid oxidation of ethanol to acetaldehyde thereby producing alcohol sensitivity symptoms. More than 90% of the Japanese and other Mongoloids possess the atypical ADH with several times higher cat-alytic activity, whereas the incidence of flushing accompanied by higher blood acetaldehyde levels is only about 50%. Hence, rapid or higher than normal pro-duction of acetaldehyde via an atypical ADH alone cannot be the major cause of intense adverse reactions to alcohol.

A positive correlation between alcohol sensitivity and elevated blood acetalde-hyde level in conjunction with ALDH2 isozyme abnormality was noted in

Japanese subjects given an acute dose of alcohol (Harada *et al.*, 1981). Subjects with variant form of ALDH2 showed a significantly higher blood acetaldehyde level than nonflushers with a normal ALDH2 isozyme profile, while blood ethanol concentrations were similar in both groups of subjects. Thus, the initial vasomotor flushing after alcohol ingestion in Orientals might be due to their inability to metabolize acetaldehyde quickly and effectively in the absence of the ALDH2 isozyme with a low Km for acetaldehyde. Apparently, slow acetalde-hyde oxidation due to an ALDH2 isozyme abnormality leads to elevated blood acetaldehyde levels (Lehmann *et al.*, 1986) resulting in catecholamine-mediated vasodilation associated with dysphoric symptoms. Indeed, a faster ethanol elimi-nation associated with a higher blood acetaldehyde level was observed in a Japanese subject homozygous for ALDH2 2 allele (Meier-Tackmann *et al.*, 1990).

ALCOHOL DRINKING HABITS, ALCOHOL FLUSHING, AND RISK FOR ALCOHOLISM

An interesting correlation exists between flushing response and alcohol drinking habits of Orientals. Generally, Orientals flush more and drink less alcohol than Caucasians (Agarwal and Goedde, 1987, 1990; Chi *et al.*, 1989; Higuchi *et al.*, 1992).
    Comparison among racial and ethnic groups shows that

- a larger proportion of Orientals than Caucasians report no use of alcohol;
- Caucasians report heavier alcohol use;
- a large proportion of Orientals who drink alcohol experiences facial flushing and associated sensitivity symptoms after drinking alcohol.

Reed and Hanna (1986) reported a between- and within-race variation in cardio-vascular responses to alcohol and usual consumption of alcohol. The alcohol response data showed similarities between Japanese and Chinese and marked differences between either of these and Europeans. The mean usual alcohol con-sumption (g/week) was 40.8, 73.0, and 135.8 for the Chinese, Japanese and Europeans, respectively.

ALDH2 ABNORMALITY, ALCOHOL SENSITIVITY AND ALCOHOL DRINKING

Individuals sensitive to alcohol because of their genetically controlled polymor-phism of ALDH2 may be discouraged from abuse of alcohol due to initial adverse reaction (Agarwal and Goedde, 1989). The frequency of the atypical ALDH2 isozyme in two Japanese districts in the regions of Sendai and Gifu differed significantly and correlated negatively with the per capita alcohol con-sumption (Harada *et al.*, 1985). The putative role of ALDH2 genotypes in affect-ing the alcohol drinking habits in Orientals has been further supported by many subsequent studies (Yamashita *et al.*, 1990; Higuchi *et al.*, 1992; Thomasson and Li, 1993).

ADH AND ALDH POLYMORPHISM, ALCOHOL USE AND EFFECTS ON END-ORGANS

It is of interest to note the role of ADH and ALDH isozyme variation in the development of alcohol dependency and alcohol-related liver disease in different populations. The presence of both ADH2[2] and ADH3[1] alleles appears to be protective against the development of alcoholism in Chinese of Taiwan (Thomasson *et al.*, 1991). Moreover, both alcoholic and non-alcoholic patients with liver disease show a lower frequency of the atypical ADH phenotype (Pares *et al.*, 1994).

The diverse adverse physiological reactions to alcohol would make it less likely for sensitive individuals to abuse alcohol. A significantly low incidence of ALDH2 isozyme abnormality was observed in a group of alcoholics as compared to the psychiatric patients, drug dependants and healthy controls in a Japanese psychiatric hospital (Harada *et al.*, 1982). Other studies (Enomoto *et al.*, 1991; Takada *et al.*, 1994) indicate that humans heterozygous for the ALDH2 alleles are at higher risk for the development of alcoholic liver disease when they drink more than a critical level of alcohol.

Therefore, the higher the prevalence of abnormal ALDH2 isozyme in racial or ethnic groups the lower the prevalence of alcohol-related problems. Taken together, the ALDH2 deficient individuals drink less, have the tendency not to become habitual drinkers, suffer less from liver disease and are rarely alcoholics. Acetaldehyde-induced aversion to alcohol drinking may represent only one aspect concerning the relation between acetaldehyde metabolism and human drinking. Thus, in future studies, the possible role of acetaldehyde and its metabolism in mediating reinforcing effects, particularly in Caucasian alcoholics has to be elucidated.

GENETIC DETERMINANTS OF ALCOHOL USE AND ABUSE

There are a series of factors which may all interact in predisposing or protecting an individual against alcohol abuse, alcoholism and alcohol-related disorders: the availability of alcohol and the price, an individual's socio-cultural, psychological, physiological, and genetic make-up. Genes have long been suspected to play a role in the aetiology of alcoholism. The phenomenon that 'alcoholism runs in families' has been described by ancient Greek scholars. The current evidence strongly suggests that alcoholism may be a genetically influenced complex multifactorial disorder (Agarwal and Goedde, 1990; Couzigou *et al.*, 1993). In order to identify discrete underlying genetic factors in the development of alcoholism one need to answer a number of questions such as: is alcoholism an inherited disorder; are there genetically distinct forms of alcoholism; is alcoholism a multifactorial disorder with a monogenic or a polygenic aetiology; what are the predisposing factors; are the genetic contributions from the environmental influences separable; what is the mode of transmission of the biological risk factors; is it possible to modulate genetic influences through prevention strategies (e.g., education, making alcohol non available via price and production policies)?

Among the most common strategies employed thus far to identify hereditary factors in alcoholism are:

- family studies (family system variables, drinking behaviour, drinking history);
- twin studies (alcohol metabolism, pattern of alcohol use and abuse);
- adoption studies (biological parents versus foster parents) and
- high risk groups (identification of biological markers of vulnerability to alcoholism).

Family studies

Because of a high degree of familial association, for many years, alcoholism was regarded as a distinct disease which may be transmitted from generation to generation. A familial association could result from cultural factors tending to encourage heavy drinking in family members. The family system may generate or promote the development of alcoholism in a family member. Children try to model their behaviour on that of their parents and doing so may also imitate their drinking habits. On the other hand, drinking may be discouraged in some families for religious, cultural or climatic grounds while in other families, constraints on heavy drinking may be virtually non-existent.

Family investigations of alcoholic probands have yielded consistently higher rates of alcoholism than would be expected in the general population regardless of the nationality of the sample (Cotton, 1979). More recent studies have reported that on average the risk for developing alcoholism is many times greater among first-degree relatives of alcoholics than among controls (Cloninger, 1987; Merikangas, 1990; Hesselbrock et al., 1991).

Twin studies

Twins have been extensively used to differentiate between genetic and environmental influences interacting in alcohol drinking behaviour, alcohol metabolism and alcoholism-related cognitive deficits. More recent twin studies indicate a modest to strong genetic predisposition (Merikangas 1990; McGue et al., 1992; Marshall and Murray, 1992). In a recent study, Kendler et al. (1992) found a significantly higher concordance rate for alcoholism in female MZ than in DZ twins (26.2% versus 11.9%). These reports strongly support the hypothesis that genetic factors play a major role in the aetiology of alcoholism in women.

The distribution of alcohol consumption in the Australian Twin Register study (Heath and Martin, 1991) showed that twin pairs were highly concordant both for teenage drinking and for early versus late onset of drinking. Environmental influences on onset of drinking appeared to be sex-specific. Among drinkers, early versus late onset of drinking was more strongly influenced by inherited factors in females, but by shared social environment (for example, family background or peer pressure) in males. Twin studies also have clearly demonstrated that alcohol metabolism rate is under genetic control. A study of psychomotor performance and subjective intoxication, in twin pairs unselected for risk of alcoholism, after intake of alcohol (0.75 g/kg BW) has shown a significant difference in alcohol-specific reactivity such as body sway and other psychomotor performances

(Martin *et al.*, 1985). In addition to these studies, the general population twin studies have also demonstrated that alcohol consumption patterns are more alike in monozygotic than in dizygotic twin pairs (Heath and Martin, 1991). In summary, these twin studies show the important role played by genetic differences in determining how much people drink and how they are affected by alcohol.

Adoption studies

Extensive adoption studies conducted in Denmark and Sweden have provided substantial evidence that alcoholism is genetically influenced and that there are distinct patterns of alcoholism with different genetic and environmental causes. When the adopted-away sons of an alcoholic parent were compared to their siblings raised by the alcoholic biologic parent, a remarkably similar rate of alcoholism was noted in both groups. Subsequent adoption studies from other countries have clearly shown that children born to alcoholic parents but adopted away during infancy were at greater risk for alcoholism than adopted-away children born to non-alcoholic parents (Merikangas, 1990). Data from a large adoptee sample in Sweden demonstrated, for the first time, the existence of two forms of inherited vulnerability to alcoholism, type I and type II (Bohman *et al.*, 1984; Cloninger, 1990).

Type I alcoholism is the more common form (75%), is less severe (mild) and affects both men and women. The genetic contribution comes from either the biological father or the mother (or both), but is sensitive to environmental factors for its expression. It is associated with adult-onset alcoholism in either biological parent. Furthermore, it is considered to be milieu limited as postnatal environment affects its occurrence and severity in genetically susceptible offspring. They start chronic drinking usually well after the age of 25, rarely have trouble with the law, and often successfully become abstinent. Their children are only twice as likely to have trouble with alcohol compared with the general population.

In contrast, type II alcoholism is characterized by severe type of susceptibility that is more influenced by genetic than environment. This form is limited to males with an early onset associated with aggressive behaviour and antisocial personality. Biological fathers of these individuals also have similar features. The alcoholics in this group (25% of the total) tend to drink heavily before the age of 25, have bad work and police records and meet with little success with treatment programmes. Drinking is a habit these alcoholics seem to pick up on their own with little encouragement from friends or other influences. Alcoholism surfaced nine times as often as in the general population when the sons of men with this form of alcoholism were followed-up. This form is called 'male limited'. Recently, a third type of alcoholism has been proposed (Hill, 1992). Like type II alcoholism, it is significantly influenced by genetic factor, but it is not associated with antisocial behaviour.

NEUROBEHAVIOURAL MARKERS OF RISK FOR ALCOHOLISM

Genetic susceptibility to alcoholism may be expressed in changes in various neurobehavioural functions as alcoholism is associated with many psychiatric problems such as childhood conduct disorder which is marked by aggression and other antisocial activities. Hyperactivity may be one aspect of the genetically mediated predisposition to alcoholism. Some differences in neuropsychological deficits were observed between healthy FHP (Family History Positive) and FHN men in a large scale study in Denmark (Drejer *et al.*, 1985). A premorbid assessment of the cohort clearly indicated that the FHP group is characterized by poor school career, verbal difficulties, impulsivity, abnormal EEG response after ethanol ingestion and poor abstracting capability compared to the FHN group (Knop *et al.*, 1988). Preliminary evidence suggests that a neurophysiological disorder may also characterize children who are at risk for alcoholism.

In addition, the sons of alcoholics presented a more neurotic personality profile than sons of non-alcoholics. When adolescent sons of alcoholics and non-alcoholics were compared on a battery of intellectual, neuropsychological, personality and behavioural parameters, the former group demonstrated certain deficits in perceptual motor ability, memory and language processing (Tarter *et al.*, 1990). They also had auditory and visual attentional impairment regarding reading comprehension. In addition, the sons of alcoholics presented a more neurotic personality profile than sons of non-alcoholics (Pihl *et al.*, 1990). Children with attention deficit hyperactivity disorder (ADHD) and de la Tourette syndrome (vocal and motor tics, learning disorder, sleep disorder etc.) are shown to be at greater risk to develop problems with alcohol and drug abuse or dependence as adults (Comings, 1993).

Electroencephalographic activity (EEG)

Electrophysiological techniques have proven valuable in detecting subtle abnormalities in alcohol-related brain function. In humans, EEG examined under resting conditions is under strong genetic control (Stassen *et al.*, 1988). Moreover, many past studies have shown that resting-state EEGs in sober, awake alcoholics contain excess beta activity and a deficiency in alpha, theta, and delta activity.

Event-Related Potentials (ERP)

ERPs represent sensitive indices of brain responsiveness to external visual, auditory, somatosensory, and olfactory stimuli. Differences in ERPs may also reflect differences in information processing thus making it a useful tool for studying inherited vulnerability to alcoholism. It was reported that abstinent alcoholics manifest a significantly reduced P300 component of ERP, which is a wave form associated with various cognitive responses of the individual, compared to matched normal control subjects (Porjesz and Begleiter, 1990).

Reaction to ethanol as a potent predictor of alcoholism

A decreased intensity of reaction to a challenge dose of ethanol was observed in male FHP subjects as compared to FHN controls (Schuckit, 1989a). The family history positive groups experienced a significantly lower levels of subjective intoxication. Despite identical blood alcohol concentrations, the FHP men rated themselves as significantly less intoxicated than the FHNs. Similarly, FHPs demonstrated significantly lower levels of prolactin and to a less extent cortisol, the two hormones thought to be released in response to ethanol ingestion

(Schuckit, 1989a). A stepwise discriminant function analysis of 30 FHP/FHN pairs revealed that the combination of subjective feelings reported after ethanol, increase in body sway and increases in several hormones correctly identify almost 80% of the two family history groups (Schuckit and Gold, 1988).

RISKS AND BENEFITS OF MODERATE DRINKING

Though there is a huge amount of literature addressing the health effects of excessive consumption, much less is known about the effects of light to moderate alcohol consumption. There is an accumulating body of evidence to suggest that low or moderate alcohol consumption is associated with a reduction in mortality from ischemic heart disease. Though moderate drinking is difficult to define, many epidemiological studies have shown an inverse correlation between low to moderate alcohol consumption, hypertension, and CHD. Such a protective effect of alcohol against atherosclerosis has been associated to the elevated concentration of HDL-cholesterol induced by alcohol (reviewed by Srivastava *et al.*, 1993).

However, the underlying mechanisms whereby alcohol drinking enhances HDL-cholesterol levels are not yet clear. Various lifestyle variables, *viz.* diet, smoking, hypertension, body mass index and exercise can affect the lipoprotein status both in users and non users of alcohol and hence these variables may have a direct relationship to CHD. Moderate alcohol may directly affect coronary artery diameter and coronary blood flow. Also, alcoholic effects on blood-clotting factor have been suggested as a possible protective mechanism. Other variables that potentially influence the development of CHD such as reduced tendency for blood coagulation and increased fibrinolytic activity are affected by alcohol and may be considered as possible additional underlying mechanisms for the cardio-protective effect of alcohol (Steinberg *et al.*, 1991). There is a general consensus that there is a U-shaped curve for the association between alcohol consumption and total cardiovascular disease. On the other hand, there is a J-shaped relationship between alcohol intake and total mortality, with increase in deaths from cirrhosis, accidents, violence, and cancer in heavier alcohol drinkers.

In France, coronary heart disease mortality is much lower than other Western countries, even if the data are corrected for such risk factors as plasma cholesterol and intake of saturated fatty acids. This apparent discrepancy is known as the 'French paradox' (Sharp, 1993). The high level of wine consumption in France, in particular the red wine, has been suggested to be the protective factor (Seigneur *et al.*, 1990). Ingestion of red wine has been shown to be associated with increased antioxidant activity in serum of healthy volunteers (Maxwell *et al.*, 1994). It is speculated that antioxidants such as polyphenolics and flavonoids present in red wine may be partly responsible for the lower incidence of CHD among French wine drinkers (Frankel *et al.*, 1993). These antioxidants apparently reduce LDL lipid oxidation leading to reduced atherogenesis.

SUMMARY

Ethanol enters rapidly into the circulation by diffusion mainly across the lining or membrane of the duodenum and jejunum, and to a lesser extent from the stomach and large intestine. Only 5–15% is excreted directly through the lungs, sweat, and urine, the remainder being metabolized by the liver, via oxidation to carbon dioxide and water. The total alcohol eliminated by the human body per hour is usually in the range of 100–130 mg/litre/hr which is equivalent to about 6–9 g alcohol per hour for a healthy subject with an average body weight. Alcohol dehydrogenase (ADH) constitutes the principal oxidative pathway for ethanol metabolism in humans. Differences in the kinetic properties of the polymorphic forms of ADH isozymes may contribute to differences in *in vivo* elimination rate of ethanol. Aldehyde dehydrogenase (ALDH) is responsible for the detoxification of acetaldehyde. Many Japanese, Chinese and Koreans have been found to have a genetically abnormal form of ALDH2 isozyme. In these individuals, ingestion of moderate amounts of alcohol exerts the so-called alcohol sensitivity symptoms such as facial flushing, and they are therefore discouraged from alcohol abuse.

Many genes are probably involved in determining a vulnerability to alcoholism, and the current evidence for a genetic role is compelling. However, the specific contribution of genetic and environmental factors on normal and abnormal drinking, on the development of organ damage, and on the mechanism of tolerance and physical dependence has yet to be fully elucidated.

CHAPTER FOURTEEN

Acute Effects of Alcohol Revealed by Event Related Potentials

Werner SOMMER and Hartmut LEUTHOLD

INTRODUCTION

This review[1] is concerned with acute effects of alcohol on even related potentials (ERP) in social drinkers. Such effects may reveal alterations of information processing that are not necessarily seen in overt behaviour (Callaway, 1983). Clues from ERPs about the functional locus or loci of alcohol effects within the cognitive system are potentially valuable because as yet behavioural evidence alone has been inconclusive. For example, whereas moderate doses of alcohol usually slow reaction time (RT) (see for review Maylor and Rabbitt, 1993a) some studies did not find any effect (Pearson, 1968; Shillito *et al.*, 1974; Tharp *et al.*, 1974) or even reported performance improvements (Maylor *et al.*, 1987; Wilkinson and Colquhoun, 1968). As pointed out by Maylor and Rabbitt (1993a), when alcohol effects are present, it is not clear whether they can be localized in particular subprocesses or stages of information processing. Maylor *et al.* (1992) even suggested that alcohol does not so much affect specific stages but rather leads to a general slowing of information processing in proportion to the complexity of the task.

ERPs provide a supplementary approach to the localization of drug effects because the successive ERP components reflect sensory transmission and processing in spinal cord or brainstem and primary cortical areas, cognitive operations such as stimulus selection and classification, and finally motor processes (for reviews, see Coles *et al.*, 1990; Hillyard and Picton, 1987; Regan, 1989). If the functional significance of an ERP component is known, its amplitude can be used as a measure for the intensity of the respective subprocess, whereas its latency reflects dynamic aspects of processing. ERPs may reflect the activity of specific modules or stages of information processing more directly than behaviour. Thus, drug actions may be revealed through ERPs even when there are no overt behavioural effects, for example at low doses or due to compensatory strategies or attempts. In addition, the spatial distribution of the ERPs across the

---

[1]To our knowledge, the latest comprehensive overview of acute alcohol effects on ERPs is that of Porjesz and Begleiter (1985, pp. 433-444); Oscar-Berman (1987) commented on several studies of mainly P_{300}.

scalp may help to localize or at least differentiate the affected neuroanatomical or functional systems.

In the present review we will start with alcohol effects on ERP components reflecting sensory processes and finish with motor processes. Because early ERP components are highly modality-specific, we will distinguish between ERPs in response to auditory (Auditory Evoked Potential: AEP), visual (Visal Evoked Potential: VEP), and somatosensory (Somatosensory Evoked Potential: SEP) stimulation. Later ERP components are strongly affected by processes under the control of the subject. Therefore, experimental tasks where the subject is passive or where her or his options are not controlled for can only provide data of very limited value. We will therefore distinguish between uncontrolled or passive and controlled or active tasks except for very early ERP components that are largely immune to cognitive influences.

AUDITORY MODALITY

Auditory brainstem responses

Auditory Brainstem Responses (ABR) (for review, see Starr and Don, 1988) reflect the activity of the auditory pathway (Figure 14.1) from the auditory nerve (wave I at about 1.5 ms) up to thalamocortical radiations (wave VII at about 9 ms). Alcohol progressively increased latencies over the various ABR waves, reaching significance from wave III or IV onward with latency shifts around 0.1 ms for wave III and 0.2 to 0.36 ms for wave VII (Campbell *et al.*, 1981; Church and Williams, 1982; Fukui *et al.*, 1981; Squires *et al.*, 1978). Church and Williams (1982) provided data which indicate that at least the ABR delays from moderate (0.5 g/kg BW)[2] to high doses (1 g/kg BW) were independent of any concomitant changes in brain temperature, a variable known to affect ABR latencies. ABR amplitude changes due to alcohol have not been reported.

Auditory evoked potentials

AEPs following ABRs up to about 50 ms are termed *mid latency components* (MLC) and reflect primary and secondary cortical activity. Whereas McRandle and Goldstein (1973) did not observe any effects at 0.4 g/kg alcohol, Squires *et al.* (1978) found the latency of the Na to increase from 18.9 ms to 20.2 ms with their, in some subjects, considerably higher dose. Amplitude effects were not observed.

The auditory P_{50} component showed no effects of moderate alcohol doses on amplitude or latency in the passive tasks of Pfefferbaum *et al.* (1979) and Freedman *et al.* (1987). However, Freedman *et al.* (1987) found alcohol to

---

[2] For the sake of comparability we attempted to express the dosages used as grams of pure alcohol per kilogram body weight (g/kg). If this was impossible, we report blood alcohol concentration (BAC g/l). We will alternatively refer to doses of up to 0.5, between 0.5 and 0.8, and above 0.8 g/kg as low, moderate, and high, respectively. Unless stated otherwise, tests were conducted during a time of maximal BAC.

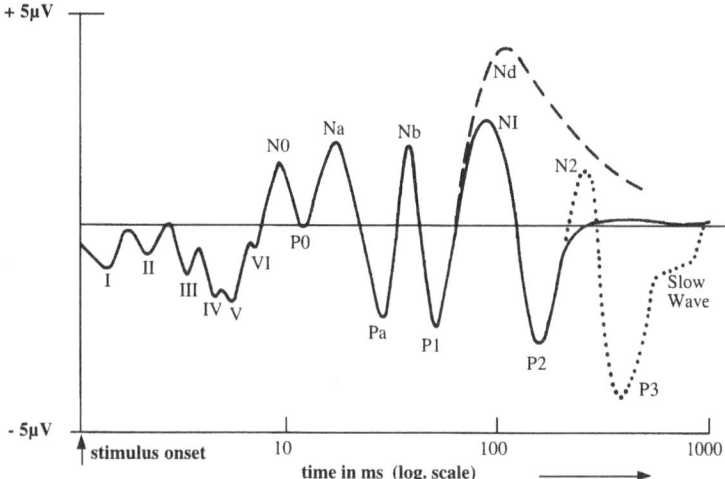

Figure 14.1 – Components of the Auditory evoked potential: The components include waves evoked from the auditory brainstem pathways (I to VII), cortical components (N_0 to P_1), late cortical components (N_1 and P_2) and additional components that vary because of attentional and cognitive processing (dashed lines Nd and dotted lines N_2, P_3 and Slow Wave) (Redrawn from Regan, 1989).

diminish P50 suppression, that is the relative decrease of P50 amplitude when elicited by two tones in rapid (0.5 s) succession.

The first studies of alcohol effects on the auditory *vertex potentials* (N_1–P_2) used *passive* or behaviourally unverified tasks. No or inconsistent alcohol effects on the *amplitudes* of the late cortical N_1 and the P_2 components were observed at doses of up to 0.4 g/kg (Campbell *et al.*, 1984; McRandle and Goldstein, 1973). Above that level, however, these components were usually found to be reduced (Campbell *et al.*, 1984; Fruhstorfer and Soveri, 1968; Gross *et al.*, 1966; Hari *et al.*, 1979; Kopell *et al.*, 1978; Krogh *et al.*, 1978; Pfefferbaum *et al.*, 1979; Wolpaw and Penry, 1978). This holds true also for the offset AEP at the end of long-lasting tones (Hari *et al.*, 1979). Inconsistent effects were observed for AEP latencies. Gross *et al.* (1966), Kopell *et al.* (1978), and Krogh *et al.* (1978) did not report alcohol effects at doses of up to 0.8 g/kg. On the other hand, Flach *et al.* (1977), and Fruhstorfer and Soveri (1968) observed latency increases at doses as low as 0.4 g/kg.

Alcohol effects during *active* tasks are somewhat in contrast with the amplitude reductions commonly reported for passive conditions. In line with passive tasks, there are no alcohol effects at doses below 0.5 g/kg (Campbell *et al.*, 1984; Pfefferbaum *et al.*, 1980; Teo and Ferguson, 1986), whereas at or above that dose, amplitude decreases are usually found (Campbell and Lowick, 1987; Campbell *et al.*, 1984; Daruna *et al.*, 1987; Hamon and Camara, 1991; Pfefferbaum *et al.*, 1980). It is noteworthy, however, that two studies reported no alcohol effects on N_1 amplitude at 0.8 g/kg (Campbell *et al.*, 1984; Roth *et al.*, 1977). Both tasks were quite demanding. Roth and associates (1977) used a memory search task and Campbell and colleagues (1984) required subjects to detect rare 1025-Hz target tones among 1000-Hz standards. When targets were more discriminable (1500 Hz), alcohol decreased AEP amplitudes as is commonly observed. Therefore it is conceivable that the adverse effects of alcohol on N_1 amplitude and the underlying cognitive processes can be compensated under increased task demands.

Unfortunately, the processes underlying the N_1 *amplitude* reduction and its possible reinstatement at higher task demands can not be unambiguously inferred. Thus, in the latency range of 50 to 150 ms several different sub components of the AEP are distinguished (Näätänen and Picton, 1987), related either to physical and temporal aspects of the stimulus or depending on the stimulus context and selective attention processes. However, attempts to specify which of these sub components may be affected by alcohol have been exceedingly rare.

Some indirect conclusions may be based on alcohol effects on the P_2, following N_1 (Campbell and Lowick, 1987). It is conceivable that alcohol negatively affects selective attention, a process that can be compensated by increased task demands or effort. In this case alcohol would reduce the processing negativity, a sub component of sufficient duration to overlap the P_2. If alcohol reduces the processing negativity it should not only diminish N_1 but might also increase P_2 by reducing overlap with this sub component. However, the studies which reported N_1 reductions under alcohol also reported P_2 to decrease (e.g., Campbell *et al.*, 1984; Campbell and Lowick, 1987, standard stimuli; Pfefferbaum *et al.*, 1980) or to be unaffected (Campbell and Lowick, 1987, targets: rare or deviating stimuli). Therefore, it appears more likely that alcohol affects sensation rather than attention-related aspects of N_1.

As for passive situations most studies with active tasks did not find latency effects on AEP vertex potentials, at doses of up to 0.9 g/kg (Campbell and Lowick, 1987; Daruna *et al.*, 1987; Hamon and Camara, 1991; Peeke *et al.*, 1980; Roth *et al.*, 1977). Only Teo and Ferguson (1986) reported latency increases in a counting task for N_1 and P_2 at 0.55 g/kg.

The N_{200} or N_2 deflection in the AEP is usually observed when a deviant or rare stimulus is presented before a background of standards. Such deviants may elicit a mismatch negativity (MMN) even when the stimuli are to be ignored. They may however also elicit a later negativity (N_2b) when the rare stimuli are task-relevant, for example in oddball tasks where rare targets have to be counted or responded to. Only two studies have provided data on auditory N_2 (Grillon *et al.*, 1994; Teo and Ferguson, 1986). The former study reported that the N_2b to rare nontargets (novel stimuli) was smaller in amplitude whereas the MMN was apparently unaffected even at a high dose of 0.85 g/kg. Teo and Ferguson (1986) found a 45-ms latency increase in N_2 at low and moderate doses.

To summarize, according to the ABRs, alcohol affects auditory transmission at loci as low as the pons (waves III and IV) and possibly already at low to moderate doses of alcohol. At moderate doses, there are also adverse effects on the processes underlying the vertex potential with the possible exception of demanding tasks. In contrast, the temporal characteristics of auditory processing (AEP latencies) beyond the brainstem appear to be relatively unaffected by alcohol.

VISUAL EVOKED POTENTIALS

In contrast to the auditory modality there are no established manifestations of subcortical processes to visual stimulation. The component structure of the VEP depends very much on stimulation. Early studies of alcohol effects on the VEP often used flash stimulation. Following single flashes, a series of VEP responses can be observed, starting at 30 to 40 ms.

In response to patterned stimulation, an occipital P_1 between 85 to 120 ms appears, preceded and followed by negative-going components. The P_1 is considered to have its principal origin in primary visual cortex and to represent the processing of mainly foveal information.

Passive situations

Studies with *passive* recording situations did not find any alcohol effects on VEP amplitudes or latencies below 0.7 to 0.8 g/kg both with flashes and patterned stimuli (Erwin and Linnoila, 1981; Jensen and Krogh, 1984; Lewis *et al.*, 1970; Pollock *et al.*, 1988; Seppäläinen *et al.*, 1981; Simpson *et al.*, 1981). Above 0.8 g/kg there are consistent reductions of amplitude and delays of latencies (Erwin and Linnoila, 1981; Lewis *et al.*, 1970; Müller and Haase, 1967; Porjesz and Begleiter, 1973a; Seppäläinen *et al.*, 1981; Simpson *et al.*, 1981).

Two studies investigated visual steady state responses where stimulation rate is so high that successive VEPs merge into a sinewave-like pattern. With checkerboard reversals Jensen and Krogh (1984) found no effects of a 0.57 g/l BAC. Spilker and Callaway (1969) modulated the intensity of an unstructured light stimulus according to a 10 Hz sinewave function. The effect of a high dose of 1.1 g/kg alcohol depended on the difference between the minimum and maximum light intensity (modulation depth). Alcohol diminished VEP amplitude at high modulation depth whereas VEP appeared to be increased at the lowest modulation depth.

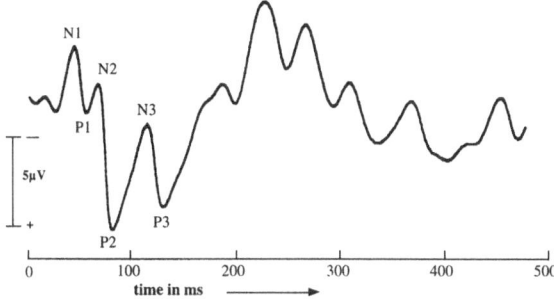

Figure 14.2 – Components of the flash Visual evoked potential: The components include waves evoked from the cortex (early components (N_1 to P_3), and late cortical components (beyond P_3) that vary as a function of attentional and cognitive processing (Redrawn from Regan, 1989).

Active tasks

For *active tasks* the main body of evidence suggests that alcohol decreases VEP components with *latencies* up to 200 ms at doses as low as 0.5 g/kg (Neill *et al.*, 1990; Rhodes *et al.*, 1975; Sommer *et al.*, 1993; Taghavy *et al.* 1976; Yamamoto and Saito, 1987). Most of these studies required simple responses to the presence of stimuli, target detection, or choice responses. Perceptual factors have been specifically manipulated by Neill *et al.* (1990) who studied depth perception with dynamic random dot stereograms, creating the impression of moving stimuli either by monocular or binocular cues. Subjects had to identify different trajectories of the apparent movements. The ERPs elicited in both conditions were decreased in amplitude by both low and moderate (0.4 and 0.8 g/kg) doses of alcohol, indicating that one effect of alcohol is on the perception of motion in depth. Interestingly, the alcohol effects were clearer in the ERPs than in the reaction times (RTs).

As concerns VEP latencies of components below 200 ms, there is little evidence of significant changes at low or moderate doses (Colrain *et al.*, 1993; Krull *et al.*, 1993; Jensen and Krogh, 1984; Rohrbaugh *et al.*, 1987; Yamamoto and Saito, 1987). The two studies that reported an increase in latency used high doses of 1.1 g/kg or more (Obitz *et al.*, 1977; Taghavy *et al.*, 1976).

VEP components in the latency range around 200 ms have been related to pattern recognition and stimulus classification processes (Ritter *et al.*, 1982) or covert orienting of attention (Näätänen and Gaillard, 1983). With the exception of Sommer *et al.* (1993) all studies that measured N_{200} latency reported significant alcohol-related delays. These delays are clear at doses of 0.7 g/kg and above (Krull *et al.*, 1993; Taghavy *et al.*, 1991) but may start even at lower doses (Colrain *et al.*, 1993; Rohrbaugh *et al.*, 1987; Yamamoto and Saito, 1987). Where both N_{200} latency and RTs are available, the alcohol effects are of comparable magnitude (Colrain *et al.*, 1993: 64 versus 45 ms; Rohrbaugh *et al.*, 1987: 37 versus ca. 40 ms). In no case were pre-N_{200} components delayed as much as the N_{200}. Therefore, the N_{200} delay seems to be quite specific to the underlying processes.

As regards visual N_{200} *amplitude*, findings are controversial. Colrain *et al.* (1993) and Taghavy *et al.* (1991) reported a decrease but used a N_{200}–P_{300} amplitude measure which confounds the two components. In response to low doses of alcohol, Rohrbaugh *et al.* (1987) and Yamamoto and Saito (1987) found N_{200} amplitude *increases* but only as a trend in the former report and measured as P_2–N_2 peak-to-peak amplitude in the latter. Rohrbaugh *et al.* (1987) suggested that the N_{200} increase may relate to similar increases seen during drowsiness in the auditory modality. One might speculate that the amplitude increases of the visual N_{200} may reflect stimulating effects of small alcohol doses. Alternatively, the apparent N_{200} increase may reflect a release from overlap by the subsequent P_{300} component. Given that P_{300} develops already during the N_{200} latency and that P_{300} is more sensitive to alcohol than N_{200}, its reduction will enhance N_{200}.

An alcohol-induced VEP reduction in the latency-range of the N_{200} found by Sommer *et al.* (1993) may relate to a different, that is, more occipital component than the findings above. Alcohol (mean BAC 6.9 g/l) virtually abolished the occipital N_{200} regardless of whether stimuli were easy or difficult to discriminate. Therefore this alcohol effect was interpreted as affecting visual spatial attention rather than pattern recognition.

In conclusion, in *passive* tasks alcohol above 0.8 g/kg reduces VEP *amplitudes* before 200 ms; in *active* tasks possibly even at lower doses. One of the specific visual processes affected may be motion perception. Whereas VEP *latency* increases are seen at 0.8 g/kg in passive tasks, higher doses appear to be necessary in active tasks. N_{200} amplitude appears to show sometimes increases rather than decreases, whereas a N_{200} latency increase seems to specifically relate to the underlying perceptual processes.

SOMATOSENSORY EVOKED POTENTIALS

Somatosensory evoked potentials comprise early components reflecting peripheral nerve and brain stem action (10–20 ms); short- and medium latency cortical components (20–100 ms), the vertex potential (100–200 ms) and an afterdischarge (200–500 ms) (Regan, 1989). These SEP components reflect sensory transmission from peripheral nerves, spinal cord, thalamus, and cortex (Desmedt, 1988). Therefore, SEPs might provide a similar, if not better, opportunity to investigate alcohol effect on sensory information processing as do AEPs.

Passive tasks

To our knowledge there are only SEP studies with passive tasks. Salamy and Williams (1973) reported conduction velocity in the medianus nerve to be unaffected even at Blood Alcohol Concentrations (BACs) of up to 1.45 g/l. Components originating in thalamus, primary and secondary cortex seem to be only affected when BAC is high or when it is maintained over several hours. Thus Salamy (1973) and Salamy and Williams (1973) found no alcohol effects on SEP components recorded from the primary somatosensory cortex between 12 and 30 ms unless BAC was raised to 1.5 g/l. On the other hand, Porjesz and Begleiter (1973b) reported decreases of the P20 (primary cortex) and of a negative component between 60 and 80 ms when BACs were maintained at 0.5 to 0.6 g/l for three hours by giving repeated drinks. We should also mention a finding by Seppäläinen *et al.* (1981) who reported an *increase* of an *early* (N_{42}–P_{65}) SEP component, specifically at a low but not at a higher dose of alcohol.

As regards alcohol effects on *late* SEP components, Lewis *et al.* (1970) found their N_{140} and P_{250} to be reduced in amplitude at 1.23 g/kg, but no effect was discernible at 0.41 g/kg. Salamy and Williams (1973) traced the effects of BACs ascending to about 1 g/l and then descending. They reported a clear inverse relationship of BAC level and N_1, P_2, and N_2 amplitudes. Interestingly, a given BAC level had the same effect regardless of whether measurement occurred during its rising or falling slope, weakening suggestions that the direction of the BAC slope over time may be more important than BAC level.

Only two studies appear to have considered SEP *latencies*. Seppäläinen *et al.* (1981) reported an increase of N_{27} latency but only at a low dose of 0.4 and not at 0.8 g/kg. Salamy and Williams (1973) found no effect on N_{100} or P_{200} latency at BACs of up to 1.0 g/l but reported an increase of N_{100} latency variability.

In sum, it appears that peripheral somatosensory conduction and also the first stages of cortical information processing are fairly robust against alcohol, whereas the later SEP components seem to be much more susceptible.

P_{300} AND SLOW POTENTIALS

P_{300}

The best investigated endogenous component in the context of alcohol effects is the P_{300} or P_3b, an electrically positive-going deflection around 300 or more milliseconds after task-relevant stimuli. P_{300} amplitude is sensitive to the subjective probability of the eliciting stimulus and to the amount of perceptual resources allocated to the stimulus (e.g., Johnson, 1988). P_{300} peak latency has been shown to increase with the time required to 'evaluate' the stimulus whereas it is much less sensitive to response-related processing (e.g., Kutas *et al.*, 1977; McCarthy and Donchin, 1981).

P_{300} *latency*

The majority of studies indicates that P_{300} latency is increased by alcohol (Campbell and Lowick 1987; Colrain *et al.*, 1993; Daruna *et al.*, 1987; Fowler and Adams, 1993; Grillon *et al.*, 1994; Krein *et al.*, 1987; Pfefferbaum *et al.*, 1980; Rohrbaugh *et al.*, 1987; Schuckit *et al.*, 1988; Taghavy *et al.*, 1991; Teo and Ferguson, 1986; Yamamoto and Saito 1987). These delays occurred at alcohol doses as low as 0.5 g/kg. In several other cases a P_{300} increase did not quite reach significance (Campbell *et al.*, 1984; Sommer *et al.*, 1993) or it was only present for old and not for young subjects when a low dose of alcohol was given (Noldy *et al.*, 1994b). Exceptional studies which did not observe an alcohol effect on P_{300} latency are those of Elmasian *et al.* (1982), Genkina (1984), Hamon and Camara (1991), Kostandov *et al.* (1982) and Roth *et al.* (1977). Possibly, the negative results of Hamon and Camara (1991) and Roth *et al.* (1977) relate to their low processing demands because the P3-eliciting event was merely a warning stimulus.

Only some of the studies above have directly compared ERP measures with performance. Pfefferbaum *et al.* (1980) investigated subjects at placebo and two moderate doses of alcohol while they responded to rare tones with button presses. Whereas there was no significant effect of alcohol on RTs, the P_{300} component was delayed by both doses of alcohol by about 20 ms. Elmasian *et al.* (1982) required reactions to rare tone bursts immediately before, immediately following and 30 min after consummation of a drink. P_{300} latency was not affected by alcohol nor was RT. Rohrbaugh *et al.* (1987) required the detection of visually degraded signal digits in a vigilance task. Detection performance deteriorated and RT increased as a function of alcohol dosage which was paralleled by a delay in P_{300} latency. Campbell and Lowick (1987) required responses to rare tones when presented to one ear whereas rare tones presented to the other ear were task-irrelevant. Alcohol significantly delayed P_{300} latency for both the distractor and target by about 16 ms, whereas the 20-ms increase in RT was not significant.

Thus the magnitude of the alcohol effect on P_{300} latency is often as large as that on RT; sometimes it may be even stronger. For example, at the high dose level of the vigilance study of Rohrbaugh *et al.* (1987) the alcohol effect on RTs was 35 ms, whereas it was 105 ms on P_{300} latency. From the 11 conditions available with values both for RTs and P_{300} latency (Campbell and Lowick, 1987; Colrain *et al.*, 1993; Fowler and Adams, 1993; Rohrbaugh *et al.*, 1987; Sommer *et al.*, 1993) we calculated a correlation between the alcohol effects (alcohol minus placebo) on RT and on P_{300} latency of r = 0.55.

Following the common interpretation of P_{300} latency one may suggest that alcohol causes a slowing of stimulus evaluation. Interestingly, the data of Rohrbaugh *et al.* (1987) and Colrain *et al.* (1993) suggest, that the delay in P_{300} latency may be only secondary to similar delays in even earlier processing stages. In both studies N_{200} latency was affected by alcohol to a similar degree as P_{300} and also RT. As suggested above, at least in these cases, alcohol may have had an effect on pattern recognition, target detection, or orienting visual attention but not on earlier or later stages.

P_{300} amplitude

As regards P_{300} amplitude, we disregard those studies which used only peak-to-peak measures (Peeke *et al.*, 1980; Taghavy *et al.*, 1991; Teo and Ferguson, 1986; Yamamoto and Saito, 1987) because, obviously, it is impossible to disentangle effects on P_{300} from those on the 'reference peak'. The majority of studies on P_{300} used variants of the oddball task, where two or more classes of stimuli are presented in random order, one class being relatively rare.

Above alcohol doses of about 0.5 g/kg, P_{300} amplitude decrements are commonly reported (Campbell and Lowick, 1987; Campbell *et al.*, 1981; Campbell *et al.*, 1984; Colrain *et al.*, 1993; Elmasian *et al.*, 1982; Genkina, 1984; Hamon and Camara, 1991; Kostandov *et al.*, 1982; Lukas *et al.*, 1990; Roth *et al.*, 1977; Rohrbaugh *et al.*, 1987). However, there are also some noteworthy exceptions at doses of up to 1.2 g/kg (Campbell and Lowick, 1987; Campbell *et al.*, 1984; Fowler and Adams, 1993; Pfefferbaum, 1980; Sommer *et al.*, 1993). A common characteristic of these studies may be their relatively high task demands over a short period of time. This clearly holds true for the 20 min duration choice RT-task of Sommer *et al.* (1993) and the 5 minutes name categorization task of Fowler and Adams (1993). Both Pfefferbaum (1980) and Campbell and Lovick (1987) required subjects to respond for 7.5 and 13 minutes respectively to one of three possible tones and found P_{300} amplitude to the targets to be unaffected but P_{300} amplitude to rare nontargets was decreased at least as a trend in both studies. Task difficulty was directly manipulated by Campbell *et al.* (1984) as described above. P_{300} amplitude was decreased in their easy but not in the difficult condition.

Because P_{300} amplitude is taken to indicate the amount of perceptual resources required by the processing of a particular stimulus one would conclude from the common amplitude decrements that alcohol tends to diminish these resources. Although more direct evidence is clearly necessary it may be suggested that the stability of P_{300} amplitude in certain demanding tasks of short duration reflects the ability of subjects to compensate for alcohol effects by recruiting additional processing resources. In our visual discrimination task study (Sommer *et al.*,

1993) the P_{300} was numerically even somewhat larger in the alcohol group (mean BAC: 0.69 g/l) than in the placebo group. The effect became highly significant for a late occipital Slow Wave with higher amplitudes by almost 5 μV in the alcohol group. Because Slow Waves have been related to processing demands (Rösler and Heil, 1991; Ruchkin *et al.*, 1988), their increase in our alcohol group who performed as well as the placebo group may reflect the high processing demands for the intoxicated subjects. Therefore P_{300} and Slow Wave components may be excellent indicators of compensatory efforts which are difficult to observe with overt measures of behaviour.

Lateralized Readiness Potential

Interestingly, one of the longest-known endogenous ERP components, the readiness potential (RP) which appears prior to voluntary actions, has never been studied under alcohol. A more recent development in ERP research is the focus on the so-called lateralized readiness potential (LRP) in bimanual choice tasks, (Coles, 1989). The LRP provides a measure of relative response preparation (preparation of the left hand to respond with versus the right hand and vice versa). The LRP onset is thought to reflect the point in time when the responding hand is selected or activated.

In the study of Sommer *et al.* (1993) the LRP revealed covert alcohol effects which were absent in overt behaviour. In a variant of the Simon task (for example, Simon, 1990), we required choice responses to geometric shapes that were either easy or hard to discriminate and presented to the left or right point of fixation. No behavioural effects of alcohol on RTs or error rates were observed, possibly due to strategic adjustments of the subjects. The LRP showed an early component indicating automatic stimulus location-dependent response activation and a late component controlled by stimulus identity. Alcohol decreased the amplitudes of both the initial and the late LRP activation components. This was seen as an indication of alcohol effects on both automatic and controlled response activation. Interestingly, alcohol left the onsets of the two LRP components unaffected. Because there was a strong trend for an alcohol-related delay in P_{300} latency, we concluded that alcohol leaves the initial flow of perceptual evidence to motor stages unimpaired whereas it appears to increase the duration of stimulus evaluation.

Contingent Negative Variation

One of the best-investigated slow brain potentials is the contingent negative variation (CNV). It is an ERP shift of negative polarity that develops between a warning signal (S_1) and a subsequent stimulus (S_2) that may require a response or deliver information (e.g., McCallum, 1988). The CNV is often seen to reflect expectation of S_2 or preparation for the required response. Alcohol effects on the CNV are surprisingly unclear. The earliest study of CNV and alcohol (Kopell *et al.*, 1972) used an amplitude measure that confounded CNV with P_{300}. CNV amplitude decreases reported by two other studies cannot be clearly attributed to alcohol because no control groups were used (Beaumanoir *et al.*, 1974; Hamon and Camara, 1991). Better controlled studies could not find alcohol effects at doses of up to 0.76 g/kg (Kopell *et al.*, 1978; Roth *et al.*, 1977; Yamamoto and Saito, 1987). However, because these studies did not demonstrate any performance effects either, their constant foreperiod RT tasks might have been too undemanding. Peeke *et al.* (1980), who did observe an alcohol effect on RTs only found a marginal CNV amplitude decrease.

In sum, both P_{300} latency and amplitude are adversely affected by even moderate doses of alcohol, indicating slowing of stimulus evaluation and depletion of perceptual resources. However, both P_{300} amplitude and subsequent Slow Wave amplitudes indicate that detrimental effects of alcohol can be compensated under certain conditions.

CONCLUSIONS

ERPs indicate clear alcohol-induced effects on the intensity and dynamics of perceptual processes. In both the visual and auditory modality, alcohol reduces the *amplitudes* of the vertex potential. Also, alcohol depletes perceptual resources as indicated by P_{300} amplitude reductions. Alcohol slows visual pattern recognition, stimulus classification, or orienting of visual attention as reflected in N_{200} *latency* increases. Stimulus evaluation as manifested in P_{300} latency is usually slowed, but, at least in the visual modality, this apparently occurs as a consequence of the effects already reflected in N_{200} latency. On the basis of behavioural studies alone, the existence of alcohol effects on perception is much more controversial (cf., Sommer *et al.*, 1993). As underscored by the moderate correlation of RT and P_3 latency, however, only part of the variance in RT can be attributed to perceptual factors.

In addition, ERPs indicate that the alcohol effects on perceptual processes may be modality-specific. Sensory transmission in the auditory modality is affected by alcohol already at the brainstem level but somatosensory transmission appears to be quite robust. Whereas latencies of VEP vertex potentials are increased by high doses of alcohol, this has not been observed for AEPs.

In *passive* tasks amplitude reductions are observed already at moderate doses for AEPs, but only at higher doses for VEPs. During *active* tasks, vulnerability for alcohol decreases in AEPs but increases in VEPs. Under increased task demands, the adverse effects of alcohol can be at least temporarily compensated by the recruitment of additional resources. Evidence for such possible compensatory effects comes from the auditory N_1 in active tasks, P_{300} amplitude and Slow Wave in short, demanding tasks.

– Evidently, ERPs may provide valuable contributions to the understanding of alcohol effects on information processing. However, as yet the great potential of ERP research has hardly been touched. Improvements might come from several sides. With the exception of very early ERPs, results from studies without manipulation of experimental factors taken to selectively influence specific cognitive stages or subprocesses and overt performance measures are exceedingly difficult to interpret on a functional level. Therefore, clearly defined and verified tasks should be elementary in cognitively oriented ERP research. However, where such tasks have been employed, these have usually been very simple or general, such as counting or simple response tasks. Therefore, the wealth of specific and well-investigated tasks in cognitive psychology might be advantageously combined with ERP recordings and alcohol.

– On the other hand, there are numerous ERP manifestations of specific cognitive subprocesses (cf., Hillyard and Picton, 1987). In the context of alcohol research few of these components have been studied in experimental set-ups designed for their elicitation. For example, possible alcohol effects on selective attention underlying the N_1 diminution could be tested by recording the processing negativity during an appropriate task (Näätänen, 1992). Recording of the MMN (Näätänen, 1992) might provide information whether also the automatic detection of deviant stimuli is affected (cf., Grillon *et al.*, 1994). In addition, by investigating visual selective attention (Mangun, 1995), one might assess the modality specificity of such findings. Such specific approaches might be combined with the investigation of other important variables for the effects of alcohol which have rarely been looked at, such as age (cf., Noldy *et al.*, 1994b) or gender.

– In addition, multi-channel analyses of ERP topography that may allow to localize or differentiate neuroanatomical and functional subsystems (e.g., Mangun, 1995) are grossly underrepresented in the field of alcohol research. These techniques might also elucidate controversial issues in previous research. Mainly on the basis of VEPs to flash stimulation in passive tasks it has been suggested that ERPs recorded over primary (occipital) areas are less sensitive to alcohol than those over non primary areas (Lewis *et al.*, 1970; Obitz *et al.*, 1977; Rhodes *et al.*, 1975). Similar results were also found in an active task by Rohrbaugh *et al.* (1987). However, Neill *et al.* (1990), Sommer *et al.* (1993) and -in a passive condition Porjesz and Begleiter (1973a) – demonstrated clear decrements of VEPs also above primary areas. Modern methods of investigating ERP topographies and relating them to possible generator systems might clarify this issue.

A similar approach might be taken to the suggestion that alcohol has a (leveling) effect on hemispheric asymmetries. A number of studies with flash stimulation found that without alcohol VEPs were larger over the right than over the left hemisphere (Erwin and Linnoila, 1981; Lewis *et al.*, 1970; Pollock *et al.*, 1988; Porjesz and Begleiter, 1973a; Rhodes *et al.*, 1975; Simpson *et al.*, 1981); and often, these asymmetries decreased with rising BAC. However, there is also evidence to the contrary (Neill *et al.*, 1990; Zuzewicz, 1981). Topographical mapping, Brain Electric Source Analysis or magnetic field recording might clarify the eliciting conditions and underlying generators of these ERP asymmetries. Conclusions from ERP asymmetries to specific hemispheric functions could be considerably strengthened by including the assessment of the lateralization of performance (cf., Schweinberger *et al.*, 1994).

SUMMARY

This review surveyed the available reports on alcohol effects on event-related potentials (ERP) in social drinkers as possible indicators of alterations in information processing. In general, alcohol appears to slow down and degrade perceptual processes and deplete perceptual resources. However, these effects are notably modality-specific and may be compensated under increased task demands. In many cases alcohol-induced alterations appear to be more directly accessible in ERPs than in overt performance. Suggestions are made, how the utilization of ERPs for the understanding of alcohol effects might be improved.

CHAPTER FIFTEEN

Acute and Chronic Effects of Social Drinking: Event Related Potentials and Quantified EEG

Nancy E. NOLDY

INTRODUCTION

Cognitive event-related potentials (ERPs) and quantified EEG (qEEG) can provide sensitive measures of alcohol-related changes in central nervous system functioning. Digitization of the EEG permits reduction of data gathered from several channels over time, as well as Fourier, topographical and statistical analyses of the effects of drugs on human brain electrical activity. ERPs are real-time indices of information processing in the brain, and are sensitive to the subtle cognitive effects of psychotropic drugs. These techniques would seem to be extremely useful in measuring the acute and chronic effects of social drinking. However, there has been little direct investigation of social drinking using ERPs or qEEG. We have knowledge from several indirect sources regarding the effects of social drinking on electrophysiological measures: studies of the acute effects of alcohol performed on social drinkers; studies examining the effects of chronic alcoholism in which social drinkers are used as controls, and ERP and qEEG studies of the hereditary aspects of alcohol intoxication among social drinkers. This chapter will draw from studies in each of these areas.

ELECTROPHYSIOLOGICAL TECHNIQUES

Quantified EEG (qEEG)

Traditional EEG methods are criticized for their subjective interpretation and time-consuming evaluations. The availability of computerized qEEG technology, and its sensitivity to drug effects has inspired interest in 'pharmaco-EEG'. Topographical mapping of multichannel recordings or 'brain mapping' effectively reduces the data and simplifies presentation. However, these techniques are also criticized. For example, analysis of frequency domains obscures transient abnormalities. In addition, statistical analysis generally involves hundreds or thousands of variables, and relatively few subjects, which may result in statistically significant results by chance alone. An additional consideration in assessing drug effects is that pharmacologically-induced alterations in the subject's state, such as changes in mood or drowsiness may themselves cause EEG changes and artifacts. Precautions such as basing investigations on hypotheses that are developed from strong neurophysiological evidence, performing a limited number of tests, rationalizing the confidence interval using a technique such as the Bonferroni procedure, replicating results, and measuring or monitoring subjective state, are therefore essential. With these precautions in mind, the qEEG may be useful in examining functional deficits associated with acute and chronic use of such drugs as cocaine (Noldy *et al.*, 1994a) and alcohol (Lukas *et al.*, 1989; Noldy and Carlen, 1990).

Event-related Potentials (ERPs)

Digitization of the resting EEG can provide quantitative information about brain functions in a global manner. The ERP accesses more specific brain functions that occur in response to physical stimuli or psychological events which occur during performance of a task. This electrical potential, which is time-locked to the stimulus presentation, is extracted from ongoing EEG activity using signal averaging procedures. As the number of stimulus presentations increases, the random EEG noise tends to cancel itself out, while the signal remains. Components of the ERP reflect the timing and quality of information processing. In general, the earlier the wave, the more likely it is generated by stimulus or sensory parameters such as intensity, frequency, presentation rate or sensory modality. The longer the latency, the greater the dependence on 'endogenous' or cognitive factors that do not directly relate to physical stimulus properties. These factors include such cognitive processes as attention, decision making and signal value. These later ERPs are useful for accessing mental events that are not amenable to behavioural measurement, and have proven sensitive to drug-induced changes in CNS functioning, including the acute and chronic effects of alcohol (see Noldy and Carlen, 1990).

ERP studies of the cognitive effects of ethanol have primarily focused on two parts of the waveform: components in the N_1/P_2 range and P_{300}. Several different negative waveforms in the 80–200 ms range reflect both physical qualities of the stimulus such as intensity and modality, and psychological factors including attention and mismatch detection (Alcaini *et al.*, 1994; Näätänen and Picton, 1987). The P_{300} component is most commonly recorded in a traditional auditory oddball task. In this task, the subject counts or indicates by button press infrequently occurring 'target' tones that differ from the more frequent 'standard' tones on some dimension (e.g., frequency). In addition to the N_1 and P_2 components that are present for both target and standard, the waveform associated with the target tones contains a large positive component, occurring at approximately 300 ms in young adults. This component is believed to reflect short-term working memory processes (Donchin, 1981). Its latency is believed to reflect stimulus evaluation time (Kutas *et al.*, 1977), and its amplitude is associated with a number of factors including allocation of processing resources, motivation and stimulus value (see reviews by Johnson, 1986; Pritchard, 1981). Since alcohol intoxication affects cognitive functions such as attention, speed of information processing and memory, these techniques may be useful indices of cognitive impairment resulting from social amounts of alcohol.

HUMAN ELECTROPHYSIOLOGY AND SOCIAL DRINKING

Acute Effects of Alcohol in Social Drinkers

qEEG Studies

In general, studies examining the effects of alcohol on the resting EEG in social drinkers report larger amplitude and slower frequency of the dominant rhythm (Davis *et al.*, 1941; Ekman *et al.*, 1964; Lukas *et al.*, 1986). However, there is wide variation in acute response to alcohol (Lehtinen *et al.*, 1978; 1985). Recent qEEG data indicate that low doses may produce predictable alterations in theta and slow alpha activity, while individual differences in the faster frequencies may be related to drinking history and EEG characteristics that precede drug administration (Ehlers *et al.*, 1989).

Lukas *et al.* (1989) measured qEEG in six female social drinkers who reported drinking 1–3 glasses of wine per week. Subjects reported their subjective impressions of intoxication and euphoria/dysphoria using a joystick during the recording. Post-ethanol (0.7 g/kg) recordings revealed the expected increased amplitude in slow and fast frequency alpha. Topographic mapping of alpha activity revealed a wider bilateral spread of alpha activity from the pre-drug occipital distribution to include the parietal, temporal and frontal areas. There was also a significant correlation between reports of euphoria and the augmented alpha distribution, which occurred 30 minutes after ethanol. The authors suggested that the alpha increases over the entire scalp were associated with the reinforcing properties of alcohol. The relationship between alpha activity and euphoria is also related to EEG characteristics that precede ethanol intake. Men with high amounts of fast alpha activity before ethanol report feeling less intoxicated or 'high' after ethanol (Ehlers *et al.*, 1989).

ERP studies

Decreases in N_1–P_2 amplitude associated with acute alcohol intake have been described by a number of investigators (Rhodes *et al.*, 1975; Kopell *et al.*, 1978; Pfefferbaum *et al.*, 1980; Wolpaw and Penry, 1978). Amplitude reductions in this portion of the waveform could reflect either depressed sensory registration or reduced attentional allocation. There is some evidence that the acute effects of ethanol on this portion of the waveform can be overcome in tasks that are more difficult than the simple auditory oddball paradigm. Campbell *et al.* (1984) recorded ERPs during auditory oddball tasks of increasing difficulty after social drinkers ingested placebo, 0.5 and 1.0 ml/kg of ethanol. N_1 and P_2 were reduced in amplitude following alcohol ingestion, but only in the easy task. P_{300} amplitude was significantly attenuated only during the easy task especially after the stronger dose. Sommer *et al.* (1993) also found alcohol-related decrements in early ERPs, but a strong enhancement of the late slow-wave in a choice reaction time task. Behaviourally, however, there were no increases in error rates or decreases in reaction times associated with alcohol use. The authors suggest that the enhanced slow-wave may reflect increased investment of processing resources to overcome the detrimental effects of alcohol. In a more complex task, Roth *et al.* (1977) recorded ERPs during a Sternberg memory retrieval paradigm in order to measure rate of scanning in short-term memory. N_1 amplitude and latency were not affected by the 1.0 ml/kg ethanol dose. If the effects of alcohol on the N_1/P_2 portion of the waveform reflected depressed sensory registration, it would be expected that N_1/P_2 amplitude in difficult tasks would be reduced at least as much as in easy task. It is therefore reasonable to suspect that the effects of alcohol are not a purely sensory phenomenon and may reflect a change in attentional allocation.

In general, alcohol tends to produce decreased amplitude and increased latency of P_{300} in the traditional oddball paradigm. Among social drinkers, Teo and Ferguson (1986) found reduced P_{300} amplitude and a dose-dependent increase in P_{300} latency for a 0.3 g/kg and a 0.54 g/kg dose of ethanol. This study also reported an increase in latency of the earlier N_1, P_2 and N_2 components for both the standard and target tones after the stronger dose. N_1–P_2 peak-to-peak amplitude also decreased only in response to the stronger dose. The authors concluded that attention (as indexed by N_1–P_2 amplitude) was only affected by the stronger dose, while stimulus categorization (as indexed by P_{300} latency) was affected by both doses. Campbell *et al* (1984) reported decreases in P_{300} amplitude for their 1.0 ml/kg dose of 94% ethanol associated with an easy, but not a more difficult oddball task. The 0.5 ml/kg dose appeared to be associated with a decrease in P_{300} amplitude in both the easy and difficult conditions. By contrast, with a similar 1.0 ml dose of 95% ethanol P_{300} amplitude was reduced in a Sternberg memory retrieval task (Roth *et al.*, 1977) even though no such decrement was found in the earlier components. In a more direct investigation of attentional allocation, Lukas *et al.* (1990) combined the traditional auditory oddball paradigm with a divided attention task. The effects of acute alcohol (0.70 g/kg) in four male social drinkers were assessed. Interestingly, the effects of the divided attention task on P_{300} were very similar to the effects of alcohol. That is, reduced amplitude, delayed latency and an apparent shift in topography from parietal maxima to more widespread activity in posterior and inferior directions. These data may indicate that alcohol effectively slows stimulus evaluation processes and fewer processing resources are allocated to the task.

Other investigators have also found changes in P_{300} topography associated with acute ethanol use in social drinkers. Daruna *et al.* (1987) examined the acute effects of ethanol in social drinkers in an auditory oddball paradigm. They found that N_1 amplitude was reduced by ethanol and P_{300} latency was increased. In addition, P_{300} topography changed after the ethanol challenge. Ethanol reduced the difference in amplitude between parietal and frontal sites. That is, compared to the placebo condition, P_{300} was larger frontally, and somewhat smaller parietally after alcohol ingestion. This was particularly evident at the left-hemisphere locations. In another study (Lukas *et al.*, 1990) the generator of the P_{300}, as estimated by topography using a computerized single source 3 shell model (brain, skull, scalp) indicated a change in activity from centro-parietal to more frontal and occipital associated with a 0.70 g/kg dose.

Preliminary data from our laboratory suggest that there may be an effect of age on ethanol effects among social drinkers (Noldy *et al.*, 1994b). We investigated the effects of ethanol on P_{300} and behavioural measures in a group of young (19–29 year old) and elderly (60–70 year old) men who were social drinkers. In this case, "social drinking" was defined as drinking no more than two drinks per day, or no more than ten drinks per week. In addition to the traditional auditory oddball task, a computerized tracking task was used to assess psychomotor functioning and a paired-associated word task was used to assess long-term memory functioning. Subjects received placebo on one day, and the other day they received several alcohol beverages such that the blood-alcohol content of both young and elderly subjects was maintained between 40 and 50 mg/100 litres (0.4 and 0.5‰) for two hours. Alcohol impaired performance on the psychomotor tracking task 2 hours after their first drink. The effects on long-term memory were more rapid, and impaired performance occurred 1 and 2 hours after the first drink for both groups. In the elderly group alone, alcohol caused a significant delay in the latency of P_{300} one hour after drinking and the topography of P_{300} became maximal frontally 1 and 2 hours after the first drink. Thus, acute intoxication seems to affect psychomotor and long-term memory abilities in similar ways for young and elderly individuals. However, in elderly male social drinkers, alcohol causes delays in P_{300} and topography changes generally associated with senescence and dementia. Alcohol may uncover or exacerbate the cognitive effects of aging in older social drinkers.

Thus, there appear to be differences in qEEG and P_{300} associated with drinking history. Moderate social drinkers appear to differ both in terms of qEEG and P_{300} from mild social drinkers even before alcohol is ingested. The source of this chronic difference has yet to be determined. Acute lectrophysiological effects of ethanol in social drinkers appear to be affected by a number of factors including personal drinking history and age. ERP data indicate that acute social doses of ethanol cause deficits in attention in easy tasks, as reflected in N_1 amplitude. With more challenging tasks it seems possible for social drinkers to counteract the ethanol-induced N_1 decrements and perhaps the P_{300} amplitude and latency effects observed in simpler tasks. However, the P_3 amplitude and latency effects are more variable. The variability in P_3 results may be attributed to many methodological and procedural differences. There is little consistency in the definition of 'social drinker'. This category seems to include all but abstainers and diagnosed chronic alcoholics. There are also differences in the definition of 'social drinkers' in each of these studies, differences in amount and procedures of alcohol ingestion, and/or methodological differences in tasks and ERP recording procedures.

For the most part, these data seem to support an attention allocation model in which acute social amounts of alcohol affect the direction in which attention is focused (Steele and Josephs, 1990). That is, alcohol reduces the amount to total processing resources available, but does not affect the ability to focus attention. Therefore if the subject is motivated and when the task requires attention, the available resources can be focused. This model would predict a decrement in performance if the number of environmental stimuli requiring attention increased (as in a divided attention task) and/or if the primary task increased in complexity to the extent that the attentional resources remaining after ethanol intake were insufficient.

Chronic Effects of Alcohol in Social Drinkers

qEEG Studies

In a comparison of social drinkers consuming moderate amounts of alcohol (average number of drinks per occasion multiplied by number of days drinking per month ≥ 40) or lower amounts (≤ 39), Ehlers *et al.* (1989) found qEEG differences in the beta range. The moderate drinkers had more power in the beta range at posterior sites both at baseline and 90 minutes after ethanol. While it is impossible to determine from this study if these electrophysiological differences predated the alcohol use, the results suggest that chronic social use of alcohol is associated with long-term electrophysiological changes.

ERP studies

In an ERP study comparing P_{300} in heavy and light social drinkers (Nichols and Martin, 1993), P_{300} was recorded during a simulated driving task in an oddball paradigm. Heavy social drinkers had longer P_{300} latencies than did low social drinkers regardless of drug condition. A correlation between P_{300} latency increases and drinking history has been reported in some investigations (Neville and Schmidt, A., 1985; Polich, 1984; Polich and Bloom, 1986) but not in others (Polich and Bloom, 1987; Polich *et al.*, 1988b). Whether this latency difference precedes or is induced by alcohol abuse is unknown.

Acute Effects of Alcohol in Social Drinkers Compared to Alcoholics

In the previous section, the effects of alcohol on social drinkers were compared to other social drinkers or themselves. It is also of interest to compare the effects of social drinking with chronic ingestion of larger amounts of alcohol.

qEEG studies

Several studies have used qEEG to demonstrate an association between EEG and cognition in alcoholics. Zilm *et al.* (1980) reported that decreased alpha was apparent in alcoholics, especially in those with greater neurological impairment. During recovery, an increase in alpha power was observed in those who demonstrated neurological improvement. Coger *et al.* (1978) found that among alcoholics, augmented fast activity was associated with greater cognitive impairment. Kaplan *et al.* (1985) reported that alcoholics demonstrated more delta and less alpha power across the scalp than did nonalcoholic social drinkers. They also found that both recency of last drink and use of detoxification medication were associated with increased beta power. A general measure of cognitive status was positively correlated with alpha power and negatively correlated with power in the delta range. Future research in this area should examine the effects of different levels of social drinking on these measures. For example, the effects of one drink per day, versus three drinks per day, and comparing this regular pattern of consumption to the weekend binger who drinks the same amount but all on one or two occasions. Variables such as sex and age may also interact with these drinking styles.

ERP studies

Pfefferbaum and associates (1980) compared ERPs in an auditory oddball paradigm in alcoholics and social drinkers. N_1 and P_2 amplitude were reduced by ethanol acutely (0.5 ml/kg and 1.0 ml/kg or 0.4 g/kg 0.8 g/kg respectively), but were not lower in the chronic alcoholics when they were not intoxicated. P_{300} latency was increased by alcohol intoxication, and was also longer in the alcoholics *before* ingesting the alcohol. While P_{300} amplitude was not affected by alcohol ingestion in the chronic alcoholics, P_{300} amplitude did seem to decrease in the non-alcoholics. Thus, N_1 and P_2 appear to be more affected by acute intoxication, while P_{300} latency may be affected by chronic exposure (also see Sandman *et al.*, 1987). By contrast, Turkkan *et al.* (1988) compared autonomic responses (heart rate and blood pressure) of alcoholics and social drinkers associated with acute ethanol ingestion. In general, the two groups did not differ in these physiological responses to alcohol and placebo, although the alcoholic subjects tended to report some tolerance to the euphoric effects of alcohol.

Interpretation of electrophysiological studies of alcoholism is inherently plagued by issues such as length of abstinence, use of psychotropic medication in detoxification, reported extent and recency of alcohol use. However, these studies are interesting in terms of the implications they have for social drinking. There appear to be long-term electrophysiological changes accompanying chronic alcoholism. However, it is as yet unknown at what level and what duration of alcohol intake these changes begin to appear.

FAMILY HISTORY OF ALCOHOLISM

Among social drinkers are those who have a family history of alcoholism (Family History Positive; FHP) and may be at greater risk for developing alcoholism. This group may also respond to the acute effects of alcohol in a different way than those who have no family history (FHN). A large proportion of the current literature on ERPs and alcohol has focused in this area. Alterations in cognitive functioning and electrophysiological activity associated with alcoholism may precede the development of alcoholism. In individuals who may be genetically predisposed, a measure that is sensitive to these cognitive changes may act as a vulnerability marker. Among FHP subjects, two types of populations have been sampled: children who have no drinking experience, and; adult sons of alcoholic fathers who are social drinkers. Family history of alcoholism is an individual difference among social drinkers which may have important implications for the way they are affected by alcohol.

qEEG studies

Gabrielli c.s. (1982) reported that EEGs of children of alcoholics (11–13 years of age) contain more relative beta activity than children without an alcoholic parent. However, in another study of youths aged 19–21 no baseline differences were found (Pollock *et al.*, 1983). After ethanol, however, the FHP youths exhibited greater decrements in slow alpha and greater enhancement of fast alpha than FHN matched controls. FHP sons of alcoholics also report less euphoria and fewer psychomotor performance deficits (Schuckit, 1980). In a ten-year follow-up, Schuckit (1991) found that the men who became dependent on alcohol had had less intense reactions to alcohol during the previous assessment.

ERP studies

In 1984, Begleiter and his colleagues recorded ERPs in two groups: 25 sons of alcoholic fathers aged 7 to 13 who had never been exposed to alcohol, and an age-matched normal control group. Subjects performed a visual task in which they discriminated between a plain oval (non-target, frequently presented stimulus), and an aerial view of the head with one 'ear' missing (targets). The heads were presented in four different orientations (target stimuli). Subjects were asked to press a button as quickly as possible to indicate whether the right ear or left ear was present. There were no differences in P_{300} latency between groups. The raw data were subjected to principal component analysis (PCA) with varimax rotation. Only the PCA results were described. A factor that was maximum at Pz and peaked at 332.5 ms appeared to reflect P_{300}. This factor discriminated between the two groups. Boys with alcoholic fathers exhibited smaller P_{300} amplitudes. The same paradigm was later used by O'Connor *et al.* (1987) with 21 to 28 year old drinkers who were biological sons of alcoholic fathers and non-alcoholic mothers (n = 24). This group drank, on average, approximately 15 g ethanol per day. Again, a positive family history was associated by reduced P_3 amplitudes, although the reduction was not as great in this study as that reported by Begleiter and his colleagues. The smaller difference between groups in this study could possibly be attributed to 1. differences in the age of the subjects or 2. the fact that O'Conner's subjects were moderate drinkers while Begleiter's had no experience with alcohol or 3. filter setting or other technical differences.

There has been little consistency in results where other tasks are used. Some studies have found latency changes, but not amplitude differences associated with positive family history (Whipple and Noble, 1986; Hill *et al.*, 1988), while others report no differences associated with family history (Polich *et al.*, 1988a; Polich *et al.*, 1988b; Polich and Bloom, 1988). These studies differ in modalities of stimulus presentation and task requirements. Hill *et al.* (1988) examined 168 FHP and FHN adult male siblings and parents. In a simple auditory oddball task and a choice reaction time task all of the brothers of alcoholic probands had longer P_{300} latency than controls. Unaffected (social drinking) siblings had shorter P_{300} latencies than affected siblings or parents, especially in the choice reaction time task. Again, it is impossible to say whether this is a result of the years of drinking of the affected siblings or a vulnerability which predates the alcohol use. The authors suggest that the unaffected sib may be protected from developing alcoholism by his increased efficiency of information processing. Polich *et al.* (1988a) reported no P_{300} differences between a group of FHP male and female students and a FHN group matched in age, height, weight, grade point average and personal drinking history. P_{300} was recorded in three auditory tasks, differing in discrimination difficulty. P_{300} amplitude decreases, however, were associated with increases in the amount of self-reported alcohol consumed for the FHP group, and only in the most difficult task situation. Polich *et al.* (1988b) also found no differences between FHP and FHN groups in P_{300} amplitude or latency in a visual discrimination task.

In a study of 21 FHP and FHN males matched on drinking history and demographics, Schuckit *et al.* (1988) found no baseline differences in the latency of P_{300} elicited in an auditory oddball task. P_{300} latency increased after 0.75 ml/kg (0.6 g/kg) and 1.1 ml/kg (0.9 g/kg) of alcohol ingestion for both groups. However, P_{300} latency in the FHP subjects returned to baseline more rapidly compared to the FHN group. Therefore, the P_{300} effects recovered more rapidly for the FHP than for the FHN subjects. Elmasian *et al.* (1982) found that FHP individuals exhibited smaller P_{300} amplitude and longer P_{300} latency compared to FHN subjects after both placebo and ethanol (0.56 and 0.94 g/kg).

Differences in P_{300} amplitude and in earlier ERPs components have also been reported in comparisons of social drinkers and alcoholics (Patterson *et al.*, 1987, Pfefferbaum *et al.*, 1991). In a relatively complex task, Patterson *et al.* (1987) presented red and green circles and 1500 Hz and 1000 Hz tones to subjects (alcoholics and social drinkers, who were either FHP or FHN). They were asked to count only one of the four stimuli, and ignore the others. The FHP alcoholics tended to have smaller visual and auditory P_{300} amplitudes than social drinkers, while there was no significant difference between the social drinkers regardless of family history and the FHN alcoholics. There were no significant differences between groups in P_{300} latency. Family history appeared to be more important than whether the subject was an alcoholic or social drinker in determining visual N_1 amplitude, but no significant differences in auditory N_1 amplitude were found. As a group, FHP subjects had a smaller amplitude visual N_1 than did FHN subjects. The effects of attention on the N_1 discriminated between alcoholics and social drinkers. The alcoholics did not exhibit as great an enhancement of N_1 to attended stimuli as did social drinkers. Pfefferbaum *et al.* (1991) also reported that FHP alcoholics have smaller amplitude P_{300}s than FHN alcoholics. This difference did not appear to be related to lifetime alcohol consumption.

CONCLUSIONS

There is some evidence that long-term neurophysiological changes are related to quantity of alcohol regularly consumed. These changes may be part of a continuum of alterations associated with mild to excessive chronic alcohol intake. The minimum amount of alcohol intake that leads to long-term changes needs to be defined. In addition, the ERP paradigms used in this research have tended to focus on attention and short-term information processing. The effects of social doses of alcohol on acquisition, consolidation, recognition and recall of information in longer-term stores can also usefully be measured with ERP techniques.

The wide variety of cognitive tasks with which this technology can be usefully combined is both its strength and weakness. It is difficult or impossible to interpret inconsistent results obtained in different paradigms. Differences in stimulus modality, presentation rate, filter settings, task difficulty and peak measurement techniques may contribute to the variability in ERP results described in this review. The number of different methodologies used renders conclusions based on the entire body of literature very difficult.

Social drinkers are a heterogeneous group, differing in dimensions such as age, sex family history, baseline electrophysiological characteristics and personal drinking history. All of these factors can reasonably be believed to have an impact on the brain's acute and chronic responses to ethanol. These factors may all contribute in complex ways to the results obtained in electrophysiological investigations, and need to be taken into consideration not only as confounding variables, but also as the focus of future investigation.

Despite the variation in sample selection and methodology, a few common threads run through this literature. Moderate amounts of alcohol (0.5 to 1.0 g/kg) in social drinkers is generally associated with slowing of the dominant rhythm, decreases in N_1 and P_{300} amplitudes in a variety of tasks. These effects are consistent with a model in which alcohol decreases information processing resources. The ability to focus attention in order to compensate for this reduction also is supported by evidence that more complex or engaging tasks eliminate the N_1 effect. The changes in P_{300} topography may indicate that different areas of the brain are recruited to perform a discrimination task when the system is challenged by alcohol.

SUMMARY

Cognitive event-related potentials (ERPs) and quantified EEG (qEEG) can provide sensitive measures of alcohol-related changes in CNS functioning. There has been little direct investigation of social drinking using these techniques. This chapter presents knowledge from studies of the acute effects of alcohol performed on social drinkers, studies examining the effects of chronic alcoholism in which social drinkers are used as controls, and studies of the hereditary aspects of alcohol intoxication among social drinkers. 'Social drinking' is not generally well-defined in these studies. Subjects vary within a range which includes all but abstainers and chronic alcoholics. Overall, there appear to be differences between moderate and mild social drinkers indicating long-term changes in qEEG characteristics and increased P_{300} latencies. Acute social doses, ranging from 0.3 to 1.0 ml/kg of ethanol have been reported to cause changes in N_1 and P_{300} associated with decreased attention, reduced processing resources and slowed information processing. However, the minimum amount of alcohol intake that leads to long-term changes has not been well investigated. Issues of subject selection, variability in methodologies, and the dearth of studies focused on social drinkers make the current state of this literature largely inconclusive, but offers a garden of opportunity for future research.

ACKNOWLEDGEMENTS

Supported in part by the Bloorview Epilepsy Program of the University of Toronto.

CHAPTER SIXTEEN

Selective Attention and Event Related Potentials in Young and Old Social Drinkers

Elly ZEEF, Jan SNEL and Bertie M. MARITZ

INTRODUCTION

Numerous reviews of neuropsychological studies have reported that compared with non-alcoholic peers, sober alcoholics perform at significantly lower levels on tests of abstracting, problem solving, new learning, memory, perceptual, and perceptual motor functions. Through complementary pneumo-encephalographic, electro-encephalograph and other neurological examinations, these deficits have likewise been associated with cortical structural changes and ventricular enlargement (Ron, 1983; Pfefferbaum *et al.*, 1990; Harper and Kril, 1991), fronto-limbic-diencephalic lesions and hypothalamic degeneration (Page and Cleveland, 1987; Ron, 1983; Harper and Kril, 1991). For example, Event Related Potential (ERP) studies report alcohol induced altered responses in different ERP-components such as N_1, N_2 and P_3, indicating impairment of brain functions that support selective attention to relevant information and that mediate perceptual discrimination (Parsons *et al*, 1990; Jääskeläinen, 1995).

Several hypotheses have been offered to account for these findings.

- The *'continuity'* hypothesis (Ryback, 1971) states that there is a single continuum of alcohol effects ranging from the profound neuropsychological impairment seen in Korsakoff patients, through the moderate deficits found in alcoholics, to effects on heavy, moderate and light social drinkers. Subtle and mild changes may be occurring in the cognitive functioning of social drinkers.
- Parker and Noble (1977) offered as their first hypothesis the *'alcohol-causal'* hypothesis, stating that alcohol consumption contributes to decrements in cognitive functioning.

  Their second hypothesis was that individuals with poorer cognitive functioning may drink more alcohol per occasion than individuals with better cognitive functioning; the *'cognitive-causal'* hypothesis. The third hypothesis was that variables such as life stress may lead to both increased drinking and decreased cognitive performance; the *'stress-emotional-causal'* hypothesis.

- Age is another issue in electrophysiological studies of alcohol brain dysfunction. Age related CNS changes are difficult to separate from CNS damage due to chronic alcohol consumption, because older alcoholics have spent more years drinking and because alcohol may interact with the aging process in such a way that older alcoholics are more susceptible to effects of alcohol. Such parallels have led to the formulation of the *'premature-aging'* hypothesis (e.g., Ryan, 1982), which suggests that alcoholism is accompanied by changes similar to those seen with advancing age. According to the *'accelerated-aging'* hypothesis (Ryan and Butters, 1979), chronic alcoholism leads to the precocious development of behavioural changes typically associated with advancing age; the alcoholic becomes old before his time. The *'age-sensitivity'* or *'increased vulnerability'* hypothesis places the timing of cognitive changes somewhat differently. According to this hypothesis, older brains are more affected by the consequences of alcohol abuse (Noonberg *et al.*, 1985). This view suggests that alcoholism does not exert a significant influence on neuropsychological function in younger individuals; however, once normal chronological aging begins to manifest itself in the middle 40s, alcoholics show increasingly greater impairments than do their peers.

SOCIAL DRINKERS

Parsons (1986) reports that 120 g of ethanol per day is the amount set as the bottom limit for defining an alcoholic level of intake by the World Health Organization. Some authors, however, set the limit at a lower level: 80 g or less. The present authors used as their definition of social drinkers an average daily intake of 80 to 120 g/alcohol.

Noble (1983) defines the term social drinking as follows: "As currently understood, a social drinker is an individual whose drinking behavior does not cause adverse consequences for himself or society. Still this term remains vague, imprecise and broad. It may well encompass an individual who consumes one drink several times a year to someone who drinks 9 to 10 drinks several days a week".

Mildly impaired cognitive function has been found in regular consumers of light and moderate amounts of alcohol by several groups of workers (Cala *et al.*, 1983: safe limits <40 g/day; Noble, 1983; Fox *et al.*, 1987; Hannon *et al.*, 1983; Parker and Noble, 1977, 1980: heavy social drinker: ≥4 drinks/occasion; Schaeffer and Parsons, 1986: 43.9 ± 26.02 ml/day; Arbuckle *et al.*, 1992: light social drinker: ≤1 drink/week, moderate social drinker 1–4 drinks/day or 7–27 drinks/week, heavy social drinker >4 drinks/day; MacVane, 1982: light to moderate 28.2 ml/occasion, heavy social drinkers: 78.1 ml alcohol/occasion or about 4.5 drinks; Nichols *et al.*, 1993: light social drinker <20 g or 2 standard glasses/week, heavy social drinker >200 g/week; Jääskeläinen, 1995: social drinkers: 3–18 standard drinks weekly.

Parsons (1986) suggested that impaired performance on neuropsychological tests by social drinkers might be due to anxiety. Schaeffer and Parsons (1986) also found anxiety to be a significant predictor of scores on neuropsychological tests for social drinkers, whereas in alcoholics the maximum consumption variables (MQF = maximum quantity × frequency) best predicted test scores. Page and Cleveland (1987) found that the effect of social drinking (≤ 90 drinks/month) on cognitive function was much less than that due to aging. Bowden (1987) concluded that the minor cognitive deficits found in social drinkers could be due to an association between drinking behaviour and innate cognitive ability, rather than a toxic effect of alcohol on the brain. The same is suggested by Emmerson *et al.* (1988) (3 times drinking/week; 4 drinks each time).

Although there is an abundance of research in this area, few studies have attempted to use matched samples of subjects. The matching by age, sex, education and socio-economic status reduced the number of confounding variables in a study by Williams and Skinner (1990) who found that their high alcohol group had significantly lower scores than the matched low alcohol group in 7 of the 10 subtests they used. They also found that the effects of age were much smaller than the effects of drinking style and that there was an effect of education levels on test scores. Waugh *et al.* (1989) divided their subjects into three groups according to their daily alcohol consumption: 40 g or less (light social drinker), 41–80 g (moderate social drinker), 81–130 g (heavy social drinker). The groups were matched for age, but no mention is made of their actual ages. They found no evidence of brain damage, as measured by NP test performance in the group who had been drinking between 41 and 80 g of alcohol per day for an average of 17 years. In the group who had been drinking up to 130 g per day for 15 years, there was a mild impairment in four NP tests in which sober alcoholics usually score poorly. Bergman (1985) found that in excessive male social drinkers (maximal 116.3 ± 97.4 g alcohol per drinking occasion; about 10 ± 9 drinks, and 47.5 ± 55.7 g alcohol consumed previous week per day) mild cognitive deficits and morphological cerebral changes (as assessed by CT scans) could be detected. This effect appeared to be the result of high recent alcohol intake, particularly during the 24 hr period prior to the investigation, representing an apparently *acute* effect of recent alcohol consumption. When excluding the acute effect of recent alcohol intake, it was possible to demonstrate mild cognitive deficits, but not morphological cerebral changes in males that were apparently due to long-term excessive drinking. In his female sample there was no association between the studied drinking variables and cognitive deficits or morphological cerebral changes. This difference may be explained by the fact that the females had significantly less advanced drinking habits, both with regard to quantity recently consumed and duration of excessive social drinking. Harper *et al.* (1988) carried out a number of volumetric and neurochemical analyses on the brains of 14 moderate drinkers (30–80 g per day) and compared the data with previous studies from controls and alcoholics. Volumetric measurements consistently suggested a loss of cerebral tissue although the differences were not statistically significant. Changes in the lipid and water content of the white matter were also noted; the water content rises and the lipid content falls in the alcoholics groups, whereas the water content appears to fall and the lipid content rises in the group of moderate drinkers.

Very few investigators have used electrophysiological measures such as ERPs to study the effects of social drinking. In Emmerson *et al.*'s study (1987) comparisons were made among groups of abstinent alcoholics, social drinkers (average of 4 ± 2.3 drinks per drinking occasion, with an average of 3 ± 1.6 drinking occasions per week) and lifetime non-drinkers. No significant impairment of ERP functioning was found for the social drinkers. All subjects were males aged 25–40 years. Jääskeläinen (1995) found ethanol-induced impairment on attention by using auditory event related potentials in 10 male and female social drinkers. Acute intake of alcohol (0.35 and 0.50 g/kg) reduced the amplitudes of early components and the mismatch negativity during a dichotic listening task, but not of later components, suggesting that especially the involuntary attention was affected.

The present study aims to determine whether effects of 'heavy' social drinking (80–120 g per day) will be reflected in the latency and amplitude of components of ERPs, reaction time (RT) and other test results of the sober subjects in such a way as is outlined in the 'accelerated aging' version of the 'premature-aging' hypothesis, namely that the heavy social drinker becomes old before his time.

Although so far the premature-aging hypothesis has not found much support, we still believe it is worth testing, because to date no ERP study has been done on social drinkers testing the premature-aging hypothesis. While we would expect the effects of alcohol to be much more widespread in alcoholics, the effects of social drinking might still be more specific, like in aging. In older people the effects of aging manifest themselves first in selective attention and fluid abilities, and since attentional impairments are also found in heavy drinkers, we focussed our attention on measuring focussed attention and memory search.

Age

With advancing age the positive deflection around 300 ms in the ERP, the so-called P_3, its amplitude is known to decrease (mainly at the posterior scalp locations), with a more even scalp distribution in old than in young adults. The latency of the P_3 generally increases with age (Ford and Pfefferbaum 1985; Looren de Jong, 1989).

Especially in complex tasks that require controlled processing, such as memory search tasks, the P_3 is diminished by overlapping negative components, the so-called processing negativities and search negativities. In memory search tasks search negativity can be pictured by subtracting the ERPs obtained in a low memory load condition from the ERPs obtained in a high memory load condition. It has been proposed that the magnitude of this subtraction wave represents the amount of controlled processing or on-task/effort. In older adults search negativities have been found more pronounced because of greater effort needed to perform the task (Zeef *et al.*, 1990).

Whereas search negativity, which overlaps with the late positive part of the ERP can be studied by varying memory load in the attended channel (non-targets – Cz), selection negativity and processing negativity, can be visualized by subtracting unattended non-targets from attended non-targets (at Cz and Fz), as both are free from motor potentials and target effects. The difference supposedly reflects only processes related to early and late selection. Early selection (selection negativity, 200–240 ms poststimulus) refers to filtering of inputs on the basis of basic, physical stimulus characteristics (that is, colour, location or pitch), late selection (processing negativity) refers to categorization of stimuli (that is, as members of a memorized target set) by further processing of semantic or functional attributes (Looren de Jong, 1989).

We expect the ERPs, RTs and error scores of the young social drinkers, as well as their test scores to be different from those of the young non-drinkers, more specifically that they will resemble those of the older non-drinkers, as compared to those of the younger controls.

We likewise expect these measures to differ for the old social drinkers as compared to the old non-drinkers. The effects of aging/social drinking should manifest themselves as: Lower scores on Digit Symbol and Digit Span Test; slower RTs in the focussed attention task; a higher percentage of errors in the focussed attention task; – ERPs: longer P_3 latency; smaller P_3 amplitude; increased search negativity (controlled memory search); later selection (slowed selective attention); more diffuse topography (less specific brain activity).

METHOD

Subjects

Because we wanted to study social drinkers with a drinking history of at least 10 years and lifetime non–drinkers of the same age, we chose the following groups: lifetime non–drinkers aged 30–40 years, social drinkers aged 30–40 years, lifetime non–drinkers aged 50–60 years, and social drinkers aged 50–60 years[1]. All drinkers reported to have started drinking alcohol on a regular basis from age 16–18.

The 47 males comprised two age groups: a 'young' group consists of 24 subjects between 30 and 40 years old, and an 'old' group of 23 subjects between 50 and 60 years old. The young group was divided into a control group of 12 lifetime non-drinkers (mean age 32, 6 years) and an experimental group of 12 social drinkers (mean age 35, 9 years).

The old group was similarly divided into a control group of lifetime non-drinkers (n = 12, mean age 57, 5 years) and an experimental group of social drinkers (n = 11, mean age 54, 9 years).

Non-drinkers were defined as those who had always averaged less than one alcoholic drink a week; social drinkers were defined as those who had an average alcohol intake of 80 to 120 g per day. The social drinkers had all started drinking alcohol on a regular basis between 16 and 18 years of age.

Subjects were paid volunteers recruited through newspapers and a radio advertisement. Beforehand they were interviewed by telephone and administered a questionnaire which sought information concerning educational background, height and body weight, neurological disease, general health, periods of unconsciousness and anaesthesia, liver disease, consumption of drugs and medication, years of drinking history, drinking related problems, use of tobacco, morningness/eveningness.

Potential subjects were not used if their medical history indicated more than 3 hours of cumulative unconsciousness or anaesthesia or neurological sequelae

---

[1] In aging research is it is more common to use age categories from 20 to 30 years and from 50 to 60 years.

resulting from injury to the head; hepatic disease, epilepsy, stroke or major psy-chiatric illness or drug abuse. Subjects who were currently taking medication that could conceivably affect their test performance were also excluded, as were those whose parents had a history of alcohol abuse dating back to the subject's birth. All subjects had completed some form of secondary education and were matched for body weight. They all had normal or corrected to normal visual acuity.

Subjects were requested to abstain from alcohol on the day of the experiment. We did not require a longer period of abstinence, since it was our intention to assess cognitive performance in the state in which they might be expected to function in their daily activities. Twenty of the 47 subjects were smokers (young non-drinkers: 4, young drinkers: 9, old non–drinkers: 3, old drinkers: 4).

Procedure

Subjects were all tested in the evening at about 7.00 p.m. The experiment took 2 to 3 hours.

During the first part of the experiment subjects were administered the follow-ing tests: the Digit Symbol Substitution Test, the Digit Span Test (forward and backward).

After having completed the above tests, there was a short break, after which subjects were prepared for ERP recordings.

ERPs were recorded from the Fz, Cz, Pz and Oz locations (10–20 system), using tin electrodes and referenced to linked earlobes. The ground was on the forehead, 3 cm anterior to Fz. ECI electrode gel was used at the interface and impedance values were kept constant at or below 5 kΩ. Horizontal eye move-ments were monitored from electrodes at the outer canthi of the right and left eyes, vertical eye movements from electrodes above and below the right eye. Trials with eye movements exceeding 100 µV and other artefacts were rejected from further analysis. Reaction times were measured in milliseconds by the computer and checked for error, omissions, premature (= <250 ms) and very late (= after 1000 ms + 2.5 sd) responses.

A Nihon Kohden 9 channel polygraph was used, low pass frequency was set to 35 Hz, time constant 5.0, sampling rate 142 Hz (1:7 ms), registration time 1200 ms, trial duration 1096 ms, prestimulus 98 ms.

Task

The ERPs were recorded in a combined focussed attention and memory search task, similar to the one used by Okita *et al.* (1985) and Kole *et al.*, Ch. 9; Figure 9.1).

Subjects were seated in a dimly lit, sound attenuated room, facing a VGA display at a distance of 90 cm. They received frame sequences of 300 ms dura-tion, with interstimulus intervals of 1096 ms. Each display-set frame consisted of four elements, two consonants and two dot masks, arranged in a square around a small, central fixation cross. The consonants and dot masks were always assigned to orthogonal diagonal positions. Either the left-up or the right-up diagonal was relevant, that is to say had to be attended to. Each experimental block began with a presentation of the fixation cross, followed by a cue frame indicating the diagonal to be attended for 5 sec, followed by a presentation of the memory set in the centre of the display lasting for 5 sec. After the presentation of these instruction frames, a sequence of 160 displays was presented. Half of the display frames belonged to the left-up diagonal category, the other half to the right-up diagonal category. Relevant targets were presented with a probability of 0.25. The frames were presented in random order. The memory set items and non-targets in each frame were randomly chosen from the set of all possible consonants except Q. The stimuli were white on a dark background. In the display frames the dimensions of the letters were 0.38° height and width, the dimensions of the display were 1.27° height and width.

Following two blocks of practice (80 trials each), one with memory load 1 and the other with memory load 4, all subjects received four experimental blocks of 160 trials each, two of the blocks with memory load 1 and the other two with memory load 4.

In two of the blocks the left-up diagonal was relevant, in the other two the right-up diagonal was to be attended. The order of administration of attended diagonal and memory load conditions was counterbalanced across subjects in each group. All subjects were presented with a different set of target letter in order to avoid any 'letter effects'. A pause of a few minutes was taken after two blocks. Subjects were told to minimize eye movement and blinking during task performance.

RESULTS

Test scores

There was no *main* effect of social drinking on test scores. Scores on the Digit Symbol Test revealed a statistically significant difference between the old groups and the young groups (t = 3.89, df = 45, p < 0.001). Within the age groups, however, there were no significant differences between the scores on the Digit Symbol Test of the drinkers and the non-drinkers.

For the Digit Span Test (forward and backward) there were no significant differences between any of the groups.

Reaction time and errors of task performance

Separate Manovas were performed on reaction time and error score with factors age, social drinking and memory load.

There was no main effect of social drinking and age on reaction times and error scores. For error scores there is an age effect for memory load 4 (F = 4.20, p < 0.05). Closer scrutiny of reaction times and error scores for the focussed attention task (Figure 16.1) reveals that contrary to the expectations the reaction times of the young non-drinkers are longest for both memory loads 1 and 4.

For memory load 4, RTs of the young non-drinkers are equal to those of the old drinkers and those of the young drinkers are equal to those of the old non-drinkers, thus cancelling out both an age effect and an alcohol effect.

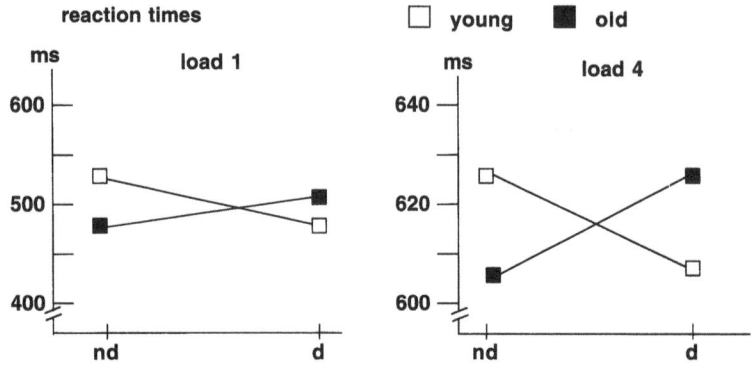

Figure 16.1 – Reaction times and error scores.

For the errors scores (Figure 16.2) we see a small effect of age for load 4, but no effect of social drinking; the young drinkers having the lowest error scores.

ERP-data analysis

Average ERPs to display-set stimuli were separately computed for each electrode location for relevant-irrelevant diagonal, target-non target frame for both memory loads 1 and 4. Relevant indicates a display frame containing consonants in the attended diagonal positions, irrelevant indicates a display frame containing consonants in the unattended diagonal. For each task condition the EEG was stimulus-locked averaged and the mean value of the pre-stimulus epoch (98 ms) was subtracted from the EEG. The averaging epoch lasted for 798 ms post-stimulus.

In order to have maximum accuracy, for all electrode locations and categories P_3 amplitude and latency at Pz were analyzed at 115 data points (areas) at intervals of 7 ms; the first at 105 ms post-stimulus and the last at 798 ms. This window was chosen on account of the F values for the factor memory load. Six way-Manovas were performed on each area with factors age, social drinking, memory load, target, relevancy and electrode location.

Results ERPs and focussed attention task

Contrary to the expectation of a more diffuse *topography* as a function of age and/or social drinking, there was no significant main effect of social drinking on grand average ERP.

Inspection of the grand average ERPs (Figure 16.2) does reveal a small but non-significant age effect on P_3 *amplitude* at Pz (a decrease of amplitude for the elderly subjects). For relevant targets a trend toward an age effect is visible, the amplitudes are larger for the young groups. For both memory loads 1 and 4 the young non-drinkers have the largest P_3 amplitudes for relevant targets.

For the other categories (irrelevant target, relevant non-target, irrelevant non-target) the P_3 amplitudes of the old groups are largest, indicating the young groups are better able to ignore irrelevant information.

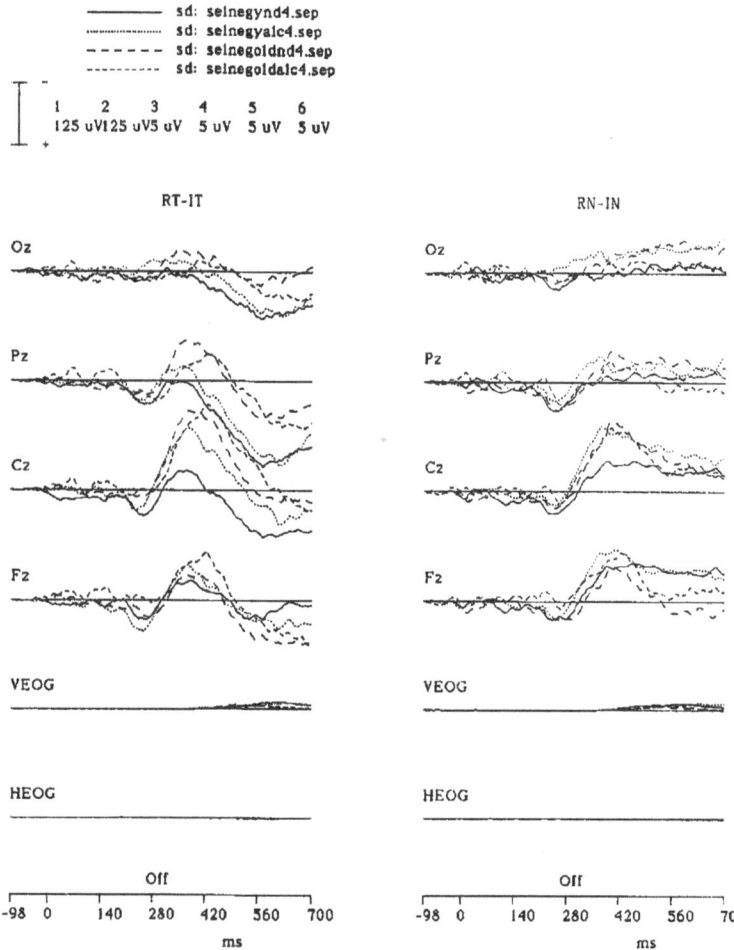

Figure 16.2 – Grand average ERPs for memory load 4.
Key: ynd = young non-drinkers; yalc = young social drinkers; oldnd = old non-drinkers; oldalc = elderly social drinkers. RT = relevant target; IT = irrelevant target; RN = relevant non target; IN = irrelevant non target.

Although the P_3 *latencies* for the old drinkers are the longest, whereas those of the old non-drinkers are the shortest, additional analyses on P_3 latency and amplitude for the Pz location did not reveal any effects of this factor either.

Memory set size affected ERPs between 252 and 301 ms poststimulus (F ≥ 6.81, p ≤ 0.05) and more strongly so from 427 ms onward (F ≥ 20.7, p < 0.0001), P_3 amplitudes being smaller for memory load 4. However, memory load did not affect P_3 latency at the Pz location. Main effects of attended versus unattended diagonal were present between 273 and 385 ms (F ≥ 7.06, p ≤ 0.01), between 413 and 511 ms (F $\geq$ 5.37, p $\leq$ 0.02) and from 560 ms onward (F $\geq$ 17.00, p < 0.0001), P_3 amplitudes being smaller for the unattended diagonal. There was a main effect of target between 357 and 476 ms (F $\geq$ 9.38, p < 0.0001) and from 532 ms onward (F $\geq$ 17.86, p < 0.0001), showing bigger P_3 amplitudes for targets.

Selection negativity concerns the difference waves obtained by subtracting the grand average ERPs to the irrelevant (unattended – irrelevant non-target) diagonal from those to the relevant (attended – relevant non-target) diagonal (at Cz); these difference waves which are free from motor potentials and target effects, supposedly reflect only processes related to early selection (selection negativity) and late selection (processing negativity at Cz and Fz). The prediction of finding a later selection negativity for social drinkers as compared to non-drinkers was not confirmed. There was only a small effect of age (age × relevancy × location) from 469 ms onward (F $\geq$ 3.03, p $\leq$ 0.05); this effect concerns processing nega-tivity, which is said to be most pronounced at Fz and Cz (Looren de Jong, 1989). This processing-related difference is more negative for the high memory load and is also more prolonged. For memory load 1 the difference between the groups is minimal, for memory load 4 however, particularly at Fz, processing negativity is most prolonged for both the young non-drinkers and the young drinkers, for the older groups it diminishes much earlier. This indicates that the older subjects process unattended information in the same way as they do attended information. The same was also found by Looren de Jong (1989).

We also expected *search negativity* (memory load × relevancy × location), resulting from subtracting the average ERPs in the lower memory load condition from the ERPs in the higher memory load condition, to be increased for social drinkers as compared to non-drinkers. This effect was not found. There was a sig-nificant interaction effect of age × drinking × memory load × relevancy × location between 343 and 581 ms (F $\geq$ 2.79, p $\leq$ 0.05) and between 609 and 700 ms (F $\geq$ 2.87, p $\leq$ 0.05). Search negativity reflecting controlled memory search, which was maximal at Cz (relevant non-target), was largest for the young drinkers and the old non-drinkers, and smallest for the young non-drinkers, while in the old drinkers falling in between, thus cancelling out a main effect of age. For the young-non-drinkers, search negativity started relatively late and was prolonged. Search negativity concerns searching processes in memory, and is supposed to be easier for young than for older people. For this reason for the young non-drinkers search negativity should have started first and should also have stopped first.

DISCUSSION

None of the expectations with regard to the accelerated aging version of the pre-mature-aging hypothesis were confirmed, neither for task performance nor for ERPs. Additionally, the effect of age as such on task performance was slight, the only test that revealed a significant age difference was the Digit Symbol Test. There was no effect of either age or social drinking on selection negativity and the effect on search negativity was small and not in the expected direction. It started relatively late for the young non-drinkers and was most prolonged for this group. This indicates that the process of searching, comparing and deciding took them longer. Because of this they should have made fewer errors than the other groups, but this was not the case. It may indicate that their processing was less efficient compared to the other, social drinking, and the elderly non-drinking group. There was also a small effect of age on processing negativity, indicating that the older subjects process attended and unattended information in the same way.

Noonberg *et al.* (1985) did find support for the accelerated aging version of the premature-aging hypothesis, however, their subjects were alcoholics, not social drinkers. Still, they note that: "While the present findings favour the accelerated aging version of the premature-aging hypothesis, they may in fact have little to do with aging at all... It is only being suggested here that while alcoholism may not produce the same pathophysiology of the brain that is typically seen among non-alcoholic elderly individuals, and while the developmental course of alcoholic deterioration may not have the same pattern found in normal aging, it nevertheless seems clear that, in certain aspects of cognitive function, alcoholics do no better than non-alcoholics 10 years their senior". Porjesz and Begleiter (1982) also conclude that, although alcoholics may exhibit similar behavioural deficits to old people, these deficits may well reflect the result of different neuronal pathology or may be general enough to be nonspecific.

In Shelton *et al.*'s study (1984) on alcoholics the premature aging hypothesis was not supported. Page and Cleveland (1987), however, studied both alcoholics and social drinkers. From their results they concluded that: "... the present investigation suggests that, on the average, alcohol usage may not accelerate the intellectual aging process to an appreciable degree". Thereby, according to Cutting in an editorial (1988) in the British Journal of Addiction, throwing the cat among the pigeons with regard to the whole area of cognitive impairment and alcoholism: "Up till now, the picture seemed fairly clear: social drinking caused moderate to severe cognitive impairment, and the older the drinker the worse the consequences. Now, one has to question the notion of whether social drinking is psychologically harmful at all".

He further mentions the hypothesis that chronic alcohol consumption may damage the cholinergic system and thus leads to dementia, but points out that an additional factor of susceptibility has to be introduced, and that only heavy drinkers with a fragile cholinergic mechanism, may develop problems. Cutting concludes that the alcohol-induced-age-acceleration hypothesis seemed sensible at the time as it fitted the few facts that were known. But that it will have to be revised in the light of new findings such as has been done by Page and Cleveland (1987) and also by Bowden (1987).

Viewed in this light it may not be surprising we did not find any alcohol-induced-aging effect, but as mentioned earlier, we thought that the effects of social drinking might be less general than those of alcoholism. As distinct from many studies on social drinkers as well as alcoholics, we have taken great care to reduce the number of confounding variables as much as possible by being very strict in our demands in the selection of our subjects with respect to age, education, health status, amount of alcohol consumed, years of drinking, family history of alcoholism and many others. All our subjects had to be in perfect health. Since the results we found were few it may plea for a less strict selection of subjects to find effects on task performance and neurophysiological parameters of social drinking independently or in interaction with other so-called confounders.

We realize that our groups are small and are heavily selected samples and therefore are maybe not representative of the general population. However, in order to rule out any (health) factors other than alcohol consumption, that might influence task performance and ERP, this seemed to us the only way.

Since the groups are small, individual differences are big, resulting in high standard deviations in ERPs, causing jitter in amplitude and flattening out results. But we do not believe this to be the cause of our scarce significant findings, which were opposite to expectations. If group size were the problem, we believe we would still have found consistent trends in the expected direction.

We also believe our test procedure was such that it could not have biased the results. All subjects were tested in the evening at about 7.00 p.m., were administered the same tests (of proven sensitivity in alcohol research) in the same order, by the same experimenter and any differences of stimulus condition in the focussed attention task were counterbalanced across subjects in each group.

For the groups of old social drinkers and young lifetime non-drinkers it took great effort to find sufficient and suitable subjects, possibly because of our strict health demands. It may well be that when heavy drinkers get older, they often develop health problems (because of weak cholinergic mechanisms?) or have to stop drinking for health reasons. Perhaps our group of old drinkers consisted of exceptionally strong, fit subjects.

The young non-drinkers may constitute an exceptional group in a different respect; their results were rather 'disappointing'. For people between 50 and 60 years of age it is not unusual if they do not drink, since when they were young, drinking alcohol was much less customary than in this epoch. The young non-drinkers (30–40 years) however, grew up in a time when drinking alcohol was quite the socially acceptable thing to do, particularly for males. It is possible that our subjects had a biological constitution (maybe a weak cholinergic mechanism) or a specific awareness of health which kept them from using alcohol; most of the young non-drinkers said they did not drink because alcohol did not agree with it.

In our health questionnaire subjects were asked about their eating habits; they all reported to have a healthy and varied diet. We did not ask them however, about the kind of food they ate. Since protein rich food increases arousal and carbohydrates have a sedating effect, this may also have been a variable we did not control for.

Since there may also be other variables that we are not aware of, which distinguish drinkers from non-drinkers, in future research it would be worthwhile to send subjects a questionnaire inquiring about diet, experienced workload/ recovery and sleep quality in order to get more insight into possible differences.

In conclusion, it is possible that an alcohol consumption between 80 and 120 g per day does not accelerate the aging process, or at least not in such a way as we have attempted to identify it in the present experiment.

It is also possible that the deleterious effects of alcohol on the brain are not yet apparent in younger subjects, as some investigators believe, and that our group of old drinkers was an exceptionally strong sample. It would therefore be interesting to test the young drinkers again in 20 years.

Another possibility is that, particularly the young non-drinkers were an exceptional group, and for the purpose of getting more information about possible differences between drinkers and non-drinkers, our subjects will be sent an additional questionnaire, and this is more or less in line with Cutting's concluding remarks. It is possible that one should look for different methods of testing to identify any alcohol related differences between drinkers and controls; what should replace it (the present research)? "This is obviously a matter for future researchers to tease out. It would appear however, that there is no simple quantitative relationship between alcohol consumption and psychological deterioration, and no particular susceptibility in older people. The true equation must take account of a large measure of individual differences in susceptibility and also take note of the issue which bedevils much neuropsychiatric research – the right direction of cause and effect if two or more factors are present".

SUMMARY

In the present study the accelerated-aging version of the premature-aging theory of chronic alcoholism was tested in four groups of healthy males: young lifetime non-drinkers (age 30–40), young social drinkers (age 30–40), elderly lifetime non-drinkers (age 50–60) and elderly social drinkers (age 50–60). The social drinkers consumed an average of 80–120 g of ethanol per day (8 to 12 units per day). ERPs were recorded during a combined focussed attention and memory search task, RTs and error scores were also measured. In addition subjects were administered the Digit Symbol Substitution Test (forward and backward) and the Digit Span Test. No evidence was found for any alcohol-related changes in the social drinkers to support the hypothesis, neither for RTs, error scores and ERPs, nor for the other test scores. It was concluded that the samples may have been exceptional, and more information is needed to get insight into possible confounding variables that may influence their performance.

CHAPTER SEVENTEEN

Social Drinking, Memory and Information Processing

Jennifer M. NICHOLS and Frances MARTIN

INTRODUCTION

Scientific research indicates that alcohol, despite its generally positive image, is a drug with neurotoxic, psychoactive and addictive properties. These properties have resulted in chronic alcohol abuse being a major social, economic, and public health problem in many parts of the world (Charness et al., 1989). Excessive alcohol consumption has been implicated as a significant cause of a number of neurological disorders including impairments in higher cortical processes. Impaired cognitive functioning, including deficits in various cerebral activities such as concentration, alertness, motivation, general ability, verbal and numerical fluency, and memory, has been reported both acutely and chronically in subjects with alcohol-related problems.

The intake of several drinks per day has long been a social pattern of use which has been regarded as harmless and even beneficial (Turner et al., 1981). At least two thirds of Western people drink more than just occasionally and evidence suggests that the average age of a person's first experience with alcohol is eleven, three years younger compared to 20 years ago (Forney et al., 1988). As the highest percentages of drinkers with the greatest per capita consumption are found between 16 and 25 years, after which consumption levels decline (Schuckit, 1985), there has been increasing concern about drinking patterns and levels in the younger population. Alcohol has been found to have a detrimental effect on the quality of campus life and generally heavy and frequent drinking is related to low academic achievement (Brennan et al., 1986) although in one study (Clark et al., 1987), medical students who abused alcohol showed evidence of better academic performance.

Recent survey studies have reported that students drinking patterns are equivalent to or worse than the general population in terms of safe alcohol intake levels (Breeze, 1985; Collier and Beales, 1989; File et al., 1990; Goddard and Ikin, 1987; West et al., 1990). File et al. (1990) found that approximately 25% of college students were drinking at levels likely to damage health as defined by the British Health Education Authority (1991). A common and perceived social use pattern is the intake of relatively large quantities (in excess of 80 g) of alcohol during any one drinking episode (File *et al.*, 1994). This consumption of excess alcohol per occasion has been commonly referred to as binge drinking. Bates and Tracy (1990) have stated that if excess alcohol consumption per occasion, that is, binge drinking, irrespective of frequency, results in persistent mild cognitive impairment then a large proportion of younger drinkers may be at risk for developing alcohol-related problems. Based on theoretical as well as practical concerns the study of whether social drinkers, particularly binge drinkers, are at risk for developing impaired cognitive functioning that persists to the sober state is of critical importance.

SOCIAL DRINKING: A CONTINUUM OF ALCOHOL-RELATED DAMAGE?

Ryback (1971) formulated the continuity hypothesis of alcohol abuse which asserts that there is a single continuum of alcohol effects which ranges from subtle deficits on social drinkers through a 'sub-clinical' amnesic disorder to full-blown Korsakoff's syndrome. The hypothesis assumes that the progressive damage of specialised structures due to the chronic abuse of alcohol results in a wide spectrum of memory and perceptual impairments and a deficit in problem-solving capacity (Brandt *et al.*, 1983; DeRenzi *et al.*, 1984). Knight and Longmore (1994) observe that "Implicit in Ryback's continuity hypothesis is the assumption that at some level of social drinking, detectable levels of impairment will become apparent" (p. 206). Research in the area of cognitive impairment due to alcohol abuse has typically focused on alcoholics with neurological evidence of Wernicke-Korsakoff's syndrome and its characteristic amnesic sequelae (see Butters, 1985; Butters and Cermack, 1980; Cutting, 1985) and on alcoholics without neurological signs (see Goldman, 1986; Loberg, 1986; Parsons *et al.*, 1987). The findings on both of these populations indicate that chronic alcohol intake results in neurological damage as well as deficits in cognitive functioning and raises the question of safe levels of alcohol consumption. There have been numerous investigations on the effect of social drinking on cognitive performance, however, the results have been inconclusive and difficult to interpret due to inconsistencies in findings, possibly as a result of methodological problems. This chapter briefly presents a number of theories on alcohol-related brain damage that have implications for social alcohol intake and then reviews the social drinking literature. The main objective of this chapter is to present a number of recent investigations that have examined the effect of social drinking on cognitive functioning, particularly memory. We conclude with a discussion of the implications of these recent findings in terms of theories and everyday functioning.

THEORIES OF ALCOHOL-RELATED BRAIN DAMAGE

Although the exact mechanisms involved in alcohol intoxication are unknown (Charness *et al.*, 1989), evidence suggests that one of the most probable sites of ethanol's intoxicating action is a complex of membrane proteins that contains a receptor for the inhibitory neurotransmitter gamma-aminobutyric acid (GABA) and an associated chloride-ion channel (Charness *et al.*, 1989). Accelerated alcohol metabolism is induced by heavy drinking which plays a role in the development of alcohol tolerance (Lieber, 1991). With the cessation of drinking, increased drug metabolism may continue for several days to weeks (Lieber, 1990, 1991). Clinical syndromes similar to that of alcohol intoxication can be induced by certain drugs which interact with receptor sites linked to this complex, that is to say, benzodiazepines and barbiturates. Chronic intake of these drugs and ethanol results in cross-tolerance (Boisse and Okamoto, 1980) and benzodiazepine receptors' inverse agonists interfere with the behavioural effects of ethanol intoxication, however, the mechanism by which this occurs is unclear (Charness *et al.*, 1989).

There has been a general acceptance of cognitive impairment as a likely concomitant of chronic alcohol abuse, however, the aetiology and course of the dysfunction have been difficult to define (Page and Cleveland, 1987). It has been postulated that ethanol and its oxidative metabolite acetaldehyde may directly damage the developing and mature nervous systems (McMullen *et al.*, 1984; King *et al.*, 1988; Arendt *et al.*, 1988; Lieber, 1988). Recent investigations suggest that ethanol-related nervous system disorders may be a result of a combination of the neurotoxic effects of ethanol or its metabolites, nutritional factors, and genetic predisposition (Charness *et al.*, 1989). Research evidence indicates that excitotoxicity plays a role in alcohol related brain damage due to changes in neurotransmission as both a direct result of alcohol and as a secondary response to thiamine deficiency (Lovinger, 1993). That is, persistent or repetitive episodes of excitatory transmission during alcohol withdrawal may result in the damage of central neurons due to 'excitotoxic' mechanisms. An early model of social drinking forwarded by Hill and Ryan (1985) suggested that the cognitive functioning deficits evident in social drinkers and alcoholics may be mediated by separate processes. The observed brain anatomy alterations in alcoholics are asserted to be due to the neurotoxic effects of alcohol consumption, either alone or in combination with nutritional deficiencies, and, they asserted, there is a threshold below which alcohol intake does not result in tissue damage. In support of this, Parker *et al.* (1985) have suggested that it is the quantity of alcohol consumed in one sitting which is associated with measurable cognitive deficits. Hill and Ryan (1985) suggest that cognitive deficits found in social drinkers are due to a transient interruption of neurophysiological functioning and represent a withdrawal phenomenon rather than neurotoxicity effects related to long-term alcohol use. They assert that the degree of disruption evident should increase with greater amounts of alcohol consumed per sitting, that is, binge drinking, and the greater the extent of withdrawal the greater cognitive processing impairment. However, this so called 'hangover hypothesis' (Parsons, 1986; Bowden *et al.*, 1988) has received limited support within the literature (e.g., Alterman and Hall, 1989; Bowden *et al.*, 1988; Carey and Maisto, 1987; Hannon *et al.*, 1987). In terms of heavy social drinking (the consumption of at least 200 g of alcohol per week including a binge session of greater than 80 g/occasion), Lovinger's (1993) assertion that repeated episodes of excitatory transmission during withdrawal may result in neuronal damage due to 'excitotoxic' mechanisms indicates that the withdrawal phenomenon itself may result in structural brain damage. Regardless of whether the cognitive functioning of social drinkers is a reflection of a transient state (Hill and Ryan, 1985) or not, the possibility that repeated episodes of increased levels of alcohol intake may lead to CNS damage or changes is important in terms of current functioning levels.

THE EFFECT OF SOCIAL DRINKING ON COGNITIVE PROCESSES

There is a large amount of neuropsychological (Parsons *et al.*, 1987), neuroradiological (e.g., Pfefferbaum *et al.*, 1992), clinical (Lishman *et al.*, 1987) and neuropathological (Harper and Kril, 1986) research on the effects of chronic alcohol intake on CNS functioning. However, in terms of public health, the effects of consumption of alcohol at levels which are perceived as social drinking requires further investigation.

Early social drinking experiments

A review of the social drinking studies (Grant, 1987; see also Knight and Longmore, 1994) revealed no consistent relationship between pattern, amount, and timing of alcohol consumption and neuropsychological findings. A number of investigations have reported a relationship between increased levels of social drinking in 30–60 year old subjects and neuropsychological impairment (e.g., MacVane *et al.*, 1982; Parker and Noble, 1977; Parker *et al.*, 1983). Parsons and Fabian (1982) were unable to replicate this finding in male and female students and Hannon *et al.* (1983; 1987) reported no specific correlations between drinking variables and cognitive variables. Hannon *et al.* (1987) suggest that the failure to replicate specific relationships demonstrates "the elusive and variable nature of the relationships being studied" (p. 506). That is, the results of numerous studies have been questionable, finding both predicted and non-predicted relations between quantity and frequency of intake and levels of functioning employing a large range of neuropsychological tests (e.g., Bergman, 1985; Hannon *et al.*, 1983, 1987; MacVane *et al.*, 1982; Parsons and Fabian, 1982; Cala *et al.*, 1983). The most consistent findings have involved research on the relationship of Average weighted Quantity per Occasion (QPO) and Shipley Institute Living Scale (SILS) performance measures (Parker and Noble, 1977; Parker *et al.*, 1980; 1983) which have received limited support from subsequent studies (Hannon *et al.*, 1983; 1985; MacVane *et al*, 1982; Parsons and Fabian, 1982; Jones and Jones, 1980). The methodological problems inherent in these early studies, such as the use of correlational analysis with no report of group means or between-group comparisons, disregard of the possibility of Type 1 errors in correlational studies, sample populations which are not representative, use of insensitive tests (e.g., SILS, Lezak, 1976), self report procedures and consumption measures employed, period of abstinence before testing, as well as the possibility that the correlations have been reduced as a result of the limited range of variables assessed may be contributory factors.

More recently, researchers have attempted to overcome some of the problems associated with the earlier studies. In support of previous null findings a number of later studies have reported no relationship between social drinking and neuropsychological test performance (Alterman and Hall, 1989; Emmerson *et al.*, 1988; Jones-Saumty and Zeiner, 1985; Page and Cleveland, 1987). For example, Bowden *et al.* (1988) found no evidence of a relationship between recent consumption and SILS scores, using multiple regression analysis, although age and education level were predictive of SILS scores. Bates and Tracy (1990) examined the relationship of drinking patterns (using 11 measures of alcohol use patterns including the QPO) to level of cognitive performance, using a large battery of cognitive tests that measured verbal and non-verbal abstracting ability, concept formation, memory, visuo-spatial skills, and visuo-motor coordination. The sample included both sexes, aged 18, 21, and 24, and was classified post-hoc to include abstainers, extremely light alcohol users, infrequent, and intensive alcohol users. The data from this study did not provide strong support for a causal relationship between alcohol and cognitive functioning and the authors suggested that subtle decreases in functioning occurring with continued alcohol use may only be detectable by the use of both within-subject comparisons over time and more sensitive cognitive tests.

The failure to replicate the earlier findings of a relationship between social drinking and an impairment in cognitive functioning (Parker and Noble, 1977; Parker *et al.*, 1983) has been suggested to be due to the variability in subject groups' drinking patterns across studies (Parker *et al.*, 1991). Parker *et al.* (1991) postulated that "the effects of quantity of alcohol consumed may be conditional upon the frequency of alcohol use" (p. 367), and that alcohol consumption within a particular population needs to reach a certain level before cognitive dysfunction is detectable using behavioural measures.

The earlier studies thus showed that, although the relationship between social drinking and cognitive performance is tenuous at best, there is some indication that social drinking may result in various forms of cognitive deficits that increase with age and consumption level. The results of a number of later studies (e.g., Bates and Tracy, 1990) were consistent with previous studies (e.g., Hannon *et al.*, 1983, 1987; Parsons and Fabian, 1982) that have found limited support to suggest that subclinical males and females are at increased risk for subtle cognitive impairments. Although, as MacVane *et al.* (1982) suggested, individual thresholds may have to be reached before alcohol affects cognitive functioning and the use of insensitive psychometric indices may have contributed to the lack of any consistent findings of mild cognitive impairment.

Neuroanatomical and neurophysiological indices

To date, magnetic resonance imaging (MRI) (e.g., Jernigan *et al.*, 1991) and positron emission tomography (PET) (e.g., Wik *et al.*, 1988) have only been used to look at the brains of alcoholics and in the case of PET scans the effect of an acute dose of alcohol on the brain metabolism of normal subjects (e.g., De Wit *et al.*, 1990). However, several investigations have employed computerised tomography (CT) (e.g., Cala *et al.*, 1978, 1983; Bergman, 1985; Mutzell and Tibblin, 1989), and autopsy reports (e.g., Harper *et al.*, 1988) to investigate the relationship between neuroanatomical indices and social drinking. Cala *et al.* (1983) using a sample of 59 male heavy social drinkers (HSDs) (54 g/day) found signs of brain atrophy. Subjects with abnormal CT scans exhibited impaired test performance on cognitive-perceptual tests, however there was no relationship between average self reported daily intake and test performance. They concluded that brain tissue damage as indexed by CT and psychometric measures studies occurs following modest alcohol consumption (40 g/day). Mutzell and Tibblin (1989) using CT scans found evidence of cerebral atrophy, particularly in the frontal lobes, in one third of subjects who drank in excess of 33 g/day. In contrast, Bergman (1985) found no evidence of brain damage in a randomly selected sample of both male and female HSDs (57 g/day). In neither sex was maximum QPO correlated with cognitive performance or CT measures, however, high alcohol consumption during the week prior to testing resulted in a negative relationship on both CT scores and cognitive test performance in the men only.

The results of these studies have shown inconsistent results with respect to social drinking, although Lishman (1987) has suggested that these methods (particularly CT scans) have reliably demonstrated shrinkage in the brains of large unselected samples of alcohol abusers. The assertion that a relationship exists between neuropsychological tests, CT results, and drinking history is not strongly supported, with the possible exception of moderately heavy alcohol intake (in excess of 40 g/week) in the weeks prior to testing (Grant, 1987; Parsons, 1986; Bergman, 1985). Knight and Longmore (1994), following a review of the CT scan findings state that, although the relationship between alcohol intake and changes in brain anatomy is not clear, there is evidence that supports the continuity hypothesis of alcohol abuse in that changes in brain anatomy are visible in the young heavy social drinker and increase in the older chronic alcoholic. These findings suggest that research on younger HSDs, i.e., particularly binge drinkers is of critical importance as even a relatively mild disruptive effect on cognitive processes, especially those involved in memory and learning may be significant.

Memory experiments

The adverse effect of alcohol intake on adolescents and young adults whose cognitive abilities are still developing has received less attention than its effect on older alcohol users. Past research of alcohol consumption in young adults has investigated the effects of heavy drinking on cognitive functioning while sober, with inconsistent findings (e.g., Hannon *et al.*, 1983; Jones-Saumty and Zeiner, 1985; Hill and Ryan, 1985). A number of the earlier studies on social drinking have investigated memory as part of a battery of neuropsychological tests in a variety of age groups (e.g., Bergman, 1985; Bowden *et al.*, 1988; Jones and Jones, 1980; MacVane *et al.*, 1982; Parker and Noble, 1977) and have generally found a significant negative relationship between memory performance and social drinking. As stated, the methodological problems associated with these studies, as well as failure to replicate significant findings, have limited any meaningful interpretations about the effect of social drinking on memory. However, in recent years a number of well designed studies have also reported negative effects of social drinking on memory performance (e.g., Salame, 1991; Waugh *et al.*, 1989; Williams and Skinner, 1990) and have indicated levels of alcohol consumption above which cognitive impairment can be detected by behavioural measures.

For example, Williams and Skinner (1990) in a study of both male and female drinkers (>500 g/week) and a group of matched controls (<200 g/week) found that impaired performance was evident in HSDs on tests which required intellectual flexibility, spatial ability, visual motor-coordination and the ability to learn. Waugh *et al.* (1989) investigated the effects of social drinking on neuropsychological performance in three groups of healthy male adults. The results indicated that the high drinking (81–130 g/day) group's performance was significantly lower than the moderate (41–80 g/day) or low (<40 g/day) drinking groups on a number of neuropsychological measures including the Rey Auditory Verbal Learning Test (RAVLT) and the Austin Maze. In contrast, no difference in neuropsychological measures was evident between low and moderate drinking groups. In addition, the quantity of alcohol consumed per drinking occasion showed no relationship with neuropsychological test scores in subjects drinking up to 80 g/day, although subjects with higher levels of intake per occasion (80–130 g) showed impaired performance on trials 2–5 and delayed recall of the RAVLT compared to lower intake groups. This lack of evidence of cognitive impairment in subjects ingesting up to 80 g/day supports previous null findings at this consumption level (Bowden *et al.*, 1988; Page and Cleveland, 1987).

Similarly, Salame (1991) reported that the frequent heavy consumption of alcohol (>50 g/day) may affect learning abilities of subclinical populations. Thirty six light social drinkers (LSDs) (estimated daily consumption 36 g) and 36 HSDs (estimated daily consumption 52.8 g; mean age 30.9 years) were selected on the basis of biological data and self report measures. The study investigated the effects of alcohol (0.7 ml/kg = 0.56 g/kg) and noise (75 dBA) on an immediate memory task (i.e., serial recall of sequences of nine digits without repetition) and choice reaction time (RT). The immediate memory results showed that HSDs were unable to learn material as quickly as LSDs and also that learning may be impaired by a moderate alcohol dose. The RT data revealed that HSDs exhibit a longer RT compared to LSDs. The results suggest an effect of past drinking history irrespective of whether the subject is sober or intoxicated. The HSDs exhibited a reduction in speed of information processing with a resultant impairment in rate of acquisition/learning of material (Salame, 1991).

Bowden (1987) has suggested that findings of a negative relationship between certain neuropsychological measures and amount of alcohol consumed per sitting are probably of limited clinical significance and may be due to pre-existing differences in cognitive ability. The recent studies reviewed employed groups matched on a number of possible confounding variables including education background (Williams and Skinner, 1990) and intelligence level (Waugh *et al.*, 1989). Therefore, the selection bias is unlikely to have affected the results. However, the possibility that low intelligence increases a person's susceptibility to alcohol damage cannot be discounted. The finding (e.g., Salame, 1991; Waugh *et al.*, 1989; Williams and Skinner, 1990) that HSDs exhibit a less severe but similar pattern of impairment to that reported in alcoholics (Bowden, 1988; Fox *et al.*, 1987) provides support for the continuity hypothesis proposed by Ryback (1971) and as Salame (1991, p. 1239) stated "extend the range of cognitive impairments to young and healthy HSDs who drink heavily and regularly but who have not yet reached. the state of chronic alcoholism which leads to admission for detoxification".

A series of studies conducted in our laboratory investigating the effect of heavy social drinking on memory functioning support this contention. We investigated the effect of heavy social drinking (binge drinking) in a group of young male university students (aged 18–30 y) on tasks that require encoding, storage and retrieval of information, in the presence and absence of a pharmacological challenge (lorazepam, 2 mg). The study of the cognitive effects of heavy alcohol intake at an early stage of abuse in combination with a pharmacological challenge may provide information on the actions/changes occurring in brain functioning due to alcohol abuse and about the processes underlying these changes. In these studies HSDs were defined as individuals who consumed at least 200 g of alcohol per week which included one episode in which in excess of 100 g of alcohol was ingested, that is, binge session, and subjects were required to be drinking at this level for at least 2 years. In comparison, LSDs were defined as subjects who drank less than 20 g of alcohol per week and had no history of heavy social drinking. The subject groups were matched for age, weight, education background, intelligence level and had no first degree relatives that were heavy drinkers. Nichols and Martin (1993a), found that lorazepam differentially affected the HSDs compared to the LSDs on the Rey Auditory Verbal Learning Test (Rey, 1964; Crawford *et al.*, 1989). In addition, the HSD's performance on a number of the tasks, that is, Controlled Oral Word Association Test (Benton and Hamsher, 1976) and copy trial of the Complex Figure Test (Rey, 1941; Taylor, 1969) was reduced compared to the LSD's irrespective of drug treatment. Verbal fluency indicates the speed and ease of verbal production and is said to be a sensitive indicator of brain dysfunction (Lezak, 1983). Hence, deficits in performance on this task may suggest an impairment in frontal lobe functioning (Miceli *et al.*, 1981; Perret, 1974). The lower scores of the HSDs may indicate that they have a reduced ability to develop strategies in order to organise verbal output or that alcohol abuse has slowed down the retrieval of information from semantic memory.

Performance on the copy trial of the complex figure was also impaired in the HSDs suggesting that their perceptual organisational ability is reduced compared to the LSDs at this stage of abuse. However, there were no differences between groups on tests that required visuo-spatial conceptualisation and manipulation, for example block design and object assembly WAIS-R (Wechsler, 1981) and NHAIS (Naylor and Harwood, 1972). In addition, lorazepam was found to differentially affect the HSDs and the LSDs on verbal recall. Anterograde impairments of immediate and delayed verbal memory were evident in the HSDs, which is consistent with the classic profile of benzodiazepine-induced amnesia (Curran *et al.*, 1987), but not in the LSDs. The finding that for LSDs immediate and delayed verbal free recall, although decreased, was not significantly impaired by the administration of lorazepam is atypical (see Lister and File, 1984; Preston *et al.*, 1988).

In a study attempting to replicate the verbal learning task findings, a new sample of 28 (14 HSDs: >200 g/week and 14 LSDs: <20 g/week) male undergraduate students were recruited. In each of the drug treatments, subjects were required to perform parallel versions of the verbal learning task presented auditorily and visually. On the delayed recall trial, lorazepam was found to reduce the number of words recalled in the LSDs only in the visually presented verbal learning task, but not in the auditorily presented verbal learning task. There was a trend for lorazepam to reduce the number of words recalled in the HSDs in both the visual and auditory verbal learning task, with the auditory verbal learning task showing a greater reduction in the number of words recalled, that is, opposite to the LSDs. These results also suggest a differential effect of lorazepam on verbal learning performance between HSDs and LSDs and suggest that the effect may be influenced by the method of presentation. These results support previous findings (Salame, 1991; Waugh *et al.*, 1989) that suggest that HSDs have changes in their cognitive functioning due to alcohol abuse placing them at an early point on the hypothesised continuum of alcohol-related impairment (Ryback, 1971).

Electrophysiological experiments

Event Related Potentials (ERPs) represent a unique approach for assessing brain functioning as they allow the study of electrophysiological manifestations of cognitive processing (Porjesz and Begleiter, 1993). They are obtained by recording the time locked brain electrical activity following presentation of a discrete stimulus to any of the sensory modalities. Psychophysiological measures of brain activity have proven to be a valuable technique for detecting subtle changes in brain function due to alcohol abuse (Emmerson *et al.*, 1987) and are also useful in indexing electrophysiological concomitants of complex cognitive tasks (Hillyard *et al.*, 1978; Donchin, 1979; Donchin *et al.*, 1978). As the impairments caused by alcohol abuse are suggested to progress on a continuum it is likely that the amplitude and latency of ERP components will reflect this progression. The component most likely to reflect this is the P_{300} component of the ERP which is a positive going waveform occurring maximally at parietal sites and 300–600 ms post stimulus presentation. It is evoked by rare or unexpected task relevant events and is independent of the physical characteristics of the eliciting stimulus. The P_{300} component has been suggested to index psychological processes associated with information processing (Donchin, 1981; Johnson, 1986).

We investigated whether heavy social drinking, including binges, causes changes in brain functioning as indexed by the P_{300} component of the ERP and RT, elicited by imminent accident scenes within a driving paradigm (Nichols and Martin, 1993). Evaluation of event-related potentials in both HSDs (n = 11) and LSDs (n = 11) revealed two major findings: HSDs showed longer P_{300} latency compared to LSDs, and HSDs and LSDs both demonstrated reduced P_{300} amplitude in the presence of a pharmacological challenge. The P_{300} latency results in conjunction with the lack of a P_{300} amplitude difference between groups suggests that the ability of HSDs to evaluate the nature of stimuli was less impaired and that it was the time taken to complete the evaluation that was more affected. Porjesz *et al.* (1987) found that alcoholics exhibited P_3 latency delays in an easy discrimination task but not in a difficult discrimination task and hence demonstrated no P_3 latency differences related to discrimination difficulty. Porjesz and Begleiter (1993) suggested that alcoholics found both tasks difficult and irrespective of task requirements adopted a similar mode of responding. Previous studies have found an increase in P_{300} latency in both alcoholics and alcoholic probands (Elmasian *et al.*, 1982; Steinhauer *et al.*, 1987) in the absence of P_{300} amplitude reduction. RT and P_{300} latency correlated positively and significantly for the HSDs under the placebo treatment and there was a trend towards a significant positive correlation for this group in the lorazepam treatment. On the other hand, there was no significant correlation between RT and P_{300} latency in the LSDs in either drug treatment. Since the correlation between RT and P_{300} latency was large and positive for the HSDs, particularly in the placebo treatment, it would seem that for this group, in spite of speed being emphasised, RT was largely determined by stimulus evaluation (Kutas *et al.*, 1977). By whatever mechanism, whether as a result of or preceding long-term alcohol ingestion, the HSDs show changes in cognitive functioning such that stimulus evaluation time is increased. This increase is present whether they are under the effects of lorazepam or not, which may indicate that, whether due to their abuse of alcohol or to differences preceding alcohol abuse, they have less resources available to evaluate stimuli and consequently stimulus evaluation takes longer. This finding is consistent with results that indicate a reduced speed of information processing ability in HSDs, with a resultant impairment in rate of acquisition/learning of material (Salame, 1991). Taken together, these findings have important implications for HSD's ability to process information quickly while performing complex tasks.

In summary, it can be seen that recent findings indicate that HSDs show impaired memory performance on behavioural tests as well as deficits in information processing ability as indexed by psychophysiological measures. Based on these findings information concerning the effect of heavy social drinking on cognitive processes involved in memory formation and retrieval could potentially be attained with the use of paradigms which investigate memory and concurrently recorded ERPs.

Electrophysiological and memory experiments

Our laboratory has investigated cognitive functioning in the presence and absence of a pharmacological challenge in groups of HSDs and LSDs. Indices of cognitive functioning came from ERPs elicited by a recognition memory paradigm (continuous recognition task) and ERP differences based on subsequent memory performance (modified Von Restorff task). The continuous recognition task required subjects to overtly discriminate between words presented for the first versus a subsequent presentation (see Rugg and Nagy, 1989). In the modified Von Restorff task subjects were required to learn a number of word lists each containing 15 words: three of which were rare or 'distinctive' words, that is, red and in upper case. Both these paradigms are designed to investigate components of the ERP which have been used to complement behavioural measures of memory. These components are N_{400} which is suggested to be a marker of lexical processing evoked by words in lists, sentences, and text (Van Petten *et al.*, 1991) and the P_{300} which has been suggested to index the updating of working memory following retrieval of information from long-term memory (Van Petten *et al.*, 1991). At a functional level, P_{300} amplitude indicates stimulus significance and it has been suggested that it may be associated with processes related to memory (Fabiani *et al.*, 1986; Paller *et al.*, 1988; Donchin, 1981). Johnson (1993) postulates that the nature of the processes performed by particular P_{300} generators is a form of memory access and while memory processes may not be the sole contributor to P_{300} amplitude (see Verleger, 1988), evidence suggests that positivity systematically fluctuates with memory processes (Van Petten *et al.*, 1991).

ERP investigations of recognition memory have reported that previously pre-sented words that are correctly identified as 'old' elicit larger positive ERPs compared to 'new' items (Friedman, 1990; Rugg and Nagy, 1989). Smith and Halgren (1989) have contended that 'old/new' ERP effects elicited in recogni-tion memory tasks largely reflect neural processes associated with episodic memory functioning. A number of studies have also shown that the P_{300} ampli-tude evoked by words on initial presentation is predictive of subsequent memory performance (Fabiani *et al.*, 1986, 1990; Paller, 1990). Even though the P_{300} dif-ference in ERPs recorded on initial presentation has led to the consensus that the 'Dm' (difference based on subsequent memory) reflects processes related to for-mation of a memory trace, there is disagreement about its relationship to the P_{300} and the functional significance of the Dm itself (see Donchin and Coles, 1988; Halgren and Smith, 1987 for differing views).

In terms of the effects of chronic alcohol intake on cognitive functioning our findings again suggest a number of differential effects between the HSDs and the LSDs. In the Von Restorff experiment a trend was found for the LSDs, but not the HSDs to have larger amplitude P_{300}s to recalled words compared to not-recalled words (Dm effect) under the placebo treatment at the parietal midline site (Pz). P_{300} amplitude to recalled words following placebo was smaller for the HSDs than for the LSDs and did not significantly differ from their results under lorazepam, whereas P_{300} amplitude to recalled words for LSDs was attenuated following lorazepam ingestion (see Figure 17.1).

This suggests (Fabiani *et al.*, 1990) that HSDs at this early stage of abuse are not encoding words or laying down a memory trace as efficiently as LSDs. However, this suggested deficit in encoding was not reflected in the behavioural data and the P_{300} amplitude data showed that HSDs were not affected to the same degree by the CNS depressant effects of lorazepam as were LSDs regardless of whether the word was later recalled or not. The differential effect of lorazepam on

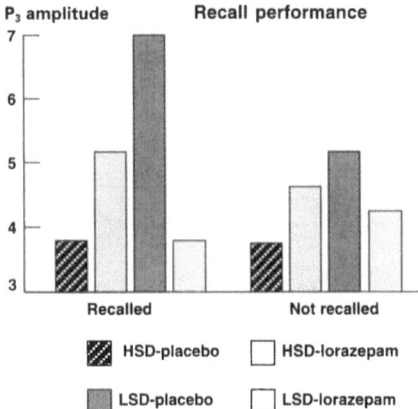

Figure 17.1 – P_{300} amplitude (μV) to recalled and not-recalled words for HSDs and LSDs for both lorazepam and placebo drug treatments at Pz.

the LSDs and HSDs was also evident in the continuous recognition task. An impairment in recognition memory, a trend towards an increase in RT, and an enhanced P_2 amplitude to 'new' words caused by lorazepam administration were only evident in the LSDs. In addition, the P_{300} amplitude of the LSDs was larger than for the HSDs to both old and new words in the placebo treatment. In the lorazepam treatment LSDs had larger P_{300} amplitudes at central (Cz) and frontal (Fz) midline sites than HSDs to old words only. In terms of the old/new effect, both groups showed larger P_{300} amplitudes to old words compared to new words in the placebo treatment. In the lorazepam treatment LSDs showed the old/new effect at Fz and Cz, whereas HSDs showed the effect at Cz and Pz. That is the old/new effect was reduced in LSDs (at Pz) under the effects of lorazepam (see Figure 17.2).

This differential effect was also evident in the behavioural data, in which only the LSDs exhibited poor discrimination/ recognition ability following lorazepam administration.

Regardless of the functional significance of the ERP component deviations evident in HSDs, the results appear to indicate that facilitation of GABA-ergic inhibition is occurring in the absence of an acute facilitation agent (e.g., alcohol or a benzodiazepine) as well as a tolerance to the effects of lorazepam. It has been suggested that a tolerance to the effects of benzodiazepine administration would reflect changes due to alcohol intake as benzodiazepines and ethanol have a similar pharmacological profile (Kril *et al.*, 1988). In the continuous recognition and Von Restorff experiments the effects of social drinking on cognitive processes involved in memory performance have been demonstrated by the differential effect of lorazepam on discrimination, Dm, early stimulus processing, and the old/new memory effect. Thus, the findings may indicate that alcohol intake in excess of 200 g per week including at least one session in which 100 g or more were consumed, has caused changes to the benzodiazepine-GABA

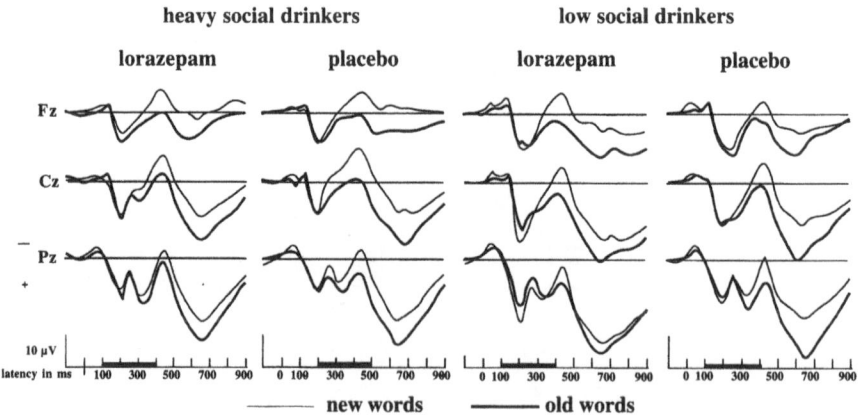

Figure 17.2 – Grand mean waveforms elicited by 'new' and 'old' words for HSDs and LSDs at each midline site (Fz, Cz, and Pz) for both the lorazepam and placebo drug treatments. Black bar indicates stimulus duration.

system which has been implicated as a mechanism involved in alcohol-related brain damage (e.g., Lovinger, 1993). Other studies (Freedman and Cermak, 1986; Kril *et al.*, 1988) add support to the contention that this system, possibly in the frontal lobes and midline di-encephalic structures, is affected by alcohol abuse and also is involved in processes related to memory formation and retrieval which appear to be affected by alcohol ingestion.

IMPLICATIONS: DOES SOCIAL DRINKING IMPAIR INFORMATION PROCESSING AND MEMORY PERFORMANCE?

In the past, recommendations of drinking levels have suggested that there is a threshold of drinking behaviour above which the individual will suffer deleterious effects. The focus has changed in recent years to educating the general public on the need for 'safe' drinking practices. Our studies and others (e.g., Nichols and Martin, 1993, Nichols *et al.*, 1993; Salame, 1991; Williams and Skinner, 1990) suggest that a daily intake of alcohol above 80 g or a regular consumption of at least 200 g/week which includes a session during which an excessive amount of alcohol (above 100 g) is consumed can result in changes occurring in CNS functioning which may affect the way we process information including memory. Since a common and perceived social use pattern is the intake of relatively large quantities of alcohol during any one drinking episode this has implications for the amount of risk people put themselves in when they drink in this fashion. The question remains, however, as to how permanent these effects are and to what extent they will be affected by future drinking behaviour.

Are the effects transient or permanent?

Evidence that supports the assertion that certain acute intoxication effects may persist to subsequent short- or long-term sober states has been inconsistent, however this issue has vital implications especially for younger drinkers (Bates and Tracy, 1990). The persistence of mild or short-term cognitive deficits to the sober state could have serious implications for everyday performance levels.

At the present time, there is insufficient data in either direction to conclude that social drinking will result in permanent brain damage. Hill and Ryan (1985) have suggested that cognitive deficits in social drinkers are a result of the disruptive effects of a withdrawal syndrome. However, this so called 'hangover hypothesis' has received limited experimental support (Bowden *et al.*, 1988; Alterman and Hall, 1989). Irrespective of whether the deficits are transient or not, it is important to investigate drinker's cognitive functioning in the typical condition in which they perform everyday activities (Waugh *et al.*, 1989). The drinkers tested in our studies had maintained the same pattern of intake over several years and according to the research of Schuckit (1985) were likely to maintain it until they reached 25, at which time statistically their intake should decrease. Hence the deficits in this population, evident in our studies following at least 2 years of binge drinking, are likely to continue until the subjects cease binge drinking. If the deficits, subtle as they appear to be are indeed transient they are recurring with every binge session and this drinking pattern, that is, persistent or repetitive episodes of withdrawal, may result in changes in brain mechanisms implicated in alcohol-related brain damage (Lovinger, 1993). The significance of these deficits and those highlighted by others (e.g., Salame, 1991) needs to be clarified, particularly in behavioural terms.

Investigations into the effects of alcohol, whether chronic or acute, have the problem inherent in all experimental research in that it is difficult to make comparisons across studies and conclusive statements about general implications. That is, studies investigating the relationship between social drinking and cognitive deficits have employed a variety of methodologies and statistical techniques leading to inconsistent findings. Issues which need to be resolved include sample or subject selection, the indices or tests, designs and statistical methods used, and subject variables such as genetic predisposition, age, sex, education, intelligence, neuromedical status, length of abstinence at time of testing, and recent drinking pattern and amount.

In recent investigations in our laboratory and others (e.g., Waugh *et al.*, 1989; Williams and Skinner, 1990) confounding factors that may influence cognitive function, for example age, educational level, occupation, other substance abuse, neurological damage, duration of alcohol consumption, number of days since last drink, were comparable across groups. However, in these studies reporting impairment in memory and information processing indices, subjects were abstinent from alcohol only for limited time periods, that is 1 week or less, prior to assessment. The possibility that the reported cognitive deficits may be due to temporary alcohol effects, particularly when binge drinking is involved, cannot be ruled out. Although studies have shown that recent alcohol intake does not influence cognitive performance in moderate drinkers (see Bowden *et al.*, 1988; Waugh *et al.*, 1989) the possibility remains that the cognitive deficits reported in HSDs may be due to 'hangover' effects.

FUTURE DIRECTIONS

Future neuropsychological research needs

- to focus on outlining the qualitative, quantitative, and temporal features of deficits associated with heavy social drinking;
- to employ current neurophysiological and neuroradiological techniques to identify the underlying functional and structural changes; and
- to investigate the relationship of these variables to neuropsychological and behavioural indices such as recent drinking pattern, age, sex, physical structure, as well as other factors.

Although the significance of the deficits reported in our investigations for everyday functioning is unclear, the HSDs (>200 g/week) show a differential effect of benzodiazepine administration which indicates that the social pattern of consuming a large quantity of alcohol per sitting in these young drinkers is causing changes in neural mechanisms that have been implicated in alcohol-related brain damage (e.g., Lovinger, 1993). In addition, this review demonstrates that pharmacological challenges can be used as tools for analysing the neurochemical mechanisms underlying behaviour. A number of research areas are likely to be advanced by the use of combinations of cognitive and pharmacological methods, for example, pharmacological modeling of neuropsychiatric disorders, the development of pharmacological challenge paradigms to help diagnostic accuracy, and the use of brain imaging techniques along with cognitive and pharmacological challenges (see Danion *et al.*, 1993).

The inconsistent findings of cognitive deficits in heavy drinkers who do not show gross clinical signs may be a consequence of unsophisticated psychometric techniques (Alderdice and Davidson, 1990). Bates and Tracy (1990) suggested that subtle decreases in functioning occurring with continued alcohol use may only be detectable by the use of both within-subject comparisons over time and more sensitive cognitive tests. ERP deviations reported in our investigations due to heavy social drinking suggests that ERP technology can be employed as a sensitive index of cognitive functioning. Even though the exact nature of the cognitive functions underlying the ERP components are yet to be delineated it is of crucial importance to establish whether subtle changes in information processing are evident at early stages of abuse as well as their persistence over time.

Advances in technology will allow a more complete picture of the effect of social drinking on CNS anatomy and functioning. Future research will need to utilise ERP measures in combination with advanced imaging techniques, for example, MRI technology, that will provide a clearer indication of cortical atrophy. PET studies to investigate metabolic changes in relation to withdrawal, reduction of alcohol consumption levels, and abstinence would also be useful. These could be employed in the context of longitudinal studies which represent an under-utilised method of investigation. That is, longitudinal studies would provide valuable information on the effect of various levels or patterns of social drinking on cognitive functioning as they allow the control of pre-existing differences in cognitive ability and would provide information on age related vulnerabilities to alcohol effects by using within-subject comparisons.

In summary, research efforts to delineate the nature of the reported alcohol related, subtle impairments in cognitive functioning as indexed by sensitive measures will need to control for confounding factors and utilise between-groups, within subjects longitudinal designs in order for conclusions to be made about the effect of various levels of social drinking on memory and information processing. In addition, the persistence of these deficits needs to be investigated by re-testing following a period of at least 3 weeks abstinence. The main aims of social drinking research should be

– to investigate whether the observed cognitive deficits become more severe with continued drinking,
– to determine whether the impairments stabilise or improve when alcohol consumption is decreased,
– to identify the significance of these impairments, whether transient or persistent, in terms current levels of everyday functioning, and
– to identify high risk patterns of drinking behaviour, for example excessive intake per occasion versus high daily intake since the possibility of a cumulative effect with each episode needs to be clarified.

In conclusion, the continued use of alcohol is virtually guaranteed due to its availability, ease of production, social acceptance, and reinforcing properties (Kalant and Khanna, 1989; McKim, 1991). Even though a definite statement still can not be made on 'safe levels' of alcohol consumption studies reported here support the need 'for a drink less per occasion' approach to alcohol intake.

SUMMARY

This chapter provides a review of the evidence suggesting that social drinking of alcohol may exert a toxic effect on brain functioning, such that memory processes and information processing abilities are impaired. A number of theories which relate to social drinking or posit a neurotoxic effect of alcohol are reviewed and consideration is given to a continuum of impairment such that social drinking occupies a place low on the continuum of alcohol related brain damage. Earlier research, particularly in relation to neuropsychological deficits caused by social drinking, was plagued by methodological inadequacies, however, with the advancement of technology, particularly CT, PET and MRI scans, and advances in experimental design, a clearer picture is emerging of the effects of heavy social drinking on brain functioning. When alcohol is consumed at levels which are regarded as socially acceptable, especially when large amounts of alcohol are consumed in one sitting, recent research indicates that impaired information processing and memory deficits may result.

CHAPTER EIGHTEEN

Limited Alcohol Consumption and Complex Task Performance

Siegfried STREUFERT and Rosanne POGASH

INTRODUCTION

The frequently observed negative impact of self administered psychoactive drugs upon behaviour has led many individuals, including several scientists, to condemn self-administration of psychoactive substances, such as alcohol, at nearly any level. A number of authors suggest universal or near universal adverse effects (e.g., Miller, 1992). Similarly, Koelega (1995) argues in favour of a 0.02 maximum allowable serum level for drivers of automobiles.

In their review of the literature, Finnigan and Hammersley (1992) suggest that prior publications could easily lead the reader to believe that *any* demanding performance may be impaired by *any* (italics ours) amount of alcohol. Yet, based on a more careful analysis of the relevant literature the same authors conclude that there is little certainty about the *universal* effects of alcohol. Findings of a *low dose* alcohol impact are especially bothersome. Where negative effects are obtained, they tend to be similar to effects of fatigue, boredom, hunger, eating and other risk factors that are part of normal life and impossible to eliminate. As Finnigan and Hammersley point out: it would be a mistake to single out small doses of alcohol as a major concern.

In addition, frequent problems inherent in alcohol research tend to place many conclusions into doubt. Publications tend to be biased toward reporting deleterious effects. Many research designs are seriously flawed. Moreover, the belief that alcohol at all or nearly *all* dose levels, and under *all* task conditions, must have universally negative effects is difficult to maintain in the face of several research publications that have reported improved functioning upon consumption of low alcohol doses (Threatt, 1976), for example improved digit symbol substitution, short term memory (cf., Dufour *et al.*, 1992), reaction time (Palva *et al.*, 1979) or visual acuity (Miyao and Ishikawa, 1994), more effective social interaction, enhanced initiative and improved communication (Mishara and Kastenbaus, 1980) or psychomotor performance (cf., Kerr and Hindmarch, 1991).

Conflicting data on the effects of alcohol upon performance can arise from a number of factors. Whether human functioning is positively, negatively or not at all affected at low treatment levels might be task specific. Different methodologies including different tests and measures may generate diverse results. Individual or population (sample) differences may influence performance (Kerr and Hindmarch, 1991). In addition, an often decried lack of experimental rigour in much alcohol research decreases confidence in published data (Finnigan and Hammersley, 1992).

The diversity of results and the differing conclusions have generated conflicting interpretations of alcohol effects upon human functioning. It is not unusual for investigators to ignore each other's conclusions or to discount data which diverge from their own. Such lack of 'inquisitive openness' to alternate research results can generate 'biased' views. For example, it has not been unusual for investigators to declare someone else's measure of performance to be 'insensitive,' whenever that measure fails to show a negative effect of alcohol intoxication. Such 'prejudgments' are questionable and may be unfortunate. Instead, we should consider and, where possible, investigate *why* obtained differences occur and what their underlying causes might be. Whenever unexpected results are *not* merely due to experimental error or poor research design, they can provide insights about alternate for example, biphasic effects of alcohol or may show how multiple variables interact to generate serendipitous effects. A greater openness to alternate interpretations would help us determine what impact alcohol consumption may have upon human functioning, not only in the somewhat sterile laboratory, but also in the complex day-to-day world (ecological validity).

Many individuals who work, many who interact with others drink alcohol from time to time (Martin, 1990). Intoxication may occur at occasions when they are called upon to function optimally. If alcohol has a detrimental effect the resulting difficulties could have serious consequences for persons with considerable responsibility. As research has shown, alcohol use among such individuals is not uncommon (Hilton, 1991). But it is not merely the fact or degree of intoxication that might have a positive or negative impact upon functioning. The outcome of alcohol consumption on effectiveness may be in part determined by an interaction of intoxication with a range of variables that include perception, individual difference and more. People's daily world not only includes multiple stimulus variables that can interact with drug effects, each with some potential importance, but it also includes multiple perceptions, habits and other person characteristics that might influence functioning during intoxication. In other words, our interpretations of research data should not only focus on alcohol effects *per se*, but should also consider covariates such as expectancy (Baum-Baicker, 1985; Dufour *et al.*, 1992), stress reduction (cf., Young *et al.*, 1990), individual differences (Hammersley *et al.*, 1994; Kerr and Hindmarch, 1991) and other environmental and person variables.

As Kerr and Hindmarch (1991) suggest, psychoactive drugs including alcohol can have biphasic dose-related effects on some but not necessarily all aspects of human functioning. In small doses some drugs may stimulate even if that stimulation occurs through selective depression of inhibition (Pohorecky, 1977; Pohorecky and Brick, 1988). In larger doses, the same drug may function as a depressant. Alcohol is generally considered a CNS depressant (Finnigan and Hammersley, 1992) yet can initially act (at low BALs) through expectancy effects (Dufour *et al.*, 1992), as a general relaxant (Finnigan and Hammersley, 1992), or by diminishing 'inhibitions' (Baum-Baiker, 1985). As a depressant, at higher doses, it likely has negative effects upon performance. Whether that effect is greater with regard to simple, moderate or complex task performance (Baylor *et al.*, 1989; Landauer and Howat, 1982; Moskowitz *et al.*, 1985), even at small doses (Billings *et al.*, 1972; Drew *et al.*, 1958; Lovibond and Bird, 1971) is a matter of debate. Available research is inconclusive. As a relaxant, alcohol might improve some forms of performance in individuals who are anxious and/or in task settings that are stressful (arousal) (cf., the extensive review of alcohol as a tension/stress reducer by Baum-Baicker, 1985).

Still, many effects of alcohol consumption (on complex functioning) may have a negative impact. For example, it has been suggested that alcohol consumption might diminish long term planning, that certain activities might decrease even though others increase (Tiplady, 1991), that reaction time in response to information and rate of information processing may slow (e.g., Maylor and Rabbitt, 1987, 1988; Maylor *et al.*, 1989).

A frequent argument (Baylor *et al.*, 1989; Landauer and Howat, 1982; Moskowitz *et al.*, 1985) that intoxication during 'complex tasks' produces a greater or universally negative impact or that alcohol will affect complex task performance more severely, or at lower BALs, appears in good part based on the finding that divided or selective attention and reaction time frequently show decrements. In tasks requiring maximal diversity of attention or maximal response speed, for example flying an aircraft under emergency conditions, that argument may be quite correct. However, settings with those demands are less common in the world outside of the laboratory. In addition, any one particular response capacity may not always be decisive. For that matter, a decrement in one particular function may either aid or hinder another (more or less important) aspect of task performance.

In other words, we should caution the reader to be somewhat circumspect in his or her judgments. It *is* possible that opposing research findings are *both* correct and *meaningful*, whenever a small variation in research design or a slight modification in a sample has occurred. Any interpretation of the impact of psychoactive drugs, such as alcohol, on human functioning should consider – differences across individuals, – differences across type and/or dose of drug, – the method or timing of administration, – the concurrent task setting and requirements, – a participant's current physiological status, as well as any other variables that differ between one research effort and another.

WHAT IS COMPLEX PERFORMANCE?

The majority of research on alcohol effects has employed controlled laboratory tasks, such as tracking, choice reaction time or signal detection. At a slightly more demanding level of functioning, divided attention tasks, driving, or aircraft simulators have been employed. Occasionally, human functioning in the day-to-day real world, such as operating an actual automobile, social interaction, or decision making[1] in challenging settings have been observed and/or scored. Certainly

---

[1] Authors of cognitive theories sometimes consider relatively simple choice tasks under the rubric 'decision making' for example, signal detection, reaction time, etc.. In this chapter we will follow views by authors of decision theory who consider decision making in a more naturalistic and complex setting, for example, in managerial or professional jobs.

the task characteristics and demands of prior research are sometimes widely dissimilar. Investigators have distinguished the 'complexity' of those settings and/or task demands in various ways. How should one optimally classify research methods if one wants to identify tasks as 'simple' versus more 'complex'?

Unfortunately, there is little agreement about the definition of 'simple' and 'complex' task performance, especially if one considers the disparate views of researchers from various branches of psychology. Better communication between scientists based in cognitive theory who, for example, tend to view divided attention tasks as 'complex' and those from social or organizational psychology (who would consider the same divided attention performance to be 'simple') is needed. The former group emphasizes the difference between a single, simple, and a multiple, complex, focus. The latter considers the potential for complexity in terms of the ecological, real-world, impact on human day-to-day endeavours, for example, complex functioning would include such behaviours as strategic social interactions, decision making about finances, etc.

To obtain a more *generically* acceptable view of 'complex' versus 'simple' tasks, we will build upon the views of Streufert and Swezey (1986) whose definition considered task simplicity vs. complexity on the basis of real-world performance demands. Selecting definitions relevant to 'ecological validity' appears meaningful: after all, much debate about the dangers of psychoactive drugs focuses upon the influence of those substances on human behaviour and effectiveness in real-world rather than laboratory settings. Nonetheless, to satisfy both cognitive and social/organizational researchers, we will extend Streufert and Swezey's distinctions into three rather than the earlier two levels on the simplicity/complexity dimension.

We will define tasks as *'simple'* whenever they require the association of a single stimulus with a single response. Many laboratory tasks fit that description, including a number of memory paradigms, many vigilance tasks (e.g., reaction time experiments where no additional stimuli are introduced), simple tracking tasks and more. In 'real-world' settings the process of hammering an already started nail into a board would also fit this description.

Tasks of *'moderate complexity'*, in contrast, would require simultaneous attention to multiple at least two *kinds* of stimuli, such as divided attention, for example, to an auditory and a visual stimulus and/or uncomplicated sequential behaviours, requiring more than one kind of action following another. Examples are laboratory experiments where additional stimuli interfere with stimuli that require primary responding, tasks where attention, differential responding, to multiple stimuli is required, or settings that require both multiple and sequential responding (where later actions are based on earlier actions) in the presence of multiple stimuli, such as driving an automobile. Note, however, that our definition of moderate task complexity focuses primarily on settings requiring *responsive* behaviours to, potentially changing, stimulus configurations. Nonetheless, tasks of moderate complexity may include some limited volitional (planning based) modifications of action which occur in *response* to given conditions, such as changing one's route home from work if the traffic on the familiar road seems too dense.

'Complex tasks' may include components of simple and intermediate task performance, but must also involve multiple volitional influences by a human actor upon his or her environment and/or sequential behaviours that reflect purposeful, strategic activities toward attempted planned outcomes. Examples may be

decisions about investment strategies, most managerial and executive actions dealing with complex and uncertain settings, or efforts to resolve interpersonal conflicts through negotiation.

With our definitions of 'simple', 'moderate' and 'complex' on the basis of task characteristics, we find ourselves in some disagreement with Finnigan and Hammersley (1992) who suggest that conflicting research data obtained from 'simple' versus 'complex' (their definition) task settings are due to a failure to distinguish *task* characteristics from the simplicity or complexity of associated *neuropsychological routines*. For example, in their view the addition of expectancy could make a 'simple' task 'complex.' In terms of our, above mentioned, definitions, the additional impact of expectancy or some other person-variable could maximally elevate a simple task to a moderate task. Yet, one may wonder whether any task, if one accepts their views, could ever be 'simple' as long as a human responder with his/her specific perceptual characteristics and cognitions is involved. Moreover, while Finnigan and Hammersley's views might help to explain potential discrepancies in research findings with any one particular task, their approach seems less helpful for conceptual distinctions: Expectancies and other subject effects may be generated by methodological (treatment) differences *but* may also originate from individual differences among research participants, sub-populations or populations. The latter differences are difficult to control on a pre-hoc basis and certainly are not interpretable post hoc when one reviews prior research.

With 'simple,' 'moderate' and 'complex' tasks defined, we must still agree on what we shall consider 'low level alcohol consumption.' Again, we should base our view of human functioning on the day-to-day real world. Persons who do not wish to become 'intoxicated' tend to limit themselves to a maximum of one or two drinks (cf., Baum-Baicker, 1985). That quantity matches the dietary guidelines of USDA and USDHHS (1990) which view 'moderate' drinking as no more than two drinks per day for men and no more than one drink for women.

However, number of drinks is not a good predictor of subsequent intoxication or of subsequent blood alcohol levels (BALs). For example, the alcohol content of drinks can vary. To allow for those differences, 'low' levels of consumption will be (somewhat arbitrarily) defined as alcohol quantities producing blood levels below 0.05, as measured in serum or via a calibrated breathalyzer.

CONTRIBUTING VARIABLES

Unfortunately, published research data on the impact of moderate or low levels of alcohol upon complex functioning as defined above, are extremely limited. As a result, we must, to some extent, consider

- the extrapolation of lower level performance *components* that frequently contribute to highly demanding (complex) real-world settings,
- performance data from moderately complex tasks, such as driving and flying an aircraft,
- effects of 'perceived intoxication,' sometimes referred to as 'expectancy variables' and

– potential joint (additive or interactive) effects of relevant variables that would or might have an impact in complex settings.

A consideration of these phenomena should hopefully aid us in interpreting the limited data on alcohol and complex functioning.

Performance components

Most laboratory data on alcohol and performance are of limited utility for human functioning in complex settings. However, a few among the more frequently researched variables may contribute to our knowledge of complex task performance. We will focus on three areas of prior research: divided attention, memory and anxiety/relaxation.

Multiple sources of information are typically present in complex task settings and may require divided attention. Research data based on moderate or high levels of intoxication, typically employing relatively uncomplicated but multiple performance requirements, suggest that effectiveness on one task component may be maintained at the cost of deteriorated performance on another component (cf., reviews by Finnigan and Hammersley, 1992). While simultaneous attention to multiple sources of information in complex task settings is *usually* of value, a more 'myopic' that is focused approach may, under specific task conditions, result in greater success. For example, managers typically overreact to too much information or overload (cf., Streufert and Swezey, 1986). Their planning efforts can easily exceed realistic projections (Streufert and Nogami, 1989). Overload may damage their capacity to handle concurrent task demands, generating superoptimal response frequency (Streufert and Streufert, 1978). If any of such 'excessive' activities are reduced, effectiveness might increase.

On the other hand, strategic behaviour, one major antecedent of success in many complex task settings, requires considerable breadth of orientation. If that breadth were reduced by limitations on divided attention, the quality of performance should deteriorate. In other words, while alcohol might, where task demands are highly complex, reduce some undesirable excesses of information processing, it might, if divided attention research data are applicable, also diminish other aspects of performance that would have been better served by increases in breadth. Whether a more broad or a more myopic approach will occur under low levels of alcohol consumption is, however, as yet unclear. Prior research has provided conflicting data: relatively low doses of alcohol may (Moskowitz *et al.*, (1985) or may not impair performance (Mills and Bisgrove, 1983).

Another important component of successful complex task performance is memory. Memory is needed in communications, in decision making and in a vast array of other activities that are carried out by managers, professionals and others. Although much research has shown that alcohol, at moderate to high BALs, has a negative impact upon certain aspects of memory (Lister *et al.*, 1987; Ryback, 1971), even at low BAL values (Roache *et al.*, 1993), recall of information received a short time prior to intoxication may improve (Lamberty *et al.*, 1990; Mueller *et al.*, 1983; Parker *et al.*, 1974, 1980). Improvements may be due to impairment of subsequent learning and indirectly by protection of contextual cues that are needed for the retrieval of episodic memory (Tyson and

Shirmuly, 1994). Better recall of *recent* information where it is more relevant than other, for example concurrent information, such as where preparation for a 'wet' business lunch with a potential client is most important, could allow an improved focus on issues at hand, possibly reducing retroactive interference, generating more optimal performance. However, where, for example in ongoing negotiations, continuous openness to new information is required, where information arriving during intoxication must be remembered and utilized, performance quality may well be diminished. In other words, whether better overall memory or specific memory would be more important should, again, depend on the complex characteristics of a task at hand.

Complex task environments are often associated with uncertainty, volatility and ambiguity, creating stress experience. Resulting anxiety may hinder effectiveness. To resolve those problems, certain individuals will consume alcohol in an attempt to reduce the stress of demanding tasks (Peyser, 1992; Young *et al.*, 1990). Limited alcohol consumption may aid anxious individuals by relieving anxiety and promoting relaxation (Netter *et al.*, 1994) or by reducing self-consciousness (Hull and Young, 1983). Perceptions of external stressors may be reduced, potentially leading to improved functioning (Patel, 1988). The degree to which alcohol might aid or hinder any one individual would, in part, depend on personality characteristics (e.g., Hammersley *et al.*, 1994) such as introversion/extraversion (Netter *et al.*, 1994), *and* would be impacted by perceived stressor levels that are generated by (inherent in) the task itself (Streufert and Swezey, 1986).

Tasks of moderate complexity

While there are important differences between tasks defined as moderately complex and those defined as complex, some effects of low levels of alcohol consumption may show limited similarities. As defined earlier, tasks of moderate complexity, such as operating a vehicle or flying an aircraft, typically do not include multiple volitional influences and sequential behaviours that reflect purposeful strategic activities toward presently planned outcomes. Rather, tasks of moderate complexity more often contain, sometimes complicated but not complex, responsive behaviours whether in reaction to current conditions or in pursuit of a 'flight plan'. Only limited sequential and volitional activities are involved. These tasks do, however, require divided attention, likely involving memory of information acquired prior to intoxication, and would likely be affected by relaxation and CNS depression if any, potentially resulting in attendant slowing of response speed, etc.

We do not intend to review performance in moderately complex settings: enough prior reviews are available (e.g., Ferrara *et al.*, 1994; Finnigan and Hammersley, 1992; Gibbons, 1988; Moskowitz and Robinson, 1988). Our approach will be limited to a brief consideration of those aspects of driving an automobile and piloting an aircraft that might have some implications for more complex functioning.

Research on driving an automobile has often employed laboratory tasks with extrapolations of data to actual functioning at the wheel. Occasionally, more applicable data were obtained in simulators and yet more useful results are based on actual driving under the influence of alcohol. The findings considered below are based on all three methods.

While some researchers have argued that even low levels of intoxication can impair driving a motor vehicle (e.g., Landauer *et al.*, 1969), published data, in general, are inconclusive and limited by multiple causation of road accidents (Ferrara *et al.*, 1994). Road accidents are often due to a secondary activity (Brewer and Sandow 1980), potentially relating to problems with divided attention, especially at higher intoxication levels. The 'popular' view that risk taking increases under the influence of alcohol appears not to hold (McMillen and Wells-Parker, 1987, McMillen *et al.*, 1989), especially not during ascending alcohol levels (Streufert *et al.*, 1992). Although occasional (partial) decrements in ability have been demonstrated for BALs as low as 0.02 (cf., Moskowitz and Robinson, 1988), a number of researchers has been unable to obtain deficits at BALs below 0.05. (e.g., Mitchell, 1985; West *et al.*, 1993). However, levels of 0.05 to 0.099 apparently generate some impairments (Hindmarch *et al.*, 1992), especially when intoxication interacts with other relevant variables, such as fatigue (Roehrs *et al.*, 1994).

Flying an aircraft under any levels of intoxication is prohibited in most countries. The inherent 'complicatedness' (not complexity!) of the pilot's task, requiring multiply divided attention, responding to potential emergencies, etc., makes optimal functioning a matter of utmost importance. But is piloting performance affected by low levels of alcohol? While some researchers argue that 'low levels' (such as a BAL of 0.04) diminish pilots' flying performance (Gibbons, 1988), especially upon approach to an airport (Davenport and Harris, 1992), others report decrements only under the most difficult task demands (Ross and Ross, 1992) or the heaviest workload conditions (Ross *et al.*, 1992).

What applicable knowledge can we gain from these data? Of course, even 'moderate' task performance should only be considered in terms of its potential component contributions it might make to complex task performance. Any reaction time based decrements that occur under moderate task conditions (such as emergency responses in piloting an aircraft) are probably rarely relevant to more complex tasks where success is not likely determined by milliseconds of reaction time. Limits on divided attention, where they do have an impact, may be more relevant. However, the few data suggesting deterioration under limited alcohol consumption hardly suggest a major impact upon settings where complex functioning is required.

Perceived intoxication: expectancy

A number of researchers and theorists have observed that a person's *impression* of having consumed alcohol or a particular amount of alcohol can have considerable influence upon subsequent behaviour and performance (Ross and Pihl, 1988; Thompson and Newlin, 1988; Tucker and Vuchinich, 1983). Perceived intoxication may diminish performance, even though the actually attained blood alcohol level may not (e.g., Nicholson *et al.*, 1992)[2].

---

[2] In research conducted in our laboratory, for example, a senior police manager showed signs of serious intoxication and finally claimed that he could not continue because he was 'much too drunk.' He was rather surprised when we showed him his breathalyzer readings of 0.00. He had participated in a placebo condition.

Expectancy at low levels of intoxication, however, does not necessarily have a negative impact upon human functioning, at least not for all individuals, or in all task settings. Where a person believes that he or she will be more effective after a drink or two, especially where increased relaxation replaces prior anxiety, expectancy could generate performance improvements. On the other hand, beliefs that alcohol will deteriorate performance can result in a negative outcome. Another rather different form of 'expectancy' occurs when perceived intoxication generates attempts to 'marshall one's energy' in order to overcome drug induced decrements. The additional effort might eliminate those decrements or, especially if decrements are minor or merely imagined, could improve functioning on affected dimensions (cf., Streufert *et al.*, 1992).

Without question, perceptions and related expectations play a considerable part in the decision to drink or not to drink alcoholic beverages. Expectations and perceptions, in turn, can affect performance levels (Nicholson *et al.*, 1992). For populations that include most managers and professionals, urban high income college graduates, who tend to drink more or more frequently than other groups (Hilton, 1991), social norms tend to permit this indulgence (Shore, 1985, 1986). In managerial and professional (complex) interpersonal settings, alcohol, in some part via expectancy mechanisms (e.g., Hull and Bond, 1986), may generate feelings of solidarity by relaxing the formality of interpersonal relationships, thereby increasing chances of consensus (Streufert *et al.*, 1994). If intoxication remains, intentionally, limited to low BAL levels, an optimal mix of consensus and control may improve a person's chances of success, such as closing a business deal (cf., MacAndrew and Edgerton, 1969).

As we have already seen in our discussions of other contributors to complex functioning, extrapolations do not generate clear predictions. The same is true for expectations of intoxication; expectancy variables. Both unfavourable and favourable effects might occur. Which impact is observed would again depend on the person, the specific task characteristics, and more.

Effects of variable interactions

Laboratory research generally attempts to eliminate or at least control variables that are extraneous to the relationship between a specific independent and a specific dependent variable. As a result, laboratory researchers can reach conclusions with some amount of certainty. However, human functioning in the 'real world' differs considerably from the laboratory, especially whenever task settings contain considerable complexity. Contributing variables may add, may cancel each other (to some extent) or may interact to generate 'unique' impacts upon performance. We have already encountered one such interaction in the section above: Expectancy effects can modify the role of alcohol in task performance. Unfortunately, *observational* research in real world settings cannot control, or may not even obtain knowledge about the impact of interactions. For example, a person who is known to have consumed alcohol may also have smoked marijuana, may have taken tranquilizers, may be consuming caffeine or may be smoking cigarettes of unknown nicotine content, that is multiple drug use is rather common (Martin *et al.*, 1994). In addition, specific characteristics of the concurrent social environment, for example social conflict, may have their own impact. Demand characteristics may shape or modify task orientation, and so on.

The most popular solution to this dilemma is a return to small scale laboratory research where such variables can be (partly!) controlled. The consequences, however, are task levels that cannot represent the day-to-day world of many persons. Another alternative is the controlled presentation of a complex task environment, as employed in quasi-experimental simulation techniques (Streufert and Swezey, 1985). Quasi-experimental simulations are to some extent a compromise between tightly controlled and applied research, but do allow considerable experimenter control and generate data with considerable ecological and statistical validity (Streufert *et al.*, 1988). However, the latter techniques have not been widely used, in part because they are relatively new, in another part because they are rather expensive.

Consequently, current knowledge about complex functioning under low levels of alcohol intoxication is limited and requires considerable future research. We are merely at the beginning. Nonetheless, in designing future research we should be aware that complex behaviour in the real world is subject to the often simultaneous impact of multiple variables and variable interactions. If we wish to study and/or predict human functioning in the real world, complex interactions are part of a behaviour that cannot (and should not) be simplified by researchers who intend to learn more about drug impact on effectiveness.

RESEARCH DATA ON ALCOHOL AND COMPLEX FUNCTIONING

As already indicated above, available data on alcohol intoxication and complex functioning is rather limited. The paucity of information is even greater for effects of low level alcohol consumption on complex functioning. As already suggested, the paucity of relevant work is easy to explain. Observational research in real-world settings is time consuming, requires consent that is not easily obtained and, in addition, suffers from design limitations that may be difficult or impossible to overcome. Potentially real-world relevant research instruments such as in-basket techniques, micro-worlds and role-play methods tend to have serious limitations (cf., Breuer *et al.*, 1996), making interpretation of obtained results difficult and uncertain. Simulations, especially quasi-experimental simulations with their considerable ecological validity, are more useful and therefore preferable, but tend to be very expensive and, consequently, are not available to most researchers.

In this section of the paper, we will report on the few available sets of data: one observational technique and three simulations. Unfortunately one of the latter, and the one with greatest relevance to low level alcohol consumption, was merely a pilot project with a limited numbers of subjects.

Interpersonal behaviour

Mishara and associates (1975) studied the effects of limited availability of alcohol, maximum 2 drinks, on older individuals. Increased interpersonal interaction, improved communication, greater initiative and more frequent group activity was observed. It is unclear whether the obtained results were generated by intoxication or by an improved social climate generated by the experimental treatment, that is by the availability of alcohol. However, to the extent to which the climate improvement was more alcohol than age or environment related, limited consumption of alcohol, probably in interaction with climate, appears to have improved some aspects of social functioning.

Complex task performance

Jobs, Fielder and Lewis (1990) conducted research on the effects of alcohol and expectancy upon business decision making. A relatively simple simulation allowed managerial decisions on pricing and product ordering. Participants received either placebo or alcohol to attain a BAL of 0.06 (actual measured peak levels were 0.04 through 0.075). According to the present definition, the attained BALs in the Jobs *et al.* research (1990) would place intoxication levels into the moderate rather than the low range. Subjects received training while consuming alcohol or placebo. Performance measures indicated decreased memory based competence. The data suggest that some learning related findings obtained from research in less complex tasks may be, at least partly, applicable to complex tasks. While learning prior to alcohol consumption might result in enhanced memory, not measured in the research, items learned during or following alcohol consumption may be subject to forgetting.

Degree of risk taking (risky versus non-risky decisions) was also assessed. Participants who were intoxicated took greater risks. However, expectancy, in the absence of intoxication, in the placebo treatment, also increased risk taking. In sum, risky behaviour in response to intoxication during complex task performance is likely due to generally held beliefs that greater risk behaviour is associated with alcohol consumption. While the data of Jobs *et al.* (1990) report on alcohol and complex functioning, we should remember that they pertain to behaviour under intoxication levels that somewhat exceed those we defined as 'low level.' The extent to which the results would apply to individuals consuming only one or at most two drinks is not known.

Streufert and associates (1994) conducted a series of validated (Streufert *et al.*, 1988) quasi-experimental simulations. A double-blind, cross-over placebo controlled design was employed. Managers made decisions in a very complex setting over a period of several hours. They received feedback which was in good part pre-programmed to maintain comparability among different participants as well as relevance to an external criterion of excellence. Two levels of intoxication were induced *and maintained* across several hours. The low BAL level in this research (0.05) is at the margin of our definition for low levels of intoxication. Results suggested considerable deterioration of performance on most measures of complex functioning under the higher intoxication level (0.10). Data for the 0.05 level showed mixed results.

At both alcohol treatment levels, participating managers became more dependent on external information. Speed of response slowed only at the higher treatment level. Note that speed is not considered in milliseconds, as is often the case in laboratory tasks, but rather in minutes, a time unit much appropriate to managerial/professional decision making. 'Breadth of approach' to the complex task increased for most, but not all, participants with increasing intoxication. Planning and strategic actions diminished in frequency for both alcohol levels, although more severely at the higher alcohol treatment level. Systematic functioning, while falling in its mean values, was not significantly affected.

Rather different results were obtained for handling of a simulated emergency. Emergency response speed *improved* under the 0.05 treatment level but deteriorated at the higher treatment level. On most other measures of emergency functioning, except for utilizing strategy in handling an emergency, the 0.05 treatment level did not differ greatly from placebo. In contrast, several measures indicated deterioration of effectiveness at the 0.10 treatment level. Strategic functioning with reference to an emergency decreased at both alcohol levels, even though that decrease was again much more pronounced at the higher treatment level.

The lesser decrements in functioning at the 0.05 treatment level, associated with one performance measure, emergency response speed, that showed an improvement over placebo treatment, led Streufert and associates to design a pilot experiment to begin exploring the impact of alcohol treatment at a yet lower level of intoxication. In a placebo-controlled, double-blind, triple-cross over design, eleven men with managerial training and experience participated in three equivalent simulation scenarios with diverse but comparable content. On one occasion, randomized order, they received placebo, on another occasion they were given alcohol to reach a BAL of 0.02 (attained maximum mean level 0.019) and on a third occasion enough alcohol to attain a BAL of 0.06 (attained maximum mean level 0.060). The detailed treatment procedures, except for alcohol treatment levels, were otherwise identical to those reported by Streufert *et al.*, 1994).

Results of this pilot experiment may be, cautiously, compared to prior data on alcohol effects at the 0.05 and 0.10 levels (reported above). Both sets of data were collected in the same simulation laboratory with equivalent research participants (managers). Differences in mean performance among the four alcohol treatment levels are presented as deviations (in percent) from their respective placebo performance. The values in Table 18.1 indicate either the percentage of decrement from placebo levels indicated by a – sign before the percent value or as a percentage of improvement over placebo levels indicated as a + sign preceding the particular percent value.

The most striking result is the number of shifts toward improved performance at a BAL of 0.02 in comparison to the much more frequent shifts toward diminished performance at all other alcohol treatment levels (0.05, 0.06 and 0.10). Exceptions to improved performance are among others, a decrease in systematic functioning, in response speed during an emergency, and in the number of actions that are related to the simulated emergency. Particularly striking (and somewhat surprising) data, however, were generated by, a frequently validated, measure for 'diversity of action.' Diversity reflects the 'breadth of approach' to the task at hand, that is, how much attention was divided among various component stimuli in the environment and how many divergent approaches were taken to the task at hand. For this measure performance decreased (compared to placebo) at the 0.02 and 0.05 treatment level but 'improved' as intoxication increased further. As a rule, greater breadth is a primary requirement for success in complex tasks. Among others, it provides the foundation for strategic thought and action, another validated predictor of success. It is, however, interesting to note, that alcohol intoxication at higher levels *increased* breadth but *diminished*

Table 18.1 – Percent improvement (+) or decrement (–) of task performance* at four levels of intoxication

Blood alcohol level %	0.02	0.05	0.06	0.10
	n = 11	n = 24	n = 11	n = 24
number of respondent decisions	+13.4	–5.1	–12.6	–71.8
unintegrated decisions	+5.1	+4.4	–13.4	–73.9
speed of response	+11.1	0.0	+7.1	–20.9
diversity of action	–5.1	–5.6	+4.3	+23.3
use of strategy	+7.2	–33.6	–14.4	–34.6
forward planning	+9.8	–45.0	–10.5	–45.1
systematic functioning	–16.7	–12.4	–9.1	–8.7
strategic actions within groups of coherent activities	+34.9	–24.8	–9.7	–17.0
emergency response speed	–14.9	+23.1	–11.4	–18.4
shift in speed upon an emergency	–31.2	–9.9	–17.4	–8.6
number of emergency actions	–21.7	–26.3	0.0	–18.9
applied emergency response	+60.7	–45.1	–72.7	–62.8
utilizing strategy in handling an emergency	+57.2	–14.7	–12.8	–42.0

*The data reported in this table are based on the same measures utilized by Streufert *et al.* (1994).

planning and strategy. Apparently two quite different processes are at work under limited and higher alcohol treatment conditions (see below).

In considering the results obtained from subjects treated at the 0.02 and 0.06 levels, we should again note that we are only dealing with *pilot data* based on 11 research participants. In other words, interpretations drawn on the basis of those data should be viewed with considerable caution until results from a larger sample have been obtained.

CONCLUSIONS

The limited quantity of research on low level alcohol intoxication upon human functioning makes it difficult to draw hard and fast conclusions. At best, we might make tentative assumptions about the phenomena at hand. Our conclusions must be preliminary for yet another reason: the multifaceted complex day-to-day world in which we live and work adds its own impact. Our complex environment is naturally, and necessarily, 'confounded' by multiple variable interactions that can add to the 'instability' of any data we collect.

Of course, even our knowledge about the impact of limited alcohol intoxication upon *simple* and *moderate* task performance is as yet incomplete and controversial. Nonetheless, cautious comparisons of data obtained from complex settings with data generated in simpler settings suggest that the impact of low level alcohol consumption on *complex* functioning may differ from general alcohol effects upon *simpler* functioning.

If we can apply the research results of Jobs *et al.* (1990), then general memory during intoxication is less effective in both simple and complex tasks. We should, however, remember that there are many complex task settings where recent, prior, preparation for an interpersonal task, such as some meetings, might be much more important than information acquisition during intoxication. Earlier we used the example of a 'wet' lunch with a business client to merely 'close a deal'. And, if the work of Mishara and associates (1975, 1980) is applicable, social interactions are enhanced and improved by limited alcohol consumption, again aiding our manager. However, tasks vary. As already suggested, the 'negotiator,' representing another kind of complex task environment, may suffer greatly from intoxication.

In the tentative comparison of research and pilot data obtained in a complex simulation by Streufert and associates (Table 18.1), we saw more potential improvements than decrements when research participants were treated at a BAL of 0.02. How should those results be interpreted? Of course, much more data is necessary before certainty is possible. We might, however, venture a somewhat 'educated' guess. The data might indicate a slightly myopic (cf., Steele and Josephs, 1990) approach to the task at the 0.02 level, as shown by decreased diversity, possibly in an attempt to overcome a perceived, potential, decrement due to alcohol. Decreased diversity (breadth) and slowed speed under emergency conditions may have resulted in less systematic functioning but may also have allowed for maintenance and even improvement in a range of 'stimulus based responses' (e.g., respondent decisions). Further, focusing on fewer task components may have allowed participants to remain involved in all-important, integrative, strategic considerations (cf., Streufert and Swezey, 1986).

But already at the 0.05 treatment level performance characteristics have changed drastically. Diversity remains reduced, but despite a potentially continuing more 'myopic' orientation, improvements in performance are generally absent. At the highest level of intoxication, only diversity (breadth) shows 'better' performance. Here, however, the increase in diversity is probably more due to confusion, trying this and that because nothing seems to work well enough, than due to an orderly and planned approach to the task at hand.

Can we draw parallels between some of the more frequent findings obtained in simpler or intermediate tasks with findings based on complex task performance? Divided attention, as measured in simpler tasks, is most similar to the measure of diversity (breadth) in complex tasks. As we have seen, some slight decrement appears to be evident for complex functioning, but only at the lowest alcohol treatment level. However, the increased breadth at higher treatment levels may reflect a different phenomenon and may not be strictly comparable. For 'divided attention' and 'breadth', in other words, some similarity might exist under some conditions.

Whether memory for complex material studied prior to intoxication is improved is not known. Material learned during intoxication may be less well remembered. However, those conclusions are based upon the effort of Jobs *et al.* (1990) which may not apply to alcohol at levels below 0.05.

Certainly, expectancy has an impact on complex functioning, although the direction of expectancy effects in complex settings may be quite variable. Diminished stress experience and decreased tension following low level alcohol consumption could, in many cases, improve performance at task demands experienced in managerial, professional and equivalent settings. Such decreases in stressor impact, if they exist at all, may be less important toward the improvement of performance in many of the simpler, laboratory, settings.

Overall, then, *if* we can trust the few data points we have to date, we might expect that limited alcohol consumption could result in both (either) improvements and decrements of complex performance. Which effect would occur may be specific to the characteristics of the task at hand and specific to the effects of several other variables that may be more or less unique to a particular person and his/her concurrent experience.

Some readers may ask whether, in the face of multi-causation, it is worthwhile to study human responding in a complex setting. Whether such efforts bear fruit will depend very much on the approach taken. To the extent to which we focus on *how* a person thinks and functions rather than on *what* he/she thinks and does (cf., Streufert and Nogami, 1989) we will obtain more meaningful and reliable data. To the extent to which we are able to control at least the more important variables while creating a research environment that is highly equivalent to the 'real world,' we will also be more successful (Streufert and Swezey, 1985). Nonetheless, it will take considerable time and many more research efforts to 'tease out' the effects of low level drug, including alcohol, treatment on complex functioning. A beginning has been made. Much more work needs to be done.

SUMMARY

Society's attitudes toward psychoactive drug consumption in general and alcohol consumption, even in limited quantities, are considered. Researchers are warned about the wholesale rejection of alcohol use at any level since prior data are, in some instances, inconclusive. To overcome disparate views by scientists from various research areas, the dimension from 'simple' to 'complex' as well as the term 'low level' alcohol consumption is redefined. Contributions of current knowledge based on variables related to simple task performance (divided attention, memory, and stressor impact) on data from moderately complex task performance (driving, piloting aircraft) on expectancy data and the impact of variable interactions, as they may affect complex task performance, are considered. Finally, the rather limited research results obtained in complex task settings to date are discussed. If any conclusions can be drawn despite the paucity of data they would suggest that limited alcohol consumption can have both favourable and unfavourable effects. Whether favourable or unfavourable outcomes are observed would depend on task characteristics, individual differences and a host of other variables.

ACKNOWLEDGEMENTS

Simulation data of Streufert and associates as discussed in this chapter were obtained with grant support by the National Institute on Drug Abuse (NIH) under grant RO1 DA 06170.

CHAPTER NINETEEN

Cognition in Social Drinkers: The Interaction of Alcohol with Nicotine and Caffeine

Robert O. PIHL, Jean-Marc ASSAAD and Kenneth R. BRUCE

INTRODUCTION

Background

According to the National Comorbidity Survey, in the U.S., 1 in 7 individuals 15–54 years old had a history of alcohol dependence, and 1 in 4 of nicotine dependence (Anthony *et al.*, 1994). Alone, and in combination, these drugs produce a variety of effects on physiology, subjective experience, emotion and cognition that affect behaviour and appear to be important determinants of both the initiation and patterns of their use. The cognitive effects of these two substances, the focus of this paper, are so varied that generalizations about them are difficult to impossible. A plethora of factors, including the environment, the cognitive task, the individual subject, as well as the time, route and rate of drug administration, all are important alone and in concert in determining possible drug effects. For example, with alcohol, factors such as the time since consumption (Westrick *et al.*, 1988), gender, specific task, and the effects of the ascending vs. descending limb of the blood-alcohol curve are important mediating factors on cognitive performance (De Wit *et al.*, 1990; Niaura *et al.*, 1987). The picture is further complicated by the potential interaction between drugs. For example, the high correlation between alcohol and nicotine use, that is, people who drink also tend to smoke, and vice versa demands for ecological validity that the effects of both drugs individually and interactively be considered. A common current belief is that nicotine's effect on cognition is opposite to that of alcohol. This position provides a strong impetus for studying the combined effects of these two drugs on cognition. Caffeine is also thought by many to be a 'cure' for the debilitating effects of alcohol and thus the study of the combined effects of caffeine and alcohol on cognitive abilities is also warranted.

The goal of this chapter is to describe some of the effects, alone and in combination, of these drugs on cognition, and to propose possible sites of action common to them. In particular, the focus will be on the effects on memory, abstraction and planning; information processing systems involving primarily the frontal and temporal lobes and related areas of the brain. These brain systems are integral in the control and execution of a variety of important motivated behaviours, not the least of which appears to be the selection and execution of appropriate adaptive responses (Pihl *et al.*, 1990). The importance of the action of these drugs on cognitive mechanisms for determining patterns of drug use will also be discussed.

This paper consists of four sections.

– The first deals with the effects of alcohol on information processing, cognition and brain activity in normal social drinkers.
– The next section examines some of the individual factors underlying the effects of alcohol on information processing, including genetic risk for alcoholism, and the role of alcohol expectancies.
– The third section details the individual effects of nicotine and caffeine, as well as the combined effects of alcohol, nicotine and caffeine on information processing and brain activity.
– Finally, the fourth section summarizes and integrates the findings, and presents theoretical implications of alcohol effects and alcohol-caffeine/nicotine interactions.

EFFECTS OF ALCOHOL ON INFORMATION PROCESSING, COGNITION AND BRAIN ACTIVITY IN NORMAL SOCIAL DRINKERS

Alcohol, perhaps not surprisingly, has heterogeneous effects on the cognitive processes in normal individuals. As described at the outset, there appear to be numerous factors that account for this, not the least of which is each individual mental task in question. Broadly defined, at mild to moderate intoxicating dosages, alcohol impairs performance on some tasks, leaves others apparently unaffected, and actually enhances still others (Loke, 1992).

Abilities diminished

Acute alcohol intoxication impairs perceptual and motor abilities, such as spatial perception, complex reaction time and sustained hand-eye co-ordination (Mongrain and Standing, 1989; Stokes *et al.*, 1991; Zacchia, *et al.*, 1991). Simple decision making is slower in intoxicated subjects who are required to give motor responses (Maylor and Rabbitt, 1993). Performance on divided attention tasks, those where the subject is performing a motor and a decision making task simultaneously, is also impaired (Zacchia, *et al.*, 1991). Intoxication impairs visual sensitivity, or the accuracy in copying complex visual stimuli such as the Rey figure (Peterson *et al.*, 1990). Judgement of facial expressions of emotion may be impaired for some emotions (anger and disgust/contempt) but not all (e.g., fear, surprise, sadness, and happiness) (Borrill *et al.*, 1987). Estimation of the passage of time is also reported to be impaired by alcohol, and by conditions where intoxication is perceived (Lapp *et al.*, 1994). Accuracy and speed of classification of word meaning and structure are impaired by alcohol, and intoxication results in more target word 'misses' (omission errors) in word recognition tasks (Maylor *et al.*, 1987).

Further, intoxication impairs verbal, difficult associative and visuospatial learning, as assessed by free recall (Peterson *et al.*, 1990). A possible explanation for the impairment by alcohol of verbal learning may be that the material is forgotten more rapidly, an explanation more plausible to some than alcohol merely disrupting attention (Maylor and Rabbitt, 1987). Also, whether the subject is intoxicated while learning, while remembering, or both is critical. Unintoxicated subjects have difficulty recalling material learned earlier while intoxicated, and intoxicated subjects have difficulty recalling material learned earlier while unintoxicated (Goodwin *et al.*, 1969; Werth and Steinbach, 1991). However, subjects recall some material (words, but not faces) better learned while intoxicated if they are remembering in an intoxicated, as compared to an unintoxicated, state (Goodwin *et al.*, 1969; but cf., Werth and Steinbach, 1991). This phenomenon has been referred to as state-dependent learning, where the idea is that some memories are more accessible in *similar* than dissimilar states (Goodwin *et al.*, 1969). Intoxicated subjects also have difficulty remembering social events (Tucker *et al.*, 1987). When this is applied to eyewitness memory, intoxication may result in the loss of memory or the production of false memories (Yuille and Tollestrup, 1990). Interestingly, alcohol has similar inhibitory effects on recall in animal studies (e.g., Castellano and Pavone, 1988), and alcohol at intoxicating concentrations has inhibitory effects on the cellular events (long-term potentiation) considered by many to be a necessary condition for some forms of memory. The induction of this mechanism, long term potentiation, is prevented at low and high doses of alcohol (Blitzer *et al.*, 1990; Morrisett and Swartzwelder, 1993; Steffensen *et al.*, 1993; Wayner *et al.*, 1993; Zhang and Morrisett, 1993). It is well known that alcohol intoxication impairs performance on some measures of abstraction, classification, and goal-directed planning. By using the Porteous Maze series, and Thurstone's word fluency, a self-directed word search task, intoxicated subjects are impaired on these abilities (Peterson *et al.*, 1990). Planning, abstracting and classifying, some of the so-called executive abilities, have been hypothesized as related primarily to the functioning of the prefrontal cortices.

As will be described later, an individual's response to alcohol that implicates these abilities, as well as memory, and those underlying structures, may be an important factor in determining motivationally significant drug responses (Peterson and Pihl, 1990; Peterson *et al.*, 1990). Most important here is the consideration that alcohol can disrupt anxiety responses, likely by interfering with the cognitive operations carried out by particular brain areas (Josephs and Steele, 1990; Sayette, 1993; Sayette *et al.*, 1989).

The important consideration that must be given to dose in assessing impairment due to alcohol is illustrated in Figure 19.1.

In this large balanced-placebo six-group, three-dosage study (active placebo; 0.66 ml parts or 1.32 ml. of 95% alcohol per kg of body weight), expectancy effects were negligible but particularly high alcohol dosage effects on tests measuring abstraction, delayed memory, and planning and problem solving were noted (Peterson *et al.*, 1990).

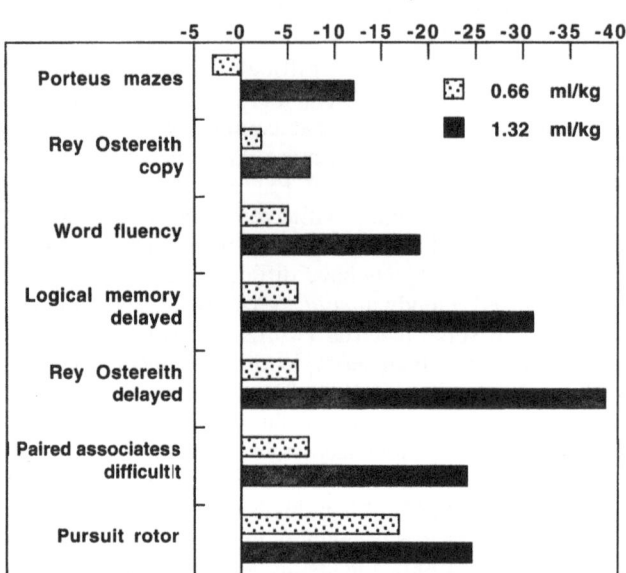

Figure 19.1 – Percent decrease in performance from placebo response under two dosages of alcohol on seven neuropsychological tests.

Abilities unaffected or inconsistently affected

Alcohol at typical dosages does not impair all cognitive abilities. Functions apparently not sensitive to alcohol's effects include the Information, Vocabulary and Digit Symbol subtests of the Wechsler Adult Intelligence Scale – Revised (Peterson *et al.*, 1990) although others (Nelson *et al.*, 1986) have reported otherwise. Reaction time is not consistently impaired by alcohol (Peterson *et al.*, 1990; Salame, 1991), and neither are planning and some indices of working memory, as indicated by tests including the Wisconsin Card Sorting Task, and easy paired associate learning (Nilsson *et al.*, 1989; Peterson *et al.*, 1990). Emotional recognition and classification from slides of human faces is not always impaired by alcohol (Baribeau *et al.*, 1986). Alcohol intoxication does not appear to affect implicit memory (Lister *et al.*, 1991; Nilsson *et al.*, 1989). Implicit memory is inferred from performance where memory is less directly assessed, and does not require the subject to have 'conscious' awareness of the procedure. Visually-presented digit memory is not consistently affected, and neither is picture recognition memory (Roache *et al.*, 1993).

Abilities strengthened

It is evident from the above that the effects of alcohol on cognition are not universally impairing. It has been proposed for memory, for example, that if alcohol-induced impairments occur they likely result indirectly from effects of the drug on other mental processes such as mood, arousal, attention and perception (Weingartner *et al.*, 1992) or the rapid decay from short term memory (Jones, 1973; Maylor and Rabbitt, 1987). Indeed, since some studies have shown alcohol to actually enhance consolidation of memories, although the role of alcohol in forgetting may be more complex than previously thought. For example, Kalin (1964) has shown that alcohol impairs memory for material written while intoxicated, but actually enhances memories for material written immediately prior to drinking. This may indicate preventing, or rapid forgetting, of *new* learning and actual enhancement of *prior* learning. Memory enhancement also occurs in post-training experiments, where subjects learn and remember while unintoxicated. In these experiments, alcohol is consumed between learning and memory testing, usually immediately after learning, and typically 24 hours prior to memory testing, the impairing psychomotor effects of alcohol have usually dissipated at this time (Lemon *et al.*, 1993). Thus, the confounding effects of alcohol on attention and retrieval are removed, and the effects of alcohol on memory consolidation can be examined in relative isolation. Such experiments have shown alcohol to enhance memory both in animals (Melia *et al.*, 1986), and in humans (Lamberty *et al.*, 1990; Mann *et al.*, 1984; Parker *et al.*, 1980) for a variety of verbal and non-verbal material. Importantly, the effect has been shown to be dose-dependent (Parker *et al.*, 1981), suggesting a direct pharmacological effect. Mechanisms that have been proposed for this post-training effect are that

- The drug may either become associated with the stimulus material motivationally (White and Milner, 1992), perhaps via brain reward systems (Mann *et al.*, 1984), or it may simply enhance (or disinhibit) pharmacologically the events underlying memory consolidation (Landauer, 1969; White and Milner, 1992; Warburton *et al*, 1992a). Underlying mechanisms here may involve hormonal and neuromodulatory systems (McGaugh, 1989).
- Another, possibility is that alcohol prevents the disruptive effects of retrograde interference, and thus indirectly protects the prior learning (Parker *et al.*, 1980).

Elucidation of operative mechanisms is important, as it would be of interest to see whether alcohol could enhance (or disinhibit) the expression of already-induced long term potentiation, or what the post-training effects of alcohol on motivationally-significant stimulus materials might be for social drinkers.

RELATED EVIDENCE OF ALCOHOL'S EFFECT ON COGNITION

Neurophysiological studies

The effects of alcohol on human cognition may also be investigated from a more basic neurophysiological level. Three technologies are important here.

• First, the electroencephalogram (EEG) can be used to investigate both resting and on-line task effects of alcohol. Alcohol produces an increase in resting alpha waves over the frontal lobes, possibly pleasure-related (Lukas *et al.*, 1990). Alcohol intoxication increases event-related potential (ERP) latency (Krull *et al.*, 1993), and decreases response amplitude (Oscar-Berman, 1987), suggesting delayed and impaired responding (Lukas *et al.*, 1990).

- Second, positron emission tomography (PET) studies have shown that alcohol affects cerebral glucose metabolism and that these effects are related to alcohol's effects on mood. Specifically, a low dose of alcohol produces changes in mood associated with larger changes in metabolism in the left parietal and right temporal areas, while a moderate dose produces mood changes associate with metabolic changes in the right frontal areas (De Wit *et al.*, 1990). Regional cerebral blood flow studies implicate alcohol in blood flow alterations (Schwartz *et al.*, 1993), changes disproportional to the temporal and frontal lobes (Sano *et al.*, 1993). *Low* doses of alcohol increase flow in frontal, and decrease flow in temporal areas disproportionately; *larger* doses produce disproportionate decrease in flow in both regions – these effects appear to be related to the biphasic arousal-sedation and mood effects of alcohol (Sano *et al.*, 1993).

- A third, relatively new and extremely exciting technology is magnetic resonance spectroscopy (MRS). MRS studies show that alcohol can be localized in human brain, and is found in highest concentrations in the ventricles, second highest in the gray and third highest in the white matter (Spielman *et al.*, 1993). Alcohol produces changes in mood associated with its presence in frontal lobe and immediately surrounding areas (Mendelson *et al.*, 1990). These changes in mood may be related to the significant relationship between alcohol and aggression which also serves as evidence for alcohol's effect on cognition.

Alcohol and aggression

Among the many consequences of alcohol intoxication, perhaps one of the most consequential is the potentiation of aggressive behaviour. Alcohol consumption immediately precedes the perpetration of approximately half of all violent crimes, homicides, assaults, rapes and instances of family violence (Murdoch *et al.*, 1981). The relationship is not simply correlational as much experimental evidence supports alcohol's role, although suggested mechanisms are arguable (Bushman and Cooper, 1990). Although it is true that situational variables are important and that aggressive individuals are more likely to drink heavily (White *et al.*, 1993), studies also show that alcohol's psychopharmacological effects *per se*, do increase the likelihood of aggressive behaviour (Pihl, *et al.*, 1997). We have speculated (Pihl *et al.*, 1993) that alcohol's effect on cognition is one mechanism responsible for this effect. Briefly, the neuroanatomical system which appears to govern our reactions to novel and/or threatening stimuli (Pihl and Peterson, 1995), is altered by alcohol and seems to potentiate aggressive behaviour. Constantly, the brain is inundated with information to be processed. A primary role of the prefrontal cortex and related structures is to provide the means to regulate behaviour after synthesizing information from the outside and inside world. When incoming stimuli are known and relevant, or irrelevant, classification in terms of motivational significance occurs and tried and true patterns of behaviour result. When, however, the unexpected happens, the threat/anxiety

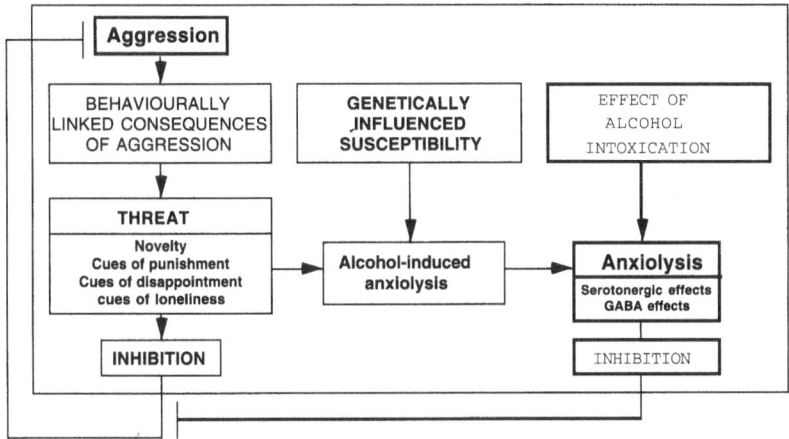

Figure 19.2 – Alcohol's hypothesized effect on the threat system and the increased likelihood of aggression.

system is involved, resulting in the inhibition of ongoing behaviour; autonomic hyperreactivity can ensue (Gray, 1982, 1987). In an evolutionary sense this response is protective, aiding in the avoidance of disappointment or punishment and provides the impetus to plan, thus allowing action with intelligence and insight (Pihl and Peterson, 1995). This inhibited state and its corresponding state of arousal does not cease until the situation is analyzed and a suitable response determined.

Alcohol's ability to cross the blood brain barrier results in the bathing of the entire brain but certain functions in certain individuals appear particularly sensitive (Pihl and Peterson, 1995). One pronounced effect is on the hippocampus/ amygdala and the resultant effect of alcohol is anxiolytic (see Pihl *et al.*, 1993 for a discussion). When the anxiety system is impaired by alcohol intoxication, the emotional valance necessary to place potentially relevant stimuli into a meaningful context, is also altered. This leads to a lack of behavioural inhibition when an individual is faced with threat. It is the absence of response to threat, historical or current, for one's actions which appears in part explanatory of the alcohol-aggression relationship. This effect also would explain the fact that close to a majority of victims of violence are also intoxicated. Thus, specifically, a diminution of threat seems to result in putting oneself in harm's way. Figure 19.2 illustrates this hypothesized effect. Recently, we compared subjects who, when sober, performed high and low on neuropsychological tests sensitive to frontal lobe function, on an aggression task under sober and intoxicated conditions (Lau *et al.*, 1995). Figure 19.3 illustrates these results. Alcohol intoxication increased aggression for high functioning individuals so that they performed like aggressive-sober low functioning subjects. These results may demonstrate how alcohol could affect cognitive functions that play a profound role in controlling aggressive behaviour.

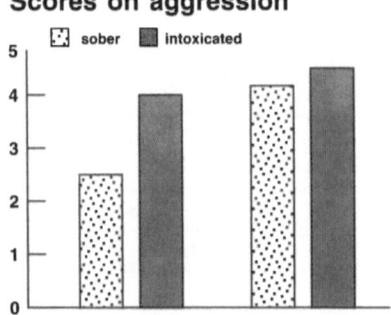

Figure 19.3 – Scores on an aggression task for high and low performers on two neuropsychological tests when sober and intoxicated.

IMPLICATIONS, MEDIATORS AND MODERATORS

Mediators and moderators of the degree of alcohol-induced performance inhibition include simultaneous food ingestion (Millar *et al.*, 1992), specifically sucrose (Zacchia *et al.*, 1991) and tryptophan and amino acid in food (Westrick *et al.*, 1988). Other substances, including drugs affecting GABA and cholinergic transmitter systems can also moderate the degree of alcohol-induced impairment (Brioni *et al.*, 1989; Castellano and Pavone, 1988; Castellano and Populin, 1990). As stated previously the dose of alcohol is also important (Jubis, 1986, 1990), with actual memory facilitation at mild intoxication, but impairment at the moderate to heavy doses described above. Intelligence (Maylor *et al.*, 1990), gender (Haut *et al.*, 1989), task repetition (Rumbold and White, 1987), and environment (Babbini *et al.*, 1991; Colbern *et al.*, 1986; Miles *et al.*, 1986) also moderate alcohol's effects on task performance. Time since drinking is important, as alcohol-induced impairment of immediate (less than 30 seconds delay) verbal memory is more pronounced on the *ascending* limb of the blood-alcohol curve than at comparable blood alcohol concentrations (BACs) on the descending limb; short- and long-term memory is *equally* disrupted at both times (30 seconds, and 15 minutes delay respectively) (Jones, 1973).

Pre-existing cognitive function is another important determinant of the more general effects of alcohol. Certain individuals, including males from families with extensive family histories of alcoholism, are 4–9 times at greater risk themselves for developing drinking problems. It has been suggested that aspects of this risk may be mediated through a neuropsychological information processing deficit in abstraction/classification, and goal-directed planning skills typical of these individuals (Peterson and Pihl, 1990; Pihl *et al.*, 1990). Deficits in these abilities have definite implications for response strategies, and for real-world behaviours (Pihl *et al.*, 1990). They result in idiosyncratic behaviour and response to alcohol (Peterson *et al.*, 1992) which in turn, are predictive of alcohol use patterns, including weekly consumption (Peterson *et al.*, 1993). Another important individual factor mediating the general effects of alcohol is expectancy. Beliefs about the effects of alcohol, are acquired through instrumental conditioning, modelling, and vicarious learning (Brown *et al.*, 1980; Oei and Baldwin, 1994; Stacy *et al.*, 1990), and are also predictive of alcohol use patterns, and response to alcohol (Fillmore and Vogel-Sprott, 1994; Rather and Goldman, 1994; Stacy *et al.*, 1994).

ALCOHOL AND NICOTINE

Background

Until recently, nicotine has been thought to unequivocally improve a wide range of cognitive abilities (such as information processing, attention and memory), leading researchers to propose that this effect contributes to its addictive properties (see Warburton, 1992, 1994 for reviews). However, West (1993) cautioned that these supposed beneficial effects on cognition are not clearly supported by indisputable scientific evidence, as the cognitive benefits observed in many studies following nicotine consumption could instead be attributed to the alleviation of withdrawal effects. An extensive review of the effects of nicotine on human performance (Heishman *et al.*, 1994) seems to support this view; specifically that, in non-abstinent smokers and non-smokers, cognitive functioning is not reliably enhanced by the administration of nicotine. This poses a problem for the study of the combined effects of nicotine and alcohol on cognitive functioning. The assumption which has guided many of the studies in this area is the moderation of the detrimental effects of alcohol by the beneficial effects of nicotine. Methodologically, however, none of these studies has been able to separate the individual effects of alcohol and nicotine from the interactive effects. The question remains whether alcohol and nicotine combinations represent a new drug effect, or merely an algebraic sum of the individual effects.

Comorbidity: alcohol and nicotine

People who drink also tend to smoke, and vice versa (Istvan and Matarazzo, 1984; Zacny, 1990). Consistently, studies have found higher rates of smoking among non-alcoholic heavy social drinkers compared to light or non-drinkers, and vice versa (Carmody *et al.*, 1985; Istvan and Matarazzo, 1984; Craig and VanNatta, 1977; Zacny, 1990). This relationship also holds for alcoholics, who smoke more cigarettes per day than non-alcoholic cigarette smokers (Ayers *et al.*, 1976; Walton, 1972). Estimates consistent since the 60s are that, on average, 91% of alcoholics smoke. This is an average computed from the following studies; Dreher and Fraser, 1967; Bobo and Gilchrist, 1983; Ayers *et al.*, 1976; Walton, 1972; Ashley *et al.*, 1981; Kozlowski *et al.*, 1986; Burling and Ziff, 1988), as compared to 24% of the general American population (Anthony *et al.*, 1994). The picture is clear, individuals who drink more, smoke more.

Further, it appears that ethanol produces increases in many measures of cigarette smoking behaviour, such as the number/frequency of cigarettes smoked and number of puffs taken by male alcoholic volunteers and social drinkers (Griffiths *et al.*, 1976; Henningfield *et al.*, 1983, 1984; Nil *et al.*, 1984; Mintz *et al.*, 1985; Mello *et al.*, 1980, 1987). These effects seem to depend on an individual's drinking history, as individuals who *drink more* are *more sensitive* to the alcohol related potentiation of nicotine intake.

It also appears that nicotine can affect alcohol consumption. Correlational data indicates that ex-smokers drink more alcohol than either smokers or non-smokers (Carmody *et al.*, 1985). This relationship does not seem to be reciprocal, as alcoholics who successfully stop drinking are reported to show no increase in smoking (McCoy and Napier, 1986). More evidence supporting the notion that nicotine consumption affects alcohol intake can be found in animal studies. Rats implanted with a slow-release nicotine pellet exhibit significantly increased ethanol consumption (Potthoff *et al.*, 1983).

POSSIBLE MECHANISMS UNDERLYING THE ALCOHOL-NICOTINE RELATIONSHIP

A plethora of theories have been proposed to explain the increased amount and rate of smoking after alcohol intake, ranging from a strong oral drive to a weakened nicotine-receptor binding, each with none or fragmentary support (Bobo, 1992). Explanations with more plausibility are:

- Alcohol increases smoking behaviours by altering craving to smoke although the mechanisms are undetermined. In abstaining smokers, although chewing nicotine gum results in a reduction in craving to smoke, drinking alcohol after the gum eliminates virtually all of the gum's beneficial effects (Mintz *et al.*, 1991).
- The link between smoking and drinking can also be explained by common personality factors, by common genetic factors, and possibly by the development of cross-tolerance over the course of chronic administration of these two substances (Zacny, 1990; Collins, 1990).
- Alcohol may weaken nicotine-receptor binding by resulting in a kind of nicotine blockade which the smoker attempts to overcome by smoking more.
- People may also smoke in order to counteract alcohol's debilitating effects on cognition or to diminish the central nervous system depressant effects of alcohol. (Michel and Bättig, 1989; Shiffman *et al.*, 1994). In addition to (or instead of) antagonism, an augmentation or enhancement effect may mediate the alcohol-tobacco relationship, such as the maintenance of a preferred mood state (Istvan and Matarazzo, 1984).
- Smoking may also come to be associated with alcohol consumption through a process of conditioned learning (Shiffman *et al.*, 1994). As both drinking and smoking tend to occur at the same locations or social contexts, a conditioned association may result, and drinking may come to elicit smoking (Diethelm and Barr, 1962; Docter and Beernal, 1964; Zacny, 1990; Istvan and Matarazzo, 1984). In addition to external (situational) cues, internal cues may also be associated. These internal cues may range from the taste to the physiological effects of alcohol.
- The relationship may reflect something about the kind of people who drink and smoke heavily. This is based on the observations that heavier drinking persists in ex-smokers (Perkins *et al.*, 1993) and heavier smoking persists in ex-drinkers (Keenan *et al.*, 1990; Maletzky and Klotter, 1974).

- Although the increased tobacco use following alcohol consumption does not seem to be due to an effect of ethanol on the rate of metabolism of nicotine, as chronic ethanol and nicotine treatment does not alter the rate of metabolism of either drug (Benowitz *et al.*, 1986; Collins, 1990; Zacny, 1990), the *disinhibition* of smoking restraint by alcohol intoxication is a plausible explanation (Shiffman *et al.*, 1994).

It is evident from this large multiple choice of alternative explanations that we are far from establishing which of these theories (if any) explain the mechanisms underlying the potentiation of tobacco consumption by alcohol.

THE COMBINED EFFECTS OF ALCOHOL AND NICOTINE ON COGNITIVE FUNCTION

Alcohol-nicotine interaction studies seem primarily based on the principle that an algebraic sum of the effects of the drugs will be demonstrated under the right conditions. The current evidence however, suggests reconsideration of the original theory and replacement with more complex positions. Simply put, the combination of these two drugs produces antagonistic, synergistic and negligible or mixed effects. As sensory-perceptual and psychomotor processes are differentially affected by the alcohol-nicotine interaction, we will review the pertinent literature separately for both processes. Studies of nicotine and alcohol's effect on sensory-perceptual processes show antagonistic, additive, synergistic, and no effects of nicotine on alcohol-induced effects. Smoking one or two cigarettes seems to produce an antagonism of alcohol's debilitating effects on processing rate in rapid visual information processing, on visual discrimination, and tends to counteract the overestimation of elapsed time caused by alcohol (Michel and Bättig, 1989; Tong *et al.*, 1974; Leigh and Tong, 1976). Nicotine has also been reported to have no effect on impairments in a divided attention task or auditory vigilance induced by alcohol (Leigh *et al.*, 1977, 1980). These effects may well be dose and task dependent. For example, Leigh (1982) in a lengthy signal detection task, found smoking six cigarettes (1.2 mg nicotine each) antagonized alcohol's debilitating effects while on a short probe test, a decrease in performance was found after subjects smoked four cigarettes (1.2 mg nicotine each) in combination with alcohol. Similarly, Jubis (1986) found that under nicotine alone, low alcohol dose alone or nicotine/low alcohol dose conditions, free recall for relevant cues was superior than under a non-drug control condition. However, at a moderate dose of alcohol, either alone or in combination with nicotine, poorer recall resulted. A recent study by Suys, Assaad, Peterson and Pihl (unpublished manuscript) further supports the notion of a dose-dependent effect of nicotine consumption on the performance of intoxicated individuals. Smoking *ad libitum* throughout an experimental procedure allowed intoxicated smokers to achieve total memory scores (based on several memory tasks) equivalent to those obtained by intoxicated non-smokers. However, semi-deprived intoxicated smokers, allowed to smoke only during alcohol consumption, achieved significantly lower scores than the intoxicated non-smokers. Further, it seems that although nicotine may antagonize (under certain conditions) alcohol's effects on sensory-perceptual tasks, psychomotor tasks are not affected (Lyon *et al.*, 1975;

Michel and Bättig, 1989; Knott and Venables, 1980). This differential cognitive/psychomotor effect may be a possible contributing factor to the observation that non-smokers are less likely to be involved in a motor vehicle accident than smokers, even when drinking habits are controlled (DiFranza *et al.*, 1986). After nicotine intake, an intoxicated individual may experience increased perceptual-sensory functioning and deems himself capable of driving a vehicle, while remaining unaware of a continued impairment in motor reaction time.

Neurophysiological studies, specifically of the combined effects of alcohol and nicotine on electrocortical activity seem to correspond to the above findings. Nicotine and alcohol produce opposing psychopharmacological effects on electrocortical activity (as measured by EEG). Nicotine has been predominantly found to stimulate EEG activity (for reviews see Conrin, 1980; Edwards *et al.*, 1985) while ethanol appears to have the opposite effect (e.g., Campbell *et al.*, 1984; Lukas *et al.*, 1986; Teo and Ferguson, 1986). Based on these findings, Michel and Bättig (1989) decided to look at EEGs after the combined consumption of both drugs. Some of the ERP (Event Related Potential) effects measured in their study led the researchers to propose that the decrease in amplitude and/or an increase in latency of the P_{300} component observed after alcohol consumption was prevented when subjects subsequently smoked a cigarette.

Implications, mediators and moderators

One source for the variability in the alcohol-nicotine interaction literature may be a dose-dependent effect, particularly in relation to effects on arousal states, which can subsequently affect attention. Arousal in this case refers to an organism's general state of alertness, which is of course influenced by a wide range of endogenous (internal) and exogenous (external) factors. Although alcohol particularly at high doses is viewed as a central nervous system depressant resulting in lowered arousal, there is much evidence that at low doses and particularly on the *rising* limb of the blood alcohol curve alcohol acts as a *stimulant* to increase arousal. Jubis (1986) using different nicotine (no nicotine, 2 cigarettes of 1.3 mg nicotine each) – alcohol (0.0 ml, 0.97 ml, 2.33 ml per kg of body weight) combinations, found that under conditions of high arousal induced by cigarette smoking or a low dose of alcohol, free recall for relevant cues is superior than under a non-drug control condition. Furthermore, low arousal brought about by a moderate dose of alcohol, either alone or in combination with nicotine, yielded poorer recall of relevant cues relative to control.

Methodological issues

Methodological problems proliferate in much of the nicotine literature which limits the extent to which firm conclusions can be drawn. For example, only 24% of experiments involving nicotine used placebo controlled conditions, only 26% maintained single or double blind testing conditions, and 70% of the studies used an imprecise method of nicotine dosing (Heishman *et al.*, 1994). This latter problem persists even if the number of cigarettes and the amount of nicotine per cigarette is precisely known. This is due to the smoker's ability to adjust nicotine intake (U.S. Department of Health and Human Services, 1988): depending on how it is smoked, the same cigarette can deliver varying amounts of nicotine to different smokers or to the same smoker at different times (Pomerleau *et al.*, 1992). The subsequent uncertainty and variability of administered doses, as well as the variability of tasks designed to test cognitive function in the alcohol-nicotine experiments, has resulted in difficulties in inter-study comparisons.

Another current issue is the need for a clear distinction between nicotine/ alcohol studies using abstinent smokers as subjects and those using non-abstinent smokers and non-smokers. The vast majority of the studies reviewed in this section used abstinent smokers as subjects. Obviously, the ethical difficulties posed by administering nicotine to non-smokers has lead to restrictions in the type of subjects available to experimenters. Nevertheless, the investigation of nicotine's antagonism of alcohol-induced cognitive deficits in individuals who may be already impaired due to nicotine withdrawal furthers our understanding of only a limited range of situations. Most commonly, people smoke and drink when not experiencing withdrawal effects from nicotine abstinence. It might be a prudent strategy to first establish the effect of alcohol intoxication alone and nicotine abstinence alone, and then compare these to the effect of alcohol intoxication following nicotine abstinence. One might then be in a position to assert whether or not deficits usually caused by alcohol intoxication are affected by blood nicotine levels. This information, in addition to knowing if any differences exist between the antagonism of alcohol in abstaining as opposed to non-abstaining individuals, will permit a clear comparison of effect sizes between studies using different methodologies.

ALCOHOL AND CAFFEINE

Background

Folklore supports the consumption of strong coffee to neutralize the debilitating effects of alcohol intoxication. For example, a clarion call for coffee is often made by the intoxicated in anticipation of driving. In 1894, Walsh asserted that "those who may desire to rescue a drunkard from his bane will find no better than strong, fresh-made coffee [which] must be administered without the addition of either milk or sugar …". Yet, although some studies have found that the consumption of caffeine following alcohol intake antagonizes some of alcohol's debilitating effects on tasks related to driving skills, caution should be employed in interpreting these results. A review of the interaction studies reveals not only antagonistic effects, but potentiation and negligible effects as well (Fudin and Nicastro, 1988). In light of these findings, it is necessary to maintain that the consumption of caffeine by legally intoxicated individuals prior to driving an automobile will most likely not antagonize alcohol-induced driving related impairments (Fudin and Nicastro, 1988)

The combined effects of alcohol and caffeine on cognitive function

Similar to nicotine, caffeine has mixed cognitive effects when combined with alcohol. Caffeine has been shown to potentiate alcohol induced impairments on a memory scanning paradigm, a perceptual-motor task and on reaction time (Lee and Lowe, 1980; Lowe, 1981; Oborne and Rogers, 1983). Basically, no effect has been reported on alcohol's debilitating effects on numerical reasoning, perceptual speed, reaction time, balance, hand steadiness and dexterity, fusion and verbal fluency, and on mathematics, verbalization and colour discrimination tasks under stress (Newman and Newman, 1956; Hughes and Forney, 1961; Forney and Hughes, 1965; Franks *et al.*, 1975; Carpenter, 1959). Caffeine has been shown to antagonize the alcohol induced deterioration of performance on a continuous arithmetic task, complex reaction time, on progressive counting and addition, and on tracking ability (Graf, 1950; Forney and Hughes, 1965; Strongin and Winsor, 1935; Franks *et al.*, 1975). A study by Fillmore and Vogel-Sprott (1995), found that caffeine (4.4 mg/kg) significantly antagonized an alcohol (0.56 g/kg) related decrement on a computerized tracking task, allowing intoxicated subjects to perform at the same level as sober subjects who did not consume caffeine. Finally, a partial counteraction by caffeine of the impairment by alcohol of performance on a simulated driving task has been reported (Rutenfranz and Jansen, 1959).

There is also one interesting study which investigated the effects of consuming caffeine prior to ingesting a moderate dose of alcohol (Hasenfratz *et al.*, 1993). These researchers used the rapid information visual processing task, which consisted of responding as rapidly as possible after the detection of a target. Qualitatively, caffeine (3.3 mg/kg) improved and alcohol (0.7 g/kg ethanol 96%) impaired both reaction time and processing rate on this task. The consumption of caffeine before alcohol led to an addition of these two effects, that is, caffeine was able to offset the debilitating effects of the alcoholic beverage.

Implications, mediators and moderators

Expectancy effects have been studied in the Caffeine-Alcohol Interaction. Fillmore and Vogel-Sprott (1995) have reported that the expectation to receive caffeine in an alcohol-caffeine interaction study did not alter psychomotor performance, as measured by a form of pursuit rotor task. However, in this study drug effect expectancies did affect performance. Subjects expecting the most impairment due to the drug combination performed the most poorly, although only in the group expecting caffeine and alcohol. The effect of caffeine on blood alcohol concentrations is also mixed. Although caffeine can sometimes alter the effects of alcohol intoxication on perceptual and motor skills, the consumption of caffeine after alcohol intoxication does not seem to effect blood alcohol concentrations (Lowe, 1981, Franks *et al.*, 1975). However, the ingestion of coffee before the ingestion of alcohol significantly reduced BACs in one study exploring the effects of caffeine consumed before alcohol; BACs increased from 0.026% to 0.031% in the alcohol/caffeine condition, as opposed to 0.038% to 0.046% in the alcohol alone condition (measured 30 and 80 minutes after the beginning of drinking) (Hasenfratz *et al.*, 1993). Finally, a state-dependent effect may be operative for the alcohol-caffeine combination. As with alcohol-nicotine, when both alcohol and caffeine are consumed to produce a specific drug state before learning, there are no recall decrements the following day when both are again ingested before recall is attempted. However, there are no recall decrements when caffeine alone is ingested before recall. The biggest recall decrements is observed with no drugs and with alcohol alone. This suggests that, if learning occurs when alcohol and caffeine are combined, the major state dependent effect is due to caffeine.

The external validity of alcohol-caffeine interaction studies depends, among other variables, on the *dose* of both drugs. Although great care is usually given to alcohol dosing and subsequent blood alcohol concentration measurements, the same cannot be said about caffeine dosing in most of the interaction studies. In fact, almost every study of alcohol-caffeine interactions use a fixed caffeine dose, regardless of body weight and alcohol dose. Other unaccounted 'confounding variables' may be partly responsible for the variability in the results of the interaction studies, as many variables are able to influence individual effects of caffeine. For example, the mean metabolic *half-life* of caffeine in smokers is 55% shorter than in non-smokers and the body clearance of unmetabolized caffeine is significantly greater in smokers. This implies that the effects of caffeine are less in smokers than in non-smokers. It is essential to take this finding into account in alcohol-caffeine interaction studies. But smoking history is not the only variable which influences individual effects of caffeine. Age, sex, mental state, personality type, degree of physical fitness and ethnic background are all variables which have been demonstrated to effect responses to caffeine. *Subject characteristics* should be recorded with great care in order to gather a clearer and more consistent body of research on the caffeine-alcohol interactions.

The *tasks* chosen to test the possible antagonism of the debilitating effects of alcohol by caffeine are another important methodological consideration. Such tasks should have previously been demonstrated to be differentially affected by both drugs. But more importantly, the joint effect of the two drugs must be assumed to be equal to the algebraic sum of their individual effects. It may therefore be unrealistic to expect caffeine to completely antagonize alcohol's strong debilitating effect on a certain task if caffeine taken alone only marginally improves performance on that task. Lastly, some of the contradictory results apparent in the alcohol/caffeine studies may be due to complex *dose related interactions* (Burns and Moskowitz, 1989–90; Oborne and Rogers, 1983). Just as some drugs have biphasic dose-dependent effects, it is not unreasonable to suppose that some drug combinations may share this pharmacological property as well. Certain doses of caffeine and alcohol may produce potentiating effects, while other doses may result in an antagonistic reaction. What can be tentatively suggested, however, is that caffeine's antagonism or potentiation of the debilitating effects of alcohol on cognitive function may vary in a dose-dependent manner, and that negligible antagonism of alcohol's effects on cognitive function by caffeine is possible under certain conditions.

CONCLUSIONS

Alcohol has varying cognitive effects, depending in part on the task at hand.

- Alcohol *impairs* many mental abilities, including complex perception, complex perceptual-motor skill, complex decision making, complex verbal associative learning, and complex verbal and visual recall. Alcohol also impairs judgement of word meanings and the passage of time. Generally speaking, alcohol appears to affect tasks that have either

    (a) a component where multiple, complex associations must be learned, or
    (b) a component where short term memory is either overextended or over-
        loaded, or
    (c) a motor component where speed or accuracy is required.

- Abilities less *reliably impaired* by alcohol include planning, and judgement of certain facial expressions.
- Alcohol does *not* appear to *affect* verbal ability *per se*, simple verbal and perceptual-motor associative learning, or implicit learning.
- Consumption of alcohol following learning results in *enhancement* of subsequent verbal and visual recall if memory is assessed when alcohol's acute effects have dissipated.

Factors moderating alcohol's effects were found to include individual (e.g., gender, IQ, etc.), as well as drug factors (e.g., dose, time, rate, and drug combinations with sucrose, tryptophan, and cholinergic and GABA-ergic drugs). These factors must all be considered when examining the cognitive effects of alcohol.

As for the interactions between alcohol and nicotine and alcohol and caffeine, one problem is a *lack of studies*. From the scarce data available to us at this time, we can tentatively claim that although caffeine and nicotine may similarly antagonize the debilitating effects of alcohol, the combination of the two does not appear to have a greater effect than when administered alone (Kerr *et al.*, 1991). A second problem is that in most relevant studies, methodological problems of discriminating between the alleviation of withdrawal symptoms caused by abstinence and absolute beneficial effects, the variations of doses of each drug across studies, and the limited sample sizes of most of these studies, make comparison of studies difficult. Seemingly contradictory findings abound across the literature. Currently, it is premature to make definite statements as to how these drugs interact.

CHAPTER TWENTY

Effects of Caffeine, Nicotine and Alcohol on Vigilance Performance

Harry S. KOELEGA

INTRODUCTION

Since ancient times the chewing of plants containing stimulants (for example, ephedrine) dates back to several thousand years B.C. and there are indications that alcohol was already used in the Palaeolithic Age, about 10,000 years ago. Although the effects of these drugs on human behaviour have been studied for more than a century, the subject has not received much recognition in psychological textbooks; likewise, effects of the drugs on human task performance are usually not to be found in books on attention and performance.

Moderate consumption for recreational purposes is a socially accepted practice: because of their euphoric properties, all three drugs are widely used for pleasure. Further, alcohol in moderate doses is man's oldest and most widely used anxiolytic, and caffeine and nicotine are considered to be able to reduce fatigue and to enhance concentration, when, for example performing protracted, sometimes boring, tasks requiring maintenance of attention in (often) monotonous situations. Low doses of alcohol have also been reported to show this stimulatory effect, whereas higher doses would impair performance (a biphasic effect). The principal aim of the present chapter is to assess whether stimulant drugs really improve, and alcohol really impairs, task performance. To that end, one particular type of task will be examined, namely vigilance or monitoring tasks.

WHY STUDY EFFECTS OF DRUGS ON VIGILANCE PERFORMANCE?

Reviewers in the field of psychopharmacology have often reported that the effects of drugs are highly inconsistent. McNair (1973), for example, concluded with some dismal comments that his catalogue remained a litany of conflicting, inconclusive, and ambiguous findings, providing no basis for generalization of drug effects. Other reviewers reached essentially the same conclusion and reported to have encountered the same bewildering chaos. The present author has suggested that most reviews of effects of drugs on 'performance' aim too high, thus precluding that one will ever arrive at generalizations of the effects of any type of drug (Koelega, 1989). Performance is not a single entity, but embraces a gamut of functions and skills, and researchers usually employ test batteries in a 'shotgun' approach, in the hope of covering as many mental functions as possible in order to provide a pharmacodynamic profile of a drug. Hindmarch (1980), after a cursory review of the literature, presented examples of more than 50 different tests of psychomotor function and stated that these are in no way exhaustive of the diversity of tasks and tests which have been used. Parrott (1991) has discussed issues of validation, reliability, and standardization of tests and test batteries and has pointed out that most tests used in psychopharmacology comprise an *ad hoc* collection of unstandardized and poorly documented procedures. Good tests should be sensitive to drug effects, should be reliable and interpretable.

In my opinion, vigilance tasks conform especially well to these requirements. In vigilance (or sustained attention) tasks, people are required to sustain a high level of attention or readiness to detect and respond to changes in the stimulus situation, occurring rarely and unpredictably. These changes may indicate a malfunction or deviation from system limits as in driving, flying or monitoring in other more or less complex systems (chemical industries, nuclear power plants, air traffic control systems, assembly line inspection, security-monitoring of luggage at airports, instrument-monitoring by nurses and anaesthetists etc.). Tasks within this definition cannot be viewed as a homogeneous class with respect to processing demands, however, different vigilance tasks impose different demands upon perceptual discrimination, memory etc. (Koelega and Brinkman, 1986; Koelega *et al.*, 1989). Vigilance tasks claim to measure (the waning of) attention but if changes due to drugs are to be interpreted in terms of the ability to sustain attention or to maintain a task set, performance should not be limited by perceptual ability, memory capacity, or processing speed, that is quality of sensory input, memory demands and processing rate should not be the main determinants of performance.

Most investigators in psychopharmacology take the view that caffeine and nicotine affect primarily tasks requiring attention and information processing, and most reviewers of the effects of alcohol concur in the view that alcohol's major impairing effects are to be found on these types of tasks (e.g., Levine *et al.*, 1975, Moskowitz and Robinson, 1987). Attention and information processing are both rather vague, catchall phrases, permitting a great variety of meanings to be associated with them, and thus a great variety of tests to measure them. Aspects of 'attention' are selectivity, including the ability to resist distractions, and switching. Another aspect concerns attentional capacity, which may not always be sufficiently taxed by performance of a single task; single-task performance may further obscure performance of other important peripheral tasks such as they occur in driving and flying and therefore it is advantageous to employ divided-attention tasks, performing several tasks at the same time. The aspect of attention to be dealt with in the present review is the ability to remain alert in boring, monotonous, deactivating situations that do not invite concentrated, compensatory effort, such as in driving and in tedious, repetitive, industrial work. Actions of drugs may manifest themselves especially in some types of tasks designed to resemble these conditions, that is to say in vigilance tasks (Koelega, 1989, 1993, 1995). There exists an extensive literature on laboratory vigilance tasks as models of information processing tasks and according to Moskowitz

(1984) these tasks are valid laboratory models of real-life situations requiring sustained alertness. Erwin *et al.* (1978) stated that the serial presentation of an endless train of meaningless (neutral) stimuli interrupted by infrequent and randomly occurring stimuli that characterizes the vigilance task, has an analogue in highway driving where neutral stimuli such as broken-line lane dividers, expansion strips etc. are present. Linnoila (1978) reported that boring vigilance tasks, based on clues from epidemiological research concerning alcohol and traffic accidents, were used. These statements testify to the validity of vigilance tasks in psychopharmacological research, in contrast to, for example, dichotic listening tasks (attending to target stimuli in one ear and ignoring different stimuli in the other ear) which have no analogue in real-life, or tasks of questionable validity and/or low sensitivity, such as the frequently used DSST (Digit Symbol Substitution Test), CFF (Critical Flicker Frequency), digit span or simple RT (Koelega, 1995). However, performance on laboratory vigilance tasks does not really predict behaviour under real-life circumstances, for example driving a car. With respect to driving performance there are no truly predictive tests, whether laboratory tasks or performance in simulators. Vigilance is only a component of many real-life activities.

Apart from validity, vigilance performance shows reasonable intraindividual reliability (test-retest measures) and the tasks are sensitive to the effects of drugs. Information on both accuracy and speed of performance is provided, as well as on stimulus- and response-related processing (perceptual sensitivity and response bias). There is further no interference from practice and learning effects, most types of vigilance task do not require practice or sophisticated skills to achieve a stable baseline, in contrast to many other tasks, for example memory tests or reaction time tasks (McClelland, 1987). Finally, it should be noted that not every repetitive, long lasting, monotonous, task is a vigilance task. Unique characteristics of this task are the lack of automaticity due to the unpredictable occurrence of target stimuli, and the inability of the subject to control or pace the work rate. Broadbent (1984) raised the possibility of a fundamental distinction in drug effects between paced and unpaced tasks. In paced tasks, the subjects cannot make compensatory efforts as is done, for example, in letter cancelling and digit symbol substitution tasks.

Before examining the effects of caffeine, nicotine and alcohol on vigilance tasks, it may be useful to provide a brief outline of the mechanisms of action of these drugs.

MECHANISMS OF ACTION OF CAFFEINE, NICOTINE AND ALCOHOL

Actions of drugs in the CNS may be specific or nonspecific. The effect of a drug is considered to be specific when it affects an identifiable molecular mechanism unique to target cells that bear receptors for that drug, as has for example, been assumed to be the case for nicotine. In contrast, caffeine and alcohol have generally been considered to be nonspecific because they produce effects on many different target cells and act by diverse molecular mechanisms. However, this distinction is not absolute: a drug that is highly specific when tested at a low concentration may exhibit nonspecific actions at substantially higher doses. And generally acting (nonspecific) drugs may not act equally on all levels of the CNS.

Caffeine belongs to the methylxanthines, together with theophylline and theobromine. These drugs have long since also been used to treat asthma, because they relax (bronchial) smooth muscle, apart from stimulating the CNS and acting on the kidney to produce diuresis. The methylxanthines block the A_1 and A_2 receptors for adenosine, so act as competitive antagonists (See Arnaud, Ch. 6). Adenosine cannot be classed properly as a neurotransmitter or hormone but is usually described as a neuromodulator. Adenosine inhibits the release of almost all neurotransmitters, whether inhibitory or excitatory.

Nicotine, isolated from leaves of tobacco, is one of two natural alkaloids (the other one is lobeline) which exhibit their primary actions by stimulating autonomic ganglia (See Le Houezec, Ch. 1). Nicotine acts with a common family of receptors which are stimulated also by the natural transmitter ACh (acetylcholine). Binding sites for nicotine have been found in many areas of the brain. The alkaloid has both stimulant and depressant phases of action (biphasic), and has been shown to stimulate the release of both inhibitory and excitatory transmitters. A recent report shows that nicotine especially enhances glutamatergic synaptic transmission (McGehee *et al.*, 1995).

Alcohol has long been considered to produce a nonspecific perturbation of neuronal membrane lipids, thus altering the permeability of the axonal membrane: the so-called 'membrane hypothesis' was already formulated some 100 years ago. A distinct CNS 'receptor' (a neural membrane protein) for alcohol has not been found, but there may be 'microdomains' functioning as specific molecular sites of action. Tabakoff and Hoffman (1993) claim that recent evidence supports the premise that alcohol in low concentrations has selective effects on particular neurotransmitters in specific brain areas: alcohol would inhibit the function of the NMDA (N-Methyl-D-Aspartate) subtype of glutamate receptor, would potentiate the actions of $GABA_A$ (gamma-amino-butyric acid) agonists and would also affect the function of $5\text{-}HT_3$ (serotonin) receptors. Alcohol may affect the activity of all known transmitters (See Agarwal, Ch. 13).

Note that the various transmitters affected by caffeine, nicotine and alcohol are not acting separately: nicotine can reduce GABA and may facilitate glutamatergic and dopaminergic transmission, GABA may in turn effect glutamate etc. etc. Nicotine has long been thought to exert its effects on EEG activation mainly by cholinergic influences, but the neural mediators of the effects of caffeine and alcohol on EEG arousal were relatively unknown. However, a recent report (Rainnie *et al.*, 1994) shows that adenosine exerts a powerful tonic inhibitory control of mesopontine cholinergic neurons important in control of EEG activation.

EFFECTS OF CAFFEINE

The first studies on the effects of caffeine were carried out in Wilhelm Wundt's laboratory during the 1880s. Early studies of the effects on human performance have been reviewed by Weiss and Laties (1962) who concluded that the stimulant can restore performance that has deteriorated by an increase in capacity and by inducing favourable attitudes and enhanced motivation. More recently, Dews

(1984b) concluded that improved performance only occurs in fatigued and sleep-deprived subjects and that the effects can only be detected in tests of long duration (at least 1 hour), a conclusion also reached by Bruce *et al*. (1986) and Fagan *et al*. (1988).

Koelega (1993) reviewed the effects of caffeine on vigilance performance: of 23 comparisons of caffeine-placebo, 14 showed an improvement in overall level of performance (correct detections or hits), 3 only under special circumstances (only in the second part of a 60-min task, only with elderly subjects etc.), and 6 comparisons showed no improvement after caffeine, two of which employed an atypical (adaptive-rate) task. Improvements were sometimes reported with very low doses, less than one cup of coffee. Response latency to hits (RT) also improved but very few studies (4 only) reported on this measure. There were no effects on false alarms or on the usually occurring vigilance decrement with time-on-task.

There was no support for earlier conclusions that effects are noticed only in fatigued subjects in protracted sessions: improved performance was also noted in 4 experiments with a session duration of 10 min or less. However, effects may more easily become manifest under fatigue conditions: in 3 studies caffeine had a beneficial effect on many measures of performance during an overnight period of work, from the afternoon until the next morning. Several other studies (Linde, 1995, Lorist *et al*., 1994b) have also reported that the effects of caffeine are dependent on the state of the subject (fatigued or well-rested) and Linde (1995) also reported an interaction with time of day and task complexity.

One study (Zwyghuizen-Doorenbos *et al*., 1990) reported a peculiar phenomenon: the caffeine group also performed better and was more alert when caffeine had already been eliminated from the system on the 3rd day of the experiment. This might be a demonstration of a conditioned effect of caffeine according to the authors, the contextual stimuli (the vehicle containing the caffeine) might elicit a conditioned alerting response. An alternative hypothesis that residual caffeine or its metabolites may have accumulated over the preceding 2 days of administration, is not likely in the light of the known pharmacokinetics (half-life = 5.4 + 2.5 h) of caffeine and metabolites, at the dose used. Caffeine might also have altered the circadian rhythm of sleepiness/alertness in some way.

The sometimes heard argument that improvement after caffeine is only a recovery of withdrawal-induced impairment cannot easily be settled because there are practically no human adults who do not use caffeine in one form or another (tea, coffee, cocoa, cola drinks). Warburton (1995) found no evidence for this alleviation-of-deficits hypothesis, but Streufert *et al*. (1995) reported decreased cognitive effectiveness upon caffeine deprivation in 25 managers. From data of 9003 adults, Jarvis (1993) concluded that there is a dose-response trend to improved performance with higher levels of coffee consumption (especially among older people), but that there is no evidence of tolerance to the performance-enhancing effects (note that the latter two studies used no vigilance tasks).

EFFECTS OF NICOTINE

Vigilance studies employing nicotine, and caffeine as well, are practically confined to the last 10–15 years which is somewhat surprising for two of the most widely and longest used drugs in the world. Of 17 nicotine-placebo comparisons discussed by Koelega (1993), nicotine (administered by cigarettes, tablets or gum) improved overall level of performance in 11 cases, and two more studies showed improvement under special conditions (only during the first half of a 20-min task or only in a 'low' nicotine group). In most cases where detectability improved, speed (RT) improved also. Of the 11 improvements, 9 have been obtained with a type of task developed by Wesnes and Warburton (e.g., 1983a) called a "Rapid Visual Information Processing" (RVIP) task (detection of sequences of three consecutive odd or even digits). In 5 out of 7 cases, using the same task, the decline with time-on-task was prevented. In some cases (e.g., Wesnes and Warburton, 1983a) it has been reported that nicotine produces *absolute* improvements in performance (above and beyond baseline levels) rather than *relative* improvement (a reduction or prevention of the decrement). A modification of the task used in Bättig's institute (e.g., Michel and Bättig, 1989) has sometimes failed to show improvements with nicotine. The Swiss modification involves presenting the digits at a subject-paced rate rather than at a fixed rate, achieved by increasing the ISI after each error and decreasing it after a hit. Performance is assessed in terms of the subject's processing rate (the intervals serving as an inverse measure of performance) rather than in terms of hits (a positive effect of caffeine on this task was reported by Bättig and Buzzi, 1986). I have earlier suggested (Koelega, 1993) that this task may measure different aspects of information processing than the more commonly used fixed-rate version; a recent experiment (Baldinger *et al.*, 1995c) produced evidence supporting this suggestion.

As with caffeine, improved performance was sometimes noted in rather brief sessions (7 minutes), for example with Jones *et al.* (1992). In the latter study, young normal, elderly normal, and Alzheimer patients were used. The normal young group and the Alzheimer group showed a dose-dependent improvement in hits, sensitivity and RT, but nicotine (intramuscular) did not affect a short-term memory task.

Sherwood (1993), reviewing the effects of nicotine on other tasks than vigilance, also concluded that nicotine has small, but robust, positive effects on the CNS which may facilitate attention, memory and sensori-motor function.

The withdrawal-deficit hypothesis (improvements are only a recovery to normal of withdrawal-induced impairment) was challenged by findings showing that *non*-smokers also improved with nicotine tablets and that there was no difference between deprived smokers and smokers smoking. There is evidence that the effects of nicotine are genuine effects above and beyond deprivation-induced impairment (Warburton and Arnall, 1994). It is noteworthy that in many studies with deprived smokers, speed (RT) was impaired but accuracy not.

EFFECTS OF ALCOHOL

Moderate consumption of alcohol may contribute to pleasure and longevity (a decreased risk for cardiovascular disease) in life, but drinking also has an effect on the unborn child and has a firmly established relationship to automobile accidents. In the area of attention and performance the effects of alcohol may be pronounced. Studies in the U.K. have shown that in about 40% of industrial accidents leading to death, alcohol was a factor, and in the Netherlands it has been estimated that almost 40% of all working employees take alcohol from time to time during working hours. Streufert *et al.* (1994) report that alcohol consumption is especially extensive among managers; at higher job levels, assess to alcohol at the workplace is easier and is often considered to be justified. The authors showed that in this group even moderate use of alcohol had an impact on activities such as the capacity to deal with novel problems. Further, alcohol bears a causal relationship to automobile accidents and is, in the USA, involved in about 50% of fatal vehicle crashes. Alcohol further plays a prominent role in fatal aircraft accidents due to pilot error (Modell and Mountz, 1990).

Experiments on the effects of alcohol on behaviour were reported as early as 1851, but systematic studies started during the 1880s in Wilhelm Wundt's laboratory. Many reviews of the effects of alcohol have appeared but it is still unclear in what way alcohol affects performance. There is no agreement on the question of whether alcohol affects some functions (motor, sensory, or cognitive) more than other functions, whether specific behavioural areas are impaired or whether alcohol has a general effect, whether automatic (effortless, highly practised) processing is more insensitive to impairing effects than controlled (effortful, capacity limited) processing etc. etc. Two major principles that seem to have been established are that the ability to time-share in a divided-attention task is seriously impaired and that information processing is slower. However, the rate of switching between sources of information or retaining information in immediate memory seem to be relatively unimpaired (Moskowitz, 1973).

Experiments with alcohol confront investigators with one of the most difficult areas of research in psychopharmacology. A host of variables may affect the outcome: sex, weight, and age of subjects as well as their own drinking history and that of their family, type of alcohol used (beer, wine, vodka etc.; each type has its own rate of absorption into the bloodstream), the time allowed for consumption, the time between ingestion and start of the experiment, the time of day, prior ingestion of food and the type of food, expectations (subjects can often discriminate between alcohol and placebo), the occurrence of tolerance, a diminished effect of alcohol at an identical BAC (Blood Alcohol Concentration), both acute tolerance within a single session (the so-called Mellanby effect, different effects on ascending and descending BAC curve) and chronic tolerance (repeated sessions) etc. etc. There are considerable individual differences in the metabolism of alcohol, in time to achieve the peak BAC (usually from 30–90 min) and in the rate of elimination. Usually, BAC and performance are non-concordant, that is peak BAC and peak impairment of performance often take place at different points of time.

Effects of alcohol on vigilance performance were reviewed by Koelega (1995). Of 38 alcohol-placebo comparisons, an effect of alcohol on level of performance was noted in 50% of the cases, which increased to about 70% when small-sized samples (n < 15) were eliminated. Only 2 out of 9 comparisons reported a precipitated performance decline with time. Speed of response (RT) was impaired in 50% of the cases as was sensitivity (d' or A'). There were no effects on false alarms or response willingness. Koelega calculated BAC values at the start of the vigilance session but concluded that generalizations cannot be made: often performance was *impaired* at 0.02–0.03% (mg alcohol per 100 ml blood) but sometimes performance was *unimpaired* at 0.1% and this could occur in the same experiment where different types of vigilance task were used. Especially tasks requiring some form of spatial (nonverbal) information processing were impaired at low doses of alcohol. Positive, facilitating, effects with low doses were not found, and there was no evidence of residual sedation when BAC had reached zero (hangover effects). The question of whether performance is more impaired during the rising limb of the BAC curve than during the falling limb, could not be answered, because most studies did not measure performance during the ascending limb.

The effects of alcohol appear to be time-dependent, are largest during the two well-known periods of sleepiness (after-midnight and mid-afternoon), the same periods that show peaks of automobile accidents. Enhanced sleepiness 'potentiates' the normally occurring impairing effects of alcohol; together with night illumination (with conditions of glare and sometimes rain), this is an extremely dangerous combination, and Koelega suggested to lower the legal limit for driving after midnight to 0.02%; for young drivers (e.g., <23 years) this should be the statutory BAC at all times, because this group is highly over represented in accident statistics.

INTERACTIONS OF THE THREE DRUGS

Positive correlations between consumption of the three drugs under consideration have been reported (e.g., Ayers *et al.*, 1976), as well as between combinations of two of the three (e.g., Adesso, 1979; Brown and Benowitz, 1989; Chait and Griffiths, 1983; Kozlowski, 1976; Marshall *et al.*, 1980; Mello *et al.*, 1980; Mintz *et al.*, 1985; Ossip and Epstein, 1981). The facilitating effect of alcohol on smoking, especially in subjects with a history of alcohol abuse, seems rather well established, but the effect of caffeine on smoking is controversial. Smoking seems to increase the clearance of caffeine (Parsons and Neims, 1978). Caffeine given alone or combined with alcohol failed to influence smoke puffing behaviour (Nil *et al.*, 1984).

Very few studies have been carried out investigating the interactions of caffeine, nicotine and alcohol during performance of vigilance tasks, an exception being some studies using the previously described subject-paced task developed by Bättig. Therefore, some reports using other tasks than vigilance will also be mentioned.

Caffeine-nicotine studies

There are very few studies investigating the effects on performance of the combination of caffeine and nicotine; in the studies known to this author, both caffeine and nicotine alone improved performance on the Bättig-task (RIVP task, subject-paced), but these effects were not additive and the positive effects of caffeine were dampened by lunch (Hasenfratz *et al.*, 1991), and on a number of tasks (choice RT, tracking, STM) caffeine and nicotine provided no greater effect than when administered alone (Kerr *et al.*, 1991). Cohen *et al.* (1994) investigated whether consumption of caffeine may have a beneficial effect while quitting smoking. A 12-hour tobacco deprivation was accompanied by changes in mood, performance (on only two out of six tasks) and physiological measures (increase in EEG theta and decrease in alpha power). The EEG changes were reversed by caffeine (not by nicotine gum) but the effects on mood and performance were not changed. The authors suggested that the 12-hour period of abstinence may not have been long enough to affect performance. The deprived smokers substituted caffeine for nicotine, consumed more coffee during tobacco-deprivation. Lane and Rose (1995) investigated the influence of caffeine on smoking behaviour in the natural environment (the smoker's normal environment rather than the artificial circumstances of the laboratory). Their study showed that caffeine intake has no effect on how much a person smokes. Pritchard *et al.* (1995) also reported that smoking and caffeine did not interact, which suggests that the epidemiological link between smoking and coffee drinking (smokers are more often coffee drinkers than non-smokers and smoking is more likely during and immediately after coffee drinking) has a non-pharmacological basis.

Caffeine-alcohol studies

There is a popular belief in the lay public that coffee can, to some extent, offset the debilitating effects of alcohol intoxication, for example before driving an automobile. Evidence has been reviewed by among others Holloway and Holloway (1979) and Fudin and Nicastro (1988). The conclusion is that legally intoxicated individuals can *not* antagonize alcohol-induced performance decrements with caffein. Note, however, that legal intoxication in most states of the U.S.A. is indicated by 0.1% BAC, whereas this is 0.05% in the Netherlands, for example. So the possibility that antagonising effects may take place at lower BAC levels cannot be excluded. Fudin and Nicastro point out that the particular function tested and the dosages of both drugs used may have a profound influence on the outcome; much more attention has been paid to the choice of alcohol dosages that produce performance decrements than to caffeine dosages that enhance performance. The authors provide many guidelines for the design of future experiments. More recent studies, not included in the aforementioned reviews, have reported that impairing effects of alcohol were antagonised by caffeine on a RVIP task (Hasenfratz *et al.*, 1993) and on some other measures such as critical flicker fusion threshold and STM (Kerr *et al.*, 1991) but in the latter study these effects may not have been statistically significant. Whereas in most studies reviewed by Fudin and Nicastro, caffeine was given after alcohol, in the study by Hasenfratz *et al.* (1993) caffeine was given *before* alcohol; caffeine pretreatment reduced BACs, so the antagonising effect may have reduced the absorption of alcohol. Fillmore and Vogel-Sprott (1995) reported that caffeine diminished alcohol-induced impairment on a pursuit rotor task, but also that participants' expectancies account for a large amount of variance in behaviour.

Nicotine-alcohol studies

In real-life situations, alcohol often increases the amount and rate of cigarette smoking in smokers, and there are reports that smoking *diminishes* the alcohol-induced performance deterioration in a subject-paced vigilance task (Michel and Bättig, 1989), in selective-attention and divided-attention tasks (Leigh *et al.*, 1977), in visual discrimination (Tong *et al.*, 1974), in choice reaction time (Kerr *et al.*, 1991; Lyon *et al.*, 1975), and in tracking (Kerr *et al.*, 1991), but in the latter study the alcohol-induced impairment on a STM (short-term memory) task was aggravated by nicotine. Perkins *et al.* (1995) reported that men and women respond differentially on subjective measures to the combination of nicotine and alcohol. Myrsten and Andersson (1973) also reported that smoking counteracted alcohol-induced impairment in simple- and choice-RT tasks, but also that heart rate increased and hand steadiness deteriorated. These findings, together with factors such as tolerance and effects of prior ingestion of food, illustrate that, at least in smokers, in operational circumstances the effects of alcohol on behaviour are hard to predict.

Can anything be concluded with respect to neurotransmitter models of attention and information processing? No single neurotransmitter can be ascribed an exclusive and prominent role in attention and information processing, probably all neurotransmitters are involved in the effects described above. Caffeine may also increase noradrenaline and dopamine synthesis, apart from the effect on ACh, and nicotine can increase the release of peptides and may also facilitate dopaminergic transmission. Dopaminergic mechanisms are involved in the negative correlation between cigarette smoking and Parkinson's disease and probably also in the prevalence of smoking among schizophrenics. Alcohol has biochemical and behavioural mechanisms in common with opiates, and although debilitating effects of alcohol on performance of some tasks have been reported to be neutralized by nicotine as we have seen, this has also been shown after ingestion of amphetamine, cocaine and the serotonin-reuptake inhibitor zimelidine, drugs with a different mechanism of action than nicotine. In most studies where alcohol interacted with other drugs in the effects on behaviour (not only with caffeine and nicotine but also with benzodiazepines, amphetamine etc.) the alcohol kinetics remained unaltered.

Finally, it should be noted that the total psychological response to a psychotropic compound is a complex interaction involving motivational factors, personality and even sociocultural habits and expectancies. Hindmarch (1980) showed for example that the intrinsic motivation of the task situation is an important determinant of performance: reaction time measured in the laboratory after temazepam was impaired, but reaction time in a car driving simulator was not, where the protocols, dose, regimen and subject populations were similar. George *et al.* (1990) showed that the impairing effects of alcohol are to some degree under volitional control via motivation (heightened concentration), although it may be questioned whether the attenuation of impairment would extend to less controllable behavioural indices such as the sway test. Further, Keister and McLaughlin (1972) reported an interaction of caffeine with introversion-extraversion in a vigilance task. Although I have earlier criticized their statistical treatment of the data, there is no doubt that groups of extreme introverts perform better than extreme extraverts on visual vigilance tasks (Koelega, 1992), and there are several reports in the literature that personality may modulate the effects of drugs.

CONCLUSIONS

The present review has shown that both caffeine and nicotine improve vigilance performance and that even low doses of alcohol impair performance on certain types of vigilance task. Experimental psychologists and psychopharmacologists have thus merely confirmed in the laboratory what people have long since been aware of. Contrary to expectations supported by folklore, however, caffeine given after alcohol often does not seem to neutralize the impairing effects of alcohol, but nicotine may do so on some but not all tasks; one isolated study reported reduced impairment when caffeine was administered before alcohol. There is no evidence of an additive effect of the two stimulants, that is to say the combination of caffeine and nicotine does not seem to facilitate performance, although consumption of the one may affect that of the other. Current information about interactions of the social drugs is incomplete, more parametric studies with multiple doses and combinations of doses and times-of-administration are needed and different measures should be examined, because caffeine and nicotine may have a more limited range in which they affect performance positively than alcohol has in impairing performance. A range of functions, some of which relevant to real-life skilled performance, such as driving and monitoring, should be investigated. For example, it may be interesting if nicotine and caffeine would appear to reduce alcohol-induced impairment of digit recall, arithmetic and letter cancelling, but if impairment of other functions on which alcohol is known to have a great impact (vestibular, oculomotor, spatial information processing) would not be reduced, the practical relevance of these antagonising effects is negligible.

References*

Abercrombie, E.A., Keefe, K.A., DiFrischia, D.A. and Zigmond, M.J. (1989). Differential effect of stress on *in vivo* dopamine release in striatum, nucleus accumbens and medial frontal cortex. *Journal of Neurochemistry*, **52**:1655–1658.

Abernethy, D.R., Todd, E.L. and Schwartz, J.B. (1985). Caffeine disposition in obesity. *British Journal of Clinical Pharmacology*, **20**:61–66.

Adan, A. (1994). Chronotype and personality factors in the daily consumption of alcohol and psychostimulants. *Addiction*, **89**:455–462.

Adesso, V.J. (1979). Some correlates between cigarette smoking and alcohol use. *Addictive Behaviors*, **4**:269–273.

Adler, T. (1993). Nicotine gives mixed results on learning and performance. *American Psychological Association Monitor*, **May**:14–15.

Agarwal, D.P. and Goedde, H.W. (1990). Alcohol metabolism, alcohol intolerance and alcoholism. Biochemical and pharmacogenetic approaches. Berlin, Heidelberg: Springer Verlag, 184 pp.

Agarwal, D.P. and Goedde, H.W. (1987). Genetic variation in alcohol metabolizing enzymes: implications in alcohol use and abuse. In: H.W. Goedde and D.P. Agarwal (Eds.), Genetics and alcoholism. New York: A.R. Liss, pp. 112–140.

Agarwal, D.P. and Goedde, H.W. (1989). Human aldehyde dehydrogenases: their role in alcoholism. *Alcohol*, **6**:517–523.

Alcaini, M., Giard, M.H., Thevent, M. and Pernier, J. (1994). Two separate frontal components in the N1 wave of the human auditory evoked response. *Psychophysiology*, **31**:611–615.

Alderdice, F.A. and Davidson, R. (1990). The effect of alcohol consumption on recency discrimination ability: an early screening test for alcohol-induced cognitive impairment. *British Journal of Addiction*, **85**:531–536.

American Psychiatric Association. (1991). DSM-IV Options Book: Work in Progress. American Psychiatric Association, Washington, D.C.

Amsel, A. (1990). Arousal, suppression, and persistence: Frustration theory, attention, and its disorders. *Cognition and Emotion*, **4**:239–268.

Anda, R.F., Williamson, D.F., Escobedo, L.G., Mast, E.E., Giovano, G.A. and Remington, P.L. (1990). Depression and the dynamics of smoking. *Journal of the American Medical Association*, **264**:1541–1545.

Anderson, K.J. and Revelle, W. (1983). The interactive effects of caffeine, impulsivity, and task demands on a visual search task. *Personality and Individual Differences*, **4**:127–134.

Anderson, K.J. (1994). Impulsivity, caffeine, and task difficulty: A within-subject test of the Yerkes-Dodson law. *Personality and Individual Differences*, **16(6)**:813–829. *[1]*.

Anderson, K.J., Revelle, W. and Lynch, M.J. (1989). Caffeine, impulsivity, and memory scanning: A comparison of two explanations for the Yerkes-Dodson effect. *Motivation and Emotion*, **13(1)**:1–20. *[2]*.

Andersson, K. and Hockey, G.R.J. (1977). Effects of cigarette smoking on incidental memory. *Psychopharmacology*, **52**:223–226.

Andersson, K. and Post, B. (1974). Effects of cigarette smoking on verbal rote learning and physiological arousal. *Scandinavian Journal of Psychology*, **15**:263–267.

---

* Number behind references in *[ ]* refer to Table 7.2 Ch. 7: Van der Stelt and Snel

Andersson, K. (1975). Effects of smoking on learning and retention. *Psychopharmacology*, **41**:1–5.

Aneshensel, C.S. and Huba, G.J. (1983). Depression, alcohol use and smoking over one year: a four-wave longitudinal causal model. *Journal of Abnormal Psychology*, **92**:134–150.

Anonymus, (1991). IARC Monographs on the Evaluation of Carcinogenic Risks to Humans, Vol. 51. Coffee, tea, maté, methylxanthines and methylglyoxal, pp. 513.

Anthony, J.C., Warner, L.A. and Kessler, R.C. (1994). Comparative epidemiology of dependence on tobacco, alcohol, controlled substances, and inhalants: Basic findings from the National Comorbidity Survey. *Experimental and Clinical Psychopharmacology*, **2**:244–268.

Arbuckle, T.Y., Chaikelson, J.S. and Gold, D.P. (1994). Social drinking and cognitive functioning revisited: the role of intellectual endowment and psychological distress. *Journal of Studies of Alcohol*, **55**:352–361.

Arci, J.B. and Grunberg, N.E. (1992). A psychophysical task to quantify smoking cessation-induced irritability: The Reactive Irritability Scale (RIS). *Addictive Behaviors*, **17**:587–601.

Arendt, T., Henning, D., Gray, J.A. and Marchbanks, R. (1988). Loss of neurons in the rat basal forebrain cholinergic projection system after prolonged intake of ethanol. *Brain Research Bulletin*, **21**:563–569.

Armitage, A.K., Hall, G.H. and Sellers, C.M. (1969). Effects of nicotine on electrocortical activity and acetylcholine release from the cat cerebral cortex. *British Journal of Pharmacology*, **35**:152–160.

Arnaud, M.J. and Enslen, M. (1992). The role of paraxanthine in mediating physiological effects of caffeine. 14th International Conference in Coffee Science, San Francisco, 14–19 July 1991, Proceedings ASIC, Paris, pp. 71–79.

Arnaud, M.J. and Welsch, C. (1979). Metabolic pathway of theobromine in the rat and identification of two new metabolites in human urine. *Journal of Agricultural and Food Chemistry*, **27**:524–527.

Arnaud, M.J. and Welsch, C. (1982). Theophylline and caffeine metabolism in man. In: N. Reitbrock, B.G. Woodcock and A.H. Staib (Eds.), *Theophylline and other methylxanthines*. Germany: Friedrich Vieweg and Sohn, pp. 135–148.

Arnaud, M.J. (1984). Products of metabolism of caffeine. In P.B. Dews (Ed.) Caffeine. Berlin: Springer Verlag, pp. 3–38.

Arnaud, M.J. (1985) Comparative metabolic disposition of [1-Me^{14}C]caffeine in rats, mice, and Chinese hamsters. *Drug Metabolism and Disposition*, **13**:471–478.

Arnaud, M.J. (1987). The pharmacology of caffeine. *Progress in Drug Research*, **31**:273–313.

Arnaud, M.J. (1993a). Metabolism of Caffeine and Other Components of Coffee. In: S. Garattini (Ed.), Caffeine, Coffee and Health. New York: Raven Press, Ltd., pp. 43–95.

Arnaud, M.J. (1993b). Caffeine. Vol. 1. In: R. Macrae, R.K. Robinson and M.J. Sadler (Eds.), *Encyclopaedia of Food Science, Food Technology and Nutrition*. London: Academic Press, pp. 566–571.

Arnaud, M.J., Wietholtz, H., Voegelin, M., Bircher, J. and Preisig, R. (1982). Assessment of the cytochrome P-448 dependent liver enzyme system by a caffeine breath test. In: R. Sato (Ed.), *Microsomes Drug Oxidation and Drug Toxicity*. New York: Wiley, pp. 443–444.

Arnold, M.E., Petros, T.V., Beckwith, B.E., Coons, G. and Gorman, N. (1987). The effects of caffeine, impulsivity, and sex on memory for word lists. *Physiology and Behavior*, **41**:25–30. *[3]*.

Ashley, M.J., Olin, J.S., Harding, L.W., Komaczewski, A. Schmidt, W. and Rankin, J.G. (1981). Morbidity patterns in hazardous drinkers: relevance of demographic, sociologic, drinking and drug use characteristics. *International Journal of the Addictions*, **16**:593–625.

Ashton, H. and Golding, J.F. (1989). Smoking: Motivation and models. In: T. Ney and A. Gale (Eds.), Smoking and human behavior. Chichester: Wiley, pp. 21–56.

Ashton, H., Millman, J.E., Telford, R. and Thompson, J.W. (1974). The effect of caffeine, nitrazepam and cigarette smoking on the contingent negative variation in man. *Electroencephalography and Clinical Neurophysiology*, **37**:59–71.

Ashton, H., Savage, R.D., Telford, R., Thompson, J.W. and Watson, D.W. (1972). The effects of cigarette smoking on the response to stress in a driving simulator. *British Journal of Pharmacology*, **45**:546–556.

Atkinson, R.C. and Shiffrin, R.M. (1968). Human memory: A proposed system and its control processes. In: K.W. Spence and J.R. Spence (Eds.), The psychology of learning and motivation: Advances in research and theory. Vol. 2. New York: Academic Press, pp. 89–195.

Atkinson, R.C. and Shiffrin, R.M. (1971). Recognition and retrieval processes in free recall. *Psychological Review*, **79**:97–123.

Ayers, J., Ruff, C.F. and Templer, D.I. (1976). Alcoholism, cigarette smoking, coffee drinking and extraversion. *Journal of Studies on Alcohol*, **37**:983–985.

Azcona, O., Barbanoj, M.J., Torrent, J. and Jané, F. (1995). Evaluation of the central effects of alcohol and caffeine interaction. *British Journal of Clinical Pharmacology*, **40**:393–400.

Babbini, M., Jones, B.L. and Alkana, R.L. (1991). Effects of post-training ethanol and group housing upon memory of an appetitive task in mice. *Behavioral and Neural Biology*, **56**:32–42.

Bachrach, H. (1966). Note on the psychological effects of caffeine. *Psychological Reports*, **18**:86.

Baddeley, A. (1983). Your memory: A user's guide. Harmondsworth, England: Penguin Books.

Baddeley, A. (1992). Working memory: The interface between memory and cognition. *Journal of Cognitive Neuroscience*, **4**:281–288.

Baddeley, A. (1993). Working memory or working attention? In: A. Baddeley and L. Weiskrantz (Eds.), Attention: Selection, Awareness and control. Oxford: Oxford University Press, pp. 152–170.

Bak, A. (1990). Coffee and cardiovascular risk; an epidemiological study. Ph.D. thesis, Ch. 2. Rotterdam: Erasmus University, pp. 13–27.

Bakan, P. (1959). Extraversion-introversion and improvement in an auditory vigilance task. *British Journal of Psychology*, **50**:325–332.

Baldinger, B., Hasenfratz, M. and Bättig, K. (1995a). Switching to ultra low nicotine cigarettes: effects of different tar yields and blocking of olfactory cues. *Pharmacology, Biochemistry and Behavior*, **50**:233–239.

Baldinger, B., Hasenfratz, M. and Bättig, K. (1995b). Effects of smoking abstinence and nicotine abstinence on heart rate, activity and cigarette craving under field conditions. *Human Psychopharmacology*, **10**:127–136.

Baldinger, B., Hasenfratz, M. and Bättig, K. (1995c). Comparison of the effects of nicotine on a fixed rate and on a subject paced version of the rapid information processing task. *Psychopharmacology*, **121**:396–400.

Baldini, F.D., Landers, D.M., Skinner, J.S. and O'Connor, J.S. (1992). Effects of varying doses of smokeless tobacco at rest and during brief, high-intensity exercise. *Military Medicine*, **157**:51–55.

Balfour, D.J.K. (1991). The influence of stress on psychopharmacological responses to nicotine. *British Journal of Addiction*, **86**:489–493.

Balfour, D.J.K. (1994). The neural mechanisms underlying the rewarding properties of nicotine. *Journal of Smoking-Related Disorders*, **5(Suppl. 1)**:141–148.

Balogh, A., Irmisch, E., Klinger, G., Splinter, F.-K. and Hoffmann, A. (1987). Untersuchungen zur Elimination von Coffein und Metamizol im Menstruationszyklus der fertilen Frau. *Zentralblatt für Gynäkologie*, **109**:1135–1142.

Baribeau, J.M., Braun, C.M. and Dube, R. (1986). Effects of alcohol intoxication on visuospatial and verbal- contextual tests of emotion discrimination in familial risk for alcoholism. *Alcoholism: Clinical and Experimental Research*, **10**:496–499.

Barlow, D.H. and Baer, D.J. (1967). Effect of cigarette smoking on the critical flicker frequency of heavy and light smokers. *Perceptual and Motor Skills*, **24**:151–155.

Barmack, J.E. (1940). The time of administration and some effects of 2 grs. of alkaloid caffeine. *Journal of Experimental Psychology*, **27**:690–698.

Baron, R.M. and Kenny, D.A. (1986). The moderator-mediator variable distinction in social psychological research: Conceptual, strategic, and statistical considerations. *Journal of Personality and Social Psychology*, **51(6)**:1173–1182.

Barraclough, S. and Foreman, N. (1994). Factors influencing recall of supraspan word lists: Caffeine dose and introversion. *Pharmacopsychoecologia*, **7**:229–236. *[4]*.

Bates, M.E. and Tracy, J.I. (1990). Cognitive functioning in young "social drinkers": Is there impairment to detect? *Journal of Abnormal Psychology*, **99**:242–249.

Bates, T.C. and Eysenck, H.J. (1994). A comparison of the information processing rates of non-smokers and cigarette-deprived smokers. *Personality and Individual Differences*, **17**:855–858.

Bates, T.C., Pellett, O., Stough, C. and Mangan, G. (1994). The effects of smoking on simple and choice reaction time. *Psychopharmacology*, **114**:365–368.

Bates, T.C., Stough, C., Mangan, G., Pellett, O. and Corballis, P. (1995). Smoking, processing speed and attention in a choice reaction time task. *Psychopharmacology*, **120**:209–212.

Batt, R.D. (1989). Absorption, distribution and elimination of alcohol. In: K.E. Crow and R.D. Batt (Eds.), Human metabolism of alcohol. Vol. I. Pharmacokinetics, medicolegal aspects, and general interest. Boca Raton: CRC Press, pp. 3–8.

Bättig, K. and Buzzi, R. (1986). Effect of coffee on the speed of subject-paced information processing. *Neuropsychobiology*, **16**:126–130. *[5]*.

Bättig, K. and Welzl, H. (1993). Psychopharmacological profile of caffeine. In: S. Garattini (Ed.), Caffeine, Coffee, and Health. New York: Raven Press, pp. 213–253.

Bättig, K. (1985). The physiological effects of coffee consumption. In: M.N. Clifford and K.C. Wilson (Eds.), Caffeine, botany, biochemistry and production of beans and beverages. London: Croom Helm, pp. 394–439.

Bättig, K. (1994). Caffeine research. *Pharmacopsychoecologia, Special Issue*, **7(2)**:231–237.

Bättig, K., Buzzi, R., Martin, J.R. and Feierabend, J.M. (1984). The effects of caffeine on physiological functions and mental performance. *Experientia*, **40(11)**:1218–1223. *[6]*.

Bättig, K., Jacober, A. and Hasenfratz, M. (1993). Cigarette smoking related variation of heart rate and physical activity with ad libitum smoking under field conditions. *Psychopharmacology*, **110**:371–373.

Bättig, K., Kos, J. and Hasenfratz, M. (1994). Smoking and food intake in a field study: Continuous actometer/heart rate recording and pocket computer assisted dietary reports, subjective self-assessments and mental performance. *Drug Development Research*, **31**:59–70.

Baum-Baicker, C. (1985). The health benefits of moderate alcohol consumption: A review of the literature. *Drug and Alcohol Dependence*, **15**:305–322.

Baylor, A.M., Layne, C.S., Mayfield, R.D, Osborne, L. and Spirduso, W.W. (1989). Effects of ethanol on human fractionated response times. *Drug and Alcohol Dependency*, **23**:31–40.

Beach, C.A., Bianchine, J.R. and Gerber, N. (1984). The excretion of caffeine in the semen of men: pharmacokinetics and comparison of the concentrations in blood and semen. *Journal of Clinical Pharmacology*, **24**:120–126.

Beaumanoir, A., Ballis, T., Nahory, A. and Genier, M. (1974). Alterations of the CNV and GSR under the effect of alcohol. *Electroencephalography and Clinical Neurophysiology*, **36**:85.

Bechtel, Y.C., Joanne, C., Grandmottet, M. and Bechtel, P.R. (1988). The influence of insulin-dependent diabetes on the metabolism of caffeine and the expression of the debrisoquin oxidation phenotype. *Clinical Pharmacology and Therapeutics*, **44**:408–417.

Beck, A.T., Wright, F.D., Newman, C.F. and Liese, B.S. (1993). Cognitive therapy of substance abuse. New York: Guilford.

Begleiter, H., Porjesz, B., Bihari, B. and Kissin, B. (1984). Event-related brain potentials in boys at risk for alcoholism. *Science*, **225**:1493–1496.

Beh, H.C. (1989). Reaction time and movement time after active and passive smoking. *Perceptual and Motor Skills*, **68**:513–514.

Behm, F.M., Schur, C., Levin, E.D., Tashkin, D.P. and Rose, J.E. (1993). Clinical evaluation of a citric acid inhaler for smoking cessation. *Drug and Alcohol Dependence*, **31**:131–138.

Behm, F.M., Schur, C., Levin, E.D., Tashkin, D.P. and Rose, J.E. (1990). Low nicotine regenerated smoke aerosol reduces desire for cigarettes. *Journal Substance Abuse*, **2**:237–247.

Benowitz, N.L. and Henningfield, J.E. (1994). Establishing a nicotine threshold for addiction. *The New England Journal of Medicine,* **331**:123–125.

Benowitz, N.L. and Jacob, P. III (1984). Daily intake of nicotine during cigarette smoking. *Clinical Pharmacology and Therapeutics*, **35**:499–504.

Benowitz, N.L. (1987). The human pharmacology of nicotine. In: H.D. Cappell (Ed.), Research advances in alcohol and drug problems. New York: Plenum, pp. 1–52.

Benowitz, N.L. (1988). Pharmacologic aspects of cigarette smoking and nicotine addiction. *New England Journal of Medicine*, **319**:1318–1330.

Benowitz, N.L. (1990). Pharmacokinetic considerations in understanding nicotine dependence. In: G. Bock and J. Marsh (Eds.), The biology of nicotine dependence (CIBA Foundation Symposium 152). Chichester: Wiley, pp. 186–209.

Benowitz, N.L., Hall, S.M. and Modin, G. (1989). Persistent increase in caffeine concentrations in people who stop smoking. *British Medical Journal*, **298**:1075–1076.

Benowitz, N.L., Hall, S.M., Herning, R.I., Jacob, P. III, Jones, R.T. and Osman, A.L. (1983). Smokers of low-yield cigarettes do not consume less nicotine. *New England Journal of Medicine*, **309**:139–142.

Benowitz, N.L., Jacob, P. III and Savanapridi, C. (1987). Determinants of nicotine intake while chewing nicotine polacrilex gum. *Clinical Pharmacology and Therapeutics*, **41**:467–473.

Benowitz, N.L., Jacob, P. III, Denaro, C. and Jenkins, R. (1991). Stable isotope studies of nicotine kinetics and bioavailability. *Clinical Pharmacology and Therapeutics*, **49**:270–277.

Benowitz, N.L., Jacob, P. III, Jones, R.T. and Rosenberg, J. (1982). Interindividual variability in the metabolism and cardiovascular effects of nicotine in man. *Journal of Pharmacology and Experimental Therapeutics*, **221**:368–372.

Benowitz, N.L., Jones, R.T. and Jacob, P. III (1986). Additive cardiovascular effects of nicotine and ethanol. *Clinical Pharmacological Therapy*, **40**:420–424.

Benowitz, N.L., Kuyt, F. and Jacob, P. III (1982). Circadian blood nicotine concentration during cigarette smoking. *Clinical Pharmacology and Therapeutics*, **32**:758–764.

Benowitz, N.L., Porchet, H. and Jacob, P. III (1990). *Pharmacokinetics, metabolism, and pharmacodynamics of nicotine*. In: S. Wonnacott, M.A.H. Russel and I.P. Stolerman (Eds.), Nicotine psychopharmacology: molecular, cellular, and behavioural aspects. Oxford: Oxford University Press, pp. 112–157.

Benowitz, N.L., Porchet, H., Sheiner, L. and Jacob, P. III (1988). Nicotine absorption and cardiovascular effects with smokeless tobacco use: comparison with cigarettes and nicotine gum. *Clinical Pharmacology and Therapeutics*, **44**:23–28.

Benton, A.L. and Hamsher, K. deS. (1976). Multilingual Aphasia Examination, Manual revised, 1978. Iowa City: University of Iowa.

Bergman, H. (1985). Cognitive deficits and morphological cerebral changes in a random sample of social drinkers. In: M. Galanter (Ed.), Recent Developments in Alcoholism. Vol. 3. Ch. 17. New York: Plenum Press, pp. 265–276.

Berthou, F., Flinois, J.-P., Ratanasavanh, D., Beaune, P., Riche, C. and Guillouzo, A. (1991). Evidence for the involvement of several cytochromes P-450 in the first steps of caffeine metabolism by human liver microsomes. *Drug Metabolism and Disposition*, **19**:561–567.

Berthou, F., Goasduff, T., Dréano, Y. and Ménez, J.-F. (1995). Caffeine increases its own metabolism through Cytochrome P4501A induction in rats. *Life Sciences*, **57**:541–549.

Bickel, W.K., Hughes, J.R., DeGrandpre, R.J., Higgins, S.T. and Rizzuto, P. (1992). Behavioral economics of drug self-Administration. IV. The effects of response requirement on the consumption of and interaction between concurrently available coffee and cigarettes. *Psychopharmacology*, **107**:211–216.

Billings, C.E., Wick, R.L., Gerke, R.J. and Chase, R.C. (1972). The effects of alcohol on pilot performance during instrument flight. (Report No. FAA-AM). Washington, DC: Federal Aviation Agency, pp. 72–74.

Birkett, D.J. and Miners, J.O. (1991). Caffeine renal clearance and urine caffeine concentrations during steady state dosing. Implications for monitoring caffeine intake during sport events. *British Journal of Clinical Pharmacology*, **31**:405–408.

Blanchard, J. and Sawers, S.J.A. (1983). The absolute bioavailability of caffeine in man. *European Journal of Clinical Pharmacology*, **24**:93–98.

Blanchard, J. (1982). Protein binding of caffeine in young and elderly males. *Journal of Pharmaceutical Science*, **71**:1415–1418.

Blanchard, J., Sawers, S.J.A., Jonkman, J.H.G. and Tang-Liu, D.-S. (1985). Comparison of the urinary metabolite profile of caffeine in young and elderly males. British *Journal of Clinical Pharmacology*, **19**:225–232.

Blitzer, R.D., Gil, O. and Laudau, E.M. (1990). Long-term potentiation in rat hippocampus is inhibited by low concentrations of ethanol. *Brain Research*, **537**:203–208.

Bobo, J.K. and Gilchrist, L.D. (1983). Urging the alcoholic client to quit smoking cigarettes. *Addictive Behaviors*, **8**:297–305.

Bobo, J.K. (1992). Nicotine dependence and alcoholism epidemiology and treatment. *Journal of Psychoactive Drugs*, **24(2)**:123–129.

Boda, D. and Németh, I. (1989). Measurement of urinary caffeine metabolites reflecting the *in vivo* xanthine oxidase activity in premature infants with RDS and in hypoxic states of children. *Biomedica Biochimica Acta*, **48**:S31–S35.

Bohman, M., Cloninger, C.R., von Knorring, A.L. and Sigvardsson, S. (1984). An adoption study of somatoform disorders. III. Crossfostering analysis and genetic relationship to alcoholism and criminality. *Archives of General Psychiatry*, **41**:872–878.

Böhme, M. and Böhme, H.-R. (1985). Der Einfluß hormonaler Kontrazeptiva und des Coffeins auf den Farnsworth-Munsell-100-Hue-Test. *Zentralblatt Gynäkologie*, **107**:1300–1306. *[7].*

Boisse, N.R. and Okamoto, M. (1980). Ethanol as a sedative-hypnotic; comparison with barbiturate and nonbarbiturate sedative-hypnotics. In: H. Rigter and J.C. Crabbe Jr. (Eds.), Alcohol tolerance and dependence. Amsterdam: Elsevier. 265–292.

Bologa, M., Tang, B., Klein, J., Tesoro, A. and Koren, G. (1991). Pregnancy-induced changes in drug metabolism in epileptic women. *Journal of Pharmacology Experimental Therapeutics*, **257**:735–740.

Bond, A. and Lader, M. (1974). The use of analogue scales in rating subjective feelings. *British Journal of Psychology*, **47**:211–218.

Bonnet, M.H. and Arand, D.L. (1994). The use of prophylactic naps and caffeine to maintain performance during a continuous operation. *Ergonomics*, **37(6)**:1009–1020. *[8].*

Borland, B.L. and Rudolph, J.P. (1975). Relative effects of low socioeconomic status, parental smoking and poor scholastic performance among high school students. *Social Science and Medicine*, **9**:27–30.

Borland, R.G., Rogers, A.S., Nicholson, A.N., Pascoe, P.A. and Spencer, M.B. (1986). Performance overnight in shiftworkers operating a day-night schedule. *Aviation, Space, and Environmental Medicine*, 57:241–249. *[9]*.

Borrill, J.A., Rosen, B.K. and Summerfield, A.B. (1987). The influence of alcohol on judgement of facial expressions of emotion. *British Journal of Medical Psychology*, 60:71–77.

Bory, C., Baltassat, P., Porthault, M., Bethenod, M., Frederich, A. and Aranda J.V. (1979). Metabolism of theophylline to caffeine in premature newborn infants. *Journal of Pediatrics*, 94:988–993.

Bosron, W.F. and Li, T.K. (1987). Catalytic properties of human liver alcohol dehydrogenase isoenzymes. *Enzyme*, 37:19–28.

Boston Collaborative Drug Surveillance Program. (1972). Coffee drinking and acute myocardial infarction. *Lancet*, 2:1278–1281.

Bowden, S.C. (1987). Brain impairment in social drinkers? No cause for concern. *Alcoholism: Clinical and Experimental Research*, 11(4):407–410.

Bowden, S.C. (1988). Learning in young alcoholics. *Journal of Clinical and Experimental Psychology*, 10:157–168.

Bowden, S.C., Walton, N.H. and Walsh, K.W. (1988). The hangover hypothesis and the influence of moderate social drinking on mental ability. *Alcoholism: Clinical and Experimental Research*, 12:25–29.

Bower, B. (1993). Smoke gets in your brain. Warning: Cigarettes may be hazardous to your thoughts. *Science News*, **January 16**:143–144.

Boyd, G.M. and Maltzman, I. (1984). Effects of cigarette smoking on bilateral skin conductance. *Psychophysiology*, 21:334–341.

Bozarth, M.A. (1987). Ventral tegmental reward system. In: L. Oreland and J. Engel (Eds.), Brain reward systems and abuse. New York: Raven Press, pp. 112–134.

Bracco, D., Ferrarra, J.-M., Arnaud, M.J., Jequier, E. and Schutz, Y. (1995). Effects of caffeine on energy metabolism, heart rate, and methylxanthine metabolism in lean and obese women. *American Journal of Physiology*, 269:E671–E678.

Brachtel, D. and Richter, E. (1988). Effect of altered gastric emptying on caffeine absorption. *Zeitschrift für Gastroenterologie*, 26:245–251.

Brandeis, D., Naylor, H., Halliday, R., Callaway, E. and Yano, L. (1992). Scopolamine effects on visual information processing, attention and event-related potential map latencies. *Psychopharmacology*, 29:315–336.

Brandt, J., Butters, N., Ryan, C. and Bayog, R. (1983). Cognitive loss and recovery in long term alcohol abusers. *Archives of General Psychiatry*, 40:435–442.

Breeze, E. (1985). Women and drinking. London: HMSO.

Brennan, A.F., Walfish, S. and AuBuchon, P. (1986). Alcohol use and abuse in college students. II. Social/environmental correlates, methodological issues, and implications for interventions. *The International Journal of the Addictions*, 21:475–493.

Breslau, N., Kilbey, M.M. and Andreski, P. (1992). Nicotine withdrawal symptoms and psychiatric disorders: Findings from an epidemiologic study of young adults. *American Journal of Psychiatry*, 149:464–469.

Breuer, K., Streufert, S. and Nogami, G.Y. (Eds.) (1996) Personalentwicklung und Auswahl mit diagnostischen Simulationen. Heidelberg: Hogrefe.

Brewer, N. and Sandow, B. (1980). Alcohol effects on driver performance under conditions of divided attention. *Ergonomics*, 23:185–190.

Brioni, J.D., McGaugh, J.L. and Izquierdo, I. (1989). Amnesia induced by short-term treatment with ethanol: Attenuation by pretest oxotremorine. *Pharmacology, Biochemistry and Behavior*, 33:27–29.

Broadbent, D.E. (1984). Performance and its measurement. *British Journal of Clinical Pharmacology*, 18:5S–9S.

Brown, C.R. and Benowitz, N.L. (1989). Caffeine and cigarette smoking: Behavioral, cardiovascular, and metabolic interactions. *Pharmacology, Biochemistry and Behavior*, 34:565–570.

Brown, C.R., Jacob, P. III, Wilson, W. and Benowitz, N.L. (1988). Changes in rate and pattern of caffeine metabolism after cigarette abstinence. *Clinical Pharmacology and Therapeutics*, **43**:488–491.

Brown, S.A., Goldman, M.S., Inn, A. and Anderson, L.R. (1980). Expectations of reinforcement from alcohol: Their domain and relation to drinking patterns. *Journal of Consulting and Clinical Psychology*, **48**:419–426.

Bruce, M., Scott, N., Lader, M. and Marks, V. (1986). The psychopharmacological and electrophysiological effects of single doses of caffeine in healthy human subjects. *British Journal of Clinical Pharmacology*, **22**:81–87. *[10]*.

Bullock, W.A. and Gilliland, K. (1993). Eysenck's Arousal theory of introversion-extraversion: A converging measures investigation. *Journal of Personality and Social Psychology*, **64(1)**:113–123. *[11]*.

Burling, T.A. and Ziff, D.C. (1988). Tobacco smoking: a comparison between alcohol and drug abuse inpatients. *Addictive Behaviors*, **13**:185–190.

Burns, M. and Moskowitz, H. (1989–90). Two experiments on alcohol-caffeine interaction. International symposium: fatigue, sleep deprivation, circadian rhythms and their interaction with alcohol and other drugs (1989, Santa Monica, California). *Alcohol, Drugs and Driving*, **5(4)** and **6(1)**:303–315.

Bushman, B.J. and Cooper, H.M. (1990). Effects of alcohol on human aggression: An integrative research review. *Psychological Bulletin*, **107**:341–354.

Busto, U., Bendayan, R. and Sellers, E.M. (1989). Clinical pharmacokinetics of non-opiate abused drugs. *Clinical Pharmacokinetics*, **16**:1–26.

Butters, N. and Cermack, L.S. (1980). Alcoholic Korsakoff's syndrome: An information processing approach to amnesia. London: Academic Press.

Butters, N. (1985). Alcoholic Korsakoff's syndrome: Some unresolved issues concerning etiology, neuropathology, and cognitive deficits. *Journal of Clinical and Experimental Neuropsychology*, **7**:181–210.

Cacioppo, J.T. and L.G. Tassinary (1990). Principles of Psychophysiology. Cambridge: Cambridge University Press.

Cala, L.A., Jones, B., Burns, P., Davis, R.E., Stenhouse, N. and Mastaglia, F.L. (1983). Results of computerized tomography, psychometric testing and dietary studies in social drinkers with emphasis on reversibility after abstinence. *Medical Journal of Australia*, **2**:264–269.

Cala, L.A., Jones, B., Mastaglia, F.L. and Wiley, B. (1978). Brain atrophy and intellectual impairment in heavy drinkers: A clinical, psychometric and computerized tomography study. *Australian and New Zealand Medical Journal*, **8**:147–153.

Callahan, M.M., Robertson. R.S., Arnaud, M.J., Branfman, A.R., McComish, M.F. and Yesair, D.W. (1982). Human metabolism of [1-methyl-^{14}C]- and [2-^{14}C]caffeine after oral administration. *Drug Metabolism and Disposition*, **10**:417–423.

Callaway, E. (1983). The pharmacology of human information processing. *Psychophysiology*, **20**:359–370.

Callaway, E., Halliday, R., Naylor, H. and Schechter, G. (1985). Effects of oral scopolamine on human stimulus evaluation. *Psychopharmacology*, **85**:133–138.

Campbell, K.B. and Lowick, B.M. (1987). Ethanol and event-related potentials: The influence of distractor stimuli. *Alcohol*, **4**:257–263.

Campbell, K.B., Marangoni, C., Walsh, C. and Báribeau-Braun, J. (1981). The effects of alcohol on the human auditory evoked potential and signal detection. *Psychophysiology*, **18**:176.

Campbell, K.B., Marois, R. and Arcand, L. (1984). Ethanol and the event-related evoked potentials: Effects of rate of stimulus presentation and task difficulty. *Annals of the New York Academy of Sciences*, **425**:551–555.

Campbell, M.E., Grant, D.M., Tang, B.K. and Kalow, W. (1987a). Biotransformation of caffeine, paraxanthine, theophylline and theobromine by polycyclic aromatic hydrocarbon-inducible cytochrome(s) P-450 in human liver microsomes. *Drug Metabolism and Disposition*, **15**:237–249.

Campbell, M.E., Spielberg, S.P. and Kalow, W. (1987b). A urinary metabolic ratio that reflects systemic caffeine clearance. *Clinical Pharmacology and Therapeutics*, **42**:157–165.

Caraco, Y., Zylber-Katz, E., Berry, E.M. and Levy, M. (1995). Caffeine pharmacokinetics in obesity and following significant weight reduction. *International Journal of Obesity*, **19**:234–239.

Caraco, Y., Zylber-Katz, E., Granit, L. and Levy, M. (1990). Does restriction of caffeine intake affect mixed function oxidase activity and caffeine metabolism. *Biopharmaceutics and Drug Disposition*, **11**:639–643.

Carey, K.B. and Maisto, S.A. (1987). Effect of a change in drinking pattern on the cognitive function of female social drinkers. *Journal of Studies on Alcohol*, **48**:236–242.

Carmelli, D., Swan, G.E., Robinette, D. and Fabsitz, M.A. (1992). Genetic influence on smoking – A study of male twins. *New England Journal of Medicine*, **327**:829–833.

Carmelli, D., Swan, G.E., Robinette, D. and Fabsitz, R.R. (1989). Heritability of substance use in the NAC-NRC Twin registry. *Acta Geneticae Medicae et Gemellogiae; Twin research*, **39**:91–98.

Carmody, T.P., Brischetto, C.S., Matarazzo, J.D., O'Donnell, R.P. and Connor, W.E. (1985). Co-occurent use of cigarettes, alcohol and coffee in healthy, community-living men and women. *Health Psychology*, **4**:323–335.

Carpenter, J.A. (1959). The effect of alcohol and caffeine on simple visual reaction time. *Journal of Comparative and Physiological Psychology*, **52**:491–496.

Carrier, O., Pons, G., Rey, E., Richard, M.-O., Moran, C., Badoual, J. and Olive, G. (1988). Maturation of caffeine metabolic pathways in infancy. *Clinical Pharmacology and Therapeutics*, **44**:145–151.

Carter, G.L. (1974). Effects of cigarette smoking on learning. *Perceptual and Motor Skills*, **39**:1344–1346.

Carton, S., Jouvent, R. and Widlöcher, D. (1994a). Nicotine dependence and motives for smoking in depression. *Journal of Substance Abuse*, **6**:67–76.

Carton, S., Jouvent, R. and Widlöcher, D. (1994b). Sensation seeking, nicotine dependence, and smoking motivation in female and male smokers. *Addictive Behaviors*, **19**:219–227.

Castellano, C. and Pavone, F. (1988). Effects of ethanol on passive avoidance behavior in the mouse: Involvement of GABA-ergic mechanisms. *Pharmacology, Biochemistry and Behavior*, **29**:321–324.

Castellano, C. and Populin, R. (1990). Effect of ethanol on memory consolidation in mice: Antagonism by the imidazobenzodiazepine Ro 15-4513 and decrement by familiarization with the environment. *Behavioural Brain Research*, **40**:67–72.

Central Bureau of Statistics. (1995). Statistisch Jaarboek 1995. CBS-publikaties's Gravenhage: SDU/Uitgeverij, p. 22.

Chait, L.D. and Griffiths, R.R. (1983). Effects of caffeine on cigarette smoking and subjective response. *Clinical Pharmacology and Therapeutics*, **34**:612–622.

Charness, M.E., Simon, R.P. and Greenberg, D.A. (1989). Ethanol and the nervous system. *The New England Journal of Medicine*, **321**:442–455.

Cheng, W.S.C., Murphy, T.L., Smith, M.T., Cooksley, W.G.E., Halliday, J.W. and Powell, L.W. (1990). Dose-dependent pharmacokinetics of caffeine in humans: relevance as a test of quantitative liver function. *Clinical Pharmacology and Therapeutics*, **47**:516–524.

Cherry, N. and Kiernon, K. (1976). Personality scores and smoking behaviour – a longitudinal study. *British Journal of Preventative Social Medicine*, **30**:123–131.

Cherry, N. and Kiernon, K. (1978). A longitudinal study of smoking and personality. In: R.E. Thornton (Ed.), Smoking Behaviour. New York: Churchill Livingstone, pp. 12–18.

Chi, I., Lubben, J.E. and Kitano, H.H. (1989). Differences in drinking behavior among three Asian-American groups. *Journal of Studies on Alcohol*, **50**:15–23.

Church, M.W. and Williams, H.L. (1982). Dose- and time-dependent effects of ethanol on brain stem auditory evoked responses in young adult males. *Electroencephalography and Clinical Neurophysiology*, **54**:161–174.

Church, R. (1989). Smoking and the human EEG. In: T. Ney and A. Gale (Eds.), Smoking and human behavior. Chichester: Wiley, pp. 115–140.

Cinciripini, P.M., Lapitsky, L., Wallfisch, M.A., Haque, W. and Van Vunakis, H. (1995, March). Smoking Cessation and Depressed Mood: Does Transdermal Nicotine Replacement Enhance Success? Paper presented at the first annual meeting of the Society for Research on Nicotine and Tobacco. San Diego, CA.

Clark, D.C., Eckenfels, E.J., Daugherty, S.R. and Fawcett, J. (1987). Alcohol-use patterns through medical school. A longitudinal study of one class. *Journal of the American Medical Association*, **257**:2921–2926.

Clark, M.S.G. and Rand, M.J. (1968). Effect of tobacco smoke on the knee-jerk reflex in man. *European Journal of Pharmacology*, **3**:294–302.

Clark, M.S.G., Rand, M.J. and Vanov, S. (1965). Comparison of pharmacological activity of nicotine and related alkaloids occurring in cigarette smoke. *Archives Internationales de Pharmacodynamie et de Thérapie*, **156**:363–379.

Clarke, P.B.S. (1990). The central pharmacology of nicotine: electrophysiological approaches In: S. Wonnacott, M.A.H. Russell and I.P. Stolerman (Eds.), Nicotine Psychopharmacology: molecular, cellular and behavioral aspects. Oxford: Oxford University Press, pp. 158–193.

Clarke, P.B.S., Quik, M., Thurau, K. and Adlkofer, F. (1994). International symposium on nicotine: the effects of nicotine on biological Systems II. The Abstracts. Basel: Birkhäuser.

Clarke, V.A. (1987). Smoking and academic achievement: A pilot study. *Psychological Reports*, **60**:259–262.

Clodoré, M., Benoit, O., Foret, J., Touitou, Y., Touron, G., Bouard, G. and Auzeby, A. (1987). Early rising or delayed bedtime: which is better for a short night's sleep. *European Journal of Applied Physiology and Occupational Physiology*, **56**:403–411.

Cloninger, C.R. (1987). Recent advances in family studies of alcoholism. In: H.W. Goedde and D.P. Agarwal (Eds.), Genetics and alcoholism. New York: Liss, A.R, pp. 47–60.

Cloninger, C.R. (1990). Genetic epidemiology of alcoholism: observations critical in the design and analysis of linkage studies. In: C.R. Cloninger and H. Begleiter (Eds.), Genetics and Biology of Alcoholism. Banbury Report 33. New York: Cold Spring Harbor Laboratory Press, pp. 105–133.

Clubley, M., Bye, C.E., Henson, T.A., Peck, A.W. and Riddington, C.J. (1979). Effects of caffeine and cyclizine alone and in combination on human performance, subjective effects, and EEG activity. *British Journal of Clinical Pharmacology*, **7**:157–163.

Coambs, R.B., Li, S. and Kozlowski, L.T. (1992). Age interacts with heaviness of smoking in predicting success in cessation of smoking. *American Journal of Epidemiology*, **135**:240–246.

Coger, R.W., Dymond, A.M., Serafetinides, E.A., Lowenstam, I. and Pearson, E. (1978). EEG signs of brain impairment in alcoholism. *Biological Psychiatry*, **13**:729–738.

Cohen, C., Pickworth, W.B., Bunker, E.B. and Henningfield, J.E. (1994). Caffeine antagonizes EEG effects of tobacco withdrawal. *Pharmacology, Biochemistry and Behavior*, **47**:919–926.

Colbern, D.L., Sharek, P. and Zimmermann, E.G. (1986). The effect of home or novel environment on the facilitation of passive avoidance by post-training ethanol. *Behavioral and Neural Biology*, **46**:1–12.

Coles, M.G.H., Gratton, G. and Fabiani, M. (1990). Event-related brain potentials. In: J.T. Cacioppo and L.G. Tassinary (Eds.), Principles of psychophysiology: Physical, social, and inferential elements. Cambridge, UK: Cambridge University Press, pp. 413–455.

Coles, M.G.H. (1989). Modern mind-brain reading: Psychophysiology, physiology, and cognition. *Psychophysiology*, **26**:251–269.

Collier, D.J. and Beales, I.L.P. (1989). Drinking among medical students, a questionnaire survey. *British Medical Journal*, **299**:19–22.

Collins, A.C. (1990). Interactions of ethanol and nicotine at the receptor level. *Recent Developments in Alcoholism*, **8**:221–231.

Collomp, K., Anselme, F., Audran, M., Gay, J.P., Chanal, J.L. and Prefaut, C. (1991). Effects of moderate exercise on the pharmacokinetics of caffeine. *European Journal of Clinical Pharmacology*, **40**:279–282.

Colrain, I.M., Mangan, G.L., Pellett, O.L. and Bates, T.C. (1992). Effects of post-learning smoking on memory consolidation. *Psychopharmacology*, **108**:448–451.

Colrain, I.M., Taylor, J., McLean, S., Buttery, R., Wise, G. and Montgomery, I. (1993). Dose dependent effects of alcohol on visual evoked potentials. *Psychopharmacology*, **112**:383–388.

Comings, D.E. (1993). Serotonin and the biochemical genetics of alcoholism: lessons from studies of attention deficit hyperactivity disorders (ADHD) and Tourette syndrome. *Alcohol and Alcoholism*, **(Suppl. 2)**:237–241.

Cone, E.J. and Henningfield, J.E. (1989). Premier 'smokeless cigarette' can be used to deliver crack. *Journal of the American Medical Association*, **261**:41.

Conner, D.P., Millora, E., Zamani, K., Nix, D., Almirez, R.G., Rhyne-Kirsch, P. and Peck, C.C. (1991). Transcutaneous chemical collection of caffeine in normal subjects: relationship to area under the plasma concentration-time curve and sweat production. *Journal of Investigative Dermatology*, **96**:186–190.

Conrin, J. (1980). The EEG effects of tobacco smoking – a review. *Clinical Electroencephalography*, **11**:180–187.

Conway, T.L., Vickers, R.R., Ward, H.W. and Rahe, R.H. (1981). Occupational stress and variation in cigarette, coffee and alcohol consumption. *Journal of Health and Social Behaviour*, **22**:155–165.

Conze, C., Scherer, G., Tricker, A.R. and Adlkofer, F. (1994). The influence of orally resorbed nicotine on smoking behavior. In: P.B.S. Clarke, M. Quik, K. Thurau and F. Adlkofer (Eds.), International Symposium on Nicotine: The effects of nicotine on biological systems II. Basel: Birkhäuser.

Cooper, P.J. and Bowskill, R. (1986). Dysphoric mood and overeating. *British Journal of Clinical Psychology*, **25**:155–156.

Cooper, R., Osselton, J.W. and Shaw, J.C. (1980). EEG technology. London: Butterworths.

Corrigall, W.A., Franklin, K.B.J., Coen, K.M. and Clarke, P.B.S. (1992). The mesolimbic dopaminergic system is implicated in the reinforcing effects of nicotine. *Psychopharmacology*, **107**:285–289.

Cotten, D.J., Thomas, J.R. and Stewart, D. (1971). Immediate effects of cigarette smoking on simple reaction time of college male smokers. *Perceptual and Motor Skills*, **33**:336.

Cotton, N.S. (1979). The familial incidence of alcoholism. A review. *Quarterly Journal of Studies on Alcohol*, **40**:89–116.

Couzigou, P., Begleiter, H., Kiianmaa, K. and Agarwal, D.P. (1993). Genetics and Alcohol: In: P.M. Verschuren (Ed.), Health issues related to alcohol consumption. Brussels: ILSI Europe, pp. 281–329.

Covey, L.S. and Tam, D. (1990). Depressive mood, the single-parent home, and adolescent cigarette smoking. *American Journal of Public Health*, **80**:1330–1333.

Covey, L.S., Glassman, A.H. and Stetner, F. (1990). Depression and depressive symptoms in smoking cessation. *Comprehensive Psychiatry*, **31**:350–354.

Cox, B.D., Blaxter, M., Buckle, A.L.J., Fenner, N.P., Golding, J.F., Gore, M., Huppert, F.A., Nickson, J., Roth, M., Stark, J., Wadsworth, M.E.J. and Whichelow, M. (1987). The Health and Lifestyle Survey. London: The Health Promotion Research Trust.

Craig, A. and Cooper, R.E. (1992). Symptoms of acute and chronic fatigue. In: A.P. Smith and D.M. Jones (Eds.), Handbook of Human Performance. Vol. 3. London: Academic Press, pp. 289–340.

Craig, T.J. and VanNatta, P.A. (1977). The association of smoking and drinking habits in a community sample. *Journal of Studies on Alcohol*, **38**:1434–1439.

Craik, F.I.M. and Tulving, E. (1975). Depth of processing and the retention of words in episodic memory. *Journal of Experimental Psychology: General*, **104**:268–294.

Crawford, J.R., Stewart, L.E. and Moore, J.W. (1989). Demonstration of savings on the AVLT and development of a parallel form. *Journal of Clinical and Experimental Neuropsychology*, **11**:975–981.

Croxton, J.S., Putz-Anderson, V.R. and Setzer, J.V. (1985). Attributional consequences of chemical exposure and performance feedback. *Journal of Applied Social Psychology*, **15(4)**:313–329. *[12]*.

Curran, H., Schiwy, W. and Lader, M. (1987). Differential amnesic properties of benzodiazepines: a dose-response comparison of two drugs with similar elimination half-lives. *Psychopharmacology*, **92**:358–3643.

Curran, S. (1990). Critical flicker fusion techniques in psychopharmacology. In: I. Hindmarch and P.D. Stonier (Eds.), Human Psychopharmacology: Methods and Measures. Vol. 3. Chichester: Wiley, pp. 21–38.

Cutler, G.H. and Barrios, F.X. (1988). Effects of deprivation on smokers' mood during the operation of a complex computer simulation. *Addictive Behaviors*, **13**:379–382.

Cutting, J.C. (1985). Korsakoff's syndrome. In: J.A.M. Frederiks (Ed.), Handbook of clinical neurology: Vol. 45 (revised series 1). Clinical Neuropsychology. Amsterdam: Elsevier, pp. 193–204.

Cutting, J.C. (1988). Alcohol cognitive impairment and aging: Still an uncertain relationship. *British Journal of Addiction*, **83**:995–997.

Cynoweth, K.R., Ternai, B., Simeral, L.S. and Maciel, G.E. (1973). NMR studies of the conformation and electron distributions in nicotine and acetylcholine. *Molecular Pharmacology*, **9**:144–151.

Daly, J.W. (1993). Mechanism of action of caffeine. In: S. Garattini (Ed.), Caffeine, coffee, and health. New York: Raven Press, pp. 97–150.

Danion, J.M., Weingartner, H., File, S.E., Jaffard, R., Sunderland, T., Tulving, E. and Warburton, D.M. (1993). Pharmacology of human memory and cognition: Illustrations from the effects of benzodiazepines and cholinergic drugs. *Journal of Psychopharmacology*, **7**:371–377.

Daruna, J.H., Goist Jr., K.C., West, J.A. and Sutker, P.B. (1987). Scalp distribution of the P3 component of event-related potentials during acute ethanol intoxication: a pilot study. In: R. Johnson Jr., J.W. Rohrbaugh and R. Parasuraman (Eds.), Current Trends in Event-Related Potential Research (EEG Suppl. 40). Amsterdam: Elsevier, pp. 521–526.

Davenport, M. and Harris, D. (1992). The effect of low blood alcohol levels of pilot performance in a series of simulated approach and landing trials. *International Journal of Aviation Psychology*, **2**:271–280.

Davidson, R.A. and Smith, B.D. (1989). Arousal and habituation: Differential effects of caffeine, sensation seeking and task difficulty. *Personality and Individual Differences*, **10(1)**:111–119. *[13]*.

Davidson, R.A. and Smith, B.D. (1991). Caffeine and novelty: Effects on electrodermal activity and performance. *Physiology and Behavior*, **49(6)**:1169–1175. *[14]*.

Davidson, R.J. (1984). Hemispheric asymmetry and emotion. In: K.R. Scherer and P. Ekman (Eds.), Approaches to Emotion. Hillsdale, New Jersey: Lawrence Erlbaum, pp. 39–57.

Davidson, R.J. (1993). Cerebral asymmetry and emotion: Conceptual and methodological conundrums. *Cognition and Emotion*, 7:115–138.

Davies, D.R. and Parasuraman, R. (1982). The psychology of vigilance. London: Academic Press.

Davis, P., Gibbs, F., Davis, H., Jetter, W. and Trowbridge, I. (1941). The effects of alcohol upon the electroencephalogram (brain waves). *Quarterly Journal of Studies in Alcohol*, 1:626–637.

De Jong, R., Wierda, M., Mulder, G. and Mulder, L.J.M. (1988). Use of partial stimulus information in response processing. *Journal of Experimental Psychology: Human Perception and Performance*, 14:682–692.

De Wijk, R. (1989). Temporal factors in human olfactory perception. Ph.D. Thesis, Utrecht: University of Utrecht, The Netherlands.

De Wit, H., Metz, J., Wagner, N. and Cooper, M. (1990). Behavioral and subjective effects of ethanol: Relationship to cerebral metabolism using PET. *Alcoholism: Clinical and Experimental Research*, 14:482–489.

Dekker, P., Van het Reven, C.G.M. and Den Hartog, A.P. (1993). Koffie verbruik en koffiegewoonten in Nederland. *Voeding*, 54:6–9.

Dembroski, T.M., MacDougall, J.M., Cardozo, S.R., Ireland, S.K. and Krug-Fite, J. (1985). Selective cardiovascular effects of stress and cigarette smoking in young women. *Health Psychology*, 4:153–167.

Denaro, C. and Benowitz, N. (1989). Diabetic patients and hepatic drug metabolism. *Clinical Pharmacology and Therapeutics*, 45:695–696.

Denaro, C.P., Brown, C.R., Wilson, M., Jacob, P. and Benowitz, N.L. (1990). Dose-dependency of caffeine metabolism with repeated dosing. *Clinical Pharmacology and Therapeutics*, 48:277–285.

DeRenzi, E., Faglioni, P., Nichelli, P. and Pignattari, L. (1984). Intellectual and memory impairment in moderate and heavy drinkers. *Cortex*, 20:525–533.

Derryberry, D. and Tucker, D.M. (1992). Neural mechanisms of emotion. *Journal of Consulting and Clinical Psychology*, 60:329–338.

Desmedt, J.E. (1988). Somatosensory evoked potentials. In: T.W. Picton (Ed.), Handbook of electroencephalography and clinical neurophysiology. Revised series. Vol. 3. Human event-related potentials. Amsterdam: Elsevier, pp. 245–360.

Desmond, P.V., Patwardhan, R.V., Johnson, R.F. and Schenker, S. (1980). Impaired elimination of caffeine in cirrhosis. *Digestive Diseases and Sciences*, 25:193–197.

Dews, P.B. (1984a). Caffeine: Perspectives from recent research. Berlin: Springer Verlag.

Dews, P.B. (1984b). Behavioral effects of caffeine. In: P.B. Dews (Ed.), Caffeine perspectives from recent research. Berlin: Springer Verlag, pp. 86–103.

Diamond, A.L. and Cole, R.E. (1970). Visual threshold as a function of test area and caffeine administration. *Psychonomic Science*, 20:109–111.

Diethelm, O. and Barr, R.M. (1962). Psychotherapeutic interviews and alcohol intoxication. *Quarterly Journal of Studies on Alcohol*, 23:243–251.

DiFranza, J.R., Winters, T.H., Goldberg, R.J., Cirrillo, L. and Biliouris, T. (1986). The relationship of smoking to motor vehicle accidents and traffic violations. *New York State Journal of Medicine*, 86:464–467.

Dimpfel, W., Schober, F. and Spüler, M. (1993). The influence of caffeine on human EEG under resting conditions and during mental loads. *Clinical Investigator*, 71:197–207. *[15].*

Docter, R.F. and Beernal, M.E. (1964). Immediate and prolonged psychophysiological effects of sustained alcohol intake in alcoholics. *Quarterly Journal of Studies on Alcohol*, 25:438–450.

Domino, E.F. (1973). Neuropsychopharmacology of nicotine and tobacco smoking. In: W.L. Dunn Jr. (Ed.), Smoking behavior: Motives and incentives. Washington, D.C.: Winston, pp. 5–31.

Domino, E.F. (1995, March). Variable Brain Localization Effects of Tobacco Smoking Depending Upon the Increment in Plasma Nicotine Levels. Paper presented at the first annual meeting of the Society for Research on Nicotine and Tobacco. San Diego, CA.

Donchin, E. and Coles, M.G.H. (1988). Is the P300 component a manifestation of context updating? *Behavioral and Brain Sciences*, **11**:357–374.

Donchin, E. (1979). Event-related brain potentials: A tool in the study of human information processing. In: H. Begleiter (Ed.), Evoked brain potentials and behavior. New York: Plenum, pp. 13–88.

Donchin, E. (1979). Event-related brain potentials: a tool in the study of human information processing. In: H. Begleiter (Ed.), Evoked potentials and behavior. New York: Plenum, pp. 13–75.

Donchin, E. (1981). Surprise! … surprise? *Psychophysiology*, **18**:493–513.

Donchin, E., Kramer, A.F. and Wickens, C. (1986). Application of brain event-related potentials to problems in engineering psychology. In: M.G.H. Coles, E. Donchin and S.W. Porges (Eds.), Psychophysiology: Systems, processes, and applications. New York: Guilford Press, pp. 702–718.

Donchin, E., Ritter, W. and McCallum, W.C. (1978). Cognitive psychophysiology: The endogenous components of the ERP. In: E. Callaway, P. Tueting and S.H. Koslow (Eds.), Event-related brain potentials in man. New York: Academic Press, pp. 349–441.

Dreher, K.F. and Fraser, J.G. (1967). Smoking habits of alcoholic outpatients. *International Journal of Addictions*, **2**:259–269.

Drejer, K., Theilgaard, A., Teasdale, T.W., Goodwin, D.W. and Schulsinger, F. (1985). A prospective study of young men at high risk for alcoholism. Neuropsychological assessment. *Alcoholism Clinical and Experimental Research*, **9**:298–302.

Dufour, M.C., Archer, L. and Gordis, E. (1992). Alcohol and the elderly. *Clinics in Geriatric Medicine*, **8**:127–141.

Dunne, M.P., MacDonald, D. and Hartley, L.R. (1986). The effects of nicotine upon memory and problem solving performance. *Physiology and Behaviour*, **37**:849–854.

Duthel, J.M., Vallon, J.J., Martin, G., Ferret, J.M., Mathieu, R. and Videman, R. (1991). Caffeine and sport: role of physical exercise upon elimination. *Medicine and Science in Sports and Exercise*, **23**:980–985.

Dye, L., Sherwood, N. and Hindmarch, I. (1991). The influence of nicotine on CNS arousal during the menstrual cycle. In: F. Adlkofer and K. Thurau (Eds.), Effects of nicotine on biological systems. Basel: Birkhäuser Verlag, pp. 527–530.

Easterbrook, J.A. (1959). The effect of emotion on cue utilization and the organization of behavior. *Psychological Review*, **66**:183–201.

Edwards, J.A. and Warburton, D.M. (1983). Smoking, nicotine and electrocortical activity. *Pharmacology and Therapeutics*, **19**:147–164.

Edwards, J.A., Wesnes, K., Warburton, D.M. and Gale, A. (1985). Evidence of more rapid stimulus evaluation following cigarette smoking. *Addictive Behavior*, **10**:113–126.

Ehlers, C.L., Wall, T.L. and Schuckit, M.A. (1989). EEG spectral characteristics following ethanol administration in young men. *Electroencephalography and Clinical Neurophysiology*, **73**:179–187.

Ekman, G., Frankenhaeuser, M., Goldber, L., Hagdahl, R. and Myrsten, A.-L. (1964). Subjective and objective effects of alcohol as functions of dosage and time. *Psychopharmacologia*, **6**:399–409.

Elbert, T. and Birbaumer, N. (1987). Hemispheric differences in relation to smoking. In: A. Glass (Ed.), Individual differences in hemispheric specialization. Tubingen: Butterworth, pp. 195–206.

Elgerot, A. (1976). Note on selective effects of short-term tobacco abstinence on complex versus simple mental tasks. *Perceptual and Motor Skills*, **42**:413–414.

Elgerot, A. (1978). Psychological and physiological changes during tobacco-abstinence in habitual smokers. *Journal of Clinical Psychology*, **34**:759–764.

Elmasian, R., Neville, H., Woods, D., Schuckit, M. and Bloom, F. (1982). Event-related brain potentials are different in individuals at high and low risk for developing alcoholism. *Proceedings of the National Academy of Science, USA*, **29**:7900–7903.

Emmerson, R.Y., Dustman, R.E., Heil, J. and Shearer, D.E. (1988). Neuropsychological performance of young nondrinkers, social drinkers, and long- and short-term sober alcoholics. *Alcoholism: Clinical and Experimental Research*, **12(5)**:625–629.

Emmerson, R.Y., Dustman, R.E., Shearer, D.E. and Chamberlin, H.M. (1987). EEG, visually evoked and event related potentials in young abstinent alcoholics. *Alcohol*, **4**:241–248.

Enomoto, N., Takase, S., Takada, N. and Takada, A. (1991). Alcoholic liver disease in heterozygotes of mutant and normal aldehyde dehydrogenase-2 gene. *Hepatology*, **13**:1071–1075.

Eriksen, C. and Schultz, D. (1979). Information processing in visual search: a continuous flow conception and experimental results. *Perception and Psychophysics*, **25**:249–263.

Erikson, G.C., Hager, L.B., Houseworth, C., Dungan, J., Petros, T. and Beckwith, B.E. (1985). The effects of caffeine on memory for word lists. *Physiology and Behavior*, **35**:47–51. *[17]*.

Erwin, C.W. and Linnoila, M. (1981). Effect of ethyl alcohol on visual evoked potentials. *Alcoholism: Clinical and Experimental Research*, **5**:49–55.

Erwin, C.W., Wiener, E.L, Linnoila, M.I. and Truscott, T.R. (1978). Alcohol-induced drowsiness and vigilance performance. *Journal of Studies on Alcohol*, **39**:505–516.

Eskenazi, B., Cain, W.S. and Friend, K. (1986). Exploration of olfactory aptitude. *Bulletin of the Psychonomic Society*, **24(3)**:20–36.

Etevenon, P., Peron-Magnan, P., Guillou, S., Toussaint, M., Gueguen, B., Deniker, P., Loo, H. and Zarifian, E. (1989). A pharmacological model of "local cerebral activation": EEG cartography of caffeine effects in normals. In: N.C. Andreasen (Ed.), Brain imaging: Applications in psychiatry. Washington, District of Columbia: American Psychiatric Press, pp. 171–180.

Ettenberg, A., Laferrière, A., Milner, P.M. and White, N. (1981). Response involvement in brain stimulation reward. *Physiology and Behavior*, **27**:641–647.

Evans, W.E., Relling, M.V., Petros, W.P., Meyer, W.H., Mirro, J. and Crom, W.R. (1989). Dextromethorphan and caffeine as probes for simultaneous determination of debrisoquin-oxidation and N-acetylation phenotypes in children. *Clinical Pharmacology and Therapeutics*, **45**:568–573.

Eysenck, H.J. and Kelley, M. J. (1987). The interaction of neurohormones with Pavlovian A and Pavlovian B conditioning in the causation of neurosis, extinction, and incubation of anxiety. In: G. Davey (Ed.), Cognitive processes and Pavlovian conditioning in humans. Chichester: Wiley.

Eysenck, H.J. and O'Connor, K. (1979). Smoking arousal and personality. In: A. Remond and C. Izard (Eds.), Electrophysiological effects of nicotine. Amsterdam: Elsevier/North Holland, pp. 147–157.

Eysenck, H.J. (1973). Personality and the maintenance of the smoking habit. In: W.L. Dunn (Ed.), Smoking behavior: Motives and incentives. Washington, D.C.: V.H. Winston, pp. 113–146.

Eysenck, H.J. (1980). The causes and effects of smoking. London: Maurice Temple Smith.

Eysenck, M.W. (1982). Attention and arousal: Cognition and performance. Berlin: Springer Verlag.

Eysenck, M.W. (1983). Memory and arousal. In: A. Gale and J. Edwards (Eds.), Physiological correlates of human behavior. London: Academic Press, pp. 97–113.

Fabiani, M., Karis, D. and Donchin, E. (1986). P300 and recall in an incidental memory paradigm. *Psychophysiology*, 23:298–331.

Fabiani, M., Karis, D. and Donchin, E. (1990). Effects of mnemonic strategy manipulation in a Von Restorff paradigm. *Electroencephalography and Clinical Neurophysiology*, 75:22–35.

Fabricant, N.D. and Rose, I.W. (1951). Effects of smoking cigarettes on the flicker fusion threshold of normal persons. *Eye Ear Mouth*, 30:541–543.

Fagan, D., Swift, C.G. and Tiplady, B. (1988). Effects of caffeine on vigilance and other performance tests in normal subjects. *Journal of Psychopharmacology*, 2(1):19–25. [18].

Fagan, D., Tiplady, B. and Scott, D.B. (1987). Effects of ethanol on psychomotor performance. *British Journal of Anaesthesia*, 59:961–965.

Fagerström, K.O. (1978). Measuring degree of physical dependence to tobacco smoking with reference to individualization of treatment. *Addictive Behaviors*, 3:235–241.

Farris, J.J. and Jones, B.M. (1978). Ethanol metabolism in male American Indians and whites. *Alcoholism Clinical and Experimental Research*, 2:77–82.

Ferrara, S.D., Zancaner, S. and Giorgetti, R. (1994). Low blood alcohol concentrations and driving impairment: A review of experimental studies and international legislation. *International Journal of Legal Medicine*, 106:169–177.

Feyerabend, C., Ings, R.M.J. and Russell, M.A.H. (1985). Nicotine pharmacokinetics and its application to intake from smoking. *British Journal of Clinical Pharmacology*, 19:239–247.

Feyerabend, C., Levitt, T. and Russell, M.A.H. (1975). A rapid gas-liquid chromatographic estimation of nicotine in biological fluids. *Journal of Pharmacy and Pharmacology*, 27:434–436.

File, S.E., Bond, A.J. and Lister, R.G. (1982). Interaction between effects of caffeine and lorazepam in performance tests and self-ratings. *Journal of Clinical Psychopharmacology*, 2:102–106.

File, S.E., Mabbutt, P.S. and Shaffer, J. (1994). Alcohol consumption and lifestyle in medical students. *Journal of Psychopharmacology*, 8:22–26.

File, S.E., Shaffer, J. and Sharma, R. (1990). UMDS student life style. *Guy's Hospital Gazette*, 154–155.

Fillmore, M.T. and Vogel-Sprott, M. (1994). Psychomotor performance under alcohol and under caffeine: Expectancy and pharmacological effects. *Experimental and Clinical Psychopharmacology*, 2(4):319–327. [19].

Fillmore, M.T. and Vogel-Sprott, M. (1994). Psychomotor performance under alcohol and under caffeine: Expectancy and pharmacological effects. *Experimental and Clinical Psychopharmacology*, 2:319–327.

Fillmore, M.T. and Vogel-Sprott, M. (1995). Behavioral effects of combining alcohol and caffeine: Contribution of drug-related expectancies. *Experimental and Clinical Psychopharmacology*, 3(1):33–38.

Fine, B.J., Kobrick, J.L., Lieberman, H.R., Marlowe, B., Riley, R.H. and Tharion, W.J. (1994). Effects of caffeine or diphenhydramine on visual vigilance. *Psychopharmacology*, 114:233–238. [20].

Finnigan, F. and Hammersley, R. (1992). The effects of alcohol on performance. In: A.P. Smith and D.M. Jones (Eds.), Handbook of Human Performance. Vol. 2. London: Academic Press, pp. 73–126.

Fiore, M.C., Smith, S.S., Jorenby, D.E. and Baker, T.B. (1994). The effectiveness of the nicotine patch for smoking cessation – A Meta-Analysis. *Journal of the American Medical Association*, 271:1940–1947.

Flach, M., Krause, D. and Hofmann, G. (1977). Gehör und Alkohol: Latenzzeitverhalten von akustisch evozierten Potentialen unter Alkoholeinwirkung. *Laryngologie und Rhinologie*, 56:863–867.

Fleishman, E.A. and Quaintance, M.K. (1984). Taxonomies of human performance. Orlando: Academic Press.

Foote, S.L., Aston-Jones, G. and Bloom, F.E. (1980). Impulse activity of locus coeruleus neurons in awake rats and monkeys is a function of sensory stimulation and arousal. *Proceedings of the National Academy of Science, USA*, **77**:303–337.

Ford, J.M. and Pfefferbaum, A. (1985). Age-related changes in event-related potentials. *Advances of Psychology*, **1**:301–339.

Foreman, H., Barraclough, S., Moore, C., Mehta, A. and Madon, M. (1989). High doses of caffeine impair performance of a numerical version of the Stroop task in men. *Pharmacology, Biochemistry and Behavior*, **32**:399–403. *[21].*

Forney, P., Forney, M.A. and Ripley, W.K. (1988). Profile of an adolescent drinker. *The Journal of Family Practice*, **27**:65–70.

Forney, R.B. and Hughes, F.W. (1965). Effect of alcohol on performance under stress of audiofeedback. *Quarterly Journal of Studies on Alcohol*, **26**:206–212.

Foulds, J., McSorley, K., Sneddon, J., Feyerabend, C., Jarvis, M.J. and Russell, M.A.H. (1994). Effect of subcutaneous nicotine injections on EEG alpha frequency in non-smokers: a placebo-controlled pilot study. *Psychopharmacology*, **115**:163–166.

Fowler, B. and Adams, J. (1993). Dissociation of the effects of alcohol and amphetamine on inert gas narcosis using reaction time and P300 latency. *Aviation, Space, and Environmental Medicine*, **64**:493–499.

Fox, G.A., Jackson M., Waugh, M. and Tuck, R.R. (1987). Brain dysfunction in social drinkers. *Australian Alcohol and Drug Review*, **6**:15–19.

Frankel, E.N., Kanner, J. and German, J.B. (1993). Inhibition of oxidation of human low density lipoprotein by phenolic substances in red wine. *Lancet*, **341**:454–457.

Frankenhaeuser, M., Myrsten, A.-L. and Post, B. (1970). Psychophysiological reactions to cigarette smoking. *Scandinavian Journal of Psychology*, **11**:237–245.

Frankenhaeuser, M., Myrsten, A.-L., Post, B. and Johansson, G. (1971). Behavioural and physiological effect of cigarette smoking in a monotonous situation. *Psychopharmacologia*, **22**:1–7.

Franks, H.M., Hagedorn, H., Hensley, V.R., Hensley, W.J. and Starmer, G.A. (1975). The effect of caffeine on human performance, alone and in combination with ethanol. *Psychopharmacologia*, **45**:177–181.

Franks, P., Harp, J. and Bell, B. (1989). Randomized, controlled trial of clonidine for smoking cessation in a primary care setting. *Journal of the American Medical Association*, **262**:3011–3013.

Fraser, H.S., Dotson, O.Y., Howard, L., Grell, G.A.C. and Knight, F. (1983). Drug metabolizing capacity in Jamaican cigarette and marijuana smokers and non-smokers. *West Indian Medical Journal*, **32**:207–211.

Frearson, W., Barrett, P. and Eysenck, H.J. (1988). Evidence of more rapid stimulus evaluation following cigarette smoking. *Addictive Behaviors*, **10**:113–126.

Frearson, W., Barrett, P. and Eysenck, H.J. (1988). Intelligence, reaction time, and the effects of smoking. *Personality and Individual Differences*, **9**:497–517.

Freedman, M. and Cermack, L.S. (1986). Semantic encoding deficits in frontal lobe disease and amnesia. *Brain and Cognition*, **5**:108–114.

Freedman, R., Waldo, M., Waldo, C.I. III and Wilson, J.R. (1987). Genetic influences on the effects of alcohol on auditory evoked potentials. *Alcohol*, **4(4)**:249–253.

Freeman, W.J. (1987). Simulation of chaotic EEG-patterns with a dynamic model of the olfactory system. *Biological Cybernetics*, **56**:136–150.

Freeman, W.J. and Baird, B. (1987). Relation of olfactory EEG to behavior: spatial analysis. *Behavioral Neuroscience*, **101(3)**:393–408.

Freeman, W.J. and Viana Di Prisco, G. (1986). Relation of olfactory EEG to behavior: time series analysis. *Behavioral Neuroscience*, **100(5)**:753–763.

Freeman, W.J. (1991). The physiology of perception. *Scientific American*, **2**:34–41.

Frewer, L.J. and Lader, M. (1991). The effects of caffeine on two computerized tests of attention and vigilance. *Human Psychopharmacology*, **6**:119–128. *[22]*.

Frezza, M., Di Padova, C., Pozzato, G., Terpin, M., Baraona, E. and Lieber, C.S. (1990). High blood alcohol levels in women: the role of decreased gastric alcohol dehydrogenase activity and first-pass metabolism. *New England Journal of Medicine*, **322**:95–99.

Friedman, D. (1990). ERPs during continuous recognition memory for words. *Biological Psychology*, **30**:61–88.

Friedman, G.D., Siegelaub, A.B. and Seltzer, C.C. (1974). Cigarettes, alcohol, coffee and peptic ulcer. *The New England Journal of Medicine*, **290**:469–473.

Friedman, J., Horvath, T. and Meares, R. (1974). Tobacco smoking and a 'stimulus barrier'. *Nature*, **248**:455–456.

Frith, C.D. (1967a). The effects of nicotine on tapping: I. *Life Sciences*, **6**:313–319.

Frith, C.D. (1967b). The effects of nicotine on tapping: II. *Life Sciences*, **6**:321–326.

Frith, C.D. (1967c). The effects of nicotine on tapping: III. *Life Sciences*, **6**:1541–1548.

Frith, C.D. (1971). Smoking behaviour and its relation to the smoker's immediate experience. *British Journal of Social and Clinical Psychology*, **10**:73–78.

Frowein, H.W. (1980). Selective effects of barbiturate and amphetamine on information processing and response execution. *Acta Psychologica*, **47**:105–115.

Frowein, H.W. (1981). Selective drugs effects on information processing. Ph.D. Thesis, Tilburg, The Netherlands: Katholieke Hogeschool Tilburg.

Frowein, H.W., Gaillard, A.W.K. and Varey, C.A. (1981). Evoked potential compounds, visual processing stages and the effects of a barbiturate. *Biological Psychology*, **13**:239–249.

Frowein, H.W., Reitsma, D. and Aquarius, C. (1981). Effects of two counteracting stresses on the reaction process. In: J. Long and A. Baddeley (Eds.), Attention and Performance IX. Hillsdale, New Jersey: Lawrence Erlbaum, pp. 575–590.

Fruhstorfer, H. and Soveri, P. (1968). Alcohol and auditory evoked responses in man. *Acta Physiologica Scandinavica*, **74**:26A–27A.

Fudin, R. and Nicastro, R. (1988). Can caffeine antagonize alcohol-induced performance decrements in humans? *Perceptual and Motor Skills*, **67**:375–391.

Fuhr, U., Wolff, T., Harder, S., Schymanski, P. and Staib, A.H. (1990). Quinolone inhibition of cytochrome P-450-dependent caffeine metabolism in human liver microsomes. *Drug Metabolism and Disposition*, **18**:1005–1010.

Fukui, Y., Mori, M., Kohga, M., Tadai, T., Tanaka, K. and Katoh, N. (1981). Reassessment of CNS effects of acute ethanol administration with auditory evoked responses: A comparative study of brain stem auditory evoked response, middle latency response and slow vertex response. *Japanese Journal of Alcohol and Drug Dependence*, **16**:9–32.

Fuller, R.G. and Forrest, D.W. (1977). Cigarette smoking under relaxation. *Journal of Psychology*, **3**:165–180.

Gabrielli, W.F., Nednick, S.A., Volavka, J., Pollack, B.E., Schulsinger, F. and Itil, T.M. (1982). Electroencephalograms in children of alcoholic fathers. *Psychophysiology*, **19(4)**:404–407.

Gaillard, A.W.K. (1988). The Evaluation of Drug Effects in Laboratory Tasks. In: I. Hindmarch, B., Aufdenbrinke and H. Ott (Eds.), Psychopharmacology and Reaction time, Chichester: Wiley, pp. 15–24.

Gale, A. and Edwards, J.A. (1983). The EEG and human behaviour. In: A. Gale and J.A. Edwards (Eds.), Physiological correlates of human behaviour. Vol. 2. Attention and performance. London: Academic Press, pp. 99–127.

Garattini, S. (1993). Caffeine, coffee, and health. New York: Raven Press.

Gasser, T., Bacher, P. and Mocks, J. (1982). Transformations towards the normal distribution of broad band spectral parameters of the EEG. *Electroencephalography and Clinical Neurophysiology*, **53**:119–124.

Geen, R.G. (1986). Preferred stimulation levels in introverts and extraverts: effects on arousal and performance. *Journal of Personality and Social Psychology*, **46**:130–312.

Gemmell, H.A. and Jacobson, B.H. (1990). Effect of selected doses of caffeine on hand steadiness following fasting. *Chiropractic Sports Medicine*, **4(3)**:93–95. *[23]*.

Genkina, O.A. (1984). Effect of alcohol on evoked electrical activity in healthy man. *Human Physiology*, **10**:189–192.

George, J., Murphy, T., Roberts, R., Cooksley, W.G.E., Halliday, J.W. and Powell, L.W. (1986). Influence of alcohol and caffeine consumption on caffeine elimination. *Clinical and Experimental Pharmacology and Physiology*, **13**:731–736.

George, W.H., Raynor, J.O. and Nochajski, T.H. (1990). Resistance to alcohol impairment of visual-motor performance II: Effects for attentional set and self-reported concentration. *Pharmacology, Biochemistry and Behavior*, **36**:261–266.

Giacobini, E. (1992). Nicotine acetylcholine receptors in the human cortex: Aging and Alzheimer's disease. In: P.M. Lippiello, A.C. Collins, J.A. Gray and J.H. Robinson (Eds.), The biology of nicotine: Current research issues. New York: Raven Press, pp. 183–194.

Gibbons, H.L. (1988). Alcohol, aviation, and safety revisited: A historical review and a suggestion. *Aviation, Space, and Environmental Medicine*, **59**:657–660.

Gilbert, D.G. and Gilbert, B.O. (1995). Personality, psychopathology, and nicotine response as mediators of the genetics of smoking. *Behavior Genetics*, **25**:133–147.

Gilbert, D.G. and Hagen, R.L. (1980). The effects of nicotine and extraversion on self-report, skin conductance, electromyographic and heart responses to emotional stimuli. *Addictive Behaviors*, **5**:247–257.

Gilbert, D.G. and Hagen, R.L. (1985). Electrodermal responses to movie stressors: Nicotine × extraversion interactions. *Personality and Individual Differences*, **6**:573–578.

Gilbert, D.G. and Meliska, C.J. (1992). Individual differences in reliability of electroencephalogram, cortisol, beta-endorphin, heart rate, and subjective responses to smoking multiple cigarettes via a quantified smoke delivery system. In: P.M. Lippiello, A.C. Collins, J.A. Gray and J.H. Robinson (Eds.), The biology of nicotine: Current research issues. New York: Raven Press, pp. 141–155.

Gilbert, D.G. and Welser, R. (1989). Emotion, anxiety and smoking. In: Ney, T. and Gale, A. (Eds.), Smoking and human behavior. Chichester: Wiley, pp. 171–196.

Gilbert, D.G. (1979). Paradoxical tranquilizing and emotion-reducing effects of nicotine. *Psychological Bulletin*, **86**:643–661.

Gilbert, D.G. (1985, March). Nicotine's effects on lateralized EEG and emotion. Paper presented at the Sixth Annual Meeting of the Society of Behavioral Medicine, New Orleans, LA.

Gilbert, D.G. (1987). Effects of smoking and nicotine on EEG lateralization as a function of personality. *Personality and Individual Differences*, **8**:933–941.

Gilbert, D.G. (1988). EEG and personality differences between smokers and nonsmokers. *Personality and Individual Differences*, **9**:659–665.

Gilbert, D.G. (1994). Why people smoke: Stress reduction, coping enhancement and nicotine. *Recent Advances in Tobacco Science*, **20**:106–161.

Gilbert, D.G. (1995). Smoking: Individual differences, psychopathology, and emotion. Washington, DC.: Taylor and Francis.

Gilbert, D.G., Estes, S.L. and Welser, R. (1997). Does noise stress modulate effects of smoking/nicotine? Mood, vigilance, and EEG responses. *Psychopharmacology*, **129**:382–389.

Gilbert, D.G., Gehlbach, B., Estes, S.L., Rabinovich, N. and Detwiler, R.J. (1994b, October). Effects of smoking deprivation and a quantified dose of tobacco smoke on EEG power and lateralization as a function of depression and habitual nicotine intake. Paper presented at the Thirty-Fourth Annual Meeting of the Society for Psychophysiological Research, Atlanta, GA.

Gilbert, D.G., Jensen, R. and Meliska, C.J. (1989a). A system for administering quantified doses of tobacco smoke to human subjects: Plasma nicotine and filter pad validation. Pharmacology, *Biochemistry and Behavior*, 31:905–908.

Gilbert, D.G., Meliska, C.J., Welser, R. and Estes, S.L. (1994c). Depression, personality, and gender influence EEG, cortisol, beta-endorphin, heart rate, and subjective responses to smoking multiple cigarettes. *Personality and Individual Differences*, 16:247–264.

Gilbert, D.G., Meliska, C.J., Welser, R., Scott, S., Jensen, R.A. and Meliska, J. (1992a, March). Individual differences in the effects of smoking cessation on EEG, mood and vigilance. Paper presented at the Thirteenth Annual Meeting of the Society of Behavioral Medicine, New York.

Gilbert, D.G., Meliska, C.J., Williams, C. and Jensen, R.A. (1992b). Subjective Correlates of smoking-induced elevations of plasma beta-endorphin and cortisol. *Psychopharmacology*, 106:275–281.

Gilbert, D.G., Robinson, J.H., Chamberlin, C.L. and Spielberger, C.D. (1989b). Effects of smoking/nicotine on anxiety, heart rate, and lateralization of EEG during a stressful movie. *Psychophysiology*, 26:311–320.

Glassman, A.H. (1993). Cigarette smoking: Implications for psychiatric illness. *American Journal of Psychiatry*, 150(86):507–510.

Glassman, A.H., Helzer, J.E., Covey, L.S., Cottler, L.B., Stetner, F., Tipp, J.E. and Johnson, J. (1990). Smoking, smoking cessation and major depression. *Journal of the American Medical Association*, 264:1546–1549.

Glassman, A.H., Jackson, W.K., Walsh, B.T., Roose, S.P. and Rosenfeld, B. (1984). Cigarette craving, smoking withdrawal and clonidine. *Science*, 226:864–867.

Goddard, I. and Ikin, C. (1988). Drinking in England and Wales in 1987. London: HMSO.

Goedde, H.W. and Agarwal, D.P. (1992) Pharmacogenetics of aldehyde dehydrogenase. In: W. Kalow (Ed.), Pharmacogenetics of drug metabolism. New York: Pergamon Press, pp. 281–311.

Goedde, H.W., Agarwal, D.P. and Fritze, G. (1992). Distribution of ADH2 and ALDH2 genotypes in different populations. *Human Genetics*, 88:344–346. .

Golding, J.F and Mangan, G.L. (1982). Arousing and de-arousing effects of cigarette smoking under conditions of stress and mild sensory isolation. *Psychophysiology*, 19:449–456.

Golding, J.F. (1988). Effects of cigarette smoking on resting EEG, visual evoked potential and photic driving. *Pharmacology, Biochemistry and Behavior*, 29:23–32.

Goldman, M.S. (1986). Neuropsychological recovery in alcoholics: Endogenous and exogenous processes. *Alcoholism: Clinical and Experimental Research*, 10:136–144.

Gonzales, M.A. and Harris, M.B. (1980). Effects of cigarette smoking on recall and categorization of written material. *Perceptual and Motor Skills*, 50:407–410.

Goodwin, D.W., Powell, B., Bremer, D., Hoine, H. and Stern, J. (1969). Alcohol and recall: State-dependent effects in man. *Science*, 163:1358–1360.

Gopher, D. and Donchin, E. (1986). Workload: An examination of the concept. In: K.R. Boff, L. Kaufman and J.P. Thomas (Eds.), Handbook of perception and human performance. Vol. 2. Cognitive processes and performance. New York: Wiley, pp. 41.1–41.49.

Gorodischer, R., Zmora, E., Ben-Zvi, Z., Warszawski, D., Yaari, A., Sofer, S. and Arnaud, M.J. (1986). Urinary metabolites of caffeine in the premature infant. *European Journal of Clinical Pharmacology*, 31:497–499.

Graf, O. (1950). Increase of efficiency by means of pharmaceutics (stimulants). In: German Aviation Medicine, World War II. *Washington, D.C.; U.S. Government Printing Office*, 2:1080–1103.

Grandinetti, A., Morens, D.M., Reed, D. and Maceachern, D. (1994). Prospective study of cigarette smoking and the risk of developing idiopathic Parkinson's disease. *American Journal of Epidemiology*, 139:1129–1138.

Grant, D.M., Tang, B.K. and Kalow, W. (1984). A simple test for acetylation phenotype using caffeine. *British Journal of Clinical Pharmacology*, **17**:459–464.

Grant, D.M., Tang, B.K. and Kalow, W. (1983a). Polymorphic N-acetylation of a caffeine metabolite. *Clinical Pharmacology and Therapeutics*, **33**:355–359.

Grant, D.M., Tang, B.K. and Kalow, W. (1983b). Variability in caffeine metabolism. *Clinical Pharmacology and Therapeutics*, **33**:591–602.

Grant, I. (1987). Alcohol and the brain: Neuropsychological correlates. *Journal of Consulting and Clinical Psychology*, **55**:310–324.

Gray, J.A. (1982). The Neuropsychology of Anxiety: An enquiry into the function of the septo-hippocampal system. New York: Oxford University Press.

Gray, J.A. (1987). The Psychology of Fear and Stress, 2nd ed. New York: Cambridge University Press.

Griffiths, R.R., Bigelow, G.E. and Liebson, I. (1976). Facilitation of human tobacco self-administration by ethanol: a behavioral analysis. *Journal of Experimental Analysis of Behavior*, **25**:279–292.

Griffiths, R.R., Evans, S.M., Heishman, S.J., Preston, K.L., Sannerud, C.A., Wolf, B. and Woodson, P.P. (1990). Low-dose caffeine physical dependence in humans. *Journal of Pharmacology and Experimental Therapeutics*, **225**:1123–1132.

Grillon, C., Sinha, R. and O'Malley, S. (1994). Effects of ethanol on P3 to target and novel stimuli. *Psychophysiology*, **31**:S54.

Grobe, J.E., Perkins, K.A., Stiller, R.L. and Jacob, R.G. (1993). Cognitive and behavioural effects of nicotine in smokers and nonsmokers. Presented at the 55th Annual Meeting of CPDD, Toronto, June.

Gross, M.M., Begleiter, H., Tobin, M. and Kissin, B. (1966). Changes in auditory evoked response induced by alcohol. *Journal of Nervous and Mental Disease*, **143**:152–156.

Gross, T.M., Jarvik, M.E. and Rosenblatt, M.R. (1993). Nicotine abstinence produces content-specific Stroop interference. *Psychopharmacology*, **110**:333–336.

Grunberg, N.E. and Raygada, M. (1991). Effects of nicotine on insulin: Actions and implications. In: F. Adlkofer and K. Thurau (Eds.), Effects of nicotine on biological systems. Basel: Birkhäuser Verlag, pp. 131–142.

Guha, D. and Pradhan, S.N. (1976). Effects of nicotine on EEG and evoked potentials and their interactions with autonomic drugs. *Neuropharmacology*, **15**:225–232.

Gunter, T.C., Van der Zande, R.D., Wiethoff, M., Mulder, G. and Mulder, L.J.M. (1987). Visual selective attention during meaningful noise and after sleep deprivation. In: R. Johnson Jr., R. Parasuraman and J.W. Rohrbaugh (Eds.), Current trends in event-related potential research (EEG Suppl. 40). Amsterdam: Elsevier, pp. 99–107.

Gupta, U. and Gupta, B.S. (1990). Caffeine differentially affects kinesthetic aftereffect in high and low impulsives. *Psychopharmacology*, **102**:102–105. *[29]*.

Gupta, U. and Gupta, B.S. (1994). Effects of caffeine on perceptual judgement: A dose-response study. *Pharmacopsychoecologia*, **7**:215–219. *[30]*.

Gupta, U. (1988a). Effects of impulsivity and caffeine on human cognitive performance. *Pharmacopsychoecologia*, **1**:33–41. *[24]*.

Gupta, U. (1988b). Personality, caffeine and human cognitive performance. *Pharmacopsychoecologia*, **1**:79–84. *[25]*.

Gupta, U. (1991). Differential effects of caffeine on free recall after semantic and rhyming tasks in high and low impulsives. *Psychopharmacology*, **105**:137–140. *[26]*.

Gupta, U. (1993). Effects of caffeine on recognition. *Pharmacology, Biochemistry and Behavior*, **44**:393–396. *[27]*.

Gupta, U., Dubey, G.P. and Gupta, B.S. (1994). Effects of caffeine on perceptual judgement. *Neuropsychobiology*, **30**:185–188. *[28]*.

Guyatt, A.R., Kirkham, A.J.T., Mariner, D.C., Baldry, A.G. and Cumming, G. (1989). Long-term effects of switching to cigarettes with lower tar and nicotine yields. *Psychopharmacology*, **99**:80–86.

Haertzen, C. and Hickey, J. (1987). Addiction research centre inventory (ARCI): measurement of euphoria and other drug effects. In: M. Bozorth (Ed.), Methods of assessing the reinforcing properties of abused drugs. New York: Springer Verlag, pp. 489–524.

Hahn, R.G., Norberg, A., Gabrielsson, J., Danielsson, A. and Jones, A.W. (1994). Eating a meal increases the clearance of ethanol given by intravenous infusion. *Alcohol and Alcoholism*, **29**:673–678.

Halgren, E. and Smith, M.E. (1987). Cognitive evoked potentials as modulatory processes in human memory formation and retrieval. *Human Neurobiology*, **6**:129–139.

Hall, S.M., Munoz, R.F. and Reus, V.I. (1991). Depression and smoking treatment: A clinical trial of an affect regulation treatment. In: Problems of drug dependence 1991. Proceedings of the 53rd Annual Scientific Meeting, The Committee on Problems of Drug Dependence, Inc. Rockville, MD: National Institute on Drug Abuse.

Halliday, R., Le Houezec, J., Benowitz, N., Naylor, H., Yano, L., Herzig, K. and Callaway, E. (1992). Nicotine speeds P_3 and RT in non-smokers. Presented at the thirty-second annual meeting of the Society for Psychophysiological Research, San Diego, October.

Halliday, R., Naylor, H., Callaway, E., Yano, L. and Walton, P. (1987). What's done can't always be undone: the effects of stimulant drugs and dopamine blockers on information processing. *Electroencephalography and Clinical Neurophysiology*, **Suppl. 40**:240–244.

Hammersley, R., Finnigan, F. and Millar, K. (1994). Individual differences in the acute response to alcohol. *Personality and Individual Differences*, **17**:497–510.

Hamon, J.F. and Camara, P. (1991). Combined effects of methanol and ethanol on event-related potentials in non alcohol dependent men. *Homeostasis*, **33**:274–279.

Hanna, J.M. (1976). Ethnic groups, human variation and alcohol use. In: M.W. Everett, J.O. Waddell and D.B. Heath (Eds.), Crosscultural approaches to the study of alcohol. An interdisciplinary perspective. Paris: Mouton, pp. 235–244.

Hannon, R., Butler, C.P., Day, C.L., Khan, S.A., Quitoriano, L.A., Butler, A.M. and Meredith, L.A. (1985). Alcohol use and cognitive functioning in men and women college students. In: M. Galanter (Ed.), Recent developments in alcoholism. Vol. 3. New York: Plenum Press, pp. 241–252.

Hannon, R., Butler, C.P., Day, C.L., Khan, S.A., Quitoriano, L.A., Butler, A.M. and Meredith, L.A. (1987). Social drinking and cognitive functioning in college students: A replication and reversibility study. *Journal of Studies on Alcohol*, **48**:502–506.

Hannon, R., Day, C.L., Butler, A.M., Larson, A. and Casey, M.B. (1983). Alcohol consumption and cognitive functioning in college students. *Journal of Studies on Alcohol*, **44**:283–298.

Hansson, L., Choudry, N.B., Karlsson, J.A. and Fuller, R.W. (1994). Inhaled nicotine in humans: effect on the respiratory and cardiovascular systems. *Journal of Applied Physiology*, **76**:2420–2427.

Harada, S., Agarwal, D.P. and Goedde, H.W. (1981). Aldehyde dehydrogenase deficiency as cause of facial flushing reaction to alcohol in Japanese. *Lancet*, **ii**:982.

Harada, S., Agarwal, D.P., Goedde, H.W., Tagaki, S. and Ishikawa, B. (1982). Possible protective role against alcoholism for aldehyde dehydrogenase isozyme deficiency in Japan. *Lancet*, **ii**:827.

Harada, S., Misawa, S., Agarwal, D.P. and Goedde, H.W. (1985). Aldehyde dehydrogenase polymorphism and alcohol metabolism in alcoholics. *Alcohol*, **2**:391–392.

Hardy, B.G., Lemieux, C., Walter, S.E. and Bartle, W.R. (1988). Interindividual and intraindividual variability in acetylation: characterization with caffeine. *Clinical Pharmacology and Therapeutics*, **44**:152–157.

Hari, R., Sams, M. and Järvilehto, T. (1979). Auditory evoked transient and sustained potentials in the human EEG: II. Effects of small doses of ethanol. *Psychiatry Research*, **1**:307–312.

Harper, C. and Kril, J. (1986). Pathological changes in alcoholic brain shrinkage. *Medical Journal of Australia*, **144**:3–4.

Harper, C. and Kril, J. (1991). If you drink your brain will shrink. Neuropathological considerations. *Alcohol and Alcoholism*, **1(Suppl.)**:375–380.

Harper, C., Kril, J. and Daly, J. (1988). Does a "moderate" alcohol intake damage the brain? *Journal of Neurology, Neurosurgery and Psychiatry*, **51**:909–913.

Hart, P., Farrell, G.C., Cooksley, W.G.E. and Powell, L.W. (1976). Enhanced drug metabolism in cigarette smokers. *British Medical Journal*, **2**:147–149.

Harter, M.R. and Previc, F.H. (1978). Size-specific information channels and selective attention: Visual evoked potential and behavioral measures. *Electroencephalography and Clinical Neurophysiology*, **45**:628–640.

Hartley, L.R. (1973). Cigarette smoking and stimulus selection. *British Journal of Psychology*, **64**:593–599.

Hasenfratz, M. and Bättig, K. (1991). Psychophysiological reactions during active and passive stress coping following smoking cessation. *Psychopharmacology*, **104**:356–362.

Hasenfratz, M. and Bättig, K. (1992). Action profiles of smoking and caffeine: Stroop effect, EEG, and peripheral physiology. *Pharmacology, Biochemistry and Behavior*, **42**:155–161. *[31]*.

Hasenfratz, M. and Bättig, K. (1993a). Effects of smoking on cognitive performance and psychophysiological parameters as a function of smoking state. *Human Psychopharmacology*, **8**:335–344.

Hasenfratz, M. and Bättig, K. (1993b). Psychophysiological interactions between smoking and stress coping? *Psychopharmacology*, **113**:37–44.

Hasenfratz, M. and Bättig, K. (1994). Acute dose-effect relationships of caffeine and mental performance, EEG, cardiovascular and subjective parameters. *Psychopharmacology*, **114**:281–287. *[32]*.

Hasenfratz, M., Baldinger, B. and Bättig, K. (1993a). Nicotine or tar titration in cigarette smoking behavior? *Psychopharmacology*, **112**:253–258.

Hasenfratz, M., Bunge, A., Dal Prá, G. and Bättig, K. (1993). Antagonistic effects of caffeine and alcohol on mental performance parameters. *Pharmacology, Biochemistry and Behavior*, **46**:463–465. *[33]*.

Hasenfratz, M., Buzzini, P., Cheda, P. and Bättig, K. (1994). Temporal relationships of the effects of caffeine and alcohol on rapid information processing. *Pharmacopsychoecologia*, **7**:87–96. *[34]*.

Hasenfratz, M., Jacober, A. and Bättig, K. (1993b). Smoking-related subjective and physiological changes: pre- to post-puff and pre- to post-cigarette. *Pharmacology, Biochemistry and Behavior*, **46**:527–534.

Hasenfratz, M., Jaquet, F., Aeschbach, D. and Bättig, K. (1991). Interactions of smoking and lunch with the effects of caffeine on cardiovascular functions and information processing. *Human Psychopharmacology*, **6**:277–284. *[35]*.

Hasenfratz, M., Michel, Ch., Nil, R. and Bättig, K. (1989a). Can smoking increase attention in rapid information processing during noise? Electrocortical, physiological, and behavioural effects. *Psychopharmacology*, **98**:75–80.

Hasenfratz, M., Nil, R. and Bättig, K. (1990). Development of central and peripheral smoking effects over time. *Psychopharmacology*, **101**:359–365.

Hasenfratz, M., Pfiffner, D., Pellaud, K. and Bättig, K. (1989b). Postlunch smoking for pleasure seeking or arousal maintenance. *Pharmacology, Biochemistry and Behaviour*, **34**:631–639.

Hasenfratz, M., Thut, G. and Bättig, K. (1992). Twenty-four-hour monitoring of heart rate, motor activity and smoking behavior including comparisons between smokers and nonsmokers. *Psychopharmacology*, **106**:39–44.

Hasher, L. and Zacks, R. (1979). Automatic and effortful processes in memory. *Journal of Experimental Psychology*, **10**:356–388.

Hatsukami, D.K., Anton, D., Keenan, R. and Callies, A. (1992). Smokeless tobacco abstinence effects and nicotine gum dose. *Psychopharmacology*, **106**:60–66.

Hatsukami, D.K., Dahlgren, L., Zimmerman, R. and Hughes, J.R. (1988). Symptoms of tobacco withdrawal from total cigarette cessation versus partial cigarette reduction. *Psychopharmacology*, **94**:242–247.

Hatsukami, D.K., Gust, S.W. and Keenan, R.M. (1987). Physiologic and subjective changes from smokeless tobacco withdrawal. *Clinical Pharmacology and Therapeutics*, **41**:103–107.

Hatsukami, D.K., Huber, M., Callies, A. and Skoog, K. (1993). Physical dependence on nicotine gum: effect of duration of use. *Psychopharmacology*, **111**:449–456.

Hatsukami, D.K., Hughes, J.R. and Pickens, R. (1985). Characterization of tobacco withdrawal: Physiological and subjective effects. In: J. Grabowski and S.M. Hall (Eds.), Pharmacological adjuncts in smoking cessation (NIDA Research Monograph 53:56–67). Rockville, MD: National Institute of Drug Abuse.

Hatsukami, D.K., Skoog, K., Huber, M. and Hughes, J. (1991). Signs and symptoms from nicotine gum abstinence. *Psychopharmacology*, **104**:496–504.

Haut, J.S., Beckwith, B.E., Petros, T.V. and Russell, S. (1989). Gender differences in retrieval from long-term memory following acute intoxication with ethanol. *Physiology and Behavior*, **45**:1161–1165.

Health Education Authority. (1991). That's the limit. A guide to sensible drinking. London: Health Education Authority.

Heath, A.C. and Martin, N.G. (1991). The inheritance of alcohol sensitivity and of patterns of alcohol use. *Alcohol and Alcoholism*, **(Suppl.1)**:141–145.

Heath, A.C. and Martin, N.G. (1993). Genetic models for the natural history of smoking: Evidence for a genetic influence on smoking persistence. *Addictive Behaviours*, **18**:19–34.

Hebener, E.S., Kagan, J. and Cohen, M. (1989). Shyness and olfactory threshold. *Personality and Individual Differences*, **10(11)**:115–963.

Heemstra, M.L. (1988). Efficiency of human information processing. A model of cognitive energetics. Unpublished Ph.D. Thesis, Amsterdam: Free University of Amsterdam.

Heimstra, N.W., Bancroft, N.R. and DeKock, A.R. (1967). Effects of smoking upon sustained performance in a simulated driving task. *Annals of the New York Academy of Sciences*, **142**:295–307.

Heimstra, N.W., Fallesen, J.J., Kinsley, S.A. and Warner, N.W. (1980). The effects of deprivation of cigarette smoking on psychomotor performance. *Ergonomics*, **23**:1047–1055.

Heishman, S.J., Snyder, F.R. and Henningfield, J.E. (1990). Effects of repeated nicotine administration. In: L. Harris (Ed.), Problems of drug dependence 1990: *Proceedings of the 52nd annual meeting of the committee on problems of drug dependence (NIDA Research Monograph*, **105**:314–315. Rockville, MD: National Institute of Drug Abuse.

Heishman, S.J., Snyder, F.R. and Henningfield, J.E. (1993). Performance, subjective, and physiological effects of nicotine in non-Smokers. *Drug and Alcohol Dependence*, **34**:11–18.

Heishman, S.J., Taylor, R.C. and Henningfield, J.E. (1994). Nicotine and smoking: a review of effects on human performance. *Experimental and Clinical Psychopharmacology*, **2(4)**:345–395.

Hempel, J., Kaiser R. and Jornvall, H. (1985). Mitochondrial aldehyde dehydrogenase from human liver: primary structure, differences in relation to the cytosolic enzyme and functional correlations. *European Journal of Biochemistry*, **153**:13–28.

Henningfield, J.E. and Keenan, R.M. (1993). Nicotine delivery kinetics and abuse liability. *Journal of Consulting and Clinical Psychology*, **61**:743–750.

Henningfield, J.E., Chait, L.D. and Griffiths, R.R. (1983). Cigarette smoking and subjective response in alcoholics: effects of pentobarbital. *Clinical Pharmacological Therapy*, **33**:806–812.

Henningfield, J.E., Chait, L.D. and Griffiths, R.R. (1984). Effects of ethanol on cigarette smoking by volunteers without histories of alcoholism. *Psychopharmacology*, **82**:1–5.

Henningfield, J.E., Kozlowski, L.T. and Benowitz, N.L. (1994). A proposal to develop meaningful labeling for cigarettes. *Journal of the American Medical Association*, **272**:312–314.

Henningfield, J.E., London, E. and Benowitz, N.L. (1990). Arterial-venous differences in plasma concentrations of nicotine after cigarette smoking. *Journal of the American Medical Association*, **263**:2049–2050.

Herman, C.P. and Polivy, J. (1975). Anxiety, restraint, and eating behavior. *Journal of Abnormal Psychology*, **84**:666–672.

Herning, R. and Pickworth, W. (1985). Nicotine gum improved stimulus processing during tobacco withdrawal. *Psychophysiology*, **22**:595.

Herning, R.I., Brigham, J., Stitzer, M.L., Glover, B.J., Pickworth, W.B. and Henningfield, J.E. (1990). The effects of nicotine on information processing: Medicating a deficit. *Psychophysiology*, **27**:S2.

Herning, R.I., Jones, R.T., Hooker, W.D., Mendelson, J. and Blackwell, L. (1985). Cocaine increases EEG beta: A replication and extension of Hans Berger's historic experiments. *Electroencephalography and Clinical Neurophysiology*, **60**:470–477.

Hertz, B.F. (1978). The effects of cigarette smoking on perception of nonverbal communications. Dissertation Abstracts International, 39/2501B–2502B.

Hesselbrock, V., Bauer, L.O., Hesselbrock, M.N. and Gillen, R. (1991) Neuropsychological factors in individuals at high risk for alcoholism. In: M. Gallanter (Ed.), Recent developments in alcoholism. New York: Plenum Press, pp. 21–40.

Higuchi, S., Muramatsu, T. and Shigemori, K. (1992). The relationship between low Km aldehyde dehydrogenase phenotype and drinking behavior in Japanese. *Journal of Studies on Alcohol*, **53**:425–433.

Hill, D. and Gray, N. (1982). Patterns of tobacco smoking in Australia. *Medical Journal of Australia*, **1**:23–25.

Hill, S.Y. and Ryan, C. (1985). Brain damage in social drinkers? Reasons for caution. In: M. Galanter (Ed.), Recent developments in alcoholism. Vol. 3. New York: Plenum Press, pp. 277–288.

Hill, S.Y. (1992). Absence of paternal sociopathy in the etiology of severe alcoholism: Is there a Type III alcoholism? *Journal of Studies on Alcohol*, **53**:161–169.

Hill, S.Y., Steinhauer, S.R., Zubin, J. and Baughman, T. (1988). Event-related potentials as markers for alcoholism risk in high density families. *Alcoholism: Clinical and Experimental Research*, **12(4)**:545–554.

Hillyard, S.A. and Kutas, M. (1983). Electrophysiology of cognitive processing. *Annual Review of Psychology*, **34**:33–61.

Hillyard, S.A. and Picton, T.W. (1987). Electrophysiology of cognition. In: V.B. Mountcastle, F. Plum and S.R. Geiger (Eds.), Handbook of physiology: Section I. The nervous system: Vol. V, Part 2. Higher functions of the brain. Bethesda, MD: American Physiological Society, pp. 519–584.

Hillyard, S.A., Picton, T.W. and Regan, D.M. (1978). Sensation, perception and attention: Analysis using ERPs. In: E. Callaway, P. Tueting and S. Koslow (Eds.), Event-related brain potentials in man. New York: Academic Press, pp. 223–322.

Hilton, M.E. (1991). The demographic distribution of drinking patterns in 1984. In: W.B. Clark and M.E. Hilton (Eds.), Alcohol in America: Drinking Practices and Problems. Albany, NY: State University of New York Press, pp. 73–86.

Hindmarch, I. (1980). Psychomotor function and psychoactive drugs. *British Journal of Clinical Pharmacology*, **10**:189–209.

Hindmarch, I., Bhatti, J.Z., Starmer, G.A., Mascord, D.J., Kerr, J.S. and Sherwood, N. (1992). The effects of alcohol on the cognitive function of males and females and on skills relating to car driving. *Human Psychopharmacology*, **7**:105–114.

Hindmarch, I., Kerr, J.S. and Sherwood, N. (1990a). Effects of nicotine gum on psychomotor performance in smokers and non-smokers. *Psychopharmacology*, **100**:535–541.

Hindmarch, I., Kerr, J.S. and Sherwood, N. (1990b). Psychopharmacological aspects of psychoactive substances. In: D.M. Warburton (Ed.), Addiction Controversies. Chur: Harwood Academic Publishers, pp. 36–44.

Hindmarch, I., Kerr, J.S. and Sherwood, N. (1991a). The comparative psychopharmacology of nicotine. In: F. Adlkofer and K. Thurau (Eds.), Effects of nicotine on biological systems. Basel: Birkhäuser Verlag, pp. 509–519.

Hindmarch, I., Kerr, J.S. and Sherwood, N. (1991b). The effects of alcohol and other drugs on psychomotor performance and cognitive function. *Alcohol and Alcoholism*, **26**:71–79.

Hines, D. (1978). Olfaction and the right cerebral hemisphere. *Journal of Altered States of Consciousness*, **3(1)**:47–59.

Hockey, G.R.J. (1986). Changes in operator efficiency as a function of environmental stress, fatigue, and circadian rhythms. In: K.R. Boff, L. Kaufman and J.P. Thomas (Eds.), Handbook of Perception and Human Performance. Vol. 2. Ch. 44, New York: Wiley, pp. 44-1–44-49.

Hockey, G.R.J., Coles, M.G.H. and Gaillard, A.W.K. (1986). Energetical issues in research on human information processing. In: G.R.J. Hockey, A.W.K. Gaillard and M.G.H. Coles (Eds.), Energetics and human information processing. Dordrecht, The Netherlands: Nijhoff, pp. 3–21.

Höfer, I. and Bättig, K. (1994). Cardiovascular, behavioral, and subjective effects of caffeine under field conditions. *Pharmacology, Biochemistry and Behavior*, **48**:899–908.

Höfer, I., Nil, R. and Bättig, K. (1991). Nicotine yield as determinant of smoke exposure indicators and puffing behavior. *Pharmacology, Biochemistry and Behavior*, **40**:139–149.

Höfer, I., Nil, R., Wyss, F. and Bättig, K. (1992). The contributions of cigarette yield, consumption, inhalation and puffing behaviour to the prediction of smoke exposure. *Clinical Investigator*, **70**:343–351.

Hollingworth, H.L. (1912). The influence of caffeine on mental and motor efficiency. Archives of Psychology, **22**:1–166.

Holloway, J.A. and Holloway, F.A. (1979). Combined effects of ethanol and stimulants on behavior and physiology. *Neuroscience and Biobehavioral Reviews*, **3**:137–148.

Horne, J.A., Brass, C.G. and Pettitt, A.N. (1980). Circadian performance differences between morning and evening 'types'. *Ergonomics*, **23(1)**:29–36.

Houston, J.P., Schneider, N.G. and Jarvik, M.E. (1978). Effects of smoking on free recall and organization. *American Journal of Psychiatry*, **135**:220–222.

Hrubec, Z. (1973). Coffee drinking and ischemic heart disease. *Lancet*, **i**:548.

Hughes, F.W. and Forney, R.B. (1961). Alcohol and caffeine in choice discrimination tests in rats. *Proceeds for the Experimental and Biological Medicine*, **108**:157–159.

Hughes, J.R. (1989). Dependence potential and abuse liability of nicotine replacements therapy. *Biomedicine and Pharmacotherapy*, **43**:11–17.

Hughes, J.R. (1991). Distinguishing withdrawal relief and direct effects of smoking. *Psychopharmacology*, **104**:409–410.

Hughes, J.R., Gust, S.W., Keenan, R., Fenwick, J.W., Skoog, K. and Higgins, S.T. (1991a). Long-term use of nicotine vs. placebo Gum. *Archives of Internal Medicine*, **151**:1993–1998.

Hughes, J.R., Gust, S.W., Skoog, K., Keenan, R.M. and Fenwick, J.W. (1991b). Symptoms of tobacco withdrawal: A replication and extension. *Archives of General Psychiatry*, **48**:52–59.

Hughes, J.R., Higgins, S. and Hatsukami, D. (1990). Effects of abstinence from tobacco: A critical review. In: L.T. Kozlowski, H.M. Annis, H.D. Cappell, F.B. Glaser, M.S. Goodstadt, Y. Israel, H. Kalant, E.M. Sellers and E.R. Vingilis (Eds.), Research advances in alcohol and drug problems. Vol. 10. New York: Plenum, pp. 317–398.

Hughes, J.R., Higgins, S.T., Bickel, W.K., Hunt, W.K., Fenwick, J.W., Gulliver, S.B. and Mireault, G.C. (1991). Caffeine self-administration, withdrawal and adverse effects among coffee drinkers. *Archives of General Psychiatry*, **48**:611–617.

Hughes, J.R., Keenan, R.M. and Yellin, A. (1989a). Brief report. Effect of tobacco withdrawal on sustained attention. *Addictive Behaviours*, **14**:577–580.

Hughes, J.R., Keenan, R.M. and Yellin, A. (1989b). Effect of tobacco withdrawal on sustained attention. *Addictive Behaviours*, **14**:577–580.

Hughes, J.R., Oliveto, A.H., Helzer, J.E., Higgins, S.T. and Bickel, W.K. (1992). Should caffeine abuse, dependence or withdrawal be added to DSM-IV and ICD-10? *American Journal of Psychiatry*, **149**:33–40.

Hull, C.L. (1975). The influence of tobacco smoking on mental and motor efficiency: An experimental investigation. Psychological Review (1924, reprinted by Greenwood Press, 1975).

Hull, J.G. and Bond Jr., C.F. (1986). Social and behavioral consequences of alcohol consumption and expectancy: A meta-analysis. *Psychology Bulletin*, **99**:347–360.

Hull, J.G. and Young, R.D. (1983). Self-consciousness, self-esteem, and success-failure as determinants of alcohol consumption in male social drinkers. *Journal of Personality and Social Psychology*, **44**:1097–1109.

Hummel, T. and Kobal, G. (1992). Differences in human evoked potentials related to olfactory or trigeminal chemosensory activation. *Electroencephalography and Clinical Neurophysiology*, **84(1)**:84–89.

Hummel, T., Hummel, C., Pauli, E. and Kobal, G. (1992b). Olfactory discrimination of nicotine-enantiomers by smokers and non-smokers. *Chemical Senses*, **17**:13–21.

Hummel, T., Livermore, A., Hummel, C. and Kobal, G. (1992a). Chemosensory event-related potentials in man: relation to olfactory and painful sensations elicited by nicotine. *Electroencephalography and Clinical Neurophysiology*, **84**:192–195.

Humphreys, M.S. and Revelle, W. (1984). Personality, motivation, and performance: A theory of the relationship between individual differences and information processing. *Psychological Review*, **91(2)**:153–184.

Hurt, R.D., Dale, L.C., Offord, K.P., Bruce, B.K., McClain, F.L. and Eberman, K.M. (1992). Inpatient treatment of severe nicotine dependence. *Mayo Clinic Proceedings*, **67**:823–828.

Hyman, G.J., Stanley, R.O., Burrows, G.D. and Horne, D.J. (1986). Treatment effectiveness of hypnosis and behaviour therapy in smoking cessation: a methodological refinement. *Addictive Behaviors*, **11**:355–365.

Ikard, F.F., Green, D. and Horn, D. (1969). A scale to differentiate between types of smoking as related to the management of affect. *International Journal of Addictions*, **4**:649–659.

Isreal, J.B., Wickens, C.D., Chesney, G.L. and Donchin, E. (1980). The event-related potential as an index of display monitoring workload. *Human Factors*, **22**:211–224.

Jääskeläinen, I.P. (1995). Acute effects of ethanol on attention as revealed by event related brain potentials and behavioral measures of performance. Academic dissertation, Faculty of Arts, Helsinki: University of Helsinki.

Jacob, P. III, Benowitz, N.L., Copeland, J.R., Risner, M.E. and Cone, E.J. (1988). Disposition kinetics of nicotine and cotinine enantiomers in rabbits and beagle dogs. *Journal of Pharmaceutical Sciences*, **77**:396–400.

Jacob, P. III, Wilson, M. and Benowitz, N.L. (1981). Improved gas chromatographic method for the determination of nicotine and cotinine in biologic fluids. *Journal of Chromatography*, **222**:61–70.

Jacober, A., Hasenfratz, M. and Bättig, K. (1993). A nicotine dependent and a nicotine independent component of smoking related pulse and activity variation. *Human Psychopharmacology*, **8**:125–132.

Jacober, A., Hasenfratz, M. and Bättig, K. (1994a). Ultralight cigarettes: activity, cardio-vascular dietary and subjective parameters. *Pharmacology, Biochemistry and Behavior*, **47**:187–195.

Jacober, A., Hasenfratz, M. and Bättig, K. (1994b). Cigarette smoking: habit or nicotine maintenance? Effects of short-term smoking abstinence and oversmoking. *Human Psychopharmacology*, **9**:117–123.

Jacober, A., Hasenfratz, M. and Bättig, K. (1994c). Circadian and ultradian rhythms in heart rate and motor activity of smokers, abstinent smokers, and nonsmokers. *Chronobiology International*, **11**:320–331.

Jacobson, B.H. and Edgley, B.M. (1987). Effects of caffeine on simple reaction time and movement time. *Aviation, Space, and Environmental Medicine*, **58**:1153–1156. *[36]*.

Jacobson, B.H. and Thurman-Lacey, S.R. (1992). Effect of caffeine on motor performance by caffeine-naive and -familiar subjects. *Perceptual and Motor Skills*, **74**:151–157. *[37]*.

Jacobson, B.H., Winter-Roberts, K. and Gemmell, H.A. (1991). Influence of caffeine on selected manual manipulation skills. *Perceptual and Motor Skills*, **72**:1175–1181. *[38]*.

James, J.E. (1991). Caffeine and Health. London: Academic Press.

James, J.E. (1994). Does caffeine enhance or merely restore degraded psychomotor performance? *Neuropsychobiology*, **30**:124–125.

Janke, W. (1983). Response variability to psychotropic drugs. New York: Pergamon Press.

Jansen, G.H. (1976). De eeuwige kroeg; hoofdstukken uit de geschiedenis van het openbaar lokaal. Meppel, the Netherlands: Boom.

Jarboe, C.H., Hurst, H.E., Rodgers, G.C. and Metaxas, J.M. (1986). Toxicokinetics of caffeine elimination in an infant. *Clinical Toxicology*, **24**:415–428.

Jarvik, M.E. (1991). Beneficial effects of nicotine. British *Journal of Addiction*, **86**:571–575.

Jarvik, M.E., Caskey, N.H., Rose, J.E., Herskovic, J.E. and Sadeghpour, M. (1989). Anxiolytic effects of smoking associated with four stressors. *Addictive Behaviors*, **14**:379–386.

Jarvis, M.J. (1993). Does caffeine intake enhance absolute levels of cognitive performance? *Psychopharmacology*, **110**:45–52.

Jennings, R. (1986a). Bodily changes during attending. In: M. Coles, E. Donchin and S. Porges (Eds.), Psychophysiology: systems, processes and applications. New York: Guilford Press, pp. 268–284.

Jennings, R. (1986b). Memory, thought and bodily processes. In: M. Coles, E. Donchin and S. Porges (Eds.), Psychophysiology: systems, processes and applications. New York: Guilford Press, pp. 290–308.

Jensen, O.L. and Krogh, E. (1984). Visual evoked response and alcohol intoxication. *Acta Opthalmologica*, **62**:651–657.

Jernigan, T., Butters, N., Di Taglia, G., Shafer, K., Smith, T., Irwin, M., Grant, I., Schuckit, M. and Cermak, L.S. (1991). Reduced cerebral grey matter observed in alcoholics using magnetic resonance imaging. *Alcoholism: Clinical and Experimental Research*, **15**:418–427.

Jobs, S.M., Fiedler, F.E. and Lewis, C.T. (1990). Impact of moderate alcohol consumption on business decision making. In: S.W. Gust, J.M. Walsh, L.B. Thomas and D.J. Crouch (Eds.), Drugs in the Workplace, Research and Evaluation Data, 2, (Research Monograph 100). Rockville, MD: National Institute on Drug Abuse.

Joeres, R., Klinker, H., Heusler, H., Epping, J., Hofstetter, G., Drost, D., Reuss, H., Zilly, W. and Richter, E. (1987). Factors influencing the caffeine test for cytochrome P 448-dependent liver function. *Archives of Toxicology*, **60**:93–94.

Joeres, R., Klinker, H., Heusler, R., Epping, J., Zilly, W. and Richter, E. (1988). Influence of smoking on caffeine elimination in healthy volunteers and in patients with alcoholic liver cirrhosis. *Hepatology*, **8**:575–579.

Johnson Jr., R. (1986). A triarchic model of P300 amplitude. *Psychophysiology*, **23(4)**:367–384.

Johnson Jr., R. (1988). The amplitude of the P300 component of the event-related potential: Review and synthesis. In: P.K. Ackles, J.R. Jennings and M.G.H. Coles (Eds.), Advances in psychophysiology. Vol. III. Greenwich: JAI Press, pp. 69–138.

Johnson Jr., R. (1993). On the neural generators of the P300 component of the event related potential. *Psychophysiology*, **30**:90–97.

Jones, A.W. (1984). Interindividual variations in the disposition and metabolism of ethanol in healthy men. *Alcohol*, **1**:385–391.

Jones, B.M. (1973). Memory impairment on the ascending and descending limbs of the blood alcohol curve. *Journal of Abnormal Psychology*, **82**:24–32.

Jones, G.M.M., Sahakian, B.J., Levy, R., Warburton, D.M. and Gray, J.A. (1992). Effects of acute subcutaneous nicotine on attention, information processing and short-term memory in Alzheimer's disease. *Psychopharmacology*, **108**:485–494.

Jones, M.K. and Jones, B.M. (1980). The relationship of age and drinking habits to the effects of alcohol on memory in women. *Journal of Studies on Alcohol*, **41**:179–186.

Jones-Saumty, D.J. and Zeiner, A.R. (1985). Psychological correlates of drinking behaviour in social-drinker college students. *Alcoholism: Clinical and experimental Research*, **9**:158–163.

Josephs, R.A. and Steele, C.M. (1990). The two faces of alcohol myopia: Attentional mediation of psychological stress. *Journal of Abnormal Psychology*, **99**:115–126.

Jost, G., Wahllander, A., Von Mandach, U. and Preisig, R. (1987). Overnight salivary caffeine clearance: a liver function test suitable for routine use. *Hepatology*, **7**:338–344.

Jubis, R.M.T. (1986). Effects of alcohol and nicotine on free recall of relevant cues. *Perceptual and Motor Skills*, **62**:363–369.

Jubis, R.M.T. (1990). Effects of alcohol and white noise on recall of relevant and irrelevant task components. *Perceptual and Motor Skills*, **71**:691–702.

Kadlubar, F.F., Talaska, G., Butler, M.A., Teitel, C.H., Massengill, J.P. and Lang, N.P. (1990). Determination of carcinogenic arylamine N-oxidation phenotype in humans by analysis of caffeine urinary metabolites. *Progress in Clinical and Biology Research*, **340**:107–114.

Kahneman, D. and Treisman, A. (1984). Changing views of attention and automaticity. In: R. Parasuraman and D.R. Davies (Eds.), Varieties of attention. London: Academic Press, pp. 29–61.

Kalant, H. and Khanna, J.M. (1989). The alcohols. In: H. Kalant and W.H.E. Roschlau (Eds.), Principles of medical pharmacology, 5th ed. Toronto: B.C. Decker, pp. 244–254.

Kalant, H., Leblanc, A. and Gibbins, R. (1971). Tolerance to and dependence on some non-opiate psychotropic drugs. *Pharmacological Review*, **23**:135–191.

Kalin, R. (1964). Effects of alcohol on memory. *Journal of Abnormal and Social Psychology*, **69**:635–641.

Kall, M.A. and Clausen, J. (1995). Dietary effect on mixed function P450 1A2 activity assayed by estimation of caffeine metabolism in man. *Human and Experimental Toxicology*, **14**:801–807.

Kalow, W. and Tang, B.K. (1991a). Use of caffeine metabolic ratios to explore CYP1A2 and xanthine oxidase activities. *Clinical Pharmacology and Therapeutics*, **50**:508–519.

Kalow, W. and Tang, B.K. (1991b). Caffeine as a metabolic probe: exploration of the enzyme-inducing effect of cigarette smoking. *Clinical Pharmacology and Therapeutics*, **49**:44–48.

Kalow, W. (1984). Pharmacoanthropology: drug metabolism. *Federation Proceedings*, **43**:2326–2331.

Kalow, W. (1985). Variability of caffeine metabolism in humans. *Arzneimittel-Forschung (Drug Research)*, **35**:319–324.

Kaminori, G.H., Somani, S.M., Knowlton, R.G. and Perkins, R.M. (1987). The effects of obesity and exercise on the pharmacokinetics of caffeine in lean and obese volunteers. *European Journal of Clinical Pharmacology*, **31**:595–600.

Kaplan, R.F., Glueck, B.C., Hesselgrock, M.N. and Reed Jr., H.B.C. (1985). Power and coherence analysis of the EEG in hospitalized alcoholics and nonalcoholic controls. *Journal of Studies on Alcohol*, **46(2)**:122–127.

Kassam, J.P., Tang, B.K., Kadar, D. and Kalow, W. (1989). *In vitro* studies of human liver alcohol dehydrogenase variants using a variety of substrates. *Drug Metabolism and Disposition*, **17**:567–572.

Keenan, R.M., Hatsukami, D.K. and Anton, D.J. (1989). The effects of short-term smokeless tobacco deprivation on performance. *Psychopharmacology*, **98**:126–130.

Keenan, R.M., Hatsukami, D.K., Pickens, R.W., Gust, S.W. and Strelow, L.J. (1990). The relationship between chronic ethanol exposure and cigarette smoking in the laboratory and the natural environment. *Psychopharmacology*, **100**:77–83.

Keiding, S., Christensen, N.J. and Damgaard, S.E. (1983). Ethanol metabolism in heavy drinkers after massive and moderate alcohol intake. *Biochemical Pharmacology*, **32**:3097–3120.

Keister, M.E. and McLaughlin, R.J. (1972). Vigilance performance related to extraversion-introversion and caffeine. *Journal of Experimental Research in Personality*, **6**:5–11.

Kendler, K.S., Heath, A.C. and Neale, M.C. (1992). A population based twin study of alcoholism in women. *Journal of American Medical Association*, **268**:1877–1882.

Kenemans, J.L. and Lorist, M.M. (1995). Caffeine and visual selective processing. *Pharmacology, Biochemistry and Behavior*, **52**:461–471.

Kenemans, J.L., Kok, A. and Smulders, F.T.Y. (1993). Event-related potentials to conjunctions of spatial frequency and orientation as a function of stimulus parameters and response requirements. *Electroencephalography and Clinical Neurophysiology*, **88**:51–63.

Kenemans, J.L., Smulders, F.T.Y. and Kok, A. (1995). Selective processing of two-dimensional visual stimuli in young and old subjects: Electrophysiological analysis. *Psychophysiology*, **32**:108–120.

Kenford, S.L., Fiore, M.C., Jorenby, D.E., Smith, S.S., Wetter, D. and Baker, T.B. (1994). Predicting smoking cessation – who will quit with and without the nicotine patch. *Journal of the American Medical Association*, **271**:589–594.

Kerkhof, G.A. (1981). Brain potentials at different times of day for morning type and evening type subjects. Ph.D. Thesis. University of Leiden, Meppel: Krips Repro.

Kerkhof, G.A. (1984). A Dutch questionnaire for the selection of morning and evening types. *Nederlands Tijdschrift voor de Psychologie*, **39**:281–294.

Kerkhof, G.A. (1985). Interindividual differences in the human circadian system: a review. *Biological Psychology*, **20**:83–112.

Kerkhof, G.A. (1991). Differences between morning types and evening types in the dynamics of EEG slow wave activity during sleep. *Electroencephalography and Clinical Neurophysiology*, **78**:197–202.

Kerr, J.S. and Hindmarch, I. (1991). Alcohol, cognitive function and psychomotor performance. *Reviews on Environmental Health*, **9**:117–122.

Kerr, J.S., Sherwood, N. and Hindmarch, I. (1991). Separate and combined effects of the social drugs on psychomotor performance. *Psychopharmacology*, **104**:113–119. *[39]*.

Khanna, N.N. and Somani, S.M. (1984). Maternal coffee drinking and unusually high concentrations of caffeine in the newborn. *Clinical Toxicology*, **22**:473–483.

Kilbane, A.J., Silbart, L.K., Manis, M., Beitins, I.Z. and Weber, W.W. (1990). Human N-acetylation genotype determination with urinary caffeine metabolites. *Clinical Pharmacology and Therapeutics*, **47**:470–477.

Kinchla, R.A. (1980). The measurement of attention. In: R.S. Nickerson (Ed.), Attention and Performance. Vol. 8. Hillsdale, New Jersey: Lawrence Erlbaum, pp. 468–511.

King, D.J. and Henry, G. (1992). The effect of neuroleptics on cognitive and psychomotor function. A preliminary study in healthy volunteers. *British Journal of Psychiatry*, **160**:647–653. *[40]*.

King, M.A., Hunter, B. and Walker, D.W. (1988). Alterations and recovery of dendritic spine density in rat hippocampus following long-term ethanol ingestion. *Brain Research*, **459**:381–385.

Kinnunen, T., Doherty, K., Militello, F.S. and Garvey, A.J. (1994). Quitting smoking and feeling blue: Nicotine replacement as an aid for the depressed. [Summary]. Proceedings of the Society of Behavioral Medicine's Fifteenth Anniversary Meeting. *Annals of Behavioral Medicine*, **16**:S068.

Kinsbourne, M. (1989). A model of adaptive behavior related to cerebral participation in emotional control. In G. Gainotti and C. Caltagirone (Eds.), Emotions and the dual brain. New York: Springer Verlag, pp. 248–260.

Kirk-Smith, M.D., Van Toller, S. and Dodd, G.H. (1983). Unconscious odour conditioning in human subjects. *Biological Psychology*, **17**:221–231.

Kirsch, I. and Weixel, L.J. (1988). Double blind versus deceptive administration of a placebo. *Behavioral Neuroscience*, **102**:319–323.

Kirschbaum, C., Strasburger, C.J. and Langkrär, J. (1993). Attenuated cortisol response to psychological stress but not to CRH or ergometry in young habitual smokers. *Pharmacology, Biochemistry and Behavior*, **44**:527–531.

Kirschbaum, C., Wüst, S. and Strasburger, C.J. (1992). 'Normal' cigarette smoking increases free cortisol in habitual smokers. *Life Sciences*, **50**:435–442.

Klatzky, R.L. (1980). Human memory. Structure and processes. San Francisco: Freeman.

Kleinsmith, L. and Kaplan, S. (1963). Paired associate learning as a function of arousal and interpolated interval. *Journal of Experimental Psychology*, **65**:190–194.

Klesges, R.C. and Klesges, L.M. (1988). Cigarette smoking as a dietary strategy in a university population. *International Journal of Eating Disorders*, **7**:413–417.

Klesges, R.C., Myers, A.W., Klesges, L.M. and LaVasque, M.E. (1989). Smoking, body weight, and their effects on smoking behavior: a comprehensive review of the literature. *Psychological Bulletin*, **106**:204–230.

Knasko, S.C. (1995). Pleasant odors and congruency: effects on approach behavior. *Chemical Senses*, **20**:479–487.

Knibbe, R.A., Drop, M.J. and Muytjens, A. (1987). Correlates of stages in the progression from everyday drinking to problem drinking. *Social Science and Medicine*, **24**:463–473.

Knight, R.G. and Longmore, B.E. (1994). Clinical Neuropsychology of Alcoholism. Hillsdale, New Jersey: Lawrence Erlbaum Associates.

Knop, J., Teasdale, T.W., Goodwin, D.W. and Schulsinger, F. (1989). Premorbid characteristics in high risk and low risk individuals of developing alcoholism. In: K. Kiianmaa, B. Tabakoff and T. Saito (Eds.), Genetic aspects of alcoholism. Helsinki: The Finish Foundation for Alcohol Studies.

Knott, V.J. and De Lugt, D. (1991). Subjective and brain-evoked responses to electrical pain stimulation: Effects of cigarette smoking and warning condition. *Pharmacology, Biochemistry and Behavior*, **39**:889–893.

Knott, V.J. and Griffiths, L. (1992). Day-long smoking deprivation: Mood, performance, and EEG/ERP comparisons with smoking and non-smoker controls. *Psychophysiology*, **29**:S45.

Knott, V.J. and Venables, P.H. (1980). Separate and combined effects of alcohol and tobacco on the amplitude of the contingent negative variation. *Psychopharmacology*, **70**:167–172.

Knott, V.J. (1980). Reaction time, noise distraction and autonomic responsivity in smokers and non-smokers. *Perceptual and Motor Skills*, **50**:1271–1280.

Knott, V.J. (1985). Effects of tobacco and distraction on sensory and slow cortical evoked potentials during task performance. *Neuropsychobiology*, **13**:136–140.

Knott, V.J. (1988). Dynamic EEG changes during cigarette smoking. *Neuropsychobiology*, **19**:54–60.

Knott, V.J. (1989a). Effects of low yield cigarettes on EEG dynamics. *Neuropsychobiology*, **21**:216–222.

Knott, V.J. (1989b). Brain electrical imaging the dose-response effects of cigarette smoking. *Neuropsychobiology*, **22**:236–242.

Knott, V.J. (1989c). Brain event-related potentials (ERPs) in smoking performance research. In: T. Ney and A. Gale (Eds.), Smoking and human behavior. Chichester: Wiley, pp. 93–114.

Knott, V.J. (1990). A neuroelectric approach to the assessment of psychoactivity in comparative substance use. In: D.M. Warburton (Ed.), Addiction controversies. Chur: Harwood Academic Publishers, pp. 66–88.

Knott, V.J. (1990). Effects of cigarette smoking on subjective and brain evoked responses to electrical pain stimulation. *Pharmacology, Biochemistry and Behavior*, **35**:341–346.

Knott, V.J. (1991). Neurophysiological aspects of smoking behaviour: a neuroelectric perspective. *British Journal of Addiction*, **86**:511–515.

Knott, V.J., Hooper, C., Lusk-Mikkelsen, S. and Kerr, C. (1995). Variations in spontaneous brain electrical (EEG) topography related to cigarette smoking: acute smoking, drug comparisons, cholinergic transmission, individual differences and psychopathology. In: E. Domino (Ed.), Brain imaging of nicotine and tobacco smoking. Ann Arbor, Michigan: NPP Books, pp. 167–189.

Kobal, G. and Hummel, T. (1991). Olfactory Evoked Potentials in Humans. In: T.V. Getchell, R.L. Doty, L.M. Bartoshuk and J.B. Snow Jr. (Eds.), Smell and Taste in Health and Disease. New York: Raven Press, pp. 255–275.

Koelega, H.S. and Brinkman, J.A. (1986). Noise and vigilance: An evaluative review. *Human Factors*, **28**:465–481.

Koelega, H.S. (1989). Benzodiazepines and vigilance performance: A review. *Psychopharmacology*, **98**:145–156.

Koelega, H.S. (1992). Extraversion and vigilance performance: 30 years of inconsistencies. *Psychological Bulletin*, **112**:239–258.

Koelega, H.S. (1993). Stimulant drugs and vigilance performance: A review. *Psychopharmacology*, **111**:1–16.

Koelega, H.S. (1995). Alcohol and vigilance performance: A review. *Psychopharmacology*, **118**:233–249.

Koelega, H.S., Brinkman, J.A., Hendriks, L. and Verbaten, M.N. (1989). Processing demands, effort, and individual differences in four different vigilance tasks. *Human Factors*, **31**:45–62.

Kok, A. (1990). Internal and external control: A two-factor model of amplitude change of event-related potentials. *Acta Psychologica*, **74**:203–236.

Kolonen, S., Tuomisto, J., Puustinen, P. and Airaksinen, M.M. (1992). Effects of smoking abstinence and chain-smoking on puffing topography and diurnal nicotine exposure. *Pharmacology, Biochemistry and Behavior*, **42**:327–332.

Kopell, B.S., Roth, W.T. and Tinklenberg, J.R. (1978). Time course effects of marijuana and ethanol on event-related potentials. *Psychopharmacology*, **56**:15–20.

Kopell, B.S., Tinklenberg, J.R. and Hollister, L.E. (1972). Contingent negative variation amplitudes. Marihuana and alcohol. *Archives of General Psychiatry*, **27**:809–811.

Kopun, M. and Propping, P. (1977). The kinetics of ethanol absorption and elimination in twins supplemented by repetitive experiments in singleton subjects. *European Journal of Clinical Pharmacology*, **11**:337–344.

Kos, J., Hasenfratz, M. and Bättig, K. (1997). Effects of a 2-day smoking abstinence on dietary, cognitive, subjective and physiological parameters among younger and older female smokers. *Physiology and Behavior* (In press).

Kosslyn, S.M. and Koenig, O. (1992). Wet mind: The new cognitive neuroscience. New York: The Free Press.

Kostandov, E.A., Arsumanov, Y.L., Genkina, O.A., Restchikova, T.N. and Shostakovich, G.S. (1982). The effects of alcohol on hemispheric functional asymmetry. *Journal of Studies on Alcohol*, **43**:411–426.

Kotake, A.N., Schoeller, D.A., Lambert, G.H., Baker, A.L., Schaffer, D.D. and Josephs, H. (1982). The caffeine CO_2 breath test: Dose response and route of N-demethylation in smokers and nonsmokers. *Clinical Pharmacology and Therapeutics*, **32**:261–269.

Kozena, L., Frantik, E. and Horvath, M. (1986). The effect on vigilance of an analgesic combination (Ataralgin) and its components, guaiphenesine and caffeine. *Activitas Nervosa Superior*, **28(2)**:153–155. *[41]*.

Kozlowski, L.T. and Heatherton, T.F. (1990). Self report issues in cigarette smoking: state of the art and future directions. *Behavioral Assessment*, **12**:53–75.

Kozlowski, L.T. (1976). Effects of caffeine consumption on nicotine consumption. *Psychopharmacology*, **47**:165–168.

Kozlowski, L.T., Jelinek, L.C. and Pope, M.A. (1986). Cigarette smoking among alcohol abusers: a continuing and neglected problem. *Canadian Journal of Public Health*, **77**:205–207.

Kozlowski, L.T., Porter, C.Q., Orleans, C.T., Pope, M.A. and Heatherton, T. (1994). Predicting smoking cessation with self-reported measures of nicotine dependence – FTQ, FTND, and HSI. *Drug and Alcohol Dependence*, **34**:211–216.

Kramer, A. and Spinks, J. (1991). Capacity views of human information processing. In: J.R. Jennings and M.G.H. Coles (Eds.), Handbook of cognitive psychophysiology: Central and autonomic nervous system approaches. New York: Wiley, pp. 179–249.

Kravkov, S.V. (1939). The influence of caffeine on the colour sensitivity. *Acta Ophthalomologica*, **17**:89–92.

Krein, S., Overton, S., Young, M., Spreier, K. and Yolton, R.L. (1987). Effects of alcohol on event-related brain potentials produced by viewing a simulated traffic light. *Journal of the American Ophtometric Association*, **58**:474–477.

Kril, J.J., Dodd, P.R., Gundlach, A.L., Davies, N., Watson, W.E.J., Johnston, G.A.R. and Harper, C.G. (1988). Necropsy study of GABA/benzodiazepine receptor binding sites in brain tissue from chronic alcoholic patients. *Clinical and Experimental Neurology*, **25**:35–141.

Krogh, H.J., Khan, M.A., Fosvig, L., Jensen, K. and Kellerup, P. (1978). N_1-P_2 component of the auditory evoked potential during alcohol intoxication and interaction of pyrithioxine in healthy adults. *Electroencephalography and Clinical Neurophysiology*, **44**:1–7.

Krull, K.R., Smith, L.T., Sinha, R. and Parsons, O.A. (1993). Simple reaction time event-related potentials: Effects of alcohol and sleep deprivation. *Alcoholism: Clinical and Experimental Research*, **17**:771–777.

Kucek, P. (1975). Effects of smoking on performance under load. *Studia Psychologica*, **17**:204–212.

Kunzendorf, R. and Wigner, L. (1985). Smoking and memory: State-specific effects. *Perceptual and Motor Skills*, **61**:558.

Kutas, M., McCarthy, G. and Donchin, E. (1977). Augmenting mental chronometry: The P300 as a measure of stimulus evaluation time. *Science*, **197**:792–795.

Kuznicki, J.T. and Turner, L.S. (1986). The effects of caffeine on caffeine users and non-users. *Physiology and Behavior*, **37**:397–408. *[42]*.

Lambert, G.H., Schoeller, D.A., Humphrey, H.E.B., Kotake, A.N., Lietz, H., Campbell, M., Kalow, W., Spielberg, S.P. and Budd, M. (1990). The caffeine breath test and caffeine urinary metabolite ratios in the Michigan cohort exposed to Polybrominated Biphenyls: a preliminary study. *Environmental Health Perspectives*, **89**:175–181.

Lamberty, G.J., Beckwith, B.E., Petros, T.V. and Ross, A.R. (1990). Posttrial treatment with ethanol enhances recall of prose narratives. *Physiology and Behavior*, **48**:653–658.

Lancet editorials. (1991). Nicotine use after the year 2000. *Lancet*, **337**:1191–1192.

Landauer, A.A. and Howat, P.A. (1982). Alcohol and the cognitive aspects of choice reaction time. *Psychopharmacology*, **78**:296–297.

Landauer, A.A., Milner, G. and Patman, J. (1969). Alcohol and amitriptyline effects on skills relating to driving behavior. *Science*, **163**:1467–1468.

Landauer, T.K. (1969). Reinforcement as consolidation. *Psychological Review*, **76**:82–96.

Landers, D.M., Crews, D.J., Boutcher, S.H., Skinner, J.S. and Gustafsen, S. (1992). The effects of smokeless tobacco on performance and psychophysiological response. *Medicine and Science in sports and Exercise*, **24**:895–903.

Landers, D.M., Lindholm, E., Crews, D.J. and Koriath, J.J. (1990). Cigarette smoking and smokeless tobacco facilitate information processing and performance. *Psychophysiology*, **27**:S47.

Landrum, R.E., Meliska, C.J. and Loke, W.H. (1988). Effects of caffeine and task experience on task performance. *Psychologia*, **31**:91–97. *[43]*.

Lane, J.D. and Rose, J.E. (1995). Effects of daily caffeine intake on smoking behavior in the natural environment. *Experimental and Clinical Psychopharmacology*, **3**:49–55.

Lang, P.J. (1979). A bio-informational theory of emotional imagery. *Psychophysiology*, **16**:495–512.

Lapp, W.M., Collins, R.L., Zywiak, W.H. and Izzo, C.V. (1994). Psychopharmacological effects of alcohol on time perception: The extended balanced placebo design. *Journal of Studies on Alcohol*, **55**:96–112.

Larson, P.S., Finnegan, J.K. and Haag, H.B. (1950). Observations on the effect of cigarette smoking on the fusion frequency of flicker. *Journal of Clinical Investigation*, **29**:483–485.

Latini, R., Bonati, M., Castelli, D. and Garattini, S. (1978). Dose-dependent kinetics of caffeine in rats. *Toxicology Letters*, **2**:267–270.

Latini, R., Bonati, M., Marzi, E. and Garattini, S. (1981). Urinary excretion of an uracilic metabolite from caffeine by rat, monkey and man. *Toxicology Letters*, **7**:267–272.

Lau, M.A., Pihl, R.O. and Peterson, J.B. (1995). Provocation, acute alcohol intoxication, cognitive performance and aggression. *Journal of Abnormal Psychology*, **104(1)**: 150–155.

Lazarus, R.S. (1991). Emotion and adaptation. New York: Oxford University Press.

Le Houezec, J. and Benowitz, N.L. (1991). Basic and clinical psychopharmacology of nicotine. *Clinics in Chest Medicine*, **12**:681–699.

Le Houezec, J., Carton, S. and Jouvent, R. (1996). Affective and cognitive effects of nicotine regarded as reinforcers of tobacco dependence. Society for Research on Nicotine and Tobacco. First annual conference. March 24–25, 1995, San Diego, California. *Addiction*, **91**:140.

Le Houezec, J., Fagerström, K.O. and Jouvent, R. (1995). Nicotine abstinence impairs cognitive performance in smokers. *Clinical Pharmacology and Therapeutics*, **57**:161.

Le Houezec, J., Halliday, R., Benowitz, N.L., Callaway, E., Naylor, H. and Herzig, K. (1994). A low dose of subcutaneous nicotine improves information processing in non-smokers. *Psychopharmacology*, **114**:628–634.

Le Houezec, J., Jacob, P. III and Benowitz, N.L. (1993). A clinical pharmacological study of subcutaneous nicotine. *European Journal of Clinical Pharmacology*, **44**:225–230.

Leathwood, P. and Pollet, P. (1983). Diet-induced mood changes in normal populations. *Journal of Psychiatric Research*, **17**:147–154.

Lee, B.L., Benowitz, N.L. and Jacob, P. III (1987). Influence of tobacco abstinence on the disposition kinetics and effects of nicotine. *Clinical Pharmacology and Therapeutics*, **41**:474–479.

Lee, B.L., Benowitz, N.L., Jarvik, M.E. and Jacob, P. III (1989). Food and nicotine metabolism. *Pharmacology, Biochemistry and Behavior*, **9**:621–625.

Lee, D.J. and Lowe, G. (1980). Interaction of alcohol and caffeine in a perceptual-motor task. *IRCS Medical Science*, **8**:420.

Lee, P.N. (1994). Smoking and Alzheimer's disease: A review of the epidemiological evidence. *Neuroepidemiology*, **13**:131–141.

Lehmann, D. and Knauss, T.A. (1976). Respiratory cycle and EEG in man and cat. *Electroencephalography and Clinical Neurophysiology*, **40**:187.

Lehmann, D. (1971). Multichannel topography of human alpha EEG fields. *Electroencephalography and Clinical Neurophysiology*, **31**:439–449.

Lehmann, W.D., Heinrich, H.C., Leonhardt, R., Agarwal, D.P., Goedde, H.W., Kneer, J. and Rating, D. (1986). 13C-Ethanol and 13C-Acetate breath test in normal and aldehyde dehydrogenase deficient individuals. *Alcohol*, **3**:227–231.

Lehtinen, I., Lang, A.H. and Keskinen, H. (1978). Acute effect of small doses of alcohol on the NSD parameter of human EEG. *Psychopharmacologia (Berl.)*, **37**:577–582.

Lehtinen, I., Nyrke, J., Lang, A., Pakkanen, A. and Keskinen, H. (1985). Individual alcohol reaction profiles. *Alcohol*, **2**:511–513.

Leigh, G. and Tong, J.E. (1976). Effects of ethanol and tobacco on time judgment. *Perceptual and Motor Skills*, **43**:899–903.

Leigh, G. (1982). The combined effects of alcohol consumption and cigarette smoking on critical flicker frequency. *Addictive Behaviors*, **7(3)**:251–259.

Leigh, G., Tong, J.E. and Campbell, J.A. (1977). Effects of ethanol and tobacco on divided attention. *Journal of Studies on Alcohol*, **38**:1233–1239.

Lelo, A., Kjellen, G., Birkett, D.J. and Miners, J.O. (1989). Paraxanthine metabolism in humans: determination of metabolic partial clearances and effects of allopurinol and cimetidine. *Journal of Pharmacology and Experimental Therapeutics*, **248**:315–319.

Lemmens, P.H.H.M. and Knibbe, R.A. (1993). Seasonal variation in survey and sales estimates of alcohol consumption. *Journal of Studies on Alcohol*, **54**:157–163.

Lemmens, P.H.H.M., Tan, E.S. and Knibbe, R.A. (1988). Bias due to non-response in a Dutch survey on alcohol consumption. *British Journal of Addiction*, **83**:1069–1077.

Lemmens, P.H.H.M., Tan, E.S. and Knibbe, R.A. (1992). Measuring quantity and frequency of drinking in a general population survey: a comparison of five indices. *Journal Studies on Alcohol*, **53**:476.

Lemon, J., Chesher, G., Fox, A., Greeley, J. and Nabke, C. (1993). Investigation of the 'hangover' effects of an acute dose of alcohol on psychomotor performance. *Alcoholism: Clinical and Experimental Research*, **17**:665–668.

Levin, E.D. (1992). Nicotinic systems and cognitive function. *Psychopharmacology*, **108**:417–431.

Levin, E.D. (1994). Nicotine effects and working memory performance. *Recent Advances in Tobacco Science*, **20**:49–66.

Levin, E.D., Briggs, S.J., Christopher, N.C. and Rose, J.E. (1993). Sertraline attenuates hyperphagia in rats following nicotine withdrawal. *Pharmacology, Biochemistry, and Behavior*, **44**:51–61.

Levin, E.D., Rose, J.E. and Behm, F. (1990). Development of a citric acid aerosol as a smoking cessation aid. *Drug and Alcohol Dependence*, **25**:273–279.

Levin, E.D., Rose, J.E., Behm, F. and Caskey, N.H. (1991). The effects of smoking-related sensory cues on psychological stress. *Pharmacology, Biochemistry and Behavior*, **39**:265–268.

Levine, J.M., Kramer, G.G. and Levine, E.N. (1975). Effects of alcohol on human performance: an integration of research findings on an abilities classification. *Journal of Applied Psychology*, **60**:285–293.

Levitsky, L.L., Schoeller, D.A., Lambert, G.H. and Edidin, D.V. (1989). Effect of growth hormone therapy in growth hormone-deficient children on cytochrome P-450-dependent 3-N-demethylation of caffeine as measured by the caffeine $^{13}CO_2$ breath test. *Development and Pharmacological Therapeutics*, **12**:90–95.

Levy, M., Granit, L. and Zylber-Katz, E. (1984). Chronopharmacokinetics of caffeine in healthy volunteers. *Annual Review of Chronopharmacology*, **1**:97–100.

Lewis, E.G., Dustman, R.E. and Beck, E.C. (1970). The effects of alcohol on visual and somato-sensory evoked responses. *Electroencephalography and clinical Neurophysiology*, **28**:202–205.

Lezak, M.D. (1976). Neuropsychological Assessment. New York: Oxford University Press.

Lezak, M.D. (1983). Neuropsychological Assessment, 2nd ed. New York: Oxford University Press.

Lieber, C.S. (1988). Biochemical and molecular basis of alcohol induced injury to liver and other tissues. *New England Journal of Medicine*, **319**:1639–1650.

Lieber, C.S. (1990). Interaction of ethanol with drugs, hepatotoxic agents, carcinogens and vitamins. *Alcohol and Alcoholism*, **25(2–3)**:157–176.

Lieber, C.S. (1991). Pathways of ethanol metabolism and related pathology. In: T.N. Palmer (Ed.), Alcoholism. A molecular perspective. New York: Plenum Press, pp. 1–26.

Lieberman, H.R. (1992). Caffeine. In: A.P. Smith and D.M. Jones (Eds.), Handbook of Human Performance. Vol. 2. Health and Performance. London: Academic Press, pp. 49–72.

Lieberman, H.R., Wurtman, R.J., Emde, G.C. and Coviella, I.L.G. (1987a). The effects of caffeine and aspirin on mood and performance. *Journal of Clinical Psychopharmacology*, **7**:315–320. *[44]*.

Lieberman, H.R., Wurtman, R.J., Emde, G.G., Roberts, C. and Coviella, I.L.G. (1987b). The effects of low doses of caffeine on human performance and mood. *Psychopharmacology*, **92**:308–312. *[45]*.

Linde, L. (1994). An auditory attention task: A note on the processing of verbal information. *Perceptual and Motor Skills*, **78**:563–570. *[46]*.

Linde, L. (1995). Mental effects of caffeine in fatigued and non-fatigued female and male subjects. *Ergonomics*, **38**:864–885.

Lindsley, D. (1960). Attention, consciousness, sleep and wakefulness. In: J. Field, H. Magoun and V. Hall (Eds.), Handbook of physiology. Section I, Neurophysiology. Vol. 3. Washington: American Physiological Society, pp. 136–195.

Linnoila, M. (1978). Psychomotor effects of drugs and alcohol on healthy volunteers and psychiatric patients. In: G. Olive (Ed.), Advances in pharmacology and therapeutics. Vol. 8. Oxford: Pergamon Press, pp. 235–249.

Lishman, W.A. (1987). Organic Psychiatry: The psychological consequences of cerebral disorder, 2nd ed. Oxford: Scientific Publications.

Lishman, W.A., Jacobson, R.R. and Acker, C. (1987). Brain damage in alcoholism: Current concepts. *Acta Medica Scandinavica*, **(Suppl. 717)**:5–17.

Lister, R.G. and File, S.E. (1984). The nature of lorazepam-induced amnesia. *Psychopharmacology*, **83**:183–187.

Lister, R.G., Eckardt, M.J. and Weingartner, H. (1987). Ethanol intoxication and memory: Recent developments and new directions. In: M. Galanter (Ed.), Recent developments in alcoholism. Vol. 5. New York: Plenum, pp. 111–126.

Lister, R.G., Gorenstein, C., Risher-Flowers, D., Weingartner, H.J. and Eckardt, M.J. (1991). Dissociation of the acute effects of alcohol on implicit and explicit memory processes. *Neuropsychologia*, **29**:1205–1212.

Loberg, T. (1986). Neuropsychological findings in the early and middle phases of alcoholism. In: I. Grant and K.M. Adams (Eds.), Neuropsychological assessment of neuropsychiatric disorders. New York: Oxford University Press, pp. 415–440.

Loke, W.H. and Meliska, C.J. (1984). Effects of caffeine use and ingestion on a protracted visual vigilance task. *Psychopharmacology*, **84**:54–57. *[53]*.

Loke, W.H. (1988). Effects of caffeine on mood and memory. *Physiology and Behavior*, **44**:367–372. *[47]*.

Loke, W.H. (1989). Effects of caffeine on task difficulty. *Psychology Belgium*, **29(1)**:51–62. *[48]*.

Loke, W.H. (1990). Effects of repeated caffeine administration on cognition and mood. *Human Psychopharmacology*, **5**:339–348. *[49]*.

Loke, W.H. (1992). Physiological and psychological effects of alcohol. *Psychologia An International Journal of Psychology in the Orient*, **35**:133–146.

Loke, W.H. (1992). The effects of caffeine and automaticity on a visual information processing task. *Human Psychopharmacology*, **7**:379–388. *[50]*.

Loke, W.H. (1993). Caffeine and automaticity in encoding prelexical tasks: Theory and some data. *Human Psychopharmacology*, **8**:77–95. [51].

Loke, W.H., Hinrichs, J.V. and Ghoneim, M.M. (1985). Caffeine and diazepam: Separate and combined effects on mood, memory, and psychomotor performance. *Psychopharmacology*, **87**:344–350. *[52]*.

Looren de Jong, H. (1989). Cognitive aging. An event-related potential study. Ph.D. Thesis Enschede: Quick Service, pp. 53–72.

Looren de Jong, H., Kok, A. and Van Rooy, J.C.G.M. (1988). Early and late selection in young and old adults: An event-related potential study. *Psychophysiology*, **25(6)**:657–671.

Lorig, T.S. and Schwartz, G.E. (1988). Brain and odor: I. Alteration of human EEG by odor administration. *Psychobiology*, **16(3)**:281–284.

Lorig, T.S. (1989). Human EEG and odor responses. *Progressions in Neurobiology*, **33**:387–398.

Lorig, T.S., Huffman, E., DeMartino, A. and DeMarco, J. (1991). The effects of low concentration odors on EEG activity and behavior. *Journal of Psychobiology*, **5**:69–77.

Lorig, T.S., Schwartz, G.E., Herman, K.B. and Lane, R.D. (1988). Brain and odor: II. EEG activity during nose and mouth breathing. *Psychobiology*, **16(3)**:285–287.

Lorig. T.S., Herman, K.B., Schwartz, G.E. and Cain, W.S. (1990). *Bulletin of the Psychonomic Society*, **28(5)**:40–58.

Lorist, M.M. (1995). Caffeine and human information processing. Ph.D. Thesis, Faculty of Psychology, University of Amsterdam.

Lorist, M.M., Snel, J. and Kok, A. (1994a). Influence of caffeine on information processing stages in well rested and fatigued subjects. *Psychopharmacology*, **113**:411–421. *[54]*.

Lorist, M.M., Snel, J., Kok, A. and Mulder, G. (1994b). Influence of caffeine on selective attention in well-rested and fatigued subjects. *Psychophysiology*, **31**:525–534. *[55]*.

Lorist, M.M., Snel, J., Kok, A. and Mulder, G. (1996). Acute effects of caffeine on selective attention and visual search processes. *Psychophysiology*, **33**:354–361.

Lorist, M.M., Snel, J., Mulder, G. and Kok, A. (1995). Aging, caffeine, and information processing: An event-related potential analysis. *Electroencephalography and Clinical Neurophysiology*, **96**:453–467.

Lovibond, S.H. and Bird, K.D. (1971). Danger level – the Warwick Farm experience. Proceedings of the International Congress on Alcoholism and Drug Dependence, Sydney, Australia.

Lovinger, D.M. (1993). Excitotoxicity and alcohol-related brain damage. *Alcoholism: Clinical and Experimental Research*, **17**:19–27.

Lowe, G. (1981). The interaction of alcohol and caffeine – some behavioral effects. *Bulletin of Behavioral Psychology and Sociology*, **34**:189.

Lucas, D., Ménez, J.-F., Berthou, F., Cauvin, J.-M. and Deitrich, R.A. (1992). Differences in hepatic microsomal Cytochrome P-450 isoenzyme induction by Pyrazole, chronic ethanol, 3-Methylcholanthrene, and Phenobarbital in high alcohol sensitivity (HAS) and low alcohol sensitivity (LAS) rats. *Alcoholism, Clinical and Experimental Research*, **16**:916–921.

Luck, S.J. and Hillyard, S.A. (1994). Electrophysiological correlates of feature analysis during visual search. *Psychophysiology*, **31**:291–308.

Luck, S.J., Heinze, H.J., Mangun, G.R. and Hillyard, S.A. (1990). Visual event-related potentials index focused attention within bilateral stimulus arrays: II. Functional dissociation of P1 and N1 components. *Electroencephalography and Clinical Neurophysiology*, **75**:528–542.

Lukas, S.E., Mendelson, J.H., Benedict, R.A. and Jones, B. (1986). EEG alpha activity increases during transient episodes of ethanol-induced euphoria. *Pharmacology, Biochemistry and Behavior*, **25**:889–895.

Lukas, S.E., Mendelson, J.H., Kouri, E., Bolduc, M. and Amass, L. (1990). Ethanol-induced alterations in EEG alpha activity and apparent source of the auditory P300 evoked response potential. *Alcohol*, **7**:471–477.

Lukas, S.E., Mendelson, J.H., Woods, B.T., Mello, N.K. and Teoh, S.K. (1989). Topographic distribution of EEG alpha activity during ethanol-induced intoxication in women. *Journal of Studies on Alcohol*, **50(2)**:176–185.

Lyon, R.J., Tong, J.E., Leigh, G. and Clare, G. (1975). The influence of alcohol and tobacco on the components of choice reaction time. *Journal of Studies on Alcohol*, **36**:587–596.

MacAndrew, C. and Edgerton, R.B. (1969). Drunken Comportment: A Social Explanation. Chicago: Aldine Publishing Co. Inc.

Mackworth, J.F. (1969). Vigilance and habituation. Middlesex: Penguin.

Mackworth, N.H. (1950). Research on the measurement of human performance. Medical Research Council special report No. 268. London: HMSO.

MacVane, J., Butters, N., Montgomery, K. and Farber, J. (1982). Cognitive functioning in men social drinkers: A replication study. *Journal of Studies on Alcohol*, **43**:81–95.

Maletzky, B.M. and Klotter, J. (1974). Smoking and alcoholism. *American Journal of Psychiatry*, **131(4)**:445–447.

Mangan, G.L. and Colrain, I.M. (1991). Relationships between photic driving, nicotine, and memory. In: F. Adlkofer and K. Thurau (Eds.), Effects of nicotine on biological systems. Basel: Birkhäuser Verlag, pp. 537–546.

Mangan, G.L. and Golding, J.F. (1978). An 'enhancement' model of smoking maintenance? In: R.E. Thornton (Ed.), Smoking behaviour. Physiological and psychological influences. New York: Churchill Livingstone, pp. 87–114.

Mangan, G.L. and Golding, J.F. (1983). The effects of smoking on memory consolidation. *Journal of Psychology*, **115**:65–77.

Mangan, G.L. (1982). The effects of cigarette smoking on vigilance performance. *The Journal of General Psychology*, **106**:77–83.

Mangan, G.L. (1983). The effects on cigarette smoking on verbal learning and retention. *Journal of General Psychology*, **108**:203–210.

Mangun, G.R. (1995). Neural mechanisms of visual selective attention. *Psychophysiology*, **32**:4–18.

Mangun, G.R., Hillyard, S.A. and Luck, S.J. (1993). Electrocortical substrates of visual selective attention. In: D. Meyer and S. Kornblum (Eds.), Attention and performance. Vol. 14. Cambridge: MIT Press, pp. 219–243.

Mann, R.E., Cho-Young, J. and Vogel-Sprott, M. (1984). Retrograde enhancement by alcohol of delayed free recall performance. *Pharmacology, Biochemistry and Behavior*, **20**:639–642.

Maritz, G.M. (1991). Position estimation of a moving object: An inquiry into age-related differences. Internal Report, Amsterdam: Faculty of Psychology, University of Amsterdam.

Marshall, E.J. and Murray, R.M. (1991). The familial transmission of alcoholism. *British Medical Journal*, **303**:72–73.

Marshall, W.R., Epstein, L.H. and Green, S.B. (1980). Coffee drinking and cigarette smoking: I. Coffee, caffeine and cigarette smoking behavior. *Addictive Behaviors*, **5**:389–394.

Marshall, W.R., Green, S.B., Epstein, L.H., Rogers, C.M. and McCoy, J.F. (1980). Coffee drinking and cigarette smoking: II. coffee, urinary ph and cigarette smoking behavior. *Addictive Behaviors*, **5**:395–400.

Martin, C.S., Clifford, P.R., Earleywine, M., Maisto, S.A. and Longabaugh, R.M. (1994). Polydrug use in alcoholics. *Alcoholism: Clinical and Experimental Research*, **18**:492. (Abstract No. 435).

Martin, E., Moll, W., Schmid, P. and Dettli, L. (1984). The pharmacokinetics of alcohol in human breath, venous and arterial blood after oral ingestion. *European Journal of Clinical Pharmacology*, **26**:619–626.

Martin, J.K. (1990). Jobs, occupations and patterns of alcohol consumption: A review of the literature. In: P.M. Roman (Ed.), Alcohol problem intervention in the workplace: Employee assistance programs and strategic alternatives. New York: Quorum Books, pp. 45–65.

Martin, N.G., Oakeshott, J.G., Gibson, J.B., Starmer, G.A., Perl, J. and Wilks, A.V. (1985). A twin study of psychomotor and physiological responses to an acute dose of alcohol. *Behavioral Genetics*, **15**:305–347.

Martin, N.G., Perl, J., Oekeshott, J.G., Gibson, J.B., Starmer, G.A. and Wilks, A.V. (1985). A twin study of ethanol metabolism. *Behavioral Genetics*, **15**:93–109.

Mascord, D., Smith, J., Starmer, G.A. and Whitfield, J.B. (1991). The effect of fructose on alcohol metabolism and on the (lactate)/(pyruvate) ratio in man. *Alcohol and Alcoholism*, **26**:53–59.

Masson, C.L. and Gilbert, D.G. (1990). Cardiovascular responses to a quantified dose of nicotine as a function of personality and nicotine tolerance. *Journal of Behavioral Medicine*, **13**:505–521.

Mattila, M., Seppala, T. and Mattila, M.J. (1988). Anxiogenic effect of yohimbine in healthy subjects: Comparison with caffeine and antagonism by clonidine and diazepam. *International Clinical Psychopharmacology*, **3**:215–229. *[56]*.

Maxwell, S., Cruischank, A. and Thorpe, G. (1994). Red wine and antioxidant activity in serum. *Lancet*, **344**:193–194.

Maxwell, S.E. and Delaney, H.D. (1990). Designing experiments and analyzing data. A model comparison perspective. Belmont: Wadsworth.

May, D.C., Jarboe, C.H., Van Bakel, A.B. and Williams, W.M. (1982). Effects of cimetidine on caffeine disposition in smokers and nonsmokers. *Clinical Pharmacology and Therapeutics*, **31**:656–661.

Maylor, E.A. and Rabbitt, P.M.A. (1987a). Effects of practice and alcohol on performance of a perceptual motor task. *Quarterly Journal of Experimental Psychology*, **39A**:777–795.

Maylor, E.A. and Rabbitt, P.M.A. (1987b). Effect of alcohol on rate of forgetting. *Psychopharmacology*, **91**:230–235.

Maylor, E.A. and Rabbitt, P.M.A. (1988). Amount of practice and degree of attentional control have no influence on the adverse effect of alcohol in word categorization and visual search tasks. *Perception and Psychophysics*, **44**:117–126.

Maylor, E.A. and Rabbitt, P.M.A. (1993). Alcohol, reaction time and memory: A meta-analysis. *British Journal of Psychology*, **84**:301–317.

Maylor, E.A., Rabbitt, P.M.A. and Kingstone, A. (1987). Effects of alcohol on word categorization and recognition memory. *British Journal of Psychology*, **78**:233–239.

Maylor, E.A., Rabbitt, P.M.A., James, G.H. and Kerr, S.A. (1990). Comparing the effects of alcohol and intelligence on text recall and recognition. *British Journal of Psychology*, **81**:299–313.

Maylor, E.A., Rabbitt, P.M.A. and Connolly, S.A.V. (1989). Rate of processing judgement of response speed: comparing the effects of alcohol and practice. *Perception and Psychophysics*, **45**:431–438.

Maylor, E.A., Rabbitt, P.M.A., James, G.H. and Kerr, S.A. (1992). Effects of alcohol, practice, and task complexity on reaction time distributions. *Quarterly Journal of Experimental Psychology*, **44A**:119–139.

Maylor, E.A., Rabbitt, P.M.A., Sahgal, A. and Wright, C. (1987). Effects of alcohol on speed and accuracy in choice reaction time and visual search. *Acta Psychological*, **65**:147–163.

McCallum, W.C. (1988). Potentials related to expectancy, preparation and motor activity. In: T.W. Picton (Ed.), Handbook of electroencephalography and clinical neurophysiology. Revised series. Vol. 3. Human event-related potentials. Amsterdam: Elsevier, pp. 427–534.

McCarthy, G. and Donchin, E. (1981). A metric for thought: A comparison of P300 latency and reaction time. *Science*, **211**:77–80.

McClelland, G.R. (1987). The effects of practice on measures of performance. *Human Psychopharmacology*, **2**:109–118.

McCoy, G.D. and Napier, K. (1986). Alcohol and tobacco consumption as risk factors for cancer. *Alcohol Health and Research World*, **10**:28–33.

McGaugh, J.L. (1989). Involvement of hormonal and neuromodulatory systems in the regulation of memory storage. *Annual Review of Neuroscience*, **12**:255–287.

McGehee, D.S., Heath, M.J.S., Gelber, S., Devay, P. and Role, L.W. (1995). Nicotine enhancement of fast excitatory synaptic transmission in CNS by presynaptic receptors. *Science*, **269**:1692–1696.

McGue, M., Pickens, R.W. and Svikis, D.S. (1992). Sex and age effects on the inheritance of alcohol problems: a twin study. *Journal of Abnormal Psychology*, **101**:3–16.

McKim, W.A. (1991). Drugs and Behavior. An Introduction to Behavioral Pharmacology, 2nd ed., Englewood Cliffs: Prentice Hall.

McMillen, D.L. and Wells-Parker, E. (1987). The effect of alcohol consumption on risk taking while driving. *Addictive Behaviours*, **12**:241–247.

McMillen, D.L., Smith, S.M. and Wells-Parker, E. (1989). The effect of alcohol, expectancy and sensation seeking on driving risk taking. *Addictive Behaviours*, **14**:477–483.

McMullen, P.A., Saint-Cyr, J. and Carlen, P.L. (1984). Morphological alterations in rat CA1 hippocampal pyramidal cell dendrites resulting from chronic ethanol consumption and withdrawal. *The Journal of Comparative Neurology*, **225**:111–118.

McNair, D.M. (1973). Antianxiety drugs and human performance. *Archives of General Psychiatry*, **29**:611–617.

McNeill, A.D., Jarvis, M. and West, R. (1987). Subjective effects of cigarette smoking in adolescents. *Psychopharmacology*, **92**:115–117.

McQuilkin, S.H., Nierenberg, D.W. and Bresnick, E. (1995). Analysis of within-subject variation of caffeine metabolism when used to determine Cytochrome P4501A2 and N-acetyltransferase-2 activities. *Cancer Epidemiology, Biomarkers and Prevention*, **4**:139–146.

McRandle, C. and Goldstein, R. (1973). Effect of alcohol on the early and late components of the averaged electroencephalographic response to clicks. *Journal of Speech and Hearing Research*, **16**:353–359.

Meier-Tackmann, D., Leonhardt, R.A., Agarwal, D.P. and Goedde, H.W. (1990). Effect of acute ethanol drinking on alcohol metabolism in subjects with different ADH and ALDH genotypes. *Alcohol*, **7**:413–418.

Mela, D.J. (1989). Caffeine ingested under natural conditions does not alter taste intensity. *Pharmacology, Biochemistry and Behavior*, **34(3)**:483–485. *[57]*.

Melia, K.F., Ehlers, C.L., LeBrun, C.J. and Koob, G.F. (1986). Post-learning ethanol effects on a water-finding task in rats. *Pharmacology, Biochemistry and Behavior*, **24**:1813–1815.

Meliska, C.J. and Gilbert, D.G. (1991). Hormonal and subjective effects of cigarette smoking in males and females. *Pharmacology, Biochemistry and Behavior*, **40**:229–235.

Mello, N.K., Mendelson, J.H. and Palmieri, S.L. (1987). Cigarette smoking by women: interactions with alcohol use. *Psychopharmacology*, **93**:8–15.

Mello, N.K., Mendelson, J.H., Sellers, M.L. and Kuehnle, J.C. (1980). Effect of alcohol and marihuana on tobacco smoking. *Clinical Pharmacology and Therapeutics*, **27**:202–209.

Mendelson, J.H., Woods, B.T., Chiu, T.M., Mello, N.K., Lukas, S.E., Teoh, S.K., Sintravanarong, P., Cochin, J., Hopkins, M. and Dobrosielski, M. (1990). *In vivo* proton magnetic resonance spectroscopy of alcohol in human brain. *Alcohol*, **7**:443–447.

Merikangas, K.R. (1990). The genetic epidemiology of alcoholism. *Psychology and Medicine*, **20**:11–22.

Miceli, G., Caltagirone, C., Gainotti, G., Masullo, C. and Silveri, M.C. (1981). Neuropsychological correlates of localized cerebral lesions in nonaphasic brain damaged patients. *Journal of Clinical Neuropsychology*, **3**:53–63.

Michel, Ch. and Bättig, K. (1988). Separate and joint effects of cigarette smoking and alcohol consumption on mental performance and physiological functions. *Activitas Nervosa Superior*, **30**:107–109.

Michel, Ch. and Bättig, K. (1989). Separate and combined psychophysiological effects of cigarette smoking and alcohol consumption. *Psychopharmacology*, **97**:65–73.

Michel, Ch., Hasenfratz, M., Nil, R. and Bättig, K. (1988). Cardiovascular, electrocortical, and behavioural effects of nicotine chewing gum. *Klinische Wochenschrift*, **66(Suppl. XI)**:72–79.

Michel, Ch., Nil, R., Buzzi, R., Woodson, P.P. and Bättig, K. (1987). Rapid information processing and concomitant event-related brain potentials in smokers differing in CO absorption. *Neuropsychobiology*, **17**:161–168.

Miles, C., Porter, K. and Jones, D.M. (1986). The interactive effects of alcohol and mood on dual task performance. *Psychopharmacology*, **89**:432–435.

Millar, K., Hammersley, R.H. and Finnigan, F. (1992). Reduction of alcohol-induced performance impairment by prior ingestion of food. *British Journal of Psychology*, **83**:261–278.

Miller, E. (1992). Effect of moderate alcohol intake on cognitive functioning: Does a little bit of what you fancy do you good? *Journal of Mental Health (UK)*, **1**:19–24.

Mills, K.C. and Bisgrove, E.Z. (1983). Body sway and divided attention performance under the influence of alcohol: Dose-response differences between males and females. *Alcoholism, Clinical and Experimental Research*, **7**:393–397.

Mintz, J., Boyd, G., Rose, J.E., Charuvastra, V.C. and Jarvik, M.E. (1985). Alcohol increases cigarette smoking: a laboratory demonstration. *Addictive Behaviors*, **10**:203–207.

Mintz, J., Phipps, C.C., Arruda, M.J., Glynn, S.M., Schneider, N.G. and Jarvik, M.E. (1991). Combined use of alcohol and nicotine gum. *Addictive Behaviors*, **16(1–2)**:1–10.

Mishara, B.L. and Kastenbaus, R. (1980). Alcohol and the elderly: Five thousand years of uncontrolled experimentation. In: Alcohol and Old Age. Seminars in Psychiatry. New York: Grune and Stratton.

Mishara, B.L., Kastenbaus, R. and Baker, F. (1975). Alcohol effects in old age: An experimental investigation. *Social Science and Medicine*, **9(10)**:535–547.

Mishra, L., Sharma, S., Potter, J.J. and Mejey, E. (1989). More rapid elimination of alcohol in women as compared to their male siblings. *Alcoholism, Clinical and Experimental Research*, **13**:752–754.

Mitchell, M.C. (1985). Alcohol induced impairment of central nervous system function: Behavioral skills involved in driving. *Journal of Studies on Alcohol*, **10 (Suppl.)**:109–116.

Mitchell, M.C., Hoyumpa, A.M., Schenker, S., Johnson, R.F., Nichols, S. and Patwardhan, R.V. (1983). Inhibition of caffeine elimination by short-term ethanol administration. *Journal of Laboratory and Clinical Medicine*, **101**:826–834.

Mitchell, P.J. and Redman, J.R. (1992). Effects of caffeine, time of day and user history on study-related performance. *Psychopharmacology*, **109**:121–126. *[58]*.

Mitchell, V.E., Ross, S. and Hurst, P.M. (1974). Drugs and placebos: effects of caffeine on cognitive performance. *Psychological Reports*, **35**:875–883.

Miyao, M. and Ishikawa, H. (1994). Effect of a low dose of alcohol on dynamic visual acuity. *Perceptual and Motor Skills*, **78**:963–967.

Mizoi, Y., Yamamoto, K., Ueno, Y., Fukunaga, T. and Harada, S. (1994). Involvement of genetic polymorphism of alcohol and aldehyde dehydrogenases in individual variation of alcohol metabolism. *Alcohol and Alcoholism*, **29**:707–710.

Modell, J.G. and Mountz, J.M. (1990). Drinking and flying. The problem of alcohol use by pilots. *New England Journal of Medicine*, **323**:455–461.

Mongrain, S. and Standing, L. (1989). Impairment of cognition, risk-taking and self-perception by alcohol. *Perceptual and Motor Skills*, **69**:199–210.

Morgan, S.F. and Pickens, R.W. (1982). Reaction time performance as a function of cigarette smoking procedure. *Psychopharmacology*, **77**:383–386.

Morgenson, G.J. (1987). Limbic-motor integration. In: A.N. Epstein and A.R. Morrison (Eds.), Progress in psychobiology and physiological psychology. Vol. 12. Orlando: Academic Press, pp. 117–170.

Morisot, C., Simoens, C., Trublin, F., Lhermitte, M., Gremillet, C., Robert, M.H. and Lequin, P. (1990). Efficacité de la caffeine transcutanée dans le traitement des apnées du prématuré. *Archives Françaises de Pediatrie*, **47**:221–224.

Morley, B.J., Farley, G.R. and Javel, E. (1983). Nicotinic acetylcholine receptors in mammalian brain. *Trends in Pharmacological Sciences*, **4**:225–227.

Morley, J.E., Levine, A.S. and Rowland, N.E. (1983). Stress induced eating. *Life Sciences*, **32**:2169–2182.

Morrisett, R.A. and Swartzwelder, H.S. (1993). Attenuation of hippocampal long-term potentiation by ethanol: a patch-clamp analysis of glutamatergic and GABA-ergic mechanisms. *Journal of Neuroscience*, **13**:2264–2272.

Moskowitz, H. and Robinson, C.D. (1988). Effects of low doses of alcohol on driving-related skills: A review of the evidence. Technical Report DOT HS 807 280. Washington, DC: US Department of Transportation, National Highway Traffic Safety Administration.

Moskowitz, H. (1973). Laboratory studies of the effects of alcohol on some variables related to driving. *Journal of Safety Research*, **5**:185–199.

Moskowitz, H., Burns, M. and Williams, A.F. (1985). Skilled performance at low blood alcohol levels. *Journal of Studies on Alcohol*, **46**:482–485.

Motokizawa, F. and Furuya, N. (1973). Neural pathway associated with the EEG arousal response by olfactory stimulation. *Electroencephalography and Clinical Neurophysiology*, **35**:83–91.

Motokizawa, F. (1974). Electrophysiological studies of olfactory projection to the mesencephalic reticular formation. *Experimental Neurology*, **44(2)**:135–144.

Mueller, C.W., Lisman, S.A. and Spear, N.E. (1983). Alcohol enhancement of human memory: Tests of consolidation and interference hypotheses. *Psychopharmacology*, **80**:226–230.

Mulder, G. and Wijers, A.A. (1991). Selective attention and mental chronometry. In: C.H.M. Brunia, G. Mulder and M.N. Verbaten (Eds.), Event related Brain Research, EEG Suppl. 42. New York, Amsterdam: Elsevier Science, pp. 228–243.

Mulder, G. (1986). The concept and measurement of mental effort. In: G.R.J. Hockey, A.W.K. Gaillard and M.G.H. Coles (Eds.), Energetics and human information processing. Dordrecht, The Netherlands: Nijhoff, pp. 175–198.

Mulder, G., Wijers, A.A., Brookhuis, K.A., Smid, H.G.O.M. and Mulder, L.J.M. (1994). Selective visual attention: Selective cuing, selective cognitive processing, and selective response processing. In: H.-J. Heinze, T.-F. Münte and G.R. Mangun (Eds.), Cognitive electrophysiology. Boston: Birkhäuser, pp. 26–80.

Müller, W. and Haase, E. (1967). Das Verhalten der corticalen Antwort unter Alkoholeinwirkung. *Albrecht von Graefes Archiv für klinische und experimentelle Ophthalmologie*, **173**:108–113.

Münte, T.-F., Heinze, H.-J., Künkel, H. and Scholz, M. (1984). Personality traits influence the effects of Diazepam and caffeine on CNV magnitude. *Neuropsychobiology*, **12**:60–67.

Murdoch, D., Pihl, R.O. and Ross, D. (1990). Alcohol and crimes of violence: Present issues. *International Journal of Addiction*, **25**:1059–1075.

Murphy, T.L., McIvor, G., Yap, A., Cooksley, W.G.E., Halliday, J.W. and Powell, L.W. (1988). The effect of smoking on caffeine elimination: implication for its use as a semiquantitative test of liver function. *Clinical and Experimental Pharmacology and Physiology*, **15**:9–13.

Murray, A.L. and Lawrence, P.S. (1984). Sequelae to smoking cessation: A review. *Clinical Psychology Review*, **4**:143–157.

Murray, S.S., Bjelke, E., Gibson, R.W. and Schumen, L.M. (1981). Coffee consumption and mortality from ischemic heart disease and other causes: Results from the Lutheran Brotherhood Study. *American Journal of Epidemiology*, **113**:661–667.

Mutzell, S. and Tibblin, G. (1989). High alcohol consumption, liver toxic drugs and brain damage-a population study. *Upsala Journal of Medical Science*, **94**:305–315.

Myrsten, A.-L. and Andersson, K. (1973). Interaction between effects of alcohol intake and cigarette smoking. Reports Psychological Laboratory University of Stockholm Nr. 402.

Myrsten, A.-L., Andersson, K., Frankenhaeuser, M. and Elgerot, A. (1975). Immediate effects of cigarette smoking as related to different smoking habits. *Perceptual and Motor Skills*, **40**:515–523.

Myrsten, A.-L., Elgerot, A. and Edgren, B. (1977). Effects of abstinence from tobacco smoking on physiological and psychological arousal levels in habitual smokers. *Psychosomatic Medicine*, **39**:25–38.

Myrsten, A.-L., Post, B., Frankenhaeuser, M. and Johansson, G. (1972). Changes in behavioural and physiological activation induced by cigarette smoking in habitual smokers. *Psychopharmacologia*, **27**:305–312.

Näätänen, R. and Gaillard, A.W.K. (1983). The orienting reflex and the N_2 deflection of the event-related potential (ERP). In: A.W.K. Gaillard and W. Ritter (Eds.), Tutorials in ERP research: Endogenous components. Amsterdam: North-Holland, pp. 119–141.

Näätänen, R. and Picton, T. (1987). The N_1 wave of the human electric and magnetic response to sound: A review and an analysis of the component structure. *Psychophysiology*, **24(4)**:375–425.

Näätänen, R. and Picton, T.W. (1986). N_2 and automatic versus controlled processes. In: W.C. McCallum, R. Zappoli and F. Denoth (Eds.), Cerebral Psychophysiology: Studies in Event Related Potentials, EEG, (Suppl. 38):169–178. Amsterdam: Elsevier.

Näätänen, R. (1992). Attention and brain function. Hillsdale, New Jersey: Lawrence Erlbaum.

Nagel, R.A., Dirix, L.Y., Hayllar, K.M., Preisig, R., Tredger, J.M. and Williams, R. (1990). Use of quantitative liver function tests-caffeine clearance and galactose elimination capacity-after orthotopic liver transplantation. *Journal of Hepatology*, **10**:149–157.

Nash, H. (1962). Alcohol and caffeine. Springfield: Thomas. *[59]*.

Naylor, G.F.K. and Harwood, E. (1972). Naylor-Harwood Adult Intelligence Scale. Victoria: Australian Council for Educational Research.

Naylor, H., Halliday, R. and Callaway, E. (1985). The effects of methylphenidate on human information processing. *Psychopharmacology*, **86**:90–95.

Nehlig, A. and Debry, G. (1993). Coffee, caffeine and the coupling between cerebral blood flow and energy metabolism. ASIC, 15ᵉ Colloque, Montpellier, France.

Neill, R.A, Delahunty, A.M. and Fenelon, B. (1990). Discrimination of motion in depth trajectory following acute alcohol ingestion. *Biological Psychology*, **31**:1–22.

Nelson, T.O., McSpadden, M., Fromme, K. and Marlatt, G.A. (1986). Effects of alcohol intoxication on meta-memory and on retrieval from long-term memory. *Journal of Experimental Psychology: General*, **115**:247–254.

Netter, P., Vogel, W. and Rammsayer, T. (1994). Extraversion as a modifying factor in catecholamine and behavioral responses to ethanol. *Psychopharmacology*, **115**:206–212.

Neve, R.J., Diederiks, J.P.M., Knibbe, R.A. and Drop, M.J. (1993). Developments in drinking behavior in the Netherlands from 1958 to 1989, a cohort analysis. *Addiction*, **88**:611–621.

Neville, H.J. and Schmidt, A. (1985). Event-related brain potentials in subjects at risk for alcoholism. In: N. Chang and H. Chao (Eds.), Early identification of alcohol abuse (Research monograph 17) Rockville, MD: National Institute on Alcohol Abuse and Alcoholism, pp. 228–239.

Newman, F., Stein, M.B., Trettau, J.R., Coppola, R. and Uhde, T.W. (1992). Quantitative electroencephalographic effects of caffeine in panic disorder. *Psychiatry Research: Neuroimaging*, **45**:105–113.

Newman, H.W. and Newman, E.J. (1956). Failure of dextrine and caffeine as practical antagonists of the depressant effect of ethyl alcohol in man. *Quarterly Journal of Studies on Alcohol*, **17**:267–298.

Newton, R., Broughton, L.J., Lind, M.J., Morrisson, P.J., Rogers, H.J. and Bradbrook, I.D. (1981). Plasma and salivary pharmacokinetics of caffeine in man. *European Journal of Clinical Pharmacology*, **21**:45–52.

Ney, T., Gale, A. and Morris, H. (1989). A critical evaluation of laboratory studies of the effects of smoking on learning and memory. In: T. Ney and A. Gale (Eds.), Smoking and human behaviour. Chichester: Wiley, pp. 239–259.

Niaura, R.S., Nathan, P. E., Frankenstein, W., Shapiro, A.P. and Brick, J. (1987). Gender differences in acute psychomotor, cognitive, and pharmacokinetic response to alcohol. *Addictive Behaviors*, **12**:345–356.

Nichols, J.M. and Martin, F. (1993). P300 in heavy social drinkers: the effect of lorazepam. *Alcohol*, **10(4)**:269–274.

Nichols, J.M., Martin, F. and Kirkby, K.C. (1993). A comparison of the effect of lorazepam on memory in heavy and low social drinkers. *Psychopharmacology*, **112**:475–482.

Nicholson, A.N., Stone, B.M. and Jones, S.J. (1984). Studies on the possible central effects in man of a neuropeptide (ACTH 4–9 Analogue). *European Journal of Clinical Pharmacology*, **27**:561–565. *[60]*.

Nicholson, M.E., Wang, M.Q., Airhihenbuwa, C.O., Mahoney, B.S. and Maney, D.W. (1992). Predicting alcohol impairment: Perceived intoxication versus BAC. *Alcoholism: Clinical and Experimental Research*, **16**:747–750.

Nil, R. and Bättig, K. (1989a). Separate effects of cigarette smoke yield and smoke taste on smoking behavior. *Psychopharmacology*, **99**:54–59.

Nil, R. and Bättig, K. (1989b). Smoking behavior: a multivariate process. In: T. Ney and A. Gale (Eds.), Smoking and human behavior. Chichester: Wiley.

Nil, R., Buzzi, R. and Bättig, K. (1984). Effects of single doses of alcohol and caffeine on cigarette smoke puffing behavior. *Pharmacology, Biochemistry and Behavior*, **20**:583–590.

Nil, R., Woodson, P.P., Michel, Ch. and Bättig, K. (1988). Effects of smoking on mental performance and vegetative functions in high and low CO absorbing smokers. *Klinische Wochenschrift*, **66(Suppl. XI)**:66–71.

Nilsson, L.G., Backman, L. and Karlsson, T. (1989). Priming and cued recall in elderly, alcohol intoxicated and sleep deprived subjects: A case of functionally similar memory deficits. *Psychological Medicine*, **19**:423–433.

Noble, E.P. (1983). Social drinking and cognitive function: A review. *Substance and Alcohol Actions/Misuse*, **4**:205–216.

Noldy, N.E. and Carlen, P.L. (1990). Acute, withdrawal, and chronic alcohol effects in man: Event-related potential and quantitative EEG techniques. *Annals of Medicine*, **22**:333–339.

Noldy, N.E., Politzer, N. and Carlen, P.L. (1994b). Alcohol and the aging brain: Age-related differences in electrophysiology and behaviour associated with alcohol intoxication. *Psychophysiology*, **31(Suppl. 1)**:S73.

Noldy, N.E., Santos, C.V., Politzer, N., Blair, R.D.G. and Carlen, P.L. (1994a). Quantitative EEG changes in cocaine withdrawal: Evidence for long-term CNS effects. *Neuropsychobiology*, **30**:189–196.

Noonberg, A., Goldstein, G. and Page, H.A. (1985). Premature aging in male alcoholics: "Accelerated aging" or increased vulnerability. *Alcoholism: Clinical and Exp. Research*, **9**:334–338.

Nordberg, A. (1994). Human nicotinic receptors – their role in aging and dementia. *Neurochemistry International*, **25**:93–97.

Norman, D.A. and Shallice, T. (1980). Attention to action: willed and automatic control of behavior. CHIP Report 99, University of California, San Diego.

Norman, D.A. and Shallice, T. (1986). Attention to action: willed and automatic control of behaviour. In: R.J. Davidson, G.E. Schwartz and D. Shapiro (Eds.), Consciousness and Self regulation. Vol. 4. New York: Plenum Press, pp. 1–18.

Norman, D.A. (1986). Reflections on cognition and parallel distributed processing. In: J.L. McClelland and D.E. Rumelhart (Eds.), Parallel distributed processing. Vol. 2. Cambridge Massachusetts: MIT Press, pp. 531–552.

Norton, R. and Howard, R. (1988). Smoking, mood and the contingent variation (CNV) in a go-no go avoidance task. *Journal of Psychophysiology*, **2**:109–118.

Norton, R., Brown, K. and Howard, R. (1992). Smoking, nicotine dose and the lateralization of electrocortical activity. *Psychopharmacology*, **108**:473–479.

Norton, R., Howard, R. and Brown, K. (1991). Nicotine dose-dependent effects of smoking on P300 and mood. *Medical Science Research*, **19**:355–356.

Nyberg, G., Panfilov, V., Sivertsson, R. and Wilhelmsen, L. (1982). Cardiovascular effects of nicotine chewing gum in healthy non-smokers. *European Journal of Clinical Pharmacology*, **50**:157–164.

O'Connor, S., Hesselbrock, V., Tasman, A. and DePalma, N. (1987). P_3 amplitudes in two distinct tasks are decreased in young men with a history of paternal alcoholism. *Alcohol*, **4**:323–330.

O'Neill, S.T. and Parrott, A.C. (1992). Stress and arousal in sedative and stimulant cigarette smokers. *Psychopharmacology*, **107**:442–446.

Obitz, F.W., Rhodes, L.E. and Creel, D. (1977). Effects of alcohol and monetary reward on visually evoked potentials and reaction time. *Journal of Studies on Alcohol*, **38**:2057–2064.

Oborne, D.J. and Rogers, Y. (1983). Interaction of alcohol and caffeine on human reaction time. *Aviation, Space and Environmental Medicine*, **54(6)**:528–534.

Oei, T.P.S. and Baldwin, A.R. (1994). Expectancy theory: A two-process model of alcohol use and abuse. *Journal of Studies on Alcohol*, **55**:525–534.

Office of Population Censuses and Surveys (1986). General Household Survey for 1984. London: HMSO.

Okita, T., Wijers, A.A., Mulder, G. and Mulder, L.J.M. (1985). Memory search and visual spatial attention: An event-related brain potential analysis. *Acta Psychologica*, **60**:263–292.

Oliveto, A.H., Hughes, J.R., Terry, S.Y., Bickel, W.K., Higgins, S.T., Pepper, S.L. and Fenwick, J.W. (1991). Effects of caffeine on tobacco withdrawal. *Clinical Pharmacology and Therapeutics*, **50**:157–164.

Oscar-Berman, M. (1987). Alcohol-related ERP changes in cognition. *Alcohol*, **4**:289–292.

Ossip, D.J. and Epstein, L.H. (1981). Relative effects of nicotine and coffee on cigarette smoking. *Addictive Behaviors*, **6**:35–39.

Ossip, D.J., Epstein, L.H. and McKnight, D. (1980). Modeling, coffee drinking, and smoking. *Psychological Reports*, **47**:408–410.

Page, R.D. and Cleveland, M.F. (1987). Cognitive dysfunction and aging among male alcoholics and social drinkers. *Alcoholism: Clinical and Experimental Research*, **11(4)**:376–384.

Paller, K.A. (1990). Recall and stem-completion priming have different electrophysiological correlates and are modified differentially by directed forgetting. *Journal of Experimental Psychology: Learning, Memory, and Cognition*, **16**:1021–1032.

Paller, K.A., McCarthy, G. and Wood, C.C. (1988). ERPs predictive of subsequent recall and recognition performance. *Biological Psychology*, **26**:269–276.

Palva, E.S., Linnoila, M., Saario, I. and Mattila, M.J. (1979). Acute and subacute effects of diazepam on psychomotor skills: Interaction with alcohol. *Acta Pharmacologica et Toxicologica*, **45**:257–264.

Pares, X., Farres, J. and Pares, A. (1994). Genetic polymorphism of liver alcohol dehydrogenase in Spanish subjects: Significance of alcohol consumption and liver disease. *Alcohol and Alcoholism*, **29**:701–705.

Parker, D.A., Parker, E.S., Brody, J.A. and Schoenberg, R. (1983). Alcohol use and cognitive loss among employed men and women. *American Journal of Public Health*, **73**:521–526.

Parker, E.S. and Noble, E.P. (1977) Alcohol consumption and cognitive functioning in social drinkers. *Journal of Studies on Alcohol*, **38**:1224–1232.

Parker, E.S., Alkana, R.L., Birnbaum, I.M., Hartley, J.T. and Noble, E.P. (1974). Alcohol and the disruption of cognitive processes. *Archives of General Psychiatry*, **31**:824–828.

Parker, E.S., Birnbaum, I.M., Boyd, R.A. and Noble, E.P. (1980). Neuropsychological decrement as a function of alcohol intake in male students. *Alcoholism: Clinical and Experimental Research*, **4**:330–334.

Parker, E.S., Birnbaum, I.M., Weingartner, H., Hartley, J.T., Stillman, R.C. and Wyatt, R.J. (1980). Retrograde enhancement of human memory with alcohol. *Psychopharmacology (Berlin)*, **69**:219–222.

Parker, E.S., Morihisa, J.M., Wyatt, R.J., Schwartz, B.L., Weingartner, H. and Stillman, R.C. (1981). The alcohol facilitation effect on memory: A dose-response study. *Psychopharmacology (Berlin)*, **74**:88–92.

Parker, E.S. and Noble E.P. (1980). Alcohol and the ageing process in social drinkers. *Journal of Studies on Alcohol*, **41**:170–178.

Parker, E.S., Parker, D.A. and Brody, J.A. (1985). The impact of fathers' drinking on cognitive loss among social drinkers. In: M. Galanter (Ed.) Recent developments in alcoholism. Vol. 3. New York: Plenum, pp. 227–240.

Parker, E.S., Parker, D.A. and Harford, T.C. (1991). Specifying the relationship between alcohol use and cognitive loss: The effects of frequency of consumption and psychological distress. *Journal of Studies on Alcohol*, **52**:366–373.

Parrott, A.C. and Craig, D. (1992). Cigarette smoking and nicotine gum (0, 2 and 4 mg): Effects upon four visual attention tasks. *Neuropsychobiology*, **25**:34–43.

Parrott, A.C. and Joyce, C. (1993). Stress and arousal rhythms in cigarette smokers, deprived smokers and non-smokers. *Human Psychopharmacology*, **8**:21–28.

Parrott, A.C. and Roberts, G. (1991). Nicotine deprivation and nicotine reinstatement: Effects upon a brief sustained attention task. In: F. Adlkofer and K. Thurau (Eds.), Effects of nicotine on biological systems. Basel: Birkhäuser Verlag, pp. 485–490.

Parrott, A.C. and Winder, G. (1989). Nicotine chewing gum (2 mg, 4 mg) and cigarette smoking: Comparative effects upon vigilance and heart rate. *Psychopharmacology*, **97**:257–261.

Parrott, A.C. (1991). Performance tests in human psychopharmacology (3): Construct validity and test interpretation. *Human Psychopharmacology*, **6**:197–207.

Parrott, A.C. (1991a). Performance tests in human psychopharmacology (1): Test reliability and standardization. *Human Psychopharmacology*, **6**:1–9.

Parrott, A.C. (1991b). Performance tests in human psychopharmacology (2): Content validity, criterion validity, and face validity. *Human Psychopharmacology*, **6**:91–98.

Parrott, A.C. (1991c). Performance tests in human psychopharmacology (3): Construct validity and test interpretation. *Human Psychopharmacology*, **6**:197–207.

Parrott, A.C. (1992). Smoking and smoking cessation: Effects upon human performance. *Smoking-Related Diseases*, **3**:43–53.

Parrott, A.C. (1994b). Acute pharmacodynamic tolerance to the subjective effects of cigarette smoking. *Psychopharmacology*, **116**:93–97.

Parrott, A.C. (1994a). Individual differences in stress and arousal during cigarette smoking. *Psychopharmacology*, **115**:389–396.

Parrott, A.C., Craig, D., Haines, M. and Winder, G. (1991). Nicotine polacrilex gum and sustained attention. In: F. Adlkofer and K. Thurau (Eds.), Effects of nicotine on biological systems (pp. 559–564). Basel: Birkhäuser Verlag.

Parsons, O.A. and Fabian, M.S. (1982). Comments. *Journal of Studies on Alcohol*, **43**:178–182.

Parsons, O.A. (1986). Cognitive functioning in sober social drinkers: A review and critique. *Journal of Studies on Alcohol*, **47(2)**:101–114.

Parsons, O.A., Butters, N. and Nathan, P.E. (1987). Neuropsychology of Alcoholism: Implications for Diagnosis and Treatment. New York: Guilford Press.

Parsons, O.A., Sinha, R. and Williams, H.L. (1990). Relationships between neuropsychological test performance and event-related potentials in alcoholic and nonalcoholic samples. *Alcoholism: Clinical Experimental Research*, **14(5)**:746–755.

Parsons, W.D. and Neims, A.H. (1978). Effect of smoking on caffeine clearance. *Clinical Pharmacology and Therapeutics*, **24**:40–45.

Passmore, A.P., Kondowe, G.B. and Johnston, G.D. (1987). Renal and cardiovascular effects of caffeine: A dose-response study. *Clinical Science*, **72**:749–756.

Patel, R.M. (1988). Ethanol's effect on human vigilance during a simple task in the presence of an auditory stressor. *Psychological Reports*, **63**:363–366.

Pátkai, P. (1970). Diurnal differences between habitual morning workers and evening workers in some psychological and physiological functions. Reports from the Psychological Laboratories, no. 311, Stockholm: University of Stockholm.

Pátkai, P. (1985). Interindividual Differences in Diurnal Variations in Alertness, Performance and Adrenaline Excretion. *Acta Physiologica Scandinavica*, **81**:35–46.

Patterson, B.W., Williams, H.L., McLean, G.A., Smith, L.T. and Schaeffer, K.W. (1987). Alcoholism and family history of alcoholism: effects on visual and auditory event-related potentials. *Alcohol*, **4**:265–274.

Pauly, J.R., Grun, E.U. and Collins, A.C. (1992). Glucocorticoid regulation of sensitivity to nicotine. In: P.M. Lippiello, A.C. Collins and J.A. Gray (Eds.), The biology of nicotine: Current Research Issues. New York: Raven Press, pp. 121–155.

Pearson, R.G. (1968). Alcohol-hypoxia effects upon operator tracking, monitoring, and reaction time. *Aerospace Medicine*, **39**:303–397.

Peeke, S.C. and Peeke, H.V.S. (1984). Attention, memory, and cigarette smoking. *Psychopharmacology*, **84**:205–216.

Peeke, S.C., Callaway, E., Jones, R.T., Stone, G.C. and Doyle, J. (1980). Combined effects of alcohol and sleep deprivation in normal young adults. *Psychopharmacology*, **76**:279–287.

Peele, S. (1985). The meaning of addiction: Compulsive experience and its interpretation. Lexington, MA: Lexington Books.

Perkins, K.A. and Grobe, J.E. (1992). Increased desire to smoke during acute stress. *British Journal of Addiction*, **87**:231–234.

Perkins, K.A. (1993). Weight gain following smoking cessation. *Journal of Consulting and Clinical Psychology*, **61**:768–777.

Perkins, K.A., Bohay, J., Meylahn, E.N., Wing, R.R., Matthews, K.A. and Kuller, L.H. (1993). Diet, alcohol intake, and physical activity as a function of smoking status in middle-aged women. *Health Psychology*, **12**:410–415.

Perkins, K.A., Epstein, L.H., Jennings, J.R. and Stiller, R. (1986b). The cardiovascular effects of nicotine during stress. *Psychopharmacology*, **90**:373–378.

Perkins, K.A., Epstein, L.H., Sexton, J.E., Solberg-Kassel, R., Stiller, R.L. and Jacob, R.G. (1992a). Effects of nicotine on hunger and eating in male and female smokers. *Psychopharmacology*, **106**:53–59.

Perkins, K.A., Epstein, L.H., Stiller, R., Jennings, J.R., Christiansen, C. and McCarthy, T. (1986a). An aerosol spray alternative to cigarette smoking in the study of the behavioral and physiological effects of nicotine. *Behavioral Research Methods, Instruments and Computers*, **18**:420–426.

Perkins, K.A., Epstein, L.H., Stiller, R.L., Sexton, J.E., Debski, T.D. and Jacob, R.G. (1990a). Behavioral performance effects of nicotine in smokers and nonsmokers. *Pharmacology, Biochemistry and Behavior*, **37**:11–15.

Perkins, K.A., Epstein, L.H., Stiller, R.L., Sexton, J.E., Marks, B.L. and Jacob, R.G. (1990b). Cardiovascular effects of nicotine during physical activity and following meal consumption. *Clinical and Experimental Pharmacology and Physiology*, **17**:327–334.

Perkins, K.A., Grobe, J.E., Fonte, C. and Breus, M. (1992c). "Paradoxical" effects of smoking on subjective stress versus cardiovascular arousal in males and females. *Pharmacology, Biochemistry and Behavior*, **42**:301–311.

Perkins, K.A., Grobe, J.E., Fonte, C., Goettler, J., Caggiula, A.R., Reynolds, W.A., Stiller, R.L., Scierka, A. and Jacob, R.G. (1994b). Chronic and acute tolerance to subjective, behavioral and cardiovascular effects of nicotine in humans. *Journal of Pharmacology and Experimental Therapeutics*, **270**:628–638.

Perkins, K.A., Grobe, J.E., Stiller, R.L., Fonte, C. and Goettler, J.E. (1992b). Nasal spray nicotine replacement suppresses cigarette smoking desire and behavior. *Clinical Pharmacology and Therapeutics*, **52**:627–634.

Perkins, K.A., Sexton, J.E., DiMarco, A., Grobe, J.E., Scierka, A. and Stiller, R.L. (1995). Subjective and cardiovascular responses to nicotine combined with alcohol in male and female smokers. *Psychopharmacology*, **119**:205–212.

Perkins, K.A., Sexton, J.E., Epstein, L.H., DiMarco, A., Fonte, C., Stiller, R.L., Scierka, A. and Jacob, R.G. (1994c). Acute thermogenic effects of nicotine combined with caffeine during light physical activity in male and female smokers. *American Journal of Clinical Nutrition*, **60**:312–319.

Perkins, K.A., Sexton, J.E., Reynolds, W.A., Grobe, J.E., Fonte, C. and Stiller, R.L. (1994a). Comparison of acute subjective and heart rate effects of nicotine intake via tobacco smoking versus nasal spray. *Pharmacology, Biochemistry and Behavior*, **47**:295–299.

Perlick, D.A. (1977). The withdrawal syndrome: Nicotine addiction and the effects of stopping smoking in heavy and light smokers. Doctoral Dissertation. Columbia University. Ann Arbor: University Microfilms International, Thesis No. 38/01-B, 409.

Pernanen, K. (1981). Theoretical aspects of the relationship between alcohol use and crime. In: J.J. Collins (Ed.), Drinking and Crime. New York: Guilford Press.

Perret, E. (1974). The left frontal lobe of man and the suppression of habitual responses in verbal categorical behaviour. *Neuropsychologia*, **12**:323–330.

Peters, R. and McGee, R. (1982). Cigarette smoking and state-dependent memory. *Psychopharmacology*, **76**:232–235.

Peterson, J.B. and Pihl, R.O. (1990). Information processing, neuropsychological function, and the inherited predisposition to alcoholism. *Neuropsychology Review*, 1:343–369.

Peterson, J.B., Finn, P.R. and Pihl, R.O. (1992). Cognitive dysfunction and the inherited predisposition to alcoholism. *Journal of Studies on Alcohol*, **53**:154–160.

Peterson, J.B., Pihl, R.O., Seguin, J.R., Finn, P.R. and Stewart, S.H. (1993). Heart-rate reactivity and alcohol consumption among sons of male alcoholics and sons of non-alcoholics. *Journal of Psychiatry and Neuroscience*, **18**:190–198.

Peterson, J.B., Rothfleisch, J., Zelazo, P.D. and Pihl, R.O. (1990). Acute alcohol intoxication and cognitive functioning. *Journal of Studies on Alcohol*, **51**:114–122.

Petrie, R.X.A. and Deary, I.J. (1989). Smoking and human information processing. *Psychopharmacology*, **99**:393–396.

Peyser, H. (1992). Stress and alcohol. In: L. Goldberger and S. Breznik (Eds.), Handbook of stress: Theoretical and clinical aspects. New York: Free Press, pp. 585–598.

Pfefferbaum, A., Ford, J.M., White, P.M. and Mathalon, D. (1991). Event-related potentials in alcoholic men: P_3 amplitude reflects family history but not alcohol consumption. *Alcoholism*, **15(5)**:839–850.

Pfefferbaum, A., Horvath, T.B., Roth, W.T., Clifford, S.T. and Kopell, B.S. (1980). Acute and chronic effects of ethanol on event-related potentials. In: H. Begleiter (Ed.), Biological effects of alcohol. New York: Plenum, pp. 625–639.

Pfefferbaum, A., Lim, K.O., Zipursky, R.B., Mathalon, D.H., Rosenbloom, M.J., Lane, B., Chung, N.H. and Sullivan, E.V. (1992). Brain grey and white matter loss accelerates with aging in chronic alcoholics: a quantitative MRI study. *Alcoholism: Clinical and Experimental Research*, **16**:1078–1089.

Pfefferbaum, A., Lim, K.O., Ha, C.N. and Zipursky, R.B. (1990). [Abstract] Changes in white and grey matter volume in alcoholism: An MRI study. The 5th Congress of the International Society for Biochemical Research on Alcoholism, p. 280.

Pfefferbaum, A., Roth, W.T., Tinklenberg, J.R., Rosenbloom, M.J. and Kopell, B.S. (1979). The effects of ethanol and meperidine on auditory evoked potentials. *Drug and Alcohol Dependence*, **4**:371–380.

Philips, C. (1971). The EEG changes associated with smoking. *Psychophysiology*, **8**:64–74.

Phillis, J.W. (1991). Adenosine and adenine nucleotides as regulators of cellular function. Boca Raton, Florida: CRC Press.

Picciotto, M.R., Zoli, M., Léna, C., Bessis, A., Lallemand, Y., Le Novère, N., Vincent, P., Merlo Pich, E., Brûlet, P. and Changeux, J.-P. (1995). Abnormal avoidance learning in mice lacking functional high-affinity nicotine receptor in the brain. *Nature*, **374**:65–67.

Pihl, R.O. and Peterson, J.B. (1995). Alcoholism: the role of different motivational systems. *Journal of Psychiatry and Neuroscience*, **20**:372–396.

Pihl, R.O., Lau, M.L. and Assaad, J.M. (1997). Aggressive disposition, alcohol and aggression. *Aggressive Behavior*, **23(1)**:11–18.

Pihl, R.O., Peterson, J.B. and Finn, P. (1990). An heuristic model for the inherited predisposition to alcoholism. *Psychology of Addictive Behaviors*, **4**:12–25.

Pihl, R.O., Peterson, J.B. and Finn, P.R. (1990). Inherited predisposition to alcoholism: characteristics of sons of male alcoholics. *Journal of Abnormal Psychology*, **99**:291–301.

Pihl, R.O., Peterson, J.B. and Lau, M.A. (1993). A biosocial model of the alcohol-aggression relationship. *Journal of Studies on Alcohol*, **11**:128–139.

Pineda, J.A., Foote, S.L. and Neville, H.J. (1989). Effects of locus coeruleus lesions on auditory, long-latency, event-related potentials in monkey. *Journal of Neuroscience*, **9**:81–93.

Pirie, P.L., Murray, D.M. and Luepker, R.V. (1991). Gender differences in cigarette smoking and quitting in a cohort of young adults. *American Journal of Public Health*, **81**:324–327.

Pohorecky, L.A. and Brick, J. (1988). Pharmacology of ethanol. *Pharmacology and Therapeutics*, **36**:335–427.

Pohorecky, L.A. (1977). Biphasic action of ethanol. *Biobehavioural Review*, **1**:231–240.

Polich, J. and Bloom, F.E. (1986). P300 and alcohol consumption in normals and individuals at risk for alcoholism. *Progress in Neuropsychopharmacology, Biology and Psychiatry*, **10**:201–210.

Polich, J. and Bloom, F.E. (1987). P300 from normals and adult children of alcoholics. *Alcohol*, **4**:301–305.

Polich, J. and Bloom, F.E. (1988). Event-related brain potentials in individuals at high and low risk for developing alcoholism: Failure to replicate. *Alcoholism: Clinical and Experimental Research*, **12(3)**:368–373.

Polich, J. (1984). P300 latency reflects personal drinking history. *Psychophysiology*, **21**:592–593.

Polich, J., Burns, T. and Bloom, F.E. (1988a). P300 and the risk for alcoholism: Family history, task difficulty, and gender. *Alcoholism: Clinical and Experimental Research*, **12(2)**:248–254.

Polich, J., Haier, R.J., Buchsbaum, M. and Bloom, F.E. (1988b). Assessment of young men at risk for alcoholism with P300 from a visual discrimination task. *Journal of Studies on Alcohol*, **49(2)**:186–190.

Pollock, V.E., Teasdale, T., Stern J. and Volavka, J. (1981). Effects of caffeine on resting EEG and response to sine wave modulated light. *Electroencephalography and Clinical Neurophysiology*, **51**:470–476.

Pollock, V.E., Volavka, J., Goodwin, D.W., Gabrielli, W.F., Mednick, S.A., Knop, J. and Schulsinger, F. (1988). Pattern reversal visual evoked potentials after alcohol administration among men at risk for alcoholism. *Psychiatry Research*, **26**:191–202.

Pollock, V.E., Volavka, J., Goodwin, D.W., Sarnoff, A.M., Gabrielli, W.F., Knop, J. and Schulsinger, F. (1983). The EEG after alcohol administration in men at risk for alcoholism. *Archives of General Psychiatry*, **40**:847–861.

Pomerleau, O.F. and Pomerleau, C.S. (1989). A biobehavioral perspective on smoking. In: T. Ney and A. Gale (Eds.), Smoking and human behavior. Chichester: Wiley, pp. 69–90.

Pomerleau, O.F., Flessland, K.A., Pomerleau, C.S. and Hariharan, M. (1992). Controlled dosing of nicotine via an intranasal nicotine aerosol delivery device (INADD). *Psychopharmacology*, **108**:519–526.

Pomerleau, O.F. and Pomerleau, C.S. (1990). Behavioural studies in humans: anxiety, stress and smoking. In: G. Bock and J. Marsh, (Eds.), The Biology of Nicotine Dependence. Chichester: Wiley.

Pomerleau, O.F., Turk, D.C. and Fertig, J.B. (1984). The effects of cigarette smoking on pain and anxiety. *Addictive Behaviors*, **9**:265–271.

Pons, G., Blais, J.-C., Rey, E., Plissonnier, M., Richard, M.-O., Carrier, O., D'Athis, P., Moran, C., Badoual, J. and Olive, G. (1988). Maturation of caffeine N-demethylation in infancy: a study using the $^{13}CO_2$ breath test. *Pediatric Research*, **23**:632–636.

Pons, G., Rey, E., Carrier, O., Richard, M.-O., Moran, C., Badoual, J. and Olive, G. (1989). Maturation of AFMU excretion in infants. *Fundamental Clinical Pharmacology*, **3**:589–595.

Pons, L., Trenque, T., Bielecki, M., Mopulin, M. and Potier, J.C. (1988). Attentional effects of caffeine in man: Comparison with drugs acting upon performance. *Psychiatry Research*, **23**:329–333. *[61]*.

Porchet, H.C., Benowitz, N.L. and Sheiner, L.B. (1988). Pharmacodynamic model of tolerance: application to nicotine. *Journal of Pharmacology and Experimental Therapeutics*, **244**:231–236.

Porchet, H.C., Benowitz, N.L., Sheiner, L.B. and Copeland, J.R. (1987). Apparent tolerance to the acute effect of nicotine results in part from distribution kinetics. *Journal of Clinical Investigation*, **80**:1466–1471.

Porjesz, B. and Begleiter, H. (1973a). Alcohol and bilateral evoked brain potentials. In: M.M. Gross (Ed.), Alcohol intoxication and withdrawal. New York: Plenum, pp. 553–567.

Porjesz, B. and Begleiter, H. (1973b). The effects of alcohol on the somatosensory evoked potentials in man. In: M.M. Gross (Ed.), Alcohol intoxication and withdrawal. New York: Plenum, pp. 345–350.

Porjesz, B. and Begleiter, H. (1982). Evoked brain potential deficits in alcoholism and aging. *Alcoholism: Clinical and Experimental Research*, **6(1)**:53–60.

Porjesz, B. and Begleiter, H. (1985). Human brain electrophysiology and alcoholism. In: R.E. Tarter and D.H. Theil (Eds.), Alcohol and the brain. Chronic effects. New York: Plenum, pp. 139–182.

Porjesz, B. and Begleiter, H. (1990). Individuals at risk for alcoholism: neurophysiologic processes. In: C.R. Cloninger and H. Begleiter (Eds.), Genetics and biology of alcoholism. Banbury Report 33. New York: Cold Spring Harbor Laboratory Press, pp. 137–157.

Porjesz, B. and Begleiter, H. (1993). Neurophysiological Factors Associated with Alcoholism. In: S.J. Nixon and W.A. Hunt (Eds.), NIAAA Research Monograph-22, Alcohol induced Brain Damage, pp. 89–120.

Porjesz, B., Begleiter, H., Bihari, B. and Kissin, B. (1987). Event related brain potentials to high incentive stimuli in abstinent alcoholics. *Alcohol*, **4**:283–287.

Posner, M.I. and Peterson, S.E. (1990). The attention system of the human brain. *Annual Review of Neurosciences*, **13**:25–42.

Posner, M.I. and Raichle, M.E. (1994). Images of mind. New York: Scientific American Library.

Posner, M.I. and Rothbart, M.K. (1986). The concept of energy in psychological theory. In: G.R.J. Hockey, A.W.K. Gaillard and M.G.H. Coles (Eds.), Energetics and human information processing. Dordrecht, The Netherlands: Nijhoff, pp. 23–40.

Potthoff, A.D., Ellison, G. and Nelson, L. (1983). Ethanol intake increases during continuous administration of amphetamine and nicotine, but not several other drugs. *Pharmacology, Biochemistry and Behavior*, **18**:489–493.

Preston, G.C., Broks, P., Traub, M., Ward, C., Poppleton, P. and Stahl, S.M. (1988). Effects of lorazepam on memory, attention and sedation in man. *Psychopharmacology*, **95**:208–215.

Pribram, K.H. and McGuinness, D. (1975). Arousal, activation, and effort in the control of attention. *Psychological Review*, **82**:116–149.

Pritchard, W.S. and Kay, D.L.C. (1993). Personality and smoking motivation of U.S. smokers as measured by the state-trait personality inventory, the Eysenck Personality Questionnaire, and Spielberger's Smoking Motivation Questionnaire. *Personality and Individual Differences*, **14**:629–637.

Pritchard, W.S. (1981). Psychophysiology of P300. *Psychological Bulletin*, **89(3)**: 506–540.

Pritchard, W.S. (1991a). The link between smoking and P: A serotonergic hypothesis. *Personality and Individual Differences*, **12**:1187–1204.

Pritchard, W.S. (1991b). Electroencephalographic effects of cigarette smoking. *Psychopharmacology*, **104**:485–490.

Pritchard, W.S., Duke, D.W., Coburn, K.L. and Robinson, J.H. (1992b). Nonlinear dynamical EEG analysis applied to nicotine psychopharmacology and Alzheimer's disease. In: P.M. Lippiello, A.C. Collins, J.A. Gray and J.H. Robinson (Eds.), The biology of nicotine: Current research issues. New York: Raven, pp. 195–214.

Pritchard, W.S., Robinson, J.H. and Guy, T.D. (1992a). Enhancement of continuous performance task reaction time by smoking in non-deprived smokers. *Psychopharmacology*, **108**:437–442.

Pritchard, W.S., Robinson, J.H., DeBethizy, J.D., Davis, R.A. and Stiles, M.F. (1995). Caffeine and smoking: subjective, performance, and psychophysiological effects. *Psychophysiology*, **32**:19–27.

Prochaska, J.O., DiClemente, C.C. and Norcross, J.C. (1992). In search of how people change: Applications to addictive behaviors. *American Psychologist*, **47**:1102–1114.

Productschap voor gedestilleerde dranken. (1993). World Drink trends. Oxford: NTC Publications.

Provost, S.C. and Woodward, R. (1991). Effects of nicotine gum on repeated administration of the Stroop test. *Psychopharmacology*, **104**:536–540.

Pryor, G.T., Steinmetz, G. and Stone, H. (1970). Changes in Absolute and in Subjective Intensity of Suprathreshold Stimuli during Olfactory Adaptation and Recovery. *Perception and Psychophysics*, **8(5b)**:331–335.

Radovanovic, Z., Ljubomir, E., Dimitrijevic, L. and Jambocic, V. (1983). Cigarette smoking among first-year medical students in Yugoslavia and their academic success. *Journal of American College Health*, **31**:253–255.

Rainnie, D.G., Grunze, H.C.R., McCarley, R.W. and Greene, R.W. (1994). Adenosine inhibition of mesopontine cholinergic neurons: implications for EEG arousal. *Science*, **263**:689–692.

Rapoport, J.L., Jensvold, M., Elkins, R., Buchsbaun, M.S., Weingartner, H., Ludlow, C., Zahn, T.P., Berg, C.J. and Neims, A.H. (1981). Behavioural and cognitive effects of caffeine in boys and adult males. *Journal of Nervous and Mental Disease*, **169**:726–732.

Rather, B.C. and Goldman, M.S. (1994). Drinking-related differences in the memory organization of alcohol expectancies. *Experimental and Clinical Psychopharmacology*, **2**:167–183.

Ray, J.J. (1985). Smoking and intelligence in Australia. *Social Science and Medicine*, **20**:1279–1280.

Reed, T.E. and Hanna, J.M. (1986). Between- and within-race variation in acute cardiovascular responses to alcohol: evidence for genetic determination in normal males in three races. *Behavioral Genetics*, **16**:585–598.

Reed, T.E. (1978). Racial comparisons of alcohol metabolism: Background problems and results. *Alcoholism, Clinical and Experimental Research*, **2**:61–69.

Reeves, W.E. and Morehouse, L.E. (1950). The acute effect of smoking upon the physical performance of habitual smokers. *Research Quarterly of American Health*, **21**:245–248.

Regan, D. (1989). Human brain electrophysiology. Evoked potentials and evoked magnetic fields in science and medicine. New York: Elsevier.

Renault, B., Kutas, M., Coles, M.G.H. and Gaillard, A.W.K. (1988). Event-related potential investigations of cognition. *Biological Psychology*, **26**:1–354.

Renault, B., Ragot, N., Lesévre, N. and Remond, A. (1982). Onset and offset of brain events as indices of mental chronometry. *Science*, **215**:1413–1415.

Renner, E., Wietholtz, H., Huguenin, P., Arnaud, M.J. and Preisig, R. (1984). Caffeine: A model compound for measuring liver function. *Hepatology*, **4**:38–46.

Revell, A.D. (1988). Smoking and performance: A puff-by-puff analysis. *Psychopharmacology*, **96**:563–565.

Rey, A. (1941). L'examen psychologique dans les cas d'encéphalopathie traumatique. *Archives de Psychologie*, **28**:286–340.

Rey, A. (1964). L'examen clinique en psychologie. Paris: Presses Universitaires de France.

Rhodes, L.E., Obitz, F.W. and Creel, D. (1975). Effect of alcohol and task on hemispheric asymmetry of visual evoked potentials in man. *Electroencephalography and Clinical Neurophysiology*, **38**:561–568.

Ritter, W., Simson, R., Vaughan Jr., H.G. and Macht, M. (1982). Manipulation of event-related potential manifestations of information processing stages. *Science*, **218**:909–911.

Ritter, W., Vaughan Jr., H.G. and Simson, R. (1983). On relating event-related potential components to stages of information processing. In: A.W.K. Gaillard and W. Ritter (Eds.), Tutorials in event-related potential research: Endogenous components. Amsterdam: North-Holland, pp. 143–158.

Roache, J.D. and Griffiths, R.R. (1987). Interactions of diazepam and caffeine: Behavioral and subjective dose effects in humans. *Pharmacology, Biochemistry and Behavior*, **26**:801–812. *[62]*.

Roache, J.D., Cherek, D.R., Bennett, R.H., Schenkler, J.C. and Cowan, K.A. (1993). Differential effects of triazolam and ethanol on awareness, memory, and psychomotor performance. *Journal of Clinical Psychopharmacology*, **13**:3–15.

Robbins, T.W. and Everitt, B.J. (1995). Arousal systems and attention. In: M.S. Gazzaniga (Ed.), The cognitive neurosciences. Cambridge: MIT Press, pp. 703–720.

Robinson, J. and Pritchard, W. (1992). The role of nicotine in tobacco use. *Psychopharmacology*, **108**:397–408.

Robinson, J.H., Pritchard, W.S. and Davis, R.A. (1992). Psychopharmacological effects of smoking a cigarette with typical 'tar' and carbon monoxide yields but minimal nicotine. *Psychopharmacology*, **108**:466–472.

Roehrs, T., Beare, D., Zorick, F. and Roth, T. (1994). Sleepiness and ethanol effects on simulated driving. *Alcoholism: Clinical and Experimental Research*, **18**:154–158.

Rogers, A.S., Spencer, M.B., Stone, B.M. and Nicholson, A.N. (1989). The influence of a 1 h nap on performance overnight. *Ergonomics*, **32**:1193–1205. *[63]*.

Rohrbaugh, J.W., Stapleton, J.M., Parasuraman, R., Zubovic, E.A., Frowein, H.W., Varner, J.L., Adinoff, B., Lane, E.A., Eckardt, M.J. and Linnoila, M. (1987). Dose-related effects of ethanol on visual sustained attention and event-related potentials. *Alcohol*, **4**:293–300.

Rohren, C.L., Coghan, I.T., Hurt, R.D., Offord, K.P., Marusic, Z. and McClain, F.L. (1994). Predicting smoking cessation outcome in a medical center from stage of readiness: contemplation versus action. *Preventive Medicine*, **23**:335–344.

Roine, R.P., Gentry, T., Lim, R.T., Baraona, E. and Lieber, C.S. (1991). Effect of concentration of ingested ethanol on blood alcohol levels. *Alcoholism, Clinical and Experimental Research*, **15**:734–738.

Ron, M.A. (1983). The alcoholic brain: CT scan and psychological findings. *Psychological Medicine, Cambridge University Press a monograph*, **Suppl. 3**:1–33.

Rose, J.E and Behm, F.M. (1994). Inhalation of vapor from black pepper extract reduces smoking withdrawal symptoms. *Drug and Alcohol Dependence*, **34**:225–229.

Rose, J.E. and Hickman, C. (1988). Citric acid aerosol as a potential smoking cessation Aid. *Chest*, **92**:1005–1008.

Rose, J.E. (1986). Cigarette smoking blocks caffeine-induced arousal. *Alcohol and Drug Research*, **7**:49–55.

Rose, J.E., Ananda, S. and Jarvik, M.E. (1983). Cigarette smoking during anxiety-provoking and monotonous tasks. *Addictive Behaviors*, **8**:353–359.

Rose, J.E., Behm, F.M. and Levin, E.D. (1993). Role of nicotine dose and sensory cues in the regulation of smoke intake. *Pharmacology, Biochemistry and Behavior*, **44**:891–900.

Rose, J.E., Behm, F.M., Westman, E.C., Levin, E.D., Stein, R.M. and Ripka, G.V. (1994). Mecamylamine combined with nicotine skin patch facilitates smoking cessation beyond nicotine patch treatment alone. *Clinical Pharmacology and Therapeutics*, **56**:86–99.

Rose, J.E., Sampson, A., Levin, E.D. and Henningfield, J.E. (1989). Mecamylamine increases nicotine preference and attenuates nicotine discrimination. *Pharmacology, Biochemistry and Behavior*, **32**:933–938.

Rose, J.E., Zinser, M.C., Tashkin, D.P., Newcomb, R. and Ertle, A. (1984). Subjective response to cigarette smoking following airway anesthetization. *Addictive Behaviors*, **9**:211–215.

Rosecrans, J. and Karan, L.D. (1993). Neurobehavioral mechanisms of nicotine action: Role in the initiation and maintenance of tobacco dependence. *Journal of Substance Abuse Treatment*, **10**:161–170.

Rosenberg, J., Benowitz, N.L., Jacob, P. III and Wilson, M. (1980). Disposition kinetics and effects of intravenous nicotine. *Clinical Pharmacology and Therapeutics*, **28**:516–522.

Rosenberg, L., Slone, D., Shapiro, S., Kaufman, D.W., Stolley, P.D. and Mittinen, O.S. (1980). Coffee drinking and myocardial infarction in young women. *American Journal of Epidemiology*, **111**:675–681.

Rosenthal, L., Roehrs, T., Zwyghuizen-Doorenbos, A., Plath, D. and Roth, T. (1991). Alerting effects of caffeine after normal and restricted sleep. *Neuropsychopharmacology*, **4(2)**:103–108. *[64]*.

Rösler, F. and Heil, M. (1991). Towards a functional categorization of slow waves: Taking into account past *and* future events. *Psychophysiology*, **28**:344–358.

Ross, D.F. and Pihl, R.O. (1988). Alcohol, self-focus and complex reaction time performance. *Journal of Studies on Alcohol*, **49**:115–125.

Ross, L.E. and Ross, S.M. (1992). Alcohol use and aviation safety. *Alcohol, Drugs and Driving*, **8**:231–239.

Ross, L.E., Yeazel, L.M. and Chau, A.W. (1992). Pilot performance with blood alcohol concentrations below 0.04%. *Aviation, Space and Environmental Medicine*, **63**:951–956.

Roth, N. and Bättig, K. (1991). Effects of cigarette smoking upon frequencies of EEG alpha rhythm and finger tapping. *Psychopharmacology*, **105**:186–190.

Roth, N., Lutiger, B., Hasenfratz, M., Bättig, K. and Knye, M. (1992). Smoking deprivation in 'early' and 'late' smokers and memory functions. *Psychopharmacology*, **106**:253–260.

Roth, W.T., Tinklenberg, J.R. and Kopell, B.S. (1977). Ethanol and marihuana effects on event-related potentials in a memory retrieval paradigm. *Electroencephalography and Clinical Neurophysiology*, **42**:381–388.

Ruchkin, D.S., Johnson Jr., R., Mahaffey, D. and Sutton, S. (1988). Toward a functional categorization of slow waves. *Psychophysiology*, **25**:339–353.

Rugg, M.D. and Nagy, M.E. (1989). Event-related potentials and recognition memory for words. *Electroencephalography and Clinical Neurophysiology*, **72**:395–406.

Rumbold, G.R. and White, J.M. (1987). Effects of repeated alcohol administration on human operant behaviour. *Psychopharmacology*, **92**:186–191.

Rush, C.R., Higgins, S.T., Bickel, W.K. and Hughes, J.R. (1994a). Acute behavioral effects of lorazepam and caffeine, alone and in combination, in humans. *Behavioural Pharmacology*, **5**:245–254. *[65]*.

Rush, C.R., Higgins, S.T., Hughes, J.R. and Bickel, W.K. (1994b). Acute behavioral effects of triazolam and caffeine, alone and in combination, in humans. *Experimental and Clinical Psychopharmacology*, **2(3)**:211–222. *[66]*.

Rush, C.R., Higgins, S.T., Hughes, J.R., Bickel, W.K. and Wiegner, M.S. (1993). Acute behavioral and cardiac effects of alcohol and caffeine, alone and in combination, in humans. *Behavioural Pharmacology*, **4**:562–572. *[67]*.

Russell, M.A., Jarvis, M., Iyer, R. and Feyerabend, C. (1980). Relation of nicotine yield of cigarettes to blood nicotine concentrations in smokers. *British Medical Journal*, **280**:972–976.

Russell, M.A.H., Jarvis, M.J., Jones, G. and Feyerabend, C. (1990). Non-smokers show acute tolerance to subcutaneous nicotine. *Psychopharmacology*, **102**:56–58.

Russell, M.A.H., Peto, J. and Patel, U.A. (1974). The classification of smoking by factorial structure of motives. *The Journal of the Royal Statistical Society, Series A (General)*, **137(3)**:313–346.

Rusted, J. and Eaton-Williams, P. (1991). Distinguishing between attentional and amnestic effects in information processing: the separate and combined effects of scopolamine and nicotine on verbal free recall. *Psychopharmacology*, **104**:363–366.

Rusted, J. (1994). Caffeine and cognitive performance: Effects on mood or mental processing? *Pharmacopsychoecologia*, **7**:49–54. *[16]*.

Rusted, J.M. and Warburton, D.M. (1992). Facilitation of memory by post-trial administration of nicotine: evidence for an attentional explanation. *Psychopharmacology*, **108**:452–455.

Rutenfranz, J. and Jansen, G. (1959). On compensation of the alcohol effect by caffeine and pervitin in psychomotor performance. *Internationale Zeitschrift für Angewandte Physiologie*, **18**:62–81.

Ryan, C. and Butters, N. (1979) Accelerated aging in chronic alcoholics: Evidence from tests of learning and memory. Paper presented at the 10th Annual National Council on Alcoholism, Medical-Scientific Conference. Washington D.C.

Ryan, C. (1982). Alcoholism and premature aging: A neuropsychological perspective. *Alcoholism: Clinical and Experimental Research*, **6**:79–96.

Ryback, R.S. (1971). The continuum and specificity of the effects of alcohol on memory. *Quarterly Journal of Studies on Alcoholism*, **32**:215–216.

Sahakian, B., Jones, G., Levy, R., Gray, J. and Warburton, D.M. (1989). The effects of nicotine on attention, information processing, and short term memory in patients with dementia of the Alzheimer type. *British Journal of Psychiatry*, **154**:797–800.

Salame, P. (1991). The effects of alcohol on learning as a function of drinking habits. *Ergonomics*, **34**:1231–1241.

Salamone, J.D. (1991). Behavioral pharmacology of dopamine systems: a new synthesis. Principles of operation. In: P. Willner and J. Scheel-Kruger (Eds.), The mesolimbic dopamine system: From motivation to action. Chichester: Wiley, pp. 599–613.

Salamy, A. and Williams, H.L. (1973). The effects of alcohol on sensory evoked and spontaneous cerebral potentials in man. *Electroencephalography and Clinical Neurophysiology*, **35**:3–11.

Salamy, A. (1973). The effects of alcohol on the variability of the human evoked response. *Neuropharmacology*, **12**:1103–1107.

Saletu, B. (1987). The use of pharmaco-EEG in drug profiling. In: I. Hindmarch and P. Stonier (Eds.), Human psychopharmacology: measures and methods. Vol. 1. Chichester: Wiley, pp. 173–200.

Saletu, B., Anderer, P., Kinsperger, K. and Grünberger, J. (1987). Topographic brain mapping of EEG in neuropsychopharmacology. Part II: Clinical applications (Pharmaco EEG Imaging). *Methods and Findings in Experimental and Clinical Pharmacology*, **9**:385–408.

Sánchez-Alcaraz, A., Ibáñez, P. and Sangrador, G. (1991). Pharmacokinetics of intravenous caffeine in critically ill patients. *Journal of Clinical Pharmacy and Therapeutics*, **16(4)**:285–289.

Sanders, A.F. (1980). Stage analysis of reaction processes. In: G. Stelmach and J. Requin (Eds.), Tutorials on motor behavior. Amsterdam: North-Holland, pp. 331–354.

Sanders, A.F. (1983). Towards a model of stress and human performance. *Acta Psychologica*, **53**:61–97.

Sanders, A.F. (1988). Relative advantages and disadvantages of various performance measures in the assessment of psychotropic drug effects. In: I. Hindmarch, B. Aufdembrinke and H. Ott (Eds.), Psychopharmacology and reaction time. Chichester: Wiley, pp. 115–124.

Sanders, A.F. (1990). Issues and trends in the debate on discrete vs. continuous processing of information. *Acta Psychologica*, **74**:123–167.

Sandman, C.A., Gerner, R., O'Halloran, J.P. and Isenhart, R. (1987). Event-related potentials and item recognition in depressed, schizophrenic and alcoholic patients. *International Journal of Psychophysiology*, **5**:215–225.

Sano, M., Wendt, P.E., Wirsen, A., Stenberg, G., Risberg, J. and Ingvar, D.H. (1993). Acute effects of alcohol on regional cerebral blood flow in man. *Journal of Studies on Alcohol*, **54**:369–376.

Sawada, K., Koyama, E., Kubota, M., Hayashi, I, Komaki, R., Inui, M. and Torii, S. (1992). Effects of odors on EEG relaxation and alpha power. *Chemical Senses*, **17**:88.

Sawynok, J. and Yaksh, T.L. (1993). Caffeine as an analgesic adjuvant: A review of pharmacology and mechanisms of action. *Pharmacological Reviews*, **45(1)**:43–85.

Sayette, M.A. (1993). An appraisal-disruption model of alcohol's effects on stress responses in social drinkers. *Psychological Bulletin*, **114**:459–476.

Sayette, M.A., Wilson, G. and Carpenter, J.A. (1989). Cognitive moderators of alcohol's effects on anxiety. *Behaviour Research and Therapy*, **27**:685–690.

Schachter, S. and Singer, J.E. (1962). Cognitive, social, and physiological determinants of emotional state. *Psychological Review*, **69**:379–399.

Schachter, S. (1973). Nesbitt's paradox. In: W.L. Dunn (Ed.), Smoking behavior: motives and incentives. Washington DC: Winston, pp. 147–155.

Schachter, S. (1978). Pharmacological and psychological determinants of smoking. In: R.E. Thornton (Ed.), Smoking behaviour, physiological and psychological influences. Edinburgh: Churchtll Livingstone, pp. 208–228.

Schachter, S. (1979). Regulation, withdrawal, and nicotine addiction. In: N.A. Krasnegor (Ed.), Cigarette Smoking as a Dependence Process. Rockville, Maryland: National Institute of Drug Abuse.

Schaeffer, K.W. and Parsons, O.A. (1986). Drinking practices and neuropsychological test performance in sober male alcoholics and social drinkers. *Alcohol*, **3**:175–179.

Scheel-Kruger, J. and Willner, P. (1991). The mesolimbic system: Principles of operation. In: P. Willner and J. Scheel-Kruger (Eds.), The mesolimbic dopamine system: From motivation to action. Chichester: Wiley, pp. 559–597.

Schicatano, E.J. and Blumenthal, T.D. (1994). Caffeine delays habituation of the human acoustic startle reflex. *Psychobiology*, **22(2)**:117–122.

Schiffman, S.S., Diaz, C. and Beeker, T.G. (1986). Caffeine intensifies taste of certain sweeteners: Role of adenosine receptor. *Pharmacology, Biochemistry and Behavior*, **24(3)**:429–432. *[68]*.

Schiffman, S.S., Gill, J.M. and Diaz, C. (1985). Methylxanthines enhance taste: Evidence for modulation of taste by adenosine receptor. *Pharmacology, Biochemistry and Behavior*, **22(2)**:195–203. *[69]*.

Schneider, N.G. (1978). The effects of nicotine on learning and short-term memory. Dissertation Abstracts international, 39/10B.

Schneider, W. and Shiffrin, R.M. (1977). Controlled and automatic human information processing: I. Detection, search, and attention. *Psychological Review*, **84**:1–66.

Schori, T.R. and Jones, B.W. (1974). Smoking and multiple-task performance. *Virginia Journal of Science*, **22**:147–151.

Schuckit, M.A. and Gold, E. (1988). A simultaneous evaluation of multiple markers of ethanol/placebo challenges in sons of alcoholics and controls. *Archives of General Psychiatry*, **45**:211–216.

Schuckit, M.A. (1980). Alcoholism and genetics: possible biological mediators. *Biological Psychiatry*, **15**:437–447.

Schuckit, M.A. (1985). Overview: Epidemiology of alcoholism. In: M.A. Schuckit (Ed.), Alcohol Patterns and Problems. New Brunswick, New Jersey: Rutgers University Press, pp. 1–42.

Schuckit, M.A. (1989a). Biomedical and genetic markers of alcoholism. In: H.W. Goedde and D.P. Agarwal (Eds.), Alcoholism: biomedical and genetic aspects. New York: Pergamon Press, pp. 290–302.

Schuckit, M.A. (1989b). Drug and alcohol abuse. A clinical guide to diagnosis and treatment. 3rd ed. New York and London: Plenum, 307 pp.

Schuckit, M.A. (1991). A 10-year follow-up of sons of alcoholics: Preliminary results. *Alcohol and Alcoholism*, **Suppl. 1**:147–149.

Schuckit, M.A., Gold, E.O, Croot, K., Finn, P. and Polich, J. (1988). P300 latency after ethanol ingestion in sons of alcoholics and in controls. *Biological Psychiatry*, **24**:310–315.

Schultze, M.J. (1982). Paradoxical aspects of cigarette smoking: physiological arousal, affect, and individual differences in body cue utilization. Dissertation Abstracts International, 42/11B.

Schwartz, G.E., Wright, K.P., Polak, E.H., Kline, J.P. and Dikman, Z. (1992). Topographic EEG mapping of conscious and unconscious odors. *Chemical Senses*, **17**:695.

Schwartz, J.A., Speed, N.M., Gross, M.D., Lucey, M.R., Bazakis, A.M., Hariharan, M. and Beresford, T.P. (1993). Acute effects of alcohol administration on regional cerebral blood flow: The role of acetate. *Alcoholism: Clinical and Experimental Research*, **17**:1119–1123.

Schwarz, B., Bischof, H.P. and Kunze, M. (1994). Coffee, tea consumption and behavioral factors including smoking, eating, drinking and physical activity. *Preventive Medicine*, **23**:377–384.

Schwarz, E., Kielholz, P., Hobi, V., Goldberg, U., Hofstetter, M., Ladewig, D., Miest, P., Reggiani, G. and Richter, R. (1981). Alcohol-induced background and stimulus-elicited EEG changes in relation to blood alcohol levels. *International Journal of Clinical and Pharmacological Therapeutics and Toxicology*, **19**:102–111.

Schweinberger, S.R., Sommer, W. and Stiller, R.M. (1994). Event-related potentials and models of performance asymmetries in face and word recognition. *Neuropsychologia*, **32**:175–191.

Schweitzer, P.K., Muehlbach, M.J. and Walsh, J.K. (1992). Countermeasures for night work performance deficits: the effect of napping or caffeine on continuous performance at night. *Work and Stress*, **6(4)**:355–365. *[70].*

Scott, N.R., Chakraborty, J. and Marks, V. (1986). Urinary metabolites of caffeine in pregnant women. *British Journal of Clinical Pharmacology*, **22**:475–478.

Scott, N.R., Stambuk, D., Chakraborty, J., Marks, V. and Morgan, M.Y. (1988). Caffeine clearance and biotransformation in patients with chronic liver disease. *Clinical Science*, **74**:377–384.

Seigneur, M., Bonnet, J. and Dorian, B. (1990). Effect of consumption of alcohol, white wine, and red wine on platelet function and serum lipids. *Journal of Applied Cardiology*, **4**:215–222.

Sekizawa, S. and Tsubone, H. (1994). Nasal receptors responding to noxious chemical irritants. *Respiration Physiology*, **96**:37–48.

Seltzer, C.C. and Oechsli, F.W. (1985). Psychosocial characteristics of adolescent smokers before they started smoking: Evidence of self-selection. *Journal of Chronic Diseases*, **38**:17–26.

Seppäläinen, A.M., Savolainen, K. and Kovala, T. (1981). Changes induced by xylene and alcohol in human evoked potentials. *Electroencephalography and clinical Neurophysiology*, **51**:148–155.

Sesardic, D., Boobi, A.R., Murray, B.P., Murray, S., Segura, J., De La Torre, R. and Davies, D.S. (1990). Furafylline is a potent and selective inhibitor of cytochrome P-450IA2 in man. *British Journal of Clinical Pharmacology*, **29**:651–663.

Shahi, G.S. and Moochhala, S.M. (1991). Smoking and Parkinson's disease – A new perspective. *Reviews on Environmental Health*, **9**:123–136.

Shallice, T. and Burgess, P. (1993). Supervisory control of action and thought selection. In: A. Baddeley and L. Weiskrantz (Eds.), Attention: selection, awareness, and control. Oxford: Oxford University Press, pp. 171–187.

Shallice, T. (1988). From neuropsychology to mental structure. Cambridge: Cambridge University Press.

Sharp, D. (1993). When Wein is red. *Lancet*, **341**:27–28.

Shelton, M.D., Parsons, O.A. and Leber, W.R. (1984). Verbal and visuospatial performance in male alcoholics: A test of the premature-aging hypothesis. *Journal of Consulting and Clinical Psychology*, **52(2)**:200–206.

Sherwood, N. and Kerr, J.S. (1993). The reliability, validity and pharmacosensitivity of four psychomotor tests. In: I. Hindmarch and P.D. Stonier (Eds.), Human Psychopharmacology: Measures and Methods. Vol. 4. Chichester: Wiley, pp. 1–14.

Sherwood, N. (1993). Effects of nicotine on human psychomotor performance. *Human Psychopharmacology*, **8**:155–184.

Sherwood, N. (1994). Cognitive and psychomotor effects of nicotine and cigarette smoking. *Recent Advances in Tobacco Science*, **20**:81–105.

Sherwood, N., Kerr, J.S. and Hindmarch, I. (1990). No differences in the psychomotor response of heavy, light and non-smokers to the acute administration of nicotine. *Medical Science Research*, **18**:839–840.

Sherwood, N., Kerr, J.S. and Hindmarch, I. (1991a). Effects of nicotine on short-term memory. *Pharmacopsychoecologia*, **4**:51–55.

Sherwood, N., Kerr, J.S. and Hindmarch, I. (1991b). Subjective ratings of CNS arousal in smokers after acute and repeated doses of nicotine gum. *Medical Science Research*, **19**:455–456.

Sherwood, N., Kerr, J.S. and Hindmarch, I. (1991c). Effects of nicotine gum on short-term memory. In: F. Adlkofer and K. Thurau (Ed.), Effects of Nicotine on Biological Systems. Basel: Birkhäuser, pp. 531–535.

Sherwood, N., Kerr, J.S. and Hindmarch, I. (1992). Psychomotor performance in smokers following single and repeated doses of nicotine gum. *Psychopharmacology*, **108**:432–436.

Shiffman, S. (1989). Tobacco 'chippers': individual differences in tobacco dependence. *Psychopharmacology*, **97**:535–538.

Shiffman, S. (1991, March). Chippers: Characteristics of non-dependent cigarette smokers. Symposium conducted at the annual meeting of the Society of Behavioral Medicine, Washington, DC.

Shiffman, S., Fischer, L.A., Paty, J.A., Gnys, M., Hickcox, M. and Kassel, J.D. (1994). Drinking and smoking: a field study of their association. *Annals of Behavioral Medicine*, **16(3)**:203–209.

Shiffman, S., Paty, J., Kassel, J. and Gnys, M. (1993). Smoking typology profiles of chippers and regular smokers. Unpublished manuscript, Pittsburgh: University of Pittsburgh.

Shiffman, S., Zettler-Segal, M., Kassel, J., Paty, J., Benowitz, N.L. and O'Brien, G. (1992). Nicotine elimination and tolerance in non-dependent cigarette smokers. *Psychopharmacology*, **109**:449–456.

Shiffrin, R.M. and Schneider, W. (1977). Controlled and automatic human information processing: II. Perceptual learning, automatic attending, and a general theory. *Psychological Review*, **84**:127–190.

Shiffrin, R.M. (1988). Attention. In: R.C. Atkinson, R.J. Herrnstein, G. Lindzey and R.D. Luce (Eds.), Stevens' handbook of experimental psychology: Vol. 2. Learning and cognition. New York: Wiley, pp. 739–811.

Shillito, M.L., King, L.E. and Cameron, C. (1974). Effects of alcohol on choice reaction time. *Quarterly Journal of Studies on Alcohol*, **35**:1023–1034.

Shore, E.R. (1985). Alcohol consumption rates among managers and professionals. *Journal of Studies on Alcohol*, **46**:153–156.

Shore, E.R. (1986). Norms regarding drinking behavior in the business environment. *Journal of Social Psychology*, **125**:735–741.

Sieber, M.F. and Angst, J. (1990). Alcohol, tobacco, and cannabis: 12-year longitudinal associations with antecedent social context and personality. *Drug and Alcohol Dependence*, **25**:281–292.

Siegel, S. (1983). Classical conditioning, drug tolerance, and drug dependence. In: R.G. Smart, F.B. Glaser, Y. Israel, H. Kalant, R.E. Popham and W. Schmidt (Eds.), Research advances in alcohol and drug problems. New York: Plenum Press, pp. 207–246.

Siegers, C.-P., Strubelt, O. and Back, G. (1974). Azione inhibitrice della caffeina sull'assorbimento di etanolo nei ratti. *Archivio per le Scienze Mediche*, **131**:173–180.

Silagy, C., Mant, D., Fowler, G. and Lodge, M. (1994). Meta-analysis on efficacy of nicotine replacement therapies in smoking cessation. *Lancet*, **343**:139–142.

Silverman, K., Evans, S.M., Strain, E.C. and Griffiths, R.R. (1992). Withdrawal syndrome after double-blind cessation of caffeine consumption. *New England Journal of Medicine*, **327**:1109–1114.

Silverman, K., Mumford, G.K. and Griffiths, R.R. (1994). Enhancing caffeine reinforcement by behavioral requirements following drug ingestion. *Psychopharmacology*, **114**:424–432.

Simon, O. (1990). The effects of an irrelevant directional cue on human information processing. In: R.W. Proctor and T.G. Reeve (Eds.), Stimulus-response compatibility. An integrated perspective. Amsterdam: North-Holland, pp. 31–86.

Simpson, D., Erwin, C.W. and Linnoila, M. (1981). Ethanol and menstrual cycle interactions in the visual evoked response. *Electroencephalography and clinical Neurophysiology*, **52**:28–35.

Smid, H.G.O.M., Mulder, G. and Mulder, L.J.M. (1987). The continuous flow model revisited: Perceptual and motor aspects. In: R. Johnson Jr., J.W. Rohrbaugh and R. Parasuraman (Eds.), Current trends in event-related potential research (EEG Suppl. 40). Amsterdam: Elsevier, pp. 270–278.

Smith, A.P. and Miles, C. (1985). The combined effects of noise and nightwork on human function. In: D. Oborne (Ed.), Contemporary Ergonomics 1985, London: Taylor and Francis, pp. 33–41.

Smith, A.P. and Miles, C. (1986a). The combined effects of nightwork and noise, on human function. In: M. Haider, M. Koller and R. Cervinka (Eds.), Studies in Industrial and Organizational Psychology, Part 3: Night and Shiftwork – Long term effects and their prevention. Frankfurt: Peter Lang, pp. 331–338.

Smith, A.P. and Miles, C. (1986b). Acute effects of meals, noise and nightwork. *British Journal of Psychology*, **77**:377–389.

Smith, A.P. and Miles, C. (1987a). The combined effects of occupational health hazards: an experimental investigation of the effects of noise, nightwork and meals. *International Archives of Occupational and Environmental Health*, **59**:83–89.

Smith, A.P. and Miles, C. (1987b). Sex differences in the effects of noise and nightwork on performance. *Work and Stress*, **1**:333–339.

Smith, A.P. and Phillips, W. (1993). Effects of low doses of caffeine in coffee on human performance and mood. 15th Colloque Scientifique International sur le café. Vol. I, Montpellier. Paris: Association Scientifique Internationale du Café, pp. 461–469.

Smith, A.P. (1988). Individual differences in the combined effects of noise and nightwork on human performance. In: O. Manninen (Ed.), Recent Advances in Researches on the Combined Effects of Environmental Factors. Tampere: Finland, pp. 365–380.

Smith, A.P. (1990). An experimental investigation of the combined effects of noise and nightwork on human function, In: B. Berglund and T. Lindwall (Eds.), Noise as a Public Health Problem. Vol. 5. New Advances in Noise Research, Part II. Stockholm: Swedish Council for Building Research, pp. 255–271.

Smith, A.P. (1991). The combined effects of noise, nightwork and meals on mood. *International Archives of Occupational and Environmental Health*, **63**:105–108.

Smith, A.P. (1992). Colds, influenza and performance. In: A.P. Smith and D.M. Jones (Eds.), Handbook of Human Performance. Vol. 2. London: Academic Press, pp. 197–218.

Smith, A.P. (1994). Caffeine, performance, mood and states of reduced alertness. *Pharmacopsychoecologia*, **7**:75–86. *[71]*.

Smith, A.P. (1995). Contextual factors modifying the effects of coffee on performance and mood. Report to P.E.C. Paris: Committee on Physiological Effects of Caffeine.

Smith, A.P., Brockman, P., Flynn, R., Maben, A. and Thomas, M. (1993a). Investigation of the effects of coffee on alertness and performance during the day and night. *Neuropsychobiology*, **27**:217–223. *[72]*.

Smith, A.P., Kendrick, A. and Maben, A. (1992b). Effects of caffeine, lunch and alcohol on human performance, mood and cardiovascular function. *Proceedings of the Nutrition Society*, **51**:325–333. *[73]*.

Smith, A.P., Kendrick, A., Maben, A. and Salmon, J. (1994a). Effects of breakfast and caffeine on cognitive performance, mood and cardiovascular functioning. *Appetite*, **22**:39–55. *[74]*.

Smith, A.P., Kendrick, A.M. and Maben, A.L. (1992a). Effects of breakfast and caffeine on performance and mood in the late morning and after lunch. *Neuropsychobiology*, **26**:198–204.

Smith, A.P., Maben, A. and Brockman, P. (1993b). The effects of caffeine and evening meals on sleep and performance, mood and cardiovascular functioning the following day. *Journal of Psychopharmacology*, **7**:203–206.

Smith, A.P., Maben, A. and Brockman, P. (1994b). Effects of evening meals and caffeine on cognitive performance, mood and cardiovascular functioning. *Appetite*, **22**:57–65. *[75]*.

Smith, A.P., Rusted, J.M., Eaton-Williams, P., Savory, M. and Leathwood, P. (1990). Effects of caffeine given before and after lunch on sustained attention. *Neuropsychobiology*, **23**:160–163. *[76]*.

Smith, A.P., Rusted, J.M., Savory, M., Eaton-Williams, P. and Hall, S.R. (1991). The effects of caffeine, impulsivity and time of day on performance, mood and cardiovascular function. *Journal of Psychopharmacology*, **5(2)**:120–128. *[77]*.

Smith, B.D. (1994). Effects of acute and habitual caffeine ingestion on physiology and behavior: Tests of a biobehavioral arousal theory. *Pharmacopsychoecologia*, **7**:151–167.

Smith, B.D., Davidson, R.A. and Green, R.L. (1993). Effects of caffeine and gender on physiology and performance: Further tests of a biobehavioral model. *Physiology and Behavior*, **54(3)**:415–422. *[78]*.

Smith, B.D., Rafferty, J., Lindgren, K., Smith, D.A. and Nespor, A. (1991). Effects of habitual caffeine use and acute ingestion: Testing a biobehavioral model. *Physiology and Behavior*, **51**:131–137. *[79]*.

Smith, C.J. and Giacobini, E. (1992). Nicotine, Parkinson's and Alzheimer's disease. *Reviews in the Neurosciences*, **3**:25–43.

Smith, D.L., Tong, J.E. and Leigh, G. (1977). Combined effects of tobacco and caffeine on the components of choice reaction time, heart rate, and hand steadiness. *Perceptual and Motor Skills*, **45**:635–639.

Smith, M.E. and Halgren, E. (1989). Dissociation of recognition memory components following temporal lobe lesions. *Journal of Experimental Psychology: Learning, Memory, and Cognition*, **15**:50–60.

Smits, P., Temme, L. and Thien, T. (1993). The cardiovascular interaction between caffeine and nicotine in humans. *Clinical Pharmacology and Therapeutics*, **54**:194–204.

Smulders, F.T.Y. (1993). The selectivity of age effects on information processing. Response times and electrophysiology. Ph.D. Thesis, University of Amsterdam, Amsterdam.

Snel, J. and Lorist, M.M. (1996). Caffeine and information processing – its role in sensory perception. In: D.M. Warburton and N. Sherwood (Eds.) Pleasure and quality of life. Proceedings of the ARISE-meeting, Amsterdam, Chichester: Wiley, pp. 97–114.

Snyder, F.R. and Henningfield, J.E. (1989). Effects of nicotine administration following 12 h of tobacco deprivation: Assessment on computerized performance tasks. *Psychopharmacology*, **97**:17–22.

Snyder, F.R., Davis, F.C. and Henningfield, J.E. (1989). The tobacco withdrawal syndrome: Performance decrements assessed on a computerized test battery. *Drug and Alcohol Dependence*, **23**:259–266.

Solomon, R.L. (1977). An opponent-process of acquired motivation: The affective dynamics of addiction. In: J.D. Maser and M.E.P. Seligman (Eds.), Psychopathology: experimental models. San Francisco: Freeman, pp. 66–103.

Sommer, W., Leuthold, H. and Hermanutz, M. (1993). Covert effects of alcohol revealed by event-related potentials. *Perception and Psychophysics*, **54(1)**:127–135.

Spielberger, C.D. (1986). Psychological determinants of smoking behavior. In: R.D. Tollison (Ed.), Smoking and society: Toward a more balanced assessment. Lexington, MA: D.C. Heath and Company, pp. 89–134.

Spielman, D.M., Glover, G.H., Macovski, A. and Pfefferbaum, A. (1993). Magnetic resonance spectroscopic imaging of ethanol in the human brain: A feasibility study. *Alcoholism: Clinical and Experimental Research*, **17**:1072–1077.

Spilich, G.J., June, L. and Renner, J. (1992). Cigarette smoking and cognitive performance. *British Journal of Addiction*, **87**:1313–1326.

Spilker, B. and Callaway, E. (1969). Effects of drugs on "Augmenting/Reducing" in averaged visual evoked responses in man. *Psychopharmacologia*, **15**:116–124.

Spring, B., Pingitore, R., Kessler, K., Mahableshwarker, A., Bruckner, E., Kohlbeck, R. and Braun, J. (1993). Fluoxetine prevents withdrawal dysphoria but not anticipatory anxiety about quitting smoking. Proceedings of the fourteenth annual meeting of the Society of Behavioral Medicine. *Annals of Behavioral Medicine*, **15**:129.

Squires, K.C., Chu, N.-S. and Starr, A. (1978). Acute effects of alcohol on auditory brainstem potentials in humans. *Science*, **201**:174–176.

Srivastava, L.M., Vasisht, S., Agarwal, D.P. and Goedde, H.W. (1994). Relation between alcohol intake, lipoproteins and coronary heart disease: the interest continues. *Alcohol and Alcoholism*, **29**:11–24.

Stacy, A.W., Leigh, B.C. and Weingardt, K.R. (1994). Memory accessibility and association of alcohol use and its positive outcomes. *Experimental and Clinical Psychopharmacology*, **2**:269–282.

Stacy, A.W., Widaman, K.F. and Marlatt, G.A. (1990). Expectancy models of alcohol use. *Journal of Personality and Social Psychology*, **58**:918–928.

Stålhandske, T. and Slanina, P. (1982). Nicotyrine inhibits *in vivo* metabolism of nicotine without increasing its toxicity. *Toxicology and Applied Pharmacology*, **65**:366–72.

Starr, A. and Don, M. (1988). Brain potentials evoked by acoustic stimuli. In: T.W. Picton (Ed.), Handbook of electroencephalography and clinical neurophysiology. Revised series: Vol. 3. Human event-related potentials. Amsterdam: Elsevier, pp. 97–157.

Stassen, H.H., Lykken, D.T., Propping, P. and Bomben, G. (1988). Genetic determination of the human EEG. *Human Genetics*, **80**:165–170.

Statland, B.E. and Demas, T.J. (1980). Serum caffeine half-lives. Healthy subjects vs. patients having alcoholic hepatic disease. *American Journal of Clinical Pathology*, **73**:390–393.

Statland, B.E., Demas, T.J. and Danis, M. (1976). Caffeine accumulation associated with alcoholic liver disease. *New England Journal of Medicine*, **295**:110–111.

Steele, C.M. and Josephs, R.A. (1990). Alcohol myopia: Its prized and dangerous effects. *American Psychologist*, **45(8)**:921–933.

Steffensen, S.C., Yeckel, M.F., Miller, D.R. and Henricksen, S.J. (1993). Ethanol-induced suppression of hippocampal long-term potentiation is blocked by lesions of the septohippocampal nucleus. *Alcoholism: Clinical and Experimental Research*, **17**:655–659.

Steinberg, D., Pearson, T.A. and Kuller, L.H. (1991). Alcohol and atherosclerosis. *Annals of Internal Medicine*, **114**:967–976.

Steinhauer, S.R., Hill, S.Y. and Zubin, J. (1987). Event-related potentials in alcoholics and their first-degree relatives. *Alcohol*, **4**:307–314.

Stepney, R. (1979). Smoking as a psychological tool. *Bulletin of the British Psychological Society*, **32**:341–345.

Sternberg, S. (1966). High-speed scanning in human memory. *Science*, **153**:652–654.

Sternberg, S. (1969). The discovery of processing stages: Extensions of Donders' method. *Acta Psychological*, **30**:276–315.

Stevens, H.A. (1976). Evidence that suggests a negative association between cigarette smoking and learning performance. *Journal of Clinical Psychology*, **32**:896–898.

Stewart, S.H. and Pihl, R.O. (1993). Effects of alcohol administration on psychophysiological and subjective-emotional responses to aversive stimulation in anxiety-sensitive women. *Psychology of Addictive Behaviors*, **8**:29–42.

Stokes, A.F., Belger, A., Banich, M.T. and Taylor, H. (1991). Effects of acute aspartame and acute alcohol ingestion upon the cognitive performance of pilots. *Aviation, Space and Environmental Medicine*, **62**:648–653.

Stough, C., Bates, T.C., Mangan, G.L. and Pellett, O.L. (1995). Smoking, string length and intelligence. *Personality and Individual Differences*, **18**:75–79.

Strain, E.C., Mumford, G.K., Silverman, K. and Griffiths, R.R. (1994). Caffeine dependence syndrome. *Journal of the American Medical Association*, **272**:1043–1048.

Streufert, S. and Nogami, G.Y. (1989). Cognitive style and complexity: Implications for I/O psychology. In: C.L. Cooper and I. Robertson (Eds.), International review of industrial and organizational psychology. London: Wiley, pp. 93–143.

Streufert, S. and Streufert, S.C. (1978). Behavior in the complex environment. Washington DC and New York: Wiley.

Streufert, S. and Swezey, R.W. (1985). Simulation and related research methods in environmental psychology. In: A. Baum and J.E. Singer (Eds.), Advances in environmental psychology. Vol. 5. Hillsdale: Lawrence Erlbaum, pp. 99–117.

Streufert, S. and Swezey, R.W. (1986). Complexity, managers and organizations. San Diego, California: Academic Press.

Streufert, S., Pogash, R. and Piasecki (1988). Simulation based assessment of complex managerial performance: Reliability and validity. *Personnel Psychology*, **41**:537–557.

Streufert, S., Pogash, R., Miller, J., Gingrich, D., Landis, R., Lonardi, L., Severs, W. and Roache, J.D. (1995). Effects of caffeine deprivation on complex human functioning. *Psychopharmacology*, **118**:377–384.

Streufert, S., Pogash, R., Roache, J., Severs, W., Gingrich, D., Landis, R., Lonardi, L. and Kantner, A. (1994). Alcohol and managerial performance. *Journal of Studies on Alcohol*, **55**:230–238.

Streufert, S., Pogash, R.M., Roache, J., Gingrich, D., Landis, R., Severs, W., Lonardi, L. and Kantner, A. (1992). Effects of alcohol intoxication on risk taking, strategy, and error rate in visuomotor performance. *Journal of Applied Psychology*, **77**:515–524.

Strongin, E.I. and Winsor, A.L. (1935). The antagonistic action of coffee and alcohol. *Journal of Abnormal Social Psychology*, **30**:301–313.

Stroop, J.R. (1935). Studies of interference in serial verbal reactions. *Journal of Experimental Psychology*, **18**:643–661.

Strubelt, O., Böhme, K., Siegers, C.-P. and Bruhn, P. (1976). Der Einfluss von Coffein auf die Resorption und einige zentrale Wirkungen von Ethanol. *Zeitschrift für Ernährungswissenschaft*, **15**:125–131.

Sturtevant, F.M. and Sturtevant, R.P. (1979). Chronopharmacokinetics of ethanol. In: E. Majchrowicz and E.P. Noble (Eds.), Biochemistry and pharmacology of ethanol. Vol. 1. New York: Plenum, pp. 27–40.

Suedfeld, P. and Ikard, F.F. (1974). Use of sensory deprivation in facilitating the reduction of cigarette smoking. *Journal of Consulting and Clinical Psychology*, **42**:888–895.

Suter, T.W., Buzzi, R., Woodson, P.P. and Bättig, K. (1983). Psychophysiological correlates of conflict solving and cigarette smoking. *Activitas Nervosa Superior*, **25**:261–271.

Sutherland, G., Russell, M.A.H., Stapleton, J., Feyerabend, C. and Ferno, O. (1992a). Nasal nicotine spray: A rapid nicotine delivery system. *Psychopharmacology*, **108**:512–518.

Sutherland, G., Stapleton, J.A., Russell, M.A.H., Jarvis, M.J., Hajek, P., Belcher, M. and Feyerabend, C. (1992b). Randomised controlled trial of nasal nicotine spray in smoking Cessation. *Lancet*, **340**:324–329.

Suys, N., Assaad, J.M., Peterson, J.B. and Pihl, R.O. The effects of ethanol and tobacco on memory. Unpublished manuscript. Montreal: Department of Psychology, McGill University.

Swanson, J.A., Lee, J.W. and Hopp, J.W. (1994). Caffeine and nicotine: a review of their joint use and possible interactive effects in tobacco withdrawal. *Addictive Behaviors*, **19**:229–256.

Swift, C.G. and Tiplady, B. (1988). The effects of age on the response to caffeine. *Psychopharmacology*, **94(1)**:29–31. *[80]*.

Tabakoff, B. and Hoffman, P.L. (1993). The neurochemistry of alcohol. *Current Opinion in Psychiatry*, **6**:388–394.

Taghavy, A., Kügler, C.F.A., Brütting, L., Taghavy, D. and Machbert, G. (1991). Dosiswirkungsbeziehungen zwischen Blutalkoholkonzentrationen und kognitiven Potentialen (visuelles P300) beim Menschen. *Zeitschrift für EEG-EMG*, **22**:83–88.

Taghavy, A., Penning, J. and Hoh, E. (1976). Gleichzeitige Ableitung visuell evozierter Potentiale (VEP) und Registrierung einfacher Reaktionszeiten (RZ) im "Maximalbereich" der Äthanolwirkung. *Arzneimittel-Forschung (Drug Research)*, **26**:1125–1126.

Takada, A., Tsutsumi, M. and Kobayashi, Y. (1994). Genotypes of ALDH2 related to liver and pulmonary disease and other genetic factors related to alcoholic liver disease. *Alcohol and Alcoholism*, **29**:719–727.

Tang, B.K., Kadar, D. and Kalo, W. (1987). An alternative test for acetylator phenotyping with caffeine. *Clinical Pharmacology and Therapeutics*, **42**:509–513.

Tang, B.K., Zubovits, T. and Kalow, W. (1986). Determination of acetylated caffeine metabolites by high-performance exclusion chromatography. *Journal of Chromatography*, **375**:170–173.

Tang-Liu, D.D.S., Williams, R.L. and Reigelman, S. (1983) Disposition of caffeine and its metabolites in man. *Journal of Pharmacology and Experimental Therapeutics*, **224**:180–185.

Tarriere, H.C. and Hartemann, F. (1964). Investigations into the effects of tobacco smoke on a visual vigilance task. In: Proceedings of the second international congress on ergonomics. Dortmund (Suppl. to Ergonomics). London: Taylor and Francis, pp. 525–530.

Tarrús, E., Cami, J., Roberts, D.J., Spickett, R.G.W., Celdran, E. and Segura, J. (1987). Accumulation of caffeine in healthy volunteers treated with furafylline. *British Journal of Clinical Pharmacology*, **23**:9–18.

Tarter, R.E., Kabene, M., Escallier, E.A., Laird, S.B. and Jacob, T. (1990). Temperament deviation and risk for alcoholism. *Alcoholism Clinical and Experimental Research*, **14**:380–382.

Taylor, E.M. (1969). Localization of cerebral lesions by psychological testing. *Clinical Neurosurgery*, **16**:269–287.

Teo, R.K.C. and Ferguson, D.A. (1986). The acute effects of ethanol on auditory event-related potentials. *Psychopharmacology*, **90**:179–184.

Terry, W.S. and Phifer, B. (1986). Caffeine and memory performance on the AVLT. *Journal of Clinical Psychology*, **42(6)**:860–863. *[81]*.

Teyler, T.J. and DiScenna, P. (1986). The hippocampal memory indexing theory. *Behavioral Neuroscience*, **100**:147–154.

Tharp Jr., V.K., Rundell Jr., O.H., Lester, B.K. and Williams, H.L. (1974). Alcohol and information processing. *Psychopharmacologia*, **40**:33–52.

Thomasson, H.R. and Li, T.K. (1993). How alcohol and aldehyde dehydrogenase genes modify alcohol drinking, alcohol flushing, and the risk for alcoholism. *Alcohol, Health and Research World*, **17**:167–172.

Thomasson, H.R., Edenberg, H.J. and Crabb, D.W. (1991). Alcohol and aldehyde dehydrogenase genotypes and alcoholism in Chinese men. *American Journal of Human Genetics*, **48**:677–681.

Thompson, J.B. and Newlin, D.B. (1988). Effects of alcohol conditioning and expectancy on a visuo-motor integration task. *Addictive Behaviours*, **13**:73–77.

Threatt, R. (1976). The influence of alcohol on work performance. Washington DC: NIAAA Publication FR 16–67.

Tiffany, S.T. (1990). A cognitive model of drug urges and drug-use behavior: Role of automatic and nonautomatic processes. *Psychological Review*, **97**:147–168.

Tiffany, S.T. (1992). A critique of contemporary urge and craving research: Methodological, psychometric and theoretical issues. *Advances in Behaviour Research and Therapy*, **14**:123–139.

Tiplady, B. (1991). Alcohol as a comparator. In: I.D. Keppler, L.D. Sanders and M. Rosen (Eds.), Ambulatory Anaesthesia and Sedation: Impairment and Recovery. Oxford: Blackwell, pp. 26–37.

Tomkins, S.S. (1966). Psychological model of smoking behavior. *American Journal of Public Health*, **56(Suppl.)**:17–20.

Tong, J.E., Booker, J.L. and Knott, V.J. (1978). Effects of tobacco, time on task, and stimulus speed on judgments of velocity and time. *Perceptual and Motor Skills*, **47**:175–178.

Tong, J.E., Henderson, P.R. and Chipperfield, B.G. (1980). Effects of ethanol and tobacco on auditory vigilance performance. *Addictive Behavior*, **5**:153–158.

Tong, J.E., Knott, V.J., McGraw, D.J. and Leigh, G. (1974). Alcohol, visual discrimination and heart rate, effects of dose, activation and tobacco. *Quarterly Journal of Studies on Alcohol*, **35**:1003–1022.

Tong, J.E., Leigh, G., Campbell, J. and Smith, D. (1977). Tobacco smoking, personality and sex factors in auditory vigilance performance. *British Journal of Psychology*, **68**:365–370.

Torii, S., Fukuda, H., Kanemoto, H., Miyanchi, R., Hamauzu, Y. and Kawasaki, M. (1991). Contingent Negative Variation and the Psychological Effects of Odour. In: S. Van Toller and G.H. Dodd (Eds.), Perfumery: the Psychology and Biology of Fragrances. London: Chapman and Hall University Press, pp. 107–120.

Treisman, A. (1988). Features and objects: The fourteenth Bartlett memorial lecture. *The Quarterly Journal of Experimental Psychology*, **40**:201–237.

Treisman, A. (1993). The perception of features and objects. In: A. Baddeley and L. Weiskrantz (Eds.), Attention: Selection, awareness, and control: A tribute to Donald Broadbent. Oxford: Clarendon Press, pp. 5–35.

Tucker, D.M. and Vuchinich, R.E. (1983). An information processing analysis of the effects of alcohol on perceptions of facial emotions. *Psychopharmacology*, **79**:215–219.

Tucker, D.M. and Williamson, P.A. (1984). Asymmetric neural control systems in human self-regulation. *Psychological Review*, **91**:185–215.

Tucker, J.A., Vuchinich, R.E. and Schonhaut, S.J. (1987). Effects of alcohol on recall of social interactions. *Cognitive Therapy and Research*, **11**:273–283.

Turkkan, J.S., Stitzer, M.L. and McCaul, M.E. (1988). Psychophysiological effects of oral ethanol in alcoholics and social drinkers. *Alcoholism: Clinical and Experimental Research*, **12(1)**:30–38.

Turner, T.B., Bennett, V.L. and Hernandez, H. (1981). The beneficial side of moderate alcohol use. *John Hopkins Medical Journal*, **148**:53–63.

Tyson, P.D. and Shirmuly, M. (1994). Memory enhancement after drinking ethanol: Consolidation, interference or response bias? *Physiology and Behavior*, **56**:933–937.

U.S. Department of Agriculture and U.S. Department of Health and Human Services (1990). Nutrition and your health: Dietary guidelines for Americans, 3rd ed. Washington DC: USDA and USDHHS. 25 p.

U.S. Department of Health and Human Services (1988). The health consequences of smoking: Nicotine addiction. A report of the Surgeon General (DHHS Publication No. CDC 88–8406). Washington, DC: U.S. Government Printing Office.

Uematsu, T., Mizuno, A., Itaya, T., Suzuki, Y., Kanamaru, M. and Nakashima, M. (1987). Psychomotor performance tests using a microcomputer: evaluation of effects of caffeine and chlorpheniramine in healthy human subjects. *Yakubutsu Seishin Kodo (Japanese Journal of Psychopharmacology)*, **7(4)**:427–432. *[82]*.

Umilta, C. (1988). The control operations of consciousness. In: A.J. Marcel and E. Bisiach (Eds.), Consciousness in contemporary science. Oxford: Oxford University Press, pp. 334.

Ungeleider, L.G. and Mishkin, M. (1982). Two cortical visual systems. In: D.J. Ingle, M.A. Goodale and R.J.W. Mansfield (Eds.), Analysis of visual behavior. Cambridge: MIT Press, pp. 549–586.

Valeriote, C., Tong, J.E. and Durding, B. (1979). Ethanol, tobacco and laterality effects on simple and complex motor performance. *Journal of Studies on Alcohol*, **40**:823–830.

Van der Heijden, A.H.C. (1993). The role of position in object selection in vision. *Psychological Research*, **56**:44–58.

Van der Molen, M., Bashore, T., Halliday, R. and Callaway, E. (1991). Chronopsychophysiology: Mental chronometry augmented by physiological time markers. In: J.R. Jennings and M.G.H. Coles (Eds.), Handbook of cognitive psychophysiology. Chichester: Wiley, pp. 9–178.

Van der Stelt, O. and Snel, J. (1993). Effects of caffeine on human information processing: A cognitive-energetic approach. In: S. Garattini (Ed.), Caffeine, coffee, and health. New York: Raven Press, pp. 291–316.

Van der Stelt, O. (1994). Caffeine and attention. *Pharmacopsychoecologia*, **7**:221–227.

Van Duren, L.L. and Sanders, A.F. (1988). On the robustness of the additive factors stage structure in blocked and mixed choice reaction designs. *Acta Psychologica*, **69**:83–94.

Van Dusseldorp, M. and Katan, M.B. (1990). Headache caused by caffeine withdrawal among moderate coffee drinkers switched to ordinary decaffeinated coffee: A 12-week double blind trial. *British Medical Journal*, **300**:1558–1559.

Van Petten, C., Kutas, M., Kluender, R., Mitchiner, M. and McIsaac, H. (1991). Fractionating the word repetition effect with event-related potentials. *Journal of Cognitive Neuroscience*, **3**:131–150.

Van Reek, J. (1986). Cigarette smoking in the USA: Sociocultural influences. *Revue Épidemiologique et Santé Publique*, **34**:168–173.

Van Toller, S., Behan, J., Howells, P., Kendal-Reed, M. and Richardson, A. (1993). An analysis of spontaneous human Cortical EEG activity to odours. *Chemical Senses*, **18**:116–227.

Varagnolo, M., Plebani, M., Mussap, M., Nemetz, L., Paleari, C.D. and Burlina, A. (1989). Caffeine as indicator of metabolic functions of microsomal liver enzymes. *Clinica Chimica Acta*, **183**:91–94.

Verbrugge, L.M. (1989). The twain meet: empirical explanations of sex differences in health and mortality. *Journal of Health and Social Behavior*, **30**:282–304.

Verleger, R. (1988). Event-related potentials and cognition: A critique of the context updating hypothesis and an alternative interpretation of P_3. *Behavioral and Brain Sciences*, **11**:343–427.

Vesell, E.S., Page, J.G. and Passananti, G.T. (1971). Genetic and environmental factors affecting ethanol metabolism in man. *Clinical Pharmacology and Therapeutics*, **12**:92–201.

Vistisen, K., Loft, S. and Poulsen, H.E. (1990). Cytochrome P450IA2 activity in man measured by caffeine metabolism: effect of smoking, broccoli and exercise. *Advances in experimental and biological biology*, **283**:407–411.

Voevoda, M.I., Avksentyuk, A.V. and Ivanova, A.V. (1994). Molecular genetic studies in the populations of native inhabitants of Chuckhee Peninsula. Analysis of polymorphism of mitochondrial DNA and of genes controlling alcohol metabolizing enzymes. *Siberian Journal of Ecology*, **2**:139–151.

Von Wartburg, J.P. and Schurch, P.M. (1968). Atypical human liver alcohol dehydrogenase. *Annals of the New York Academy of Sciences*, **151**:936–946.

Von Wartburg, J.P. (1989). Pharmacokinetics of alcohol. In: K. E. Crow and R.D. Batt (Eds.), Human metabolism of alcohol. Vol. I. Pharmacokinetics, medicolegal aspects, and general interests. Boca Raton: CRC Press, pp. 9–22.

Waller, D. and Levander, S. (1980). Smoking and vigilance: the effects of tobacco on CFF as related to personality and smoking habits. *Psychopharmacology*, **70**:131–136.

Walsh, J.M. (1894). Coffee: its History, Classification and Description. Philadelphia: Winston.

Walton, R.G. (1972). Smoking and alcoholism: a brief report. *American Journal of Psychiatry*, **128(11)**:1455–1456.

Wang, T., Kleber, G., Stellaard, F. and Paumgartner, G. (1985). Caffeine elimination: a test of liver function. *Klinische Wochenschrift*, **63**:1124–1128.

Warburton, D.M. and Arnall, C. (1994). Improvements in performance without nicotine withdrawal. *Psychopharmacology*, **115**:539–542.

Warburton, D.M. and Rusted, J.M. (1989). Memory assessment. In: I. Hindmarch and P.D. Stonier (Eds.), Human psychopharmacology: Measures and Methods. Vol. 2. Chichester: Wiley, pp. 155–178.

Warburton, D.M. and Rusted, J.M. (1993). Cholinergic control of cognitive resources. *Neuropsychobiology*, **28**:43–46.

Warburton, D.M. and Thompson, D.H. (1994). An evaluation of the effects of caffeine in terms of anxiety, depression and headache in the general population. *Pharmacopsychoecologia*, **7**:55–61.

Warburton, D.M. and Walters, A.C. (1989). Attentional processing. In: T. Ney and A. Gale (Eds.), Smoking and human behavior. Chichester: Wiley, pp. 223–237.

Warburton, D.M. and Wesnes, K. (1978). Individual differences in smoking and attentional performance. In: R.E. Thornton (Ed.), Smoking Behaviour: Physiological and Psychological Influences. Edinburgh: Churchill Livingstone, pp. 19–43.

Warburton, D.M. (1981). Neurochemistry of behaviour. *British Medical Bulletin*, **37**:121–125.

Warburton, D.M. (1987). The functions of smoking. In: W.R. Martin, G.R. Van Loon G., E.T. Iwamoto and D.L. Davis (Eds.), Tobacco Smoke and Nicotine: A Neurobiologic Approach. New York: Plenum, pp. 51–61.

Warburton, D.M. (1988a). The puzzle of nicotine use. In: M. Lader (Ed.), The Psychopharmacology of Addiction. Oxford: Oxford University Press, pp. 27–49.

Warburton, D.M. (1988b). The functional use of nicotine. In: N.J. Wald and P. Froggatt (Eds.), Nicotine, smoking, and the low tar programme. Oxford: Oxford University Press, pp. 182–199.

Warburton, D.M. (1989a). Is nicotine use an addiction? *The Psychologist: Bulletin of the British Psychological Society*, **4**:166–170.

Warburton, D.M. (1989b). The functional use of nicotine. In: N. Wald and P. Froggatt (Eds.), Nicotine, smoking and the low tar programme. Oxford: Oxford University Press, pp. 182–199.

Warburton, D.M. (1990a). Heroin, cocaine and now nicotine. In: D.M. Warburton (Ed.), Addiction controversies. Chur: Harwood Academic Publishers, pp. 21–35.

Warburton, D.M. (1990b). Psychopharmacological aspects of nicotine. In: S. Wonnacott, M.A.H. Russell and I.P. Stolerman (Eds.), Nicotine psychopharmacology. Oxford: Oxford University Press, pp. 77–111.

Warburton, D.M. (1991). From molecules to cognitions. *The Psychologist*, **4**:533–536.

Warburton, D.M. (1992). Nicotine as a cognitive enhancer. *Progress in Neuropsychopharmacology and Biological Psychiatry*, **16**:181–191.

Warburton, D.M. (1993). Letter to the editor. Science News, **143**:147,159.

Warburton, D.M. (1994). The reality of the beneficial effects of nicotine. In: Comments on West's editorial "Beneficial effects of nicotine: fact or fiction?" *Addiction*, **89**:135–146.

Warburton, D.M. (1995). The effects of caffeine on cognition and mood without caffeine abstinence. *Psychopharmacology*, **119**:66–70.

Warburton, D.M., Revell, A.D. and Walters, A.C. (1988). Nicotine as a resource. In: M. Rand and K. Thurau (Eds.), The pharmacology of nicotine. London: IRL Press, pp. 359–373.

Warburton, D.M., Rusted, J.M. and Fowler, J.A. (1992a). Comparison of the attentional and consolidation hypotheses for the facilitation of memory by nicotine. *Psychopharmacology*, **108**:443–447.

Warburton, D.M., Rusted, J.M. and Müller, C. (1992b). Patterns of facilitation of memory by nicotine. *Behavioural Pharmacology*, **3**:375–378.

Warburton, D.M., Wesnes, K. and Revell, A. (1984). Smoking and academic performance. *Current Psychological Research and Reviews*, **3**:25–31.

Warburton, D.M., Wesnes, K., Shergold, K. and James, M. (1986). Facilitation of learning and state dependence with nicotine. *Psychopharmacology*, **89**:55–59.

Ward, M.M., Swan, G.E. and Jack, L.M. (1994). Subjective complaints after smoking cessation: Transient vs. offset effects. Proceedings of The Society of Behavioral Medicine's Fifteenth Anniversary Meeting. *Annals of Behavioral Medicine*, **16**:S043.

Warm, J.S., Dember, W.N. and Parasuraman, R. (1991). Effects of olfactory stimulation on performance and stress in a visual sustained attention task. *Journal of the Society of Cosmetic Chemists*, **42**:199–210.

Warwick, K.M. and Eysenck, H.J. (1963). The effects of smoking on CFF threshold. *Life Sciences*, **4**:219–225.

Warwick, K.M. and Eysenck, H.J. (1968). Experimental studies of the behavioural effects of nicotine. *Pharmarkopsychiatrie Neuro-Psychopharmakologie*, **1**:145–169.

Waugh, M., Jackson, M., Fox, G.A., Hawke, S.H. and Tuck, R.R. (1989). Effect of social drinking on neuropsychological performance. *British Journal of Addiction*, **84**:659–667.

Wayner, M.J., Armstrong, D.L., Polan-Curtain, J.L. and Denny, J.B. (1993). Ethanol and diazepam inhibition of hippocampal LTP is mediated by angiotensin II and AT1 receptors. *Peptides*, **14**:441–444.

Webster, D.D. (1964). The dynamic quantification of spasticity with automated integrals of passive motion resistance. *Clinical Pharmacology and Therapeutics*, **5**:900–908.

Wechsler, D. (1981). Wechsler Adult Intelligence Scale-Revised. Psychological Corporation, New York.

Wedel, M., Pieters, J.E., Pikkar, N.A. and Ockhuizen, Th. (1991). Application of a three-compartment model to a study of the effects of sex, alcohol dose, and concentration, exercise and food consumption on the pharmacogenetics of ethanol in healthy volunteers. *Alcohol and Alcoholism*, **26**:329–336.

Weingartner, H., Eckardt, M., Molchan, S.E. and Sunderland, T. (1992). Measurement and interpretation of changes in memory in response to drug treatments. *Psychopharmacology Bulletin*, **28**:331–340.

Weiss, B. and Laties, V.G. (1962). Enhancement of human performance by caffeine and the amphetamines. *Pharmacological Review*, **14**:1–36.

Welford, A.T. (1968). Fundamentals of skill. London: Methuen.

Werth, R. and Steinbach, T. (1991). Symptoms of prosopagnosia in intoxicated subjects. *Perceptual and Motor Skills*, **73**:399–412.

Wesnes, K. and Revell, A. (1984). The separate and combined effects of scopolamine and nicotine on human information processing. *Psychopharmacology*, **84**:5–11.

Wesnes, K. and Warburton, D.M. (1978). The effect of cigarette smoking and nicotine tablets upon human attention. In: R.E. Thornton (Ed.), Smoking behaviour: physiological and psychological influences. London: Churchill-Livingstone, pp. 131–147.

Wesnes, K. and Warburton, D.M. (1983a). The effects of smoking on rapid information processing performance. *Neuropsychobiology*, **9**:223–229.

Wesnes, K. and Warburton, D.M. (1983b). Smoking, nicotine and human performance. *Pharmacology and Therapeutics*, **21**:184–208.

Wesnes, K. and Warburton, D.M. (1984a). Smoking, nicotine, and human performance. In: D.J.K. Balfour (Ed.), Nicotine and the tobacco smoking habit. Oxford, UK: Pergamon Press, pp. 133–152.

Wesnes, K. and Warburton, D.M. (1984b). The effects of cigarettes of varying yield on rapid information processing performance. *Psychopharmacology*, **82**:338–342.

Wesnes, K. and Warburton, D.M. (1984c). Effects of scopolamine and nicotine on human rapid information processing performance. *Psychopharmacology*, **82**:147–150.

Wesnes, K. (1987). Nicotine increases mental efficiency: but how? In: W. Martin, G. Van Loon, E. Iwamoto and L. Davis (Eds.), Tobacco smoking and nicotine. New York: Plenum Press, pp. 63–80.

Wesnes, K., Warburton, D.M. and Matz, B. (1983). The effects of nicotine on stimulus sensitivity and response bias in a visual vigilance task. *Neuropsychobiology*, **9**:41–44.

West, R. (1990). Nicotine pharmacodynamics: Some unresolved issues. In: G. Bock and J. Marsh (Eds.), The Biology of Nicotine Dependence. Chichester: Wiley, pp. 210–224.

West, R. (1993). Beneficial effects of nicotine: fact or fiction? *Addiction*, **88**:589–590.

West, R., Drummond, C. and Eames, K. (1990). Alcohol consumption, problem drinking and anti-social behaviour in a sample of college students. *British Journal of Addiction*, **85**:479–486.

West, R.J. and Hack, S. (1991a). Effects of nicotine cigarettes on memory search rate. In: F. Adlkofer and K. Thurau (Eds.), Effects of nicotine on biological systems. Basel: Birkhäuser Verlag, pp. 547–557.

West, R.J. and Hack, S. (1991b). Effects of cigarettes on memory search and subjective ratings. *Pharmacology, Biochemistry and Behaviour*, **38**:281–286.

West, R.J. and Jarvis, M.J. (1986). Effects of nicotine on finger tapping rate in non-smokers. *Pharmacology, Biochemistry and Behavior*, **25**:727–731.

West, R.J. and Lennox, S. (1992). Function of cigarette smoking in relation to examinations. *Psychopharmacology*, **108**:456–459.

West, R.J. and Russell, M. (1986). Loss of acute nicotine tolerance and severity of cigarette withdrawal. *Psychopharmacology*, **94**:563–565.

West, R.J. (1990). Nicotine pharmacodynamics: Some unresolved issues. In: G. Bock and J. Marsh (Eds.), The biology of nicotine dependence. Chichester: Wiley, pp. 210–224.

West, R.J. (1993). Editorial. Beneficial effects of nicotine: Fact or fiction? *Addiction*, **88**:589–590.

West, R.J., Hajek, P. and Belcher, M. (1987). Time course of cigarette withdrawal symptoms during four weeks of treatment with nicotine chewing gum. *Addictive Behaviors*, **12**:199–203.

West, R.J., Hajek, P. and Belcher, M. (1989). Time course of cigarette withdrawal symptoms while using nicotine gum. *Psychopharmacology*, **99**:143–145.

West, R.J., Jarvis, M.J., Russell, M.A.H., Carruthers, M.E. and Feyerabend, C. (1984a). Effect of nicotine replacement on the cigarette withdrawal syndrome. *British Journal of Addiction*, **79**:215–219.

West, R.J., Russell, M.A.H., Jarvis, M.J. and Feyerabend, C. (1984b). Does switching to an ultra-low nicotine cigarette induce nicotine withdrawal effects? *Psychopharmacology*, **84**:120–123.

West, R.J., Wilding, J.M., French, D.J. and Kemp, R. (1993). Effect of low and moderate doses of alcohol on driving, hazard perception, latency and driving speed. *Addiction*, **88**:527–532.

Westrick, E.R., Shapiro, A.P., Nathan, P.E. and Brick, J. (1988). Dietary tryptophan reverses alcohol-induced impairment of facial recognition but not verbal recall. *Alcoholism: Clinical and Experimental Research*, **12**:531–533.

Whipple, S. and Noble, E.P. (1986). The effects of familial alcoholism on visual event-related potentials. *Psychophysiology*, **23**:470.

White, H.R., Brick, J. and Hansell, S. (1993). A longitudinal investigation of alcohol use and aggression in adolescence. *Journal of Studies on Alcohol*, **11**:62–77.

White, N.M. and Milner, P.M. (1992). The psychobiology of reinforcers. *Annual Review of Psychology*, **43**:443–471.

Whitfield, J.B. and Martin, N.G. (1994). Alcohol consumption and alcohol pharmacokinetics: interactions within the normal population. *Alcoholism Clinical and Experimental Research*, **18**:238–243.

Wickens, C.D. (1991). Processing resources and attention. In: D.L. Damos (Ed.), Multiple-task performance. London: Taylor and Francis, pp. 3–34.

Widmark, E.M.P. (1930). Les lois cardinales de la distribution et du metabolisme de l'alcool ethylique dans l'organisme humain. Kungl. Fysiograf. Sallskapets Hanlinger, N.F. 41:1.

Wietholtz, H., Voegelin, M., Arnaud, M.J., Bircher, J. and Preisig, R. (1981). Assessment of the cytochrome P-448 dependent liver enzyme system by a caffeine breath test. *European Journal of Pharmacology*, **21**:53–59.

Wijers, A.A. (1989). Visual selective attention, an electrophysiological approach. Ph.D. Thesis, Groningen: University of Groningen.

Wijers, A.A., Okita, T., Mulder, G., Mulder, L.J.M., Lorist, M.M., Poiesz, R. and Scheffers, M.K. (1987). Visual search and spatial attention: ERPs in focussed and divided attention conditions. *Biological Psychology*, **25**:33–60.

Wijers, A.A., Stowe, L.A., Mulder, G., Willemsen, A.T.M., Pruim, J., Paans, A.M.J. and Vaalburg, W. (in press). PET [0–15]H_2O investigation of visual selective attention. In. H.-J. Heinze, T.F. Münte and G.R. Mangun (Eds.), Mapping Cognition in time and space.

Wik, G., Borg, S., Sjogren, I., Wiesel, F.A., Blomqvist, G., Borg, J., Greitz, T., Nyback, H., Sedvall, G., Stone-Elander, S. and Widen, L. (1988). PET determination of regional cerebral glucose metabolism in alcohol dependent men and healthy controls using sup-1-sup-1c-glucose. *Acta Psychiatrica Scandinavica*, **78**:234–241.

Wikswo, J.P., Gevins, A. and Williamson, S.J. (1993). The future of the EEG and MEG. *Electroencephalography and Clinical Neurophysiology*, **87**:1–9.

Wilbert, J. (1987). Tobacco and Shamanism in South America. In: R.E. Schultes and R.F. Raffauf (Eds.) Psychoactive plants of the world. New Haven and London: Yale University Press.

Wilkinson, R.T. and Colquhoun, W.P. (1968). Interaction of alcohol with incentive and with sleep deprivation. *Journal of Experimental Psychology*, **76**:623–629.

Wilkinson, R.T. (1968). Sleep deprivation: Performance tests for partial and selective sleep deprivation. *Progress in Clinical Psychology*, **8**:28–43.

Wilkinson, R.T. (1970). Methods for research on sleep deprivation and sleep function. In: E. Hartmann (Ed.), Sleep and dreaming. Boston: Little Brown, pp. 369–381.

Williams, C.M. and Skinner, A.E.G. (1990). The cognitive effects of alcohol abuse: A controlled study. *British Journal of Addiction*, **85**:911–917.

Williams, D.G. (1980). Effects of cigarette smoking on immediate memory and performance in different kinds of smoker. *British Journal of Psychology*, **71**:83–90.

Williams, D.G., Tata, P.R. and Miskella, J. (1984). Different 'types' of cigarette smokers received similar effects from smoking. *Addictive Behaviours*, **9**:207–210.

Willner, P., Ahlenius, S., Muscat, R. and Scheel-Kruger, J. (1991). The mesolimbic dopamine system. In: P. Willner and J. Scheel-Kruger (Eds.), The mesolimbic dopamine system: From motivation to action. Chichester: Wiley, pp. 3–15.

Wise, R.A. (1988). The neurobiology of craving: Implications for the understanding and treatment of addiction. *Journal of Abnormal Psychology*, **97**:118–132.

Wittenborn, J. (1988). Assessment of the effects of drugs on memory. *Psychopharmacology Series*, **6**:67–78.

Wolff, P.H. (1973). Vasomotor sensitivity to alcohol in diverse Mongoloid populations. *American Journal of Human Genetics*, **25**:193–199.

Wolpaw, J.R. and Penry, J.K. (1978). Effects of ethanol, caffeine, and placebo on the auditory evoked response. *Electroencephalography and Clinical Neurophysiology*, **44**:568–574.

Wonnacott, S. (1990). The paradox of nicotinic acetylcholine receptor upregulation by nicotine. *Trends in Pharmacological Sciences*, **11**:216–219.

Woods, C.C. and Jennings, J.R. (1976). Speed-accuracy tradeoff functions in choice reaction time: Experimental designs and computational procedures. *Perception and Psychophysics*, **19**:92–101.

Woodson, P.P. and Griffiths, R.R. (1992). Control of cigarette smoking topography: smoke filtration and draw resistance. *Behavioural Pharmacology*, **3**:99–111.

Woodson, P.P., Buzzi, R., Nil, R. and Bättig, K. (1986). Effects of smoking on vegetative reactivity to noise in women. *Psychophysiology*, **23**:272–282.

World Health Organization. (1992). The ICD-10 classification of mental and behavioral disorders. Geneva: WHO.

Yamamoto, T. and Saito, Y. (1987). Acute effects of alcohol on ERPs and CNV in a paired stimulus paradigm. In: R. Johnson Jr., J.W. Rohrbaugh and R. Parasuraman (Eds.), Current Trends in Event-Related Potential Research (EEG Suppl. 40). Elsevier: Amsterdam, pp. 562–569.

Yamashita, I., Ohmori, T. and Koyama, T. (1990). Biological study of alcohol dependence syndrome with reference to ethnic difference: Report of a WHO collaborative study. *Japanese Journal of Psychiatry*, **44**:79–84.

Yerkes, R.M. and Dodson, J.D. (1908). The relation of strength of stimuli to rapidity of habit-formation. *Journal of Comparative Neurology and Psychology*, **18**:459–482.

Yesair, D.W., Branfman, A.R. and Callahan, M.M. (1984). Human disposition and some biological aspects of methylxanthines. In: The Methylxanthine Beverages and Foods: Chemistry, Consumption and Health Effects. *Progression in Clinical and Biological Research*, **158**:215–233.

Yin, S.J. and Li, T.-K. (1989). Genetic polymorphism and properties of human alcohol and aldehyde dehydrogenases: implications for ethanol metabolism. In: G.Y. Sun, P.K. Rudeen, W.G. Wood, Y.H. Wei and A.Y. Sun (Eds.), Molecular mechanisms of alcohol. Neurobiology and metabolism. New Jersey: Humana Press, pp. 227–247.

Yin, S.J., Bosron, W.F, Magnes, L.J. and Li, T.K. (1984). Human liver alcohol dehydrogenase: purification and kinetic characterization of the $\beta_2\beta_2$, $\beta_2\beta_1$, $a\beta_2$, and $\beta_2\gamma_1$ "Oriental" isoenzymes. *Biochemistry*, **23**:5847–5853.

Young, R. McD, Oei, T.P.S. and Knight, R.G. (1990). The tension reduction hypothesis revisited: an alcohol expectancy perspective. *British Journal of Addiction*, **85**:31–40.

Yu, G., Maskray, V., Jackson, S.H.D., Swift, C.G. and Tiplady, B. (1991). A comparison of the central nervous system effects of caffeine and theophylline in elderly subjects. *British Journal of Clinical Pharmacology*, **32**:341–345. *[83]*.

Yuille, J.C. and Tollestrup, P.A. (1990). Some effects of alcohol on eyewitness memory. *Journal of Applied Psychology*, **75**:268–273.

Zacchia, C., Pihl, R.O., Young, S.N. and Ervin, F.R. (1991). Effect of sucrose consumption on alcohol-induced impairment in male social drinkers. *Psychopharmacology*, **105**:49–56.

Zacny, J.P. and Stitzer, M.L. (1988). Cigarette brand-switching: Effects on smoke exposure and smoking behavior. *Journal of Pharmacology and Experimental Therapeutics*, **246**:619–627.

Zacny, J.P. (1990). Behavioral aspects of alcohol-tobacco interactions. *Recent Developments in Alcoholism*, **8**:205–219.

Zahn, T.P. and Rapoport, J.L. (1987). Autonomic nervous system effects of acute doses of caffeine in caffeine users and abstainers. *International Journal of Psychophysiology*, **5**:33–41. *[84]*.

Zeef, E.J., Kok, A. and Looren de Jong, H. (1990). Cognitive aging from a psychophysiological perspective. In: W. Koops, H.J.C. Soppe, J.L. Van der Linden, P.C.M. Molenaar and J.J.F. Schroots (Eds.), Developmental Psychology behind the dykes. Delft: Eburon, pp. 73–88.

Zhang, G. and Morrisett, R.A. (1993). Ethanol inhibits tetraethylammonium chloride-induced synaptic plasticity in area CA1 of rat hippocampus. *Neuroscience Letters*, **156**:27–30.

Zilm, D.H., Huszar, L., Carlen, P.L., Kaplan, H.L. and Wilkinson, D.A. (1980). EEG correlates of the alcohol-induced organic brain syndrome in man. *Clinical Toxicology*, **16**:345–358.

Zinser, M.C., Baker, T.B., Sherman, J.E. and Cannon, D.S. (1992). Relation between self-reported affect and drug urges and cravings in continuing and withdrawing smokers. *Journal of Abnormal Psychology*, **101**:617–629.

Zuzewicz, W. (1981). Ethyl alcohol effect on the visual evoked potential. *Acta Physiologica Polonica*, **32**:93–98.

Zwyghuizen-Doorenbos, A., Roehrs, T.A., Lipschutz, L., Timms, V. and Roth, T. (1990). Effects of caffeine on alertness. *Psychopharmacology*, **100**:36–39. *[85]*.

Zysset, T. and Wietholtz, H. (1991). Pharmacokinetics of caffeine in patients with decompensated Type I and Type II diabetes mellitus. *European Journal of Clinical Pharmacology*, **41**:449–452.

SUBJECT INDEX

After a page reference, 'f' indicates a figure and 't' a table.

AUTHOR INDEX

After a page reference, 'f' indicates a figure, 'n' a note and 't' a table.

1

Jernigan, T., Butters, N., Di Taglia, G., Shafer, K., Smith, T., Irwin, M., Grant, I., Schuckit, M. and Cermak, L.S. 319
Jobs, S.M., Fiedler, F.E. and Lewis, C.T. 341, 343–4
Joeres, R., Klinker, H., Heusler, H., Epping, J., Hofstetter, G., Drost, D., Reuss, H., Zilly, W. and Richter, E. 163
Joeres, R., Klinker, H., Heusler, R., Epping, J., Zilly, W. and Richter, E. 157
Johnson Jr., R. 282, 290, 323–4
Jones, A.W. 260
Jones, B.M. 351, 354
Jones, G.M.M., Sahakian, B.J., Levy, R., Warburton, D.M. and Gray, J.A. 9, 17, 26n, 251–3, 368
Jones, M.K. and Jones, B.M. 318, 320
Jones-Saumty, D.J. and Zeiner, A.R. 319–20
Josephs, R.A. and Steele, C.M. 349
Jost, G., Wahllander, A., Von Mandach, U. and Preisig, R. 157
Jubis, R.M.T. 75t, 76, 354, 357–8

Kadlubar, F.F., Talaska, G., Butler, M.A., Teitel, C.H., Massengill, J.P. and Lang, N.P. 161
Kahneman, D. and Treisman, A. 177
Kalant, H. and Khanna, J.M. 330
Kalant, H., Leblanc, A. and Gibbins, R. 121
Kalin, R. 351
Kall, M.A. and Clausen, J. 164
Kalow, W. 160, 255
Kalow, W. and Tang, B.K. 161–4
Kaminori, G.H., Somani, S.M., Knowlton, R.G. and Perkins, R.M. 153, 156
Kaplan, R.F., Glueck, B.C., Hesselgrock, M.N. and Reed Jr., H.B.C. 294
Kassam, J.P., Tang, B.K., Kadar, D. and Kalow, W. 265
Keenan, R.M., Hatsukami, D.K. and Anton, D.J. 55t, 57
Keenan, R.M., Hatsukami, D.K., Pickens, R.W., Gust, S.W. and Strelow, L.J. 356
Keiding, S., Christensen, N.J. and Damgaard, S.E. 261
Keister, M.E. and McLaughlin, R.J. 372
Kendler, K.S., Heath, A.C. and Neale, M.C. 270

Kenemans, J.L., Kok, A. and Smulders, F.T.Y. 190
Kenemans, J.L. and Lorist, M.M. 185, 190, 192, 198–9
Kenford, S.L., Fiore, M.C., Jorenby, D.E., Smith, S.S., Wetter, D. and Baker, T.B. 108
Kerkhof, G.A. 201–2, 210
Kerr, J.S. and Hindmarch, I. 42, 174, 331–2
Kerr, J.S., Sherwood, N. and Hindmarch, I. 29, 44, 48–9t, 50, 59n, 61, 170–1, 176–7, 253, 362, 370–1
Khanna, N.N. and Somani, S.M. 154
Kilbane, A.J., Silbart, L.K., Manis, M., Beitins, I.Z. and Weber, W.W. 160
Kinchla, R.A. 246
King, D.J. and Henry, G. 170–1, 176
King, M.A., Hunter, B. and Walker, D.W. 317
Kinnunen, T., Doherty, K., Militello, F.S. and Garvey, A.J. 134
Kinsbourne, M. 146
Kirk-Smith, M.D., Van Toller, S. and Dodd, G.H. 202
Kirsch, I. and Weixel, L.J. 203
Kirschbaum, C., Strasburger, C.J. and Langkrär, J. 100
Kirschbaum, C., Wüst, S. and Strasburger, C.J. 100
Klatzky, R.L. 70
Kleinsmith, L. and Kaplan, S. 128
Klesges, R.C. and Klesges, L.M. 141
Klesges, R.C., Myers, A.W., Klesges, L.M. and LaVasque, M.E. 101–2
Knasko, S.C. 202
Knibbe, R.A., Drop, M.J. and Muytjens, A. 231
Knight, R.G. and Longmore, B.E. 316, 318, 320
Knop, J., Teasdale, T.W., Goodwin, D.W. and Schulsinger, F. 272
Knott, V.J. 35t, 38, 51, 54t, 95, 116, 118, 140, 146–7
Knott, V.J. and De Lugt, D. 140
Knott, V.J. and Griffiths, L. 39
Knott, V.J., Hooper, C., Lusk-Mikkelsen, S. and Kerr, C. 116, 118
Knott, V.J. and Venables, P.H. 54t, 57, 358
Kobal, G. and Hummel, T. 210, 213
Koelega, H.S. 21, 246–7, 250, 331, 364–5, 367–70, 372
Koelega, H.S. and Brinkman, J.A. 364